HEALTH, SAFETY, AND ACCIDENT MANAGEMENT IN THE CHEMICAL PROCESS INDUSTRIES

CHEMICAL INDUSTRIES

A Series of Reference Books and Textbooks

Consulting Editor

HEINZ HEINEMANN

1. *Fluid Catalytic Cracking with Zeolite Catalysts,* Paul B. Venuto and E. Thomas Habib, Jr.
2. *Ethylene: Keystone to the Petrochemical Industry,* Ludwig Kniel, Olaf Winter, and Karl Stork
3. *The Chemistry and Technology of Petroleum,* James G. Speight
4. *The Desulfurization of Heavy Oils and Residua,* James G. Speight
5. *Catalysis of Organic Reactions,* edited by William R. Moser
6. *Acetylene-Based Chemicals from Coal and Other Natural Resources,* Robert J. Tedeschi
7. *Chemically Resistant Masonry,* Walter Lee Sheppard, Jr.
8. *Compressors and Expanders: Selection and Application for the Process Industry,* Heinz P. Bloch, Joseph A. Cameron, Frank M. Danowski, Jr., Ralph James, Jr., Judson S. Swearingen, and Marilyn E. Weightman
9. *Metering Pumps: Selection and Application,* James P. Poynton
10. *Hydrocarbons from Methanol,* Clarence D. Chang
11. *Form Flotation: Theory and Applications,* Ann N. Clarke and David J. Wilson
12. *The Chemistry and Technology of Coal,* James G. Speight
13. *Pneumatic and Hydraulic Conveying of Solids,* O. A. Williams
14. *Catalyst Manufacture: Laboratory and Commercial Preparations,* Alvin B. Stiles
15. *Characterization of Heterogeneous Catalysts,* edited by Francis Delannay
16. *BASIC Programs for Chemical Engineering Design,* James H. Weber
17. *Catalyst Poisoning,* L. Louis Hegedus and Robert W. McCabe
18. *Catalysis of Organic Reactions,* edited by John R. Kosak
19. *Adsorption Technology: A Step-by-Step Approach to Process Evaluation and Application,* edited by Frank L. Slejko
20. *Deactivation and Poisoning of Catalysts,* edited by Jacques Oudar and Henry Wise
21. *Catalysis and Surface Science: Developments in Chemicals from Methanol, Hydrotreating of Hydrocarbons, Catalyst Preparation, Monomers and Polymers, Photocatalysis and Photovoltaics,* edited by Heinz Heinemann and Gabor A. Somorjai
22. *Catalysis of Organic Reactions,* edited by Robert L. Augustine

ADDITIONAL VOLUMES IN PREPARATION

HEALTH, SAFETY, AND ACCIDENT MANAGEMENT IN THE CHEMICAL PROCESS INDUSTRIES

Ann Marie Flynn
Louis Theodore

Manhattan College
Riverdale, New York

CRC Press
Taylor & Francis Group
Boca Raton London New York

CRC Press is an imprint of the
Taylor & Francis Group, an **informa** business

CRC Press
Taylor & Francis Group
6000 Broken Sound Parkway NW, Suite 300
Boca Raton, FL 33487-2742

First issued in paperback 2019

© 2002 by Taylor & Francis Group, LLC
CRC Press is an imprint of Taylor & Francis Group, an Informa business

No claim to original U.S. Government works

ISBN-13: 978-0-367-39689-3

Visit the Taylor & Francis Web site at
http://www.taylorandfrancis.com

and the CRC Press Web site at
http://www.crcpress.com

To my husband Kevin for his understanding and wisdom
and to our children Taylor and Kevin – reach for the stars.
(A.M.F.)

To my students at Pfizer of whom I am so proud, particularly
Joe Hettenbach
Greg Hounsell
Kevin Keating
Dan O'Shea
(L.T.)

Preface

The rapid growth and expansion of the chemical industry has been accompanied by a spontaneous rise in human, material, and property losses because of fires, explosions, hazardous and toxic spills, equipment failures, other accidents, and business interruptions. Concern over the potential consequences of catastrophic accidents, particularly at chemical and petrochemical plants, has sparked interest at both the industrial and regulatory levels in obtaining a better understanding of the subject of this book: Health, Safety, and Accident Management (HS&AM). The writing of this book was undertaken, in part, as a result of this growing concern.

In recent years, the engineering profession has expanded its responsibilities to society to include HS&AM, with particular emphasis on accidents arising at industrial sources. Increasing numbers of engineers, technicians, and maintenance personnel are now confronting problems in this most important area. To cope with these challenges, the engineer and scientist of today and tomorrow must develop both a proficiency in HS&AM and an improved understanding of the subject.

Because HS&AM is a highly sophisticated and complex endeavor, many company administrators and regulatory officials are seeking highly trained and professionally educated personnel to fill positions in this area. Thus, companies and government agencies have acquired an interest in the continuing education of employees and students. It was also in the spirit of responding to this particular concern that this book was undertaken.

Regarding students, the Accreditation Board for Engineering and Technology (ABET) requires that engineering graduates understand the engineer's responsibility to protect both occupational and public health and safety. Traditionally, engineering schools have done a superb job of educating their students on the fundamental laws of nature governing their fields, and on the application of these laws to engineering problems. Unfortunately, they have been less successful in conveying to the students the importance of occupational and environmental safety in the design of chemical processes. This concern also served as a driving force for the writing of this book.

This book is intended primarily for regulatory officials, company administrators, engineers, technicians, industry maintenance personnel, and both undergraduate and first-year graduate students. It is assumed that the reader has taken basic courses in physics and chemistry; only a minimum background in mathematics is required (though calculus is desirable). The authors' aims are to offer the reader the fundamentals of HS&AM with appropriate practical applications in the chemical process industries and to provide an introduction to the specialized and reference books in this and related areas. The reader is encouraged to use the works cited in the bibliography to continue development beyond the scope of this book.

As is usually the case in preparing a manuscript, the decisions of what to include and what to omit have been difficult. However, every attempt has been made to offer engineering and science course material to readers at a level that will enable them to better cope with some of the complex problems encountered in HS&AM.

This book is divided into five parts: the problem, accidents, health risk, hazard risk, and hazard risk analysis. Part I, an introduction to HS&AM, presents legal considerations, emergency planning, and emergency response. This Part basically serves as an overview to the more technical topics covered in the remainder of the book. Part II treats the broad subject of accidents, discussing fires, explosions and other accidents. The chapters in Parts III and Part IV provide introductory material to health and hazard risk assessment, respectively. Part V examines hazard risk analysis in significant detail. The three chapters in this final part include material on fundamentals of applicable statistics theory, and the applications and calculations of risk analysis for real systems.

For some readers, particularly students, the book may serve as a starting point that will allow them to become acquainted with the HS&AM field. For others, who would classify themselves as experts, the book will serve as a reference text. It may also be useful as a tool for training in industry, government, or academia. The book should be valuable to engineers in agencies and industry, to technicians, and to maintenance personnel. It may also be of value to individuals involved with the general field of environmental management. The aim of the authors is to provide, in a thorough and clear manner, a book covering both the fundamentals of health, safety and accident management and their application to real-world problems. It is hoped that it will serve both industry and government (as well as academia) in attempting to reduce and/or eliminate accidents that can result in the loss of human and animal life, materials, vegetation, and property.

During the preparation of this book, the authors were ably assisted in many ways by a number of Manhattan College graduate students and practicing engineers with expertise in this field. We gratefully acknowledge their invaluable assistance; the names of these individuals are listed on the opening page of the chapters to which they contributed. In addition, several illustrative examples and problems were drawn, in part, from contributions of several faculty who participated in an earlier National Science Foundation funded College Faculty Workshop that was conducted at Manhattan College.

The authors also wish to acknowledge the contributors to the first generation John Wiley (1989) text titled "Accident and Emergency Management".

The following were listed as contributors:

Chapter 1: Past History, John O'Byrne
Chapter 2: Legislation, Gaetano LaVigua
Chapter 3: Emergency Planning and Response, Elizabeth Shoen
Chapter 5: Fires, Explosions and Other Accidents, Nat Federici and Isabella
　　　　　　Schroeder
Chapter 6: Accident Prevention in Process Facilities, Carol Earle (Conti)
Chapter 7: Process Applications, Anthony Gardetto and Chassam Koderska.

The following was a partial contributor:
Chapter 4: Process Fundamentals and Plant Equipment, Michael Venezia

Drs. Joseph Reynolds and Frank Taylor served as co-authors (with Dr. Theodore) for the original text.

Somehow the editor usually escapes acknowledgment. We were particularly fortunate to have B.J. Clark serve as the editor for our book. He had the vision early on to realize the need for a project of this nature. He and his staff's (including Brian Black and Ted Allen) patient handling and understanding of the numerous problems that arose during the preparation of the manuscript was particularly appreciated. Thanks, B.J.

Ann Marie Flynn
Louis Theodore

Contents

Contents

Contents

Part I

Introduction

Accidents can occur in many ways. There may be a chemical spill, an explosion, or a runaway reaction in a nuclear plant. There are often accidents in transport: trucks overturning, trains derailing, or ships capsizing. There are "acts of God" such as earthquakes and storms. It is painfully clear that accidents are a fact of life. The one common thread through all of these situations is that accidents are rarely expected and, unfortunately, they are frequently mismanaged.

Development of plans for handling accidents and emergencies must precede the actual occurrence of these events. In recent years incidents related to the chemical, petrochemical, and refinery industries have caused particular concern. Since the products of these industries are essential in a modern society, every attempt must be made to identify and reduce the risk of accidents or emergencies in these areas.

Part I of this book serves as an introduction to the general subject of *Health, Safety and Accident Management*. The more technical aspects of this subject, equipment and accident details, health risk assessment, hazard/risk assessment, and hazard/risk analysis, are covered in Parts II, III, IV, and V, respectively. There are three chapters in Part I. In Chapter 1, the histories of accidents are examined from early incidents to recent catastrophes. The evolution of safety precautions, particularly as they apply to chemical plants, is also reviewed. Chapter 2 is concerned with legislation. The major applicable pieces of legislation, the *Clean Air Act*, the *Clean Water Act*, the *Resource Conservation and Recovery Act* (RCRA), the *Comprehensive Environmental Response, Compensation, and Liability Act* (CERCLA), and the *Superfund Amendments and Reauthorization Act* (SARA), are discussed. Increased public awareness is the major thrust of the new legislation. Title III, which is the heart of SARA, establishes requirements for emergency planning and "community right to know" for federal, state, and local governments as well as for industry. Title III is a major stepping-stone in the protection of the environment, but its major principal thrust is to facilitate planning for possible catastrophes. A more comprehensive examination of Title III is provided in Chapter 3. State initiatives are also discussed in Chapter 3.

1

Past History

1.1 INTRODUCTION

Whether a careless mishap at home, an unavoidable collision on the freeway, or a miscalculation at a chemical plant, accidents are a fact of life. Even in prehistoric times, long before the advent of technology, a club-wielding caveman swings at his prey and inadvertently topples his friend in what can only be classified as an "accident." As humanity progressed, so did the severity of these misfortunes. The "modern era" has brought about assembly lines, chemical manufacturers, nuclear power plants, and other technological complexities, all carrying the capability of disaster. To keep pace with the changing times, safety precautions must constantly be upgraded. It is no longer sufficient, as with the caveman, to shout a warning "Watch out with that thing!" Today's problems require more elaborate systems of warnings and controls to minimize the chances of serious accidents.

This chapter examines the history of accidents from early incidents to recent catastrophes. In conjunction with this review, the material will study the evolution of safety precautions, particularly as they apply to chemical plants. A crucial part of any design project is the inclusion of safety controls. Whether the plans involve a chemical plant, a nuclear reactor, or a thruway, steps must be taken to minimize the likelihood, or consequences, of accidents. It is also important to realize how accident planning has improved in order to monitor today's advanced technologies. This chapter reviews a variety of actual accidents in order to provide an understanding of these phenomena, which will supplement the subsequent chapters that deal with these subjects in significant detail.

The authors gratefully acknowledge the assistance of Jason Ragona in researching, reviewing, and editing this chapter.

1.2 EARLY ACCIDENTS

Accidents have occurred since the birth of civilization and were just as damaging in early times as they are today. Anyone who crosses a street or swims in a pool runs the risk of injury through carelessness, poor judgment, ignorance, or other circumstances. This has not changed through history. In the following pages, a number of accidents and disasters that took place before the advances of modern technology are examined. Catastrophic explosions have been reported as early as 1769, when one-sixth of the city of Frescia, Italy, was destroyed by the explosion of 100 tons of gunpowder stored in the state arsenal. More than 3000 people were killed in this, the second deadliest explosion in history. The worst explosion in history occurred in 1856 on the Greek island of Rhodes. A church, which had gunpowder stored in its vaults, was struck by lightning. The resulting blast is estimated to have killed 4000 people. This remains the highest death toll for a single explosion.[1]

The Great Chicago Fire

One of the most legendary disasters occurred in Chicago in October 1871. The "Great Chicago Fire," as it is now known, is alleged to have started in a barn owned by Patrick O'Leary, when one of his cows overturned a lantern. The O'Leary house escaped unharmed, since it was upwind of the blaze, but the barn was destroyed as well as 2124 acres of Chicago real estate.

The Chicago fire may be blamed on a farmer's cow; however, the reason for the extent of the damage was the city's shoddy construction. Almost all buildings and houses were built of wood, and many of the sidewalks and roads were also of wooden construction. The streets were dangerously narrow, allowing flames to leap across easily to neighboring structures. The water supply proved to be inadequate, despite the location of Chicago on the banks of Lake Michigan. Add to this the extended drought conditions of the summer of 1871, which caused numerous smaller fires, and all the elements of a catastrophe were present.

Chicago was clearly unprepared for the events that took place in October of 1871. The city was planned and constructed with little apparent regard to safety or fire prevention. No emergency plan was implemented, and panic was the order of the day, with people grabbing whatever possessions they could carry and fleeing the city. Looters and thieves broke store windows and helped themselves. All told, between 200 and 250 people were killed, another 200 were reported missing, and 100,000 were left homeless. The total loss of property was estimated in excess of $200 million, forcing more than 60 insurance companies into bankruptcy. Whether or not O'Leary's cow actually caused this massive fire, the fact remains that accidents often occur under the most unlikely circumstances.[2]

South Fork Dam – Johnstown, Pennsylvania

On May 31, 1889, an accident that took the lives of 2209 people occurred. The site was Johnstown, Pennsylvania, and the incident was a flood that followed the collapse of a dam. The South Fork Dam was originally constructed to provide water for a proposed canal system between Johnstown and Pittsburgh. After the dam was completed, however, the project was abandoned. The dam changed ownership several times and eventually became the property of a group of rich industrialists who stocked the lake with fish and installed wire-mesh grates over the drainage canals to prevent the fish from escaping. These grates soon became clogged, decreasing the drainage capacity of the dam. Structural engineers warned that the dam was destined to collapse, but the warning went unheeded.

In the spring of 1889, unusually heavy rains caused the water level of the lake to rise, which caused the dam to rupture on May 31. A wall of water 40 feet high crashed down into the valley and the city. The lake drained at a rate of 200,000 cubic feet per second, and the entire lake was emptied in 36 minutes. The rushing wave picked up houses, trees, people, and debris. The flood was stopped when a bridge over the Conemaugh River caught most of the debris, forming a natural dam. Eventually, the rushing water sputtered out, and the danger was over. Although this disaster appears to be a clear case of negligence, it was deemed an "act of providence" and no damages were ever paid by the dam owners.

In the Johnstown flood, there was no adequate evacuation or safety plan. Once the dam had broken, the water reached the city in a matter of minutes - hardly enough time for a city to react. However, the warnings of inspectors should have been taken seriously. Johnstown, a city of 30,000 people, was destroyed because of the carelessness and ignorance of a few.[1]

Oppau, Germany

As technology advanced, large factories began growing throughout the world. Chemical manufacturers were called upon to supply an increasing number of products. These changes brought about new potential dangers. One such danger became a reality in Oppau, Germany, on September 21, 1921. Early in the morning, two massive explosions shook the surrounding area. Damage was reported 53 miles away in Frankfurt, while the shock was felt as far as 145 miles away. More than 700 houses were destroyed in the village of Oppau, and approximately 500 people died. Many lives were spared because the blast blew walls and roofs of houses away rather than knocking them down on the occupants. The cause of the blast was the sudden explosion of 4500 tons of ammonium sulfonitrate, an ammonium salt.

Specific details regarding the accident and its causes are sparse because, as is usually the case in such a mammoth disaster, none of the plant operators present at the time survived. The large buildings containing the ammonium sulfonitrate disappeared entirely, and nothing was left in their place except a crater 250 feet in diameter and 30 feet deep. Ammonium sulfonitrate had never before been known to explode or to ignite spontaneously. This compound forms a hard mass when stored in large quantities, which was usually blasted apart with dynamite to prepare it for transportation. However, this technique, which had been used thousands of times without serious incident, was apparently responsible for the explosion in Oppau. It had been noted that the salt had changed from its original white color to a slightly yellow color and that the temperature of the storage rooms had risen 20°C above normal. These conditions may have led to a partial decomposition of the stored material to ammonium nitrate, which caused spontaneous ignition of the mass.

The Oppau explosion came about because of a lack of understanding of the chemical being manufactured. Although there had been no earlier reported incidents of ammonium sulfonitrate exploding, a similar compound, ammonium nitrate is known to be highly explosive. Thus, a more careful examination of this compound's characteristics should have been conducted. In addition to performing more research, management could have developed a method of breaking up the hardened mass other than blasting with dynamite. Many lives were lost because a company decided to cut corners.[3]

East Ohio Gas Company – Cleveland, Ohio

Some accidents can be attributed to structural failures. On October 20, 1944, one of the four liquid gas tanks at the East Ohio Gas Company in Cleveland began to leak. The plant converted natural gas to the liquid form, which was stored for emergency use in holding tanks. If needed, the liquefied product could be reconverted to its gaseous state and fed into the city distribution lines. The tanks were constructed in 1941 and had a capacity of more than 400,000 cubic feet of liquid.

At the time of the disaster, the leaking liquid escaped into the plant grounds and beyond, vaporizing as it traveled. A spark or flame ignited the highly combustible gas, causing a drawback and fire which quickly engulfed the tanks and plant itself. Almost before the 80 gas company workers knew what was taking place, an explosion blasted out walls and flattened structures of the plant, killing the entire work force. The fire spread quickly through the district, destroying 52 homes and damaging 200 others. Four industrial plants were leveled and 20 others were damaged; 131 people were killed, 72 being burned beyond identification and buried in a common grave. The extent of the disaster was greatly increased by the location of the plant, which was close to a residential area. In fact, this case is still used as an argument against the siting of liquefied natural gas facilities in urban areas.[2,4]

As was the case in Oppau, Germany, in 1921, everyone with firsthand knowledge of the disaster was killed on the site, leaving the cause of the incident unknown. It is reasoned that a structural weakness in one of the tanks caused the gas leak, although the liquefied gas was not under any pressure. The tanks were only 3 years old when the accident occurred and certainly were not expected to develop any leaks. Although it is difficult to say how often plant structures should be inspected, one would reason that it should be possible to use a new tank with the assurance that it is structurally sound.

Texas City, Texas

Catastrophic accidents have occurred at sea as well as on land. In 1947 an unusual incident involved both. The French freighter *Grandcamp* arrived at Texas City, Texas, to be loaded with 1400 tons of ammonium nitrate fertilizer. During the night, a fire broke out in the hold of the vessel, but apparently fearing that water would damage the rest of the cargo, crew made only limited attempts to fight the flames. Since the *Grandcamp* was docked only 700 feet from the Monsanto chemical plant, which produced styrene, a highly combustible ingredient of synthetic rubber, the *Grandcamp* was ordered to be towed away from the harbor. As tug boats prepared to hook up their lines, the *Grandcamp* exploded in a flash of fire and steel fragments. The blast rattled windows 150 miles away, registered on a seismograph in Denver, and killed many people standing on the dock. The Monsanto plant exploded minutes later, killing many survivors of the first blast, destroying most of the Texas City business district, and setting fires throughout the rest of the city. As the fires burned out of control, the freighter *High Flyer*, also loaded with nitrates, exploded in the harbor.

This third explosion was too much for the people of Texas City, who had responded efficiently to the initial two blasts. Hundreds were forced to leave the city, letting the fire burn itself out. The series of explosions had killed approximately 500 people and seriously injured 1000 others. The final death toll may have been as high as 1000 because the dock area contained a large population of migrant workers without permanent address or known relatives. It was reported that this disaster probably was caused by careless smoking aboard the *Grandcamp*. [1,2]

1.3 RECENT MAJOR ACCIDENTS

The advances of modern technology have brought about new problems. Perhaps the most serious of these is the threat of a nuclear power plant accident known as a meltdown. In this section several of this era's most infamous accidents are examined; some possible explanations are also offered.

Flixborough, England

An explosion at the Nypro Ltd. caprolactam factory at Flixborough, England, on June 1, 1974, was one of the most serious in the history of the chemical industry and the most serious in the history of the United Kingdom. Of those working on the site, 28 were killed and 36 others injured. Outside the plant, 53 people were reported injured, while 1821 houses and 167 shops suffered damage. The estimated cost of the damage was well over $100 million.

The cyclohexane oxidation plant at Flixborough, which stored this solvent at about 120 psig and 145°C, contained six reactors in series (see Figure 1.3.1). One reactor had been removed for repairs, and the resulting gap was bridged by a temporary 20-in. pipe, connected by a bellows at each end and supported on temporary scaffolding. This pipe collapsed, and in the minute or so that elapsed before ignition, about 35 tons of cyclohexane escaped. The extensive damage that occurred could have resulted from the deflagration of as little as 10 to 20 tons.

It was later determined that no calculations had been done to ascertain whether the bellows or pipe could withstand the strain. Instead, only the capacity of the assembly to carry the required flow was calculated. No reference was made to any accepted standard, nor to the designer's guide issued by the manufacturer of the bellows. Neither the pipe nor the complete assembly was pressure tested before being fitted. Apparently no one realized that the pressurized assembly would be subject to a turning moment, imposing on the bellows shear forces for which the system was not designed. Nor did anyone appreciate that the hydraulic thrust on the bellows would tend to make the pipe buckle at the joints. This temporary pipe connection functioned satisfactorily after the initial installation. It was never closely inspected, however, even though (when at operating temperature and pressure) it was observed to lift off the scaffolding that had been put in to support it. [5,6]

What happened on the final shift will never be known because all those in the control room were killed and all instrumentation and records were destroyed. The equivalent force of the explosion was estimated to have been at least 15 tons of TNT.

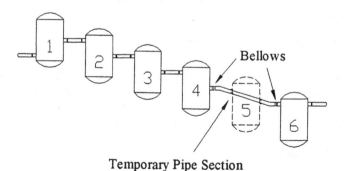

Temporary Pipe Section

Figure 1.3.1 Reactor Sequence.

Seveso, Italy

At least 220 persons were evacuated on July 26-28, 1976, from Milan's northern industrial suburb of Seveso, where a chemical factory explosion released into the atmosphere a cloud of trichlorophenol gas, containing the highly toxic defoliant, dioxin. Approximately 5 pounds of dioxin was released. Used by U.S. forces in Vietnam, dioxin is known to cause liver and kidney damage, and genetic alteration. Eighteen days after the leak, hundreds of rabbits, birds, cats, dogs, and chickens died. Plants withered, and nearly 30 people were hospitalized with skin rashes and internal disorders. A 172-acre contamination zone was enclosed with barbed wire while emergency plans were drawn up to evacuate an additional 15,000 people if the cloud spread. A decontaminant consisting of water and olive oil was field-tested on August 28 at the site. Within 48 hours of spraying the decontaminant, 70% of the toxin had been eliminated. A government spokesman estimated total damages at more than $48 million.[7]

Details on the accident are somewhat sketchy, but the following details are now available. The process system included a reactor where 2,3,5-trichlorophenol(TCP) was being synthesized from 1,2,4,5-tetrachlorobenzene in a liquid mixture of NaOH, ethylene glycol, and xylene. As the reaction neared completion, the plant crew decided to leave for the weekend before completing the final processing step that includes the addition of quench water. Because of this, the reactor system over-pressurized and the pressure relief system expelled 2 kg 2,3,7,8-tetrachlorodibenzopara-dioxin (the most toxic of all dioxins) from the reactor. The pressure relief system worked exactly as designed by preventing the explosion of the reactor; however, the relief went directly into the atmosphere. It was then dispersed by a wind across the country side where children playfully thought it was "snowing". A disaster later ensued.

Three Mile Island, Pennsylvania

Moving to more complex technology, the best known of the U.S. nuclear accidents, which occurred in 1979 on Three Mile Island in Pennsylvania, is now considered. A series of breakdowns in the cooling system of the plant's number 2 reactor led to a major accident in the early morning hours of March 28. Two days later the Nuclear Regulatory Commission (NRC) warned of a possible core meltdown, a catastrophic event that could involve major loss of life. The possible explosion of a hydrogen gas bubble that had formed in the overheated reactor vessel of the crippled plant was also a major threat. Because of concern over the continued emission of radioactive gases, pregnant women and preschool children within a 5-mile radius of the plant were advised to leave the area. On April 2 there was a dramatic reduction in the size of the dangerous gas bubble, as well as further cooling of the reactor core. A week later the bubble had been eliminated.

The accident at Three Mile Island unfortunately threatened the future of nuclear power in the United States and called into question the safety systems regulated by the NRC and used by the nuclear power industry. At the time of the accident, 72 nuclear reactors provided 13% of the nation's electrical power.

The cause of the incident has been hotly contested by the plant's owners, Metropolitan Edison Company, the NRC, the state of Pennsylvania, and companies that had constructed elements of the reactor system. Apparently, what had happened was the failure of a valve in a pump in the primary core cooling system. In addition, there was human error. This interrupted the flow of water used to cool the reactor, which caused the steam turbine to stop and consequently shutdown the reactor. However, the reactor continued to generate heat and, as a result, the emergency cooling system began operating automatically. At some point during the switchover from the primary cooling system to the emergency core cooling system, a plant operator turned off the emergency system and, after a period of time, turned it back on. During that time the core was damaged. Some of the pellets of enriched uranium fuel became so heated that they either melted through or ruptured the zirconium-clad tubes that held them. Some of the water used to cool the core spilled onto the floor of the reactor building. When some of that radioactive water became steam, it was vented into the atmosphere to relieve pressure.

Vented steam was not the only source of radiation leakage. Radiation had also been traced directly to nuclear materials within the plant that had escaped by penetrating 4-foot-thick walls. The uranium fuel in the core remained so hot that the plant's managers had to vent more steam into the atmosphere to prevent an explosion in the containment building. The direct result of the venting was the release of small amounts of radioactive iodine, krypton, and xenon. The levels were described as "quite low" and not dangerous to humans. Fortunately, there was no *apparent* serious exposure for the plant's workers. The long-range effects, however, are not yet known. According to Ernest Sternglass, professor of radiology at the University of Pittsburgh, "It's not a disaster where people are going to fall down like flies. It's a creeping thing."

Federal safety investigators reported that a series of human, mechanical, and design errors had contributed to the Three Mile Island accident. Metropolitan Edison had closed three auxiliary cooling pumps for maintenance two weeks before the accident and had kept them closed - a major violation of federal regulations. Several other errors contributed significantly to the incident: electrical magnetic relief valves that had opened to release a buildup of water pressure in the reactor had failed to close as planned; plant operators received incorrect readings from the pressure level indicator about the amount of water in the reactor; and, on two occasions after the accident, operators prematurely shut off the emergency core cooling system. Also, the release of slightly radioactive water into the Susquehanna River and the venting of steam into the air had been done without NRC approval.[7]

Chernobyl, Russia

Everyone's worst fears about nuclear power became a reality in the later part of April 1986. A large Soviet reactor - unit number 4 at Chernobyl, 80 miles north of Kiev, and only 3 years old - blew out and burned, spewing radioactive debris over much of Europe. Radiation levels increased from Sweden to Britain, through Poland, and as far south as Italy. The damage caused to the environment far surpassed that due to the accident at Three Mile Island.

The sequence of events at the Soviet reactor seems to have been as follows: First, the reactor suffered a loss of cooling water, which caused the uranium fuel elements to become overheated. The reactor had no containment building to keep in radioactive releases. Therefore, all the radioactivity generated in this part of the accident entered the atmosphere. Eventually, the temperature of the fuel rose to a point at which the graphite casing holding the uranium caught fire. Water could not be used to put the fire out because it would have evaporated, causing plumes of radioactivity to escape.

The explosion was the result of a series of errors by plant operators who were conducting an unauthorized experiment after having shut down the emergency cooling system. Operators were attempting to prove that if a turbine tripped, in the event of a power outage, and was disconnected from the steam supply, they would be able to draw kinetic energy from the still spinning rotor blades to operate emergency coolant pumps until the backup diesel generators began operating. Operators began to reduce the power output on April 25. Tests were conducted at 7% power, a level at which the plant is subject to automatic shutdown. At such low power levels, xenon gas builds up to absorb neutrons and slow the fission process. When this occurred at Chernobyl, it caused the loss of control of the reactor. Power dipped to as low as 1% before it was finally stabilized back to 7%. To increase the flow of water to the reactor, two cooling pumps, in addition to the six normally used, were engaged. The cooling water inside the reactor's pressure tubes was already close to boiling or had reached that point because the drop in pressure from the low power output had heated the coolant. When the turbine was tripped, the coolant turned to steam. Unfortunately, heat could not escape because emergency systems had been shut off. Power began to surge as the water dissipated in the reactor. A heat buildup caused the zirconium casing to react with the water, releasing hydrogen. Two explosions occurred, blowing the roof off the reactor building, destroying the cooling system, and severely damaging the core.

The accident reportedly killed 31 people, injured 299 others, and caused the evacuation of 135,000 from the site. The full extent of the damage from this incident probably will not be known for years. It is the long-term effects from exposure to radiation that frighten most people, and these fears may still become a horrible reality. [8-11]

Nuclear accidents, while being the most frightening, have not occurred often. In fact, there have only been a handful of fatal incidents since an understanding of nuclear energy and radiation has been developed. However pioneers of radiation research, including Marie Curie, are known to have died from radiation poisoning because they neglected to effectively control their exposure to this powerful energy source. Today, a better understanding of the risks associated with radioactive materials has led to fewer careless deaths. However, as Chernobyl proved, industry is far from having perfected the science of using nuclear energy.

Although much still remains to be learned about the interaction between ionizing radiation and living matter, more is known about the mechanism of radiation damage on the molecular, cellular, and organ system level than most other environmental hazards. A vast amount of quantitative dose-response data has been accumulated throughout years of studying the different applications of radionuclides. This information has allowed the nuclear technology industry to continue at risks that are no greater than any other technology.

Bhopal, India

As discussed earlier, nuclear accidents have not been the only accidents to occur in recent times. Other disasters at chemical plants have been responsible for a much greater loss of life. The worst disaster in the recent history of the chemical industry occurred in Bhopal, in central India, on December 3, 1984. A leak of methyl isocyanate (MIC) from a chemical plant, where it was used as an intermediate in the manufacture of a pesticide, spread into the adjacent city and caused the poisoning death of more than 2500 people; approximately 20,000 others were injured.

The owner of the plant, Union Carbide Corporation, reported that the accident was "the result of a unique combination of unusual events." The methyl isocyanate (MIC) was driven out of a storage tank by pressure generated from a water-induced runaway polymerization reaction. The last batch of MIC put into the tank before the accident contained more chloroform than product specifications allowed. Chloroform promotes the polymerization of MIC although chloroform alone cannot react without the high temperatures caused by the presence of water. The excess chloroform is one of the "unusual events" preceding the disaster, although the presence of water appears to be the primary culprit.

The most intriguing question is why the plant's safety equipment and controls, which should have been designed to cope with a chemical that is known to be subject to violent reactions, did not work. The MIC tank had three safety devices. A pipe leading from the tank contained a valve that was set to rupture if the pressure in the tank exceeded 40 psi. Beyond the valve there were two other safety devices. One was a flare (tower) to burn escaping gas. The other was a vent gas scrubber, a tower packed with loose material through which a solution of caustic soda could be poured onto rising MIC to decompose it. On December 3, 1984, the operator of the factory's control room noticed that the pressure in the MIC tank (see Figure 1.3.2) had risen from 2 to 30 psi, and in a

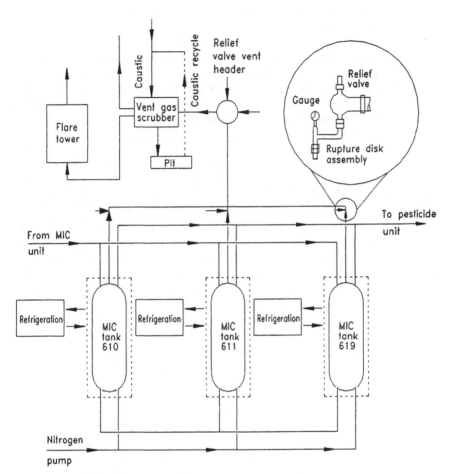

Figure 1.3.2 Bhopal Methyl Isocyanate Plant.

matter of a few minutes, it had risen above the gauge maximum of 55 psi. The tank was hot and rumbling, its concrete cladding was cracking, and the safety valve was screeching. The flare was shut down for maintenance, so the operator turned on the vent gas scrubber. The flow meter failed to indicate that a flow of caustic soda had started. However, the morning after the accident, the scrubber's runoff solution was hot, which indicated that the scrubber had worked and the pump had operated properly. When the flow meter was subsequently cleaned, it started working properly. This is one indication that maintenance at the plant was poor. Other evidence indicates that the concentration of the caustic solution had not been tested since the previous October. Despite this, the scrubber seems to have worked normally on the night of the accident. However, the scrubber was not able to control the MIC. After it had been operating for 45 minutes, the plant superintendent verified that MIC was being released into the atmosphere from the scrubber stack.

In addition to the failings of the safety system, the most pressing question was how the water made its way into the MIC tank, causing the fatal reaction. Obviously, the plant could not contain a runaway reaction, and an investigation proved that water alone caused the Bhopal disaster. The Union Carbide Corporation has cited some evidence that points to sabotage, and not plant mismanagement, as the cause of the disaster.[12] Their scenario, based on this evidence, is that a disgruntled worker sneaked into the deserted storage area, removed the pressure gauge on one tank, attached a water hose to the opening, and turned on the faucet. A chain of chemical reactions ensued and the pressure and temperature in the tank began to build. The MIC vaporized at the higher temperature and forced its way past the relief valve, through the pipeline, and into the vent tower. Several hours later, some workers noticed the odor of MIC gas and, while searching for a leak in the storage area, found the hose. It is believed that they attempted to siphon off the water by transferring it to another unit. Ten minutes later, the gas began leaking more rapidly. In an attempt to cover up the error, the workers' logs were altered to show that the transfer took place before the water was introduced into the tank. Evidence for this scenario exists in the testimony of a former employee who stated that a pressure gauge was missing from the storage tank the morning after the disaster. Additionally, shortly after the accident, several workers mentioned that water had been intentionally added to the tank. It was difficult to follow up this claim because the workers were dismissed and could not be found after the plant closed.[12]

At the time of the accident many people wondered why a chemical plant that produces such a dangerous compound was located so close to a residential area. The plant was originally built 1.5 miles from the nearest housing, but a small town grew up next to it. Zoning laws in many countries would have prevented such development. The Bhopal tragedy reinforces the need for controls to prevent the siting of plants that produce hazardous chemicals close to residential areas and to prevent residential construction close to these plants.

The chemical produced at Bhopal, methyl isocyanate, is an intermediate in the manufacturing of Carbaryl, which accounts for only 3% of all the pesticides used today. This accident suggests that perhaps, whenever feasible, industry should develop plants that use and produce less hazardous raw materials. Most plants are made safe by adding protective equipment, which, as in the Bhopal incident, may sometimes fail or prove inadequate. However, if a plant does not use or produce hazardous materials, the probability of such an accident is very low. Developing new and safer designs can, in the long run, be more economical than trying to control the hazards associated with older designs.[13-16] This is discussed in more detail in Part II.

Ashland Oil, Pennsylvania

An accident occurred recently that did not involve the loss of human life but nevertheless can be considered to be a disaster. On January 2, 1988, a 48-foot-high fuel tank ruptured at the Ashland Oil terminal in southeastern Pennsylvania. Nearly 3.9 million gallons of number 2 distillate fuel poured out. The force caused the tank to jump backward 100 feet and sent a wave 35 feet high crashing into another tank, 100 feet away. A containment dike trapped much of the spilled fuel. However, 600,000 gallons escaped into the Monongahela River at Floreffe, about 25 miles upstream from Pittsburgh. Soon after the spill, a rumor began circulating, that there was a possible gasoline leak as well. This raised concerns about a fire, leading to the evacuation of 250 homes.

Although the containment dike was large enough to hold the entire contents of the tank, the tank ruptured so suddenly that a torrent of 3.8 million gallons generated a wave of oil that splashed over the embankment. The tank, 56 feet high and 120 feet in diameter, was erected in 1986 on a newly enlarged concrete foundation that previously had supported a smaller receptacle. The tank was constructed of steel 40 years old, which presumably contains less carbon and was more brittle than steel manufactured more recently.

In addition, the tank had been moved from a terminal near Cleveland and reassembled. To facilitate the move to Pittsburgh, the tank was cut vertically and horizontally at the original welds, then restocked in eight rings above a concave steel floor at the new site. This procedure has been questioned. Some experts contend that, for reasons of structural integrity, the tank should not have been cut horizontally.

Ashland admitted that the usual testing procedures were not employed before putting the tank into service. Hydrostatic testing is recommended by the American Petroleum Institute, but pumps large enough to get water to the top of the tank were not available at the site. An alternate procedure was performed in which the tank was sprayed with diesel fuel and subjected to a vacuum. However, the tank was not inspected by any outside agency prior to use.

The fuel that had spilled into the frigid water began to emulsify and sink and the extremely cold weather caused ice to form on the river. It is nearly impossible to recover oil that sinks below the skirts of the recovery booms or that becomes trapped in ice. However, various methods were used to remove the oil from the Monongahela. Chemists developed a method that mixed the contaminated water with powdered carbon and bentonite, which gives the slurry higher absorbency. The mixture is then pumped to a treatment plant, where other chemicals are added to balance acidity and make the oil coagulate in a settling tank. This treatment is not new, but the chemists had to come up with the right combinations of chemicals to handle the oil. At one point, the EPA allowed the use of a substance called *Elastol* for the first time. *Elastol* congeals spilled hydrocarbons into a mass that can be easily recovered.

Although an accurate assessment could not be made immediately, the Ashland spill took a heavy toll on the wildlife of the Ohio River Valley. More than 5000 waterfowl were killed when their feathers became contaminated with oil. Birds that experience oil contamination lose their natural insulation and buoyancy and either drown or freeze to death. In addition, a massive fish kill was expected in the spring when the river entered a new cycle.

The Pittsburgh area was the hardest hit by the spill. Two communities with a combined population of 23,000 were completely without water for 5 days, and many others were left with critically short supplies. To alleviate the problem, workers constructed a 12-inch-diameter line to connect the unaffected city water system, which draws from the Allegheny River, with the devastated Western Pennsylvania Water Company system.

This massive spill of oil jeopardized the water supplies of more than a million people as it moved downstream into the Ohio River through Pennsylvania, Ohio, and West Virginia. While 19,000 spills are reported each year, there has been only one other complete tank collapse in the past 20 years. Many still question whether more stringent regulation is needed. They point to more site-specific plans for spill prevention and control. The plans that facilities are required to file now tend to be more "generic". New legislation regulating a tank's proximity to storm sewers, waterways, and other facilities, as well as possible weather conditions and configurations of terrain is needed.[17-19]

Guatemala

Not all of the recent major accidents have been "man-made." There have been numerous tragic floods, landslides, volcanic eruptions, and earthquakes. One such earthquake struck central Guatemala on February 4, 1976, killing more than 24,000 people and injuring another 50,000. As much as one-sixth of the country's population may have been left homeless. The death toll makes this the second worst disaster in Western Hemisphere history, exceeded only by the 1970 Peruvian earthquake that took between 50,000 and 70,000 lives.

The first major shock occurred at 3:04 A.M., when most Guatemalans were in their homes asleep. The tremor, which measured 7.5 on the Richter scale, collapsed thousands of poorly constructed buildings, burying their sleeping occupants. Had the shock occurred during the day when many of these people were outside at work, the death toll would have been much lower. By contrast, an earthquake of similar intensity at the same hour in the United States has a much lower potential for disaster. The reason for the difference lies in the construction of the buildings. As in many underdeveloped countries, the people of Guatemala are safest outside their homes, which are usually flimsy structures of adobe or stone erected over loose wood frames, often topped with clay tiles or corrugated metal sheets. The newer office buildings, luxury hotels, and apartment houses in Guatemala City suffered some damage during the earthquake but remained essentially intact. However, the traditional housing of the older, poor districts collapsed completely. The greatest damage occurred in the small villages outside the capital, some of which were completely destroyed.

A second series of tremors, most likely aftershocks associated with the original quake, struck the battered area on February 6, killing thousands more and hindering relief efforts. The full loss in property and industrial output, including the replacement of some 200,000 dwellings, was estimated at more than $3 billion. The quake also had a depressing effect on foreign tourism, a vital segment of that small nation's economy.[1]

Trans World Airlines - Long Island, New York

Accidents have also occurred in the "air", an area that would be classified as a transportation accident. An example of this is the Trans World Airlines Flight 800 from New York to Paris. On July 17, 1996, a Boeing 747 sat on the runway at J.F.K. airport for an extended period of time. The heat generated from the idle plane was sufficient to vaporize residual fuel in what was considered an "empty" fuel tank. This created a very dangerous, flammable mixture of fuel vapor and air. Approximately ten minutes after departure from J.F.K. airport, TWA Flight 800 exploded in mid-air. The most likely source of ignition for the flammable mixture of fuel vapor and air was an electrical wiring problem. All 230 people aboard perished when the jet crashed off the coast of Long Island, New York, However, this tragedy could have been avoided by preventing the flammable mixture in the fuel tank from developing.

In response to the TWA Flight 800 crash, the Federal Aviation Administration (FAA) has moved towards requiring airlines to pump nitrogen gas into empty fuel tanks to keep the pressure inside the tanks above the vapor pressure of the fuel and, hence, keep the fuel from vaporizing. This prevents a flammable mixture from developing inside the tanks. Unfortunately, it took the deaths of the 230 people aboard TWA Flight 800 to bring prior FAA philosophies that allowed aircraft to fly with flammable fuel tanks under scrutiny.[20]

Y2K Computer Bug

A recent example of the effects of poor planning is described by what has become known as the Y2K (Year 2000) computer bug. Computers work with numbers and a very important and widely used number in both computer hardware and software is the date, particularly the year. This data, like any other data used by a computer, is stored in the computer's memory. When computers first started being used and programs were being written for them, computer storage – random access memory (RAM) and disk memory – was very expensive. In an effort to reduce the amount of memory required to store the year, only the last two digits of the year were used. This system worked fine until the year 2000. On December 31, 1999, as the digits were to change from 99 to 00, the year was actually going to change from 1999 to 1900 (instead of

2000). This would affect every business that was linked to a computer and the commotion and hysteria that surrounded the Y2K bug were due to the present age's dependence on computers. For example, it was feared that the government's computer system would print social security checks with amounts seen in the year 1900 and that banks would calculate dividends based on interest rates from the year 1900. The problems could have been catastrophic. Fortunately, the response to the "potential" Y2K disaster was effective in that most corrections or "patches" were made before the year 2000. However, these corrections were costly and required large amounts of money to be spent.[21]

1.4 OTHER ACCIDENTS

The preceding sections have given only a sampling of the major accidents that have occurred throughout history. However, there have been numerous other incidents, which were perhaps less renowned but certainly not less tragic to the families of the victims. Examples of some less renowned accidents are:

Iraq

In September 1971, a huge shipment of American barley and Mexican wheat arrived at the port of Al Basrah, Iraq. The grain had been treated with a mercury solution to prevent spoilage during shipment and storage. Although high levels of mercury are poisonous, the grain was intended for use only as a seed. It was stolen from loading docks, however, and was repackaged and sold as food. Warnings had been attached in English and Spanish, but not in Arabic. Consequently, people ate the grain without realizing that it was unfit for consumption; hundreds died, and many thousands were crippled, blinded, and suffered brain damage. Although the Iraqi government did not make public news of the poisoning, it was estimated several years later that 1000 people may have died and that 10,000 others may have been injured.[2]

Boston

On January 19, 1919, a large molasses storage tank, owned by the Purity Distilling Company in Boston, ruptured. The tank was 50 feet high and 90 feet wide, and it held 2 million gallons of molasses, used in making rum. When the tank burst, a wave of molasses two stories high swept through the North End of Boston and out to sea at a speed of 35 miles per hour. Office buildings, stores, and houses were crushed under the heavy liquid. Many people escaped the area on foot, but dozens were killed and about 150 were injured in one of the most bizarre catastrophes in history.[2]

Caracas

In Caracas, Venezuela on April 9, 1952, a large crowd had gathered at a church at the beginning of Holy Week. Apparently a pickpocket, wishing to create confusion, shouted "Fire!", whereupon the worshipers rushed toward exits at the rear of the church. Many people fell and were trampled by their fellow parishioners who were rushing to escape the imaginary fire. Fifty-three people were killed, nearly half of whom were small children and infants.[2]

Glasgow

On January 2, 1971, at the Ibrox soccer stadium in Glasgow, Scotland, 66 persons were killed and 145 injured when a reinforced steel barrier collapsed under the weight of a surging crowd. The tragedy came after 80,000 spectators had thronged the exits near the end of a hotly contested game between the Glasgow Rangers and the Glasgow Celtics, traditional rivals. A group of Rangers supporters at an exit stairway reportedly attempted to reenter the stadium when they heard that their team had scored to tie the game at 1-1. A massive human pileup was created, causing the loss of many lives.[7]

Vail

Four persons died and eight others were injured on March 26, 1976, when two gondola cars fell more than 100 feet to the slopes of Vail Mountain in Vail, Colorado. The accident occurred because an automatic shutoff mechanism failed to respond to the partial derailment of a car ahead of the two that crashed. The cause was traced to five strands of steel sheath encasing the cable. The strands had begun to unravel at a point about two-thirds of the way up the mountain. The frayed cable caused the cars bearing the victims to derail and jam up, which should have activated an electrical overload switch designed to shut down the gondola. There is no explanation as to why this safety device did not function. [7]

Montgomery County, Pennsylvania

According to a coroner's report on September 10, 1977, accidentally mislabeled supply lines that carried oxygen and nitrous oxide to a new hospital emergency room caused six "premature" deaths in Montgomery County, Pennsylvania. The crossed lines were discovered on July 6 by the hospital's anesthesiologist nearly 7 months after the opening of the new facility. During the period that the error went undetected, some 300 patients had been mistakenly administered laughing gas, a substance that is usually not harmful but is potentially fatal to victims of heart attacks and emphysema. Thirty-five people had been pronounced dead in the emergency room between December and July. The death certificates of those whom the coroner determined had died as a result of the mislabeling were later changed to indicate their deaths were "accidental" - due to a lack of oxygen.

Willow Island

Scaffolding on a power plant cooling tower under construction at Willow Island, St. Mary's, West Virginia collapsed on April 27, 1978, killing all 51 workers who were on it. The men, who had been pouring cement, died instantly when they, the platform, and tons of concrete fell 170 feet to the ground. Eleven of the men were from one family. Although more than 1000 workers were employed at the site, only a few were on the ground inside the tower, and all were able to make it to safety. Witnesses said the scaffolding, which had hung on the top of both the inside and the outside of the tower (and which encircled it), peeled off "like a bunch of dominoes failing." There was no opportunity for those on the platform to reach safety at the one stairtower, which was more than 500 feet from some of the victims.

An investigation showed that three interdependent factors had led to the accident: (1) the concrete that was being poured had not been tested to determine whether it had cured sufficiently before the supporting forms were removed; (2) the scaffold was unable to withstand the weight on it because some necessary bolts were not used; and, (3) the beam sections supporting the concrete lifting system had not been anchored and maintained in a way that would have permitted it to hold the loads that were being lifted. As a result of these failures, the inner scaffold structure and the cathead cranes on top of the tower tipped inward, pulling the outer scaffolding over the top of the wall and causing it and the workers to fall to the ground. The federal Occupational Safety and Health Administration (OSHA) was criticized for not having inspected the site in more than a year before the accident, although the agency had been made aware of "possible disastrous consequences" if the scaffolding on the cooler tower was not used properly.[2,7]

Mount St. Helens

On March 20, 1980, a regional seismic network operated by the U.S. Geological Survey and the University of Washington recorded an earthquake of Richter magnitude 4.0 from a point north of the summit of Mount St. Helens, a dormant volcano that had last erupted in 1857. Two days later, the intensity of the seismic activity increased. Geologists suspected that magma, or melted rock, was moving up inside the mountain. The activity continued to increase. On March 27 a plume of steam and ash was emitted from Mount St. Helens and rose about 6600 feet above the mountain. At a point 1 mile north of the summit crater, a large bulge was observed to be forming in the mountain's side. By early May, this bulge had grown to a length of 1 mile and a width of 0.6 mile. Volcanologists watched this bulge for signs that it might split open, extruding magma. On May 18, without any warning, Mount St. Helens suddenly exploded.

Apparently triggered by an earthquake of Richter magnitude 5.0, the entire north slope burst open along the upper edge of the bulge, releasing the bottled up gases and magma. Up to 3 cubic kilometers of rock and ash were blown away from the mountain laterally. The blast spewed 1.3 billion cubic yards of material into the atmosphere. Almost everything within 5 miles of the volcano was destroyed. Tons of choking ash and dust were dropped on central and eastern Washington, northern Oregon and Idaho, and even parts of western Montana. The volcano continued to erupt during the remainder of 1980 and throughout the summer and fall of 1981. The death toll was confirmed in 1981 as 34, and 27 remain missing. Wildlife officials estimate that approximately 10,000 wild and domestic animals may also have been killed. [1]

Hilton Head

On June 27, 1978, a freak accident at a fountain pool on Hilton Head Island, South Carolina, claimed four victims. A young man and woman apparently broke one of the lights illuminating the fountain as they jumped into it at 8 P.M., sending several hundred volts through the water. The woman's roommate entered the pool, either unaware of the danger or attempting to save the pair. A neighbor then jumped in to attempt a rescue. All four were electrocuted.[7]

Brazil

The mishandling of a radioactive source in Brazil was the least publicized, but perhaps the most interesting of the nuclear accidents. The uncontrolled radiotherapy source was overlooked in an abandoned medical clinic, and was eventually discarded as a scrap. The stainless steel jacket and the platinum capsule surrounding the radioactive cesium were compromised by scavengers in a junkyard. The cesium was distributed among the people for use as "carnival glitter," because of its luminescent properties. The material was spread directly onto individuals' skin and face, as well as their clothing. Severe illness was immediately evident to most of the exposed victims. Four people died from exposure by the spring of 1988, and it was estimated that an additional five persons would die over the next five years. Over 40 tons of material, including clothing, shoes, and housing materials, were contaminated from the release of less than 1 gram of radioactive cesium.

1.5 ADVANCES IN SAFETY FEATURES

Today's sophisticated equipment and technologies require equally sophisticated means of accident prevention. Unfortunately, the existing methods of detection and prevention are often assumed to be adequate until proven otherwise. This approach to determining a technology's effectiveness sometimes is costly and often leads to loss of life. Chemical manufacturers and power plants are businesses, and thus are not as likely to "unnecessarily" update their present controls. Table 1.5.1 lists accidents, all of which have resulted in substantial loss of life, and the changes in safety-related practices they brought about.

TABLE 1.5.1   Disasters and Their Effects on Safety Measures[a]

Type	Location and Date	Total Deaths	Results
Fire	City of Chicago, Illinois October 9, 1871	250	Building codes prohibiting building structures; water reserve
Flood	Johnstown, Pennsylvania May 31, 1889	2209	Inspections
Tidal wave	Galveston, Texas September 8, 1900	6000	Sea wall built
Fire	Iroquois Theatre, Chicago December 30, 1903	575	Stricter theater safety standards
Marine	*General Slocum,* burned in the East River, New York June 15, 1904	1021	Stricter ship inspections; revision of statutes (life preservers, experienced crew, fire extinguisher)
Earthquake and fire	San Francisco, California April 18, 1906	452	Widened streets; limited heights of buildings; steel frame and fire resistant buildings
Mine	Monongah, West Virginia December 6, 1907	361	Creation of Federal Bureau of Mines; stipend mine inspections
Fire	North Collinwood School, Cleveland, Ohio March 8, 1908	176	Need realized for fire drills and planning of school structures
Fire	Triangle Shirt Waste Company, New York March 25, 1911	145	Strengthening of laws concerning alarm signals, sprinklers, fire escapes, fire drills
Marine	*Titanic* struck iceberg, Atlantic Ocean April 15, 1912	1517	Regulation regarding number of lifeboats; all passenger ships equipped for around the clock radio watch; International Ice Patrol

TABLE 1.5.1 Disasters and Their Effects on Safety Measures[a] (cont'd)

Type	Location and Date	Total Deaths	Results
Explosion	New London School, Texas March 18, 1937	294	Odorants put in natural gas
Fire	"Coconut Grove," Boston, Massachusetts November 28, 1942	492	Ordinances regarding aisle space, electrical wiring flameproofing of decorations, overcrowding; signs indicating maximum number of occupants; administration of blood plasma to prevent shock and the use of penicillin
Plane	Two-plane air collision over Grand Canyon, Arizona June 30, 1956	128	Controlled airspace expanded; use of infrared as a warning indicator

[a] From Thygerson.[23]

Before the advent of technology, there was still a need for safety features and warnings; yet these did not exist. Many accidents occurred because of a lack of knowledge of the system, process, or substance being dealt with. Many of the pioneers of modern science were sent to an early grave by their experiments. Karl Wilhelm Scheele, the Swedish chemist who discovered many chemical elements and compounds, often sniffed or tasted his finds. He died of mercury poisoning. As noted earlier, Marie Curie died of leukemia contracted from overexposure to radioactive elements. Had either of these brilliant scientists an accurate idea of the properties of their materials, their methods certainly would have been significantly different. In those days, safety precautions often were devised by trial and error; if inhaling a certain gas was found to make someone sick, the prescribed precaution was not to smell it. Today since the physical and chemical properties of most known compounds are readily found in handbooks, proper care can be exercised when working with these chemicals.[22] Labs are equipped with exhaust hoods and fans to minimize a buildup of gases. In addition, safety glasses and eye-wash stations are required, and gloves and smocks must be worn.

Many natural disasters are now accurately predicted, buying precious time in which warnings can be made and possible evacuation plans implemented. Radar equipment commonly track storms, and seismographs detect slight rumblings in the earth, which can provide early warning of potential earthquakes. Volcanic eruptions can be predicted by using seismic event counters and aerial scanning of anomalies detected in the infrared region. Where natural disasters often occurred unexpectedly in the past, similar occurrences today are more predictable. Thus, there is more time for preparation, and less likelihood of loss of life.

The use of computers and modern instrumentation has greatly enhanced plant safety. System overloads, uncontrollable reactions, and unusual changes in temperature or pressure can be detected, with the information being relayed to a computer. The computer can then shut down the system or take the steps necessary to minimize the danger. Industry has come a long way from sniffing and tasting its way to safety. The general subject of accident prevention and plant safety is addressed in more detail in Part II of this book. It is anticipated that further advances will occur in the following five areas:

1. Help develop inherently safer processes
2. Develop monitoring/detection devices that are more durable, cost effective, reliable, and accurate
3. Develop new and improved secondary/backup containment systems
4. Development of more reliable "continuous" containment systems
5. The development of more reliable "discontinuous" containment systems

1.6 ILLUSTRATIVE EXAMPLES

Example 1.1

Briefly discuss the lessons learned from the accident in Flixborough, England.

Solution 1.1

As described in the Chapter, the incident occurred on June 7, 1974. Three major lessons learned were:

1. Although the design of a process is important, retrofit and temporary installations can play an important role in accident management.
2. Reduce inventories at plant site.
3. The plant should not be located near populated areas (because of dispersion effects).

In 1985, the Canadian Chemical Producers Association (CCPA) released a pamphlet entitled, "Essential Components of Safety Assessment Systems". Modifications to process or plant was one of the topics discussed in this pamphlet. CCPA recommended a 12 element program (listed below) to formally examine and approve process conditions whether permanent or temporary prior to implementation.

1. Does the change involve any different chemicals which could react with other chemicals, including diluents, solvents and additives already in the process?

2. Does the new proposal encourage the production of undesirable byproducts either through the primary reactions, through the side reactions or through the impurities with the new chemical component?
3. Does the rate of heat generation and/or the reaction pressure increase as a result of the new scheme?
4. Does the proposed change encourage or require the operation of equipment outside of the approved operating or design limits of the chemical processing equipment?
5. Does the proposal consider the capability of the new chemical component and its impurities with material of construction?
6. Has the occupational health and environmental impact of the change been considered?
7. Has the design for modifying the process facilities or conditions been reviewed by a qualified individual using effective techniques for analyzing process hazards, particularly when the modifications are being made in rush situations or emergency conditions?
8. Has there been an on-site inspection by qualified personnel to ensure that the new equipment is installed in accordance with specifications and drawings?
9. Have the operating instructions and engineering drawings been revised to take into account the modifications?
10. Have proper communications been made for the training of chemical process operators, maintenance people, and supervisors who may be affected by the modification?
11. Have proper revisions been made to the process control logic, instrumentation set points and alarm points, especially for computer control systems to properly respond to the modification?
12. Have provisions been made to remove or completely isolate obsolete facilities in order to eliminate the chances for operator errors involving abandoned equipment?

These 12 elements indicate the types of questions that should be asked when making changes. The exact questions are not as important as realizing that many questions have to be asked.

Example 1.2

Briefly describe the lessons learned from the accident in Bhopal, India.

Solution 1.2

As described in the Chapter, the incident occurred on December 3, 1984. Four major lessons learned were:

1. Investigations are essential.

2. Protection systems must be fully operational.
3. All process and chemistry alternatives should be examined.
4. Plants should not be located near populated areas.

Example 1.3

Two major chemical reactions dominate the methyl isocyanate (MIC) reaction to form carbaryl that were occurring in the Bhopal plant. Briefly describe these two reactions.

Solution 1.3

The two major chemical reactions that were occurring in the Bhopal plant produced carbaryl using methyl isocyanate (MIC) as an intermediate. The first of the consecutive reactions combined methylamine and phosgene to produce MIC and hydrochloric acid. Then the MIC reacts with a-naphthol to form the final product, carbaryl. There is another possible route to produce carbaryl that does not make use of the MIC intermediate. Instead, phosgene is first reacted with a-naphthol to form a-naphthol chloroformate - the intermediate for the reaction - and hydrochloric acid. Then a-naphthol chloroformate reacts with methylamine to form carbaryl and hydrochloric acid. Both reaction schemes (depicted below) have the same overall stoichiometric equation.

Methyl Isocyanate Route:

$$CH_3NH_2 \quad + \quad COCl_2 \quad \longrightarrow \quad CH_3N=C=O \quad + \quad 2HCl$$

Methylamine   Phosgene   Methyl isocyanate

a – Naphthol   Carbaryl

Nonmethyl Isocyanate Route:

$$\text{Naphthol–OH} + COCl_2 \longrightarrow \text{Naphthyl–O–}\underset{\underset{\displaystyle \|}{O}}{C}\text{–Cl} + HCl$$

a – Naphthol Chloroformate

$$\text{Naphthyl–O–}\underset{\underset{\displaystyle \|}{O}}{C}\text{–Cl} + CH_3NH_2 \longrightarrow \text{Naphthyl–O–}\underset{\underset{\displaystyle \|}{O}}{C}NHCH_3 + HCl$$

Example 1.4

Briefly describe radioactive transformations, particularly as they apply to beta particle emissions.

Solution 1.4

Radioactive transformations are accomplished by several different mechanisms, most importantly alpha particle, beta particle, and gamma ray emissions. Each of these mechanisms are spontaneous nuclear transformations. The result of these transformations is the formation of different and more stable elements.

The kind of transformation that will take place for any given radioactive element is a function of the type of nuclear instability as well as the mass/energy relationship. The nuclear instability is dependent on the ratio of neutrons to protons; a different type of decay will occur to allow for a more stable daughter product. The mass/energy relationship states that for any radioactive transformation(s) the laws of conservation of mass and the conservation of energy must be followed.

Beta particle emission occurs when an ordinary electron is ejected from the nucleus of an atom. The electron (e) appears when a neutron (n) is transformed into a proton within the nucleus.

$$^1_0n \rightarrow {}^1_1H + {}^0_{-1}e \qquad\qquad (1.6.1)$$

Note that the proton is shown as a hydrogen (H) nucleus. This transformation must conserve the overall charge of each of the resulting particles. Contrary to alpha emission, beta emission occurs in elements that contain a surplus of neutrons. The daughter product of a beta emitter remains at the same atomic mass number, but is one atomic number higher than its parent. Many elements that decay by beta emission also release a gamma ray at the same instant. These elements are known as betagamma emitters. Strong beta radiation is an external hazard because of its ability to penetrate body tissue.[24]

Example 1.5

With reference to Section 1.5 "Advances in Safety Features", describe the difference between *continuous* containment/treatment systems and *discontinuous* ones.

Solution 1.5

1. The continuous containment/treatment systems are required for health problems that exist on a normal, day to day basis. In effect, it could be viewed as a "chronic" problem.
2. The discontinuous containment/treatment systems are required for accidents that are severe but of a short duration. In effect, it could be viewed as an "acute" problem.
3. The difference between these two processes and hazard systems is provided in Parts III and IV of this book.

1.7 SUMMARY

1. Accidents are a fact of life. Their frequency have increased with technology and time. Today's problems require more elaborate systems of warnings and controls to minimize the chances of serious accidents.
2. Many of the pre-technological-era accidents, such as those in Chicago (1871), Johnstown (1889), and Oppau (1921), were exacerbated by poor planning and construction. These accidents often resulted in the tightening of safety controls.
3. Advances in technology have brought about new problems. Nuclear power plant accidents (Three Mile Island and Chernobyl) have been the most frightening, perhaps because no one really knows what to expect from them.
4. There have been numerous "less publicized" accidents, many under unusual or unlikely circumstances. It is important to remember that an accident can occur at any time.

5.   Along with the rise of technology, industry has improved its accident prevention measures. Further improvements are expected in the future. Unfortunately, many improvements are not made until after an accident has occurred.

PROBLEMS

1.   What were the factors contributing to the spread and severity of the Chicago fire of 1871?
2.   What steps could have been taken to avoid the disaster at Flixborough, England, in 1974?
3.   What were the events leading to the accident at Chernobyl?
4.   How could the Ashland Oil accident have been prevented?
5.   Make a list of factors that contributed to the Bhopal disaster.
6.   What is a *core meltdown*? What was the probable cause of the Three Mile Island incident?

REFERENCES

1.   J. Cornell, *The Great International Disaster Book*, 3rd ed., Scribner, New York, 1976.
2.   *Catastrophe! When Man Loses Control*, prepared by the editors of the Encyclopaedia Britannica, Bantam Books, New York, 1979.
3.   I.C. Commentz, "The Explosion of the Nitrate Plant at Oppau," *Chem. Metall. Eng.*, 25 (18) 818-822, 1921.
4.   *Fire Eng.*, 97 (11) 795-799, 1944.
5.   F. Warner, "The Flixborough Disaster," *Chem. Eng. Prog.*, 71 (9), 77-84, 1975.
6.   T. A. Kletz, "The Flixborough Cyclohexane Disaster," *Loss Prev.*, 9, 106-110, 1975.
7.   *The Disaster File: The 1970s*, edited by Grace M. Ferrara, Facts on File, New York, 1979.
8.   C. Norman, "Chernobyl: Errors and Design Flaws," *Science*, 233, 1029-1031, 1986.
9.   T. Wilke and R. Milne, "The World's Worst Nuclear Accident," *New Sci.*, pp. 17-19, May 1, 1986.
10. E. Marshall, "Reactor Explodes Amid Soviet Science," *Science*, 232, 814-815, 1986.
11. S. Cooke, P. Galuszka, and J. J. Kosowatz, "Human Failures Led to Chernobyl," *Eng. News Record*, pp. 10, 11, August 28, 1986.
12. L. Hays and R. Koenic, "Dissecting Disaster: How Union Carbide Flushed Out Its Theory of Sabotage at Bhopal," Wall Street Journal, p. 1, July 7, 1988.
13. T. A. Kletz, *What Went Wrong! Case Histories of Process Plant Disasters*, Gulf Publishing, Houston, TX, 1985.

14. D. Mackenzie, "Design Failures that Caused Bhopal Disaster," *New Sci.*, p. 3, March 28, 1985.

15. D. Mackenzie, "The Chemistry Behind Bhopal's Disaster," *New Sci.*, pp. 3-4, Dec. 13, 1984.

16. D. Mackenzie, "Water Leak Caused Bhopal Disaster," *New Sci.*, p. 3, Jan. 10, 1985.

17. J. Campbell, "An Oil Spill May Spur New Storage Rules," *Chem. Week*, pp. 28-29, Feb. 3, 1987.

18. M. Cunningham, "An Act to Cap Chemical Leaks," *Insight*, pp. 24-25, Feb. 29, 1988.

19. J. Campbell and A. Soast, "Reassembled Oil Tank Collapse Sending Slick Far Downstream," *Eng. News Record*, pp. 12-14, Jan. 14, 1988.

20. S. Adcock, "It's Official: Spark Led to Flight 800 Explosion," *Newsday*, August 24, 2000.

21. A. Kim, "What is the deal with the Year 2000 bug? (Y2k – demystified)," Weeno.com, 1999.

22. J. Spero, B. Devito and L. Theodore, "Regulatory Chemicals Handbook", Marcel Dekker, New York, 2000.

23. A. L. Thygerson, *Accidents and Disasters: Causes and Countermeasures*, Prentice-Hall, Englewood Cliffs, NJ, 1977.

24. M. K. Theodore and L. Theodore, *Major Environmental Issues Facing The 21st Century*, (originally published by Simon and Shuster), Theodore Tutorials, East Williston, NY, 1995.

2

Legislation

2.1 INTRODUCTION

An accident is an event that has the potential to become a catastrophe. There has been a growing concern about emergency planning because hazardous and toxic substances are increasingly being created and used by industry. The fear of accidents is probably the major reason for the promulgation of emergency planning and response legislation. Increased public awareness is also a goal of the new legislation.

The concern for emergency planning and response is reflected in the legislation summarized in this chapter. Section 2.2 discusses the early air and water legislation (pre-1970). The water legislation attempted to address the issue of accidents, but it was not very successful. Section 2.3 gives an overview of the *Clean Air Act* and addresses the specific sections relevant to emergency planning and response to emergencies. Although the *Clean Air Act* does not cover emergency planning and response in a clear and comprehensive manner, certain elements of the act are particularly significant. These include implementation plans and national emission standards for hazardous air pollutants. Section 2.4 examines the *Clean Water Act* as well as other legislation pertaining to water pollution. It will be apparent to the reader that legislation for emergency planning and response is more developed for water than it is for air. Section 2.5 is devoted to the *Resource Conservation and Recovery Act* (RCRA) and the *Comprehensive Environmental Response, Compensation, and Liability Act* (CERCLA). These two important pieces of legislation are concerned with preventing releases and with the requirements for the cleanup of hazardous and toxic sites. RCRA and CERCLA contain specific sections that address emergency planning and response. Section 2.6 discusses the *Superfund*

Amendments and Reauthorization Act (SARA), an important piece of legislation. SARA deals with the cleanup of hazardous waste sites as well as emergency planning and response. Title III, which is the heart of SARA, establishes requirements for emergency planning and "community right to know" for federal, state, and local government, as well as industry. Title III is a major stepping-stone in the protection of the environment, but its principal thrust is to facilitate planning in the event of a catastrophe. A more comprehensive examination of Title III is provided in Chapter 3.

The remaining Sections examine three important topics as they relate to the subject title of this book. Section 2.7 reviews the details of the *U.S. Environmental Protection Agency's* (USEPA's) *Risk Management Program* while Section 2.8 provides information on the *Occupational Health and Safety Administration* (OSHA). The chapter continues with a short Section (2.9) on potential environmental violations and then concludes with a Section (2.10) on the *Pollution Prevention Act of 1990*.

2.2 EARLY LEGISLATION

Before the 1970s, there was little legislation regarding the prevention of air and water pollution. Although some of the early laws approached the issue of pollution prevention, none of this legislation provided for emergency planning and response in the event of an accident.

The *Rivers and Harbors Act of 1899*, also known as the *Refuse Act*, was the first piece of legislation directed toward the prevention of water pollution. It stated that:

> "It shall not be lawful to throw, discharge, or deposit, or cause, suffer, or procure to be thrown, discharged, or deposited either from or out of any ship, barge, or other floating craft of any kind, or from the shore, wharf, manufacturing establishment, or mill of any kind, any refuse matter of any kind or description whatever other than that floating from streets and sewers and passing therefrom in a liquid state, into any navigable water of the United States."[1]

Violators of the *Refuse Act* were subject to a fine of not more than $2,500 and/or imprisonment for not more than one year. Although not drafted to curb oil pollution, the *Refuse Act* has been widely used to prosecute those who discharge oil into navigable waters.

The *Oil Pollution Act*, which prohibits the discharge of oil from any vessel into navigable waters, was passed in 1924. The penalties for violators were the same as those for the *Refuse Act* of 1899 with one addition: the Coast Guard had the authority to suspend or revoke licenses held by the officers of vessels found to be in violation of the act.

The *Federal Water Pollution Control Act of 1948* was enacted to enhance the quality and value of our natural resources and to establish a national policy for the prevention, control, and abatement of water pollution. This act required the states to establish enforceable water quality standards that would protect public health and welfare, and improve the quality of the water. While these standards were adequate for preventing pollution on a continuous basis, they were not suited for nonrecurring violations such as oil spillage.

The *Oil Pollution Act of 1961* was primarily concerned with the regulation of the deliberate discharge of oil or oil wastes into the water. As a result of this act, the owners of transport ships were to take precautions to minimize the effect of pollution. Oily ballast water separators were required on ships, and owners had to maintain oil record books. Violators of the *Oil Pollution Act* were subject to fines of not more than $1,000 for improper record keeping.

The *Clean Water Restoration Act of 1966* broadened the *Oil Pollution Act of 1924*. The new law included not only coastal waters but all rivers and lakes of the United States. It also made violators responsible for the cleanup of any oil discharge and increased the fine to not more than $10,000. However, the 1966 act also narrowed the definition of "discharge," thereby restricting enforcement and reducing the effectiveness of the *Oil Pollution Act of 1924*. As a result, the nineteenth-century *Refuse Act* evoked renewed interest as a means of prosecuting those discharging oil into navigable waters.

All the early water legislation addressed the issue of pollution, but the issue of emergency planning and response was overlooked. The *Oil Pollution Act of 1961* and the *Clean Water Restoration Act of 1966* leaned toward prevention and response, but the efforts fell short of providing meaningful legislation.

The *Solid Waste Disposal Act* was passed in 1965 to address the problem of what to do with the increasing amounts of solid wastes being generated. It was designed to:

1. Promote the demonstration, construction, and application of solid waste management and resource recovery systems, which preserve and enhance the quality of air, water, and land resources
2. Provide technical and financial assistance to state and local governments and interstate agencies in the planning and development of resource recovery and solid waste disposal programs
3. Promote a national research and development program for improved management techniques, more effective organizational arrangements, and new and improved methods of collection, separation, recovery, and recycling of solid wastes, and the environmentally safe disposal of nonrecoverable residues
4. Provide for the promulgation of guidelines for solid waste collection, transport, separation, recovery, and disposal systems
5. Provide training grants in occupations involving the design, operation, and maintenance of solid waste disposal systems[2]

Although this act was directed toward the research and development of technologies to deal with the solid waste problem, it resembled the other laws discussed in this section in that it was not concerned with emergency planning and response.

2.3 AIR POLLUTION LEGISLATION

The *Clean Air Act* and its amendments were enacted to deal with the growing problem of air pollution. The purposes of the *Clean Air Act* are to:

1. Protect and enhance the quality of the nation's air resources so as to promote the public health and welfare and the productive capacity of its population
2. Initiate and accelerate a national research and development program to achieve the prevention and control of air pollution
3. Provide technical and financial assistance to state and local governments in connection with the development and execution of their air pollution prevention and control programs
4. Encourage and assist the development and operation of regional air pollution control programs

The areas of interest in the field of emergency planning and response are the sections concerning *State Implementation Plans* (SIPS) and those for the *National Emission Standards for Hazardous Air Pollutants* (NESHAPs).

Under the *Clean Air Act*, the states are required to propose implementation plans, SIPs, for the prevention of air pollution emergency episodes. These implementation plans do not apply to all toxic releases; rather, they are directed toward preventing excessive buildup of air pollutants that are known to be harmful to the population and the environment when concentrations exceed certain limits. The compounds affected under the implementation plans are sulfur dioxide, particulate matter, carbon monoxide, nitrogen dioxide, and ozone. Each implementation plan must include a contingency plan, which will outline the steps to be taken in the event that a particular pollutant concentration reaches the level at which it can be considered to be harmful.

All contingency plans must address each of the three different levels that can occur in a pollution emergency episode: air pollution alert, air pollution warning, and air pollution emergency. Table 2.3.1 lists the limits set early for the air pollutants at each level. Each episode level must have preplanned abatement strategies for the reduction and/or the elimination of these air pollutants and will be effective at the start of each level. The contingency plan must also provide for public announcement whenever any episode stage has been determined to exist. The implementation plans under the Clean Air Act are solely based on the continuous emission of the previously stated air pollutants. They do not mandate any actions to be taken in the event of an accidental toxic release.

Table 2.3.1 Clean Air Act Pollution Limits

Pollutant[a]		Air Pollution Alert	Air Pollution Warning	Air Pollution Emergency
SO_2:	24-hr avg	0.3 ppm	0.6 ppm	0.8 ppm
PM_{10}:	24-hr avg	350 $\mu g/m^3$	420 $\mu g/m^3$	500 $\mu g/m^3$
CO:	8-hr avg	15 ppm	30 ppm	40 ppm
O_3:	1-hr avg	0.2 ppm	0.4 ppm	0.5 ppm
NO_2:	1-hr avg	0.6 ppm	1.2 ppm	1.6 ppm
	24-hr avg	0.15 ppm	0.3 ppm	0.4 ppm

[a] SO_2, sulfur dioxide; PM_{10}, particulate matter, ten microns and over; CO, carbon monoxide; O_3, ozone; NO_2, nitrous oxide.

The *National Emission Standards for Hazardous Air Pollutants* (NESHAPs) is a section of the *Clean Air Act* that concerns a number of hazardous volatile organic compounds which, if emitted, can be very dangerous to the population and the environment. Under the requirements of NESHAPs, industry must develop and maintain leak detection and elimination programs to protect against fugitive emissions. A leak detection and elimination program must contain the quantitative definition of a "leak which is acceptable" when compared to the background concentration at a facility. A facility can have more than one definition of a leak if the background concentration varies. With the establishment of an understanding of what constitutes a leak, the facility must initiate and maintain a reliable and accurate volatile hazardous air pollutant monitoring system that will be operated for the detection of major leaks. Such a monitoring system involves obtaining air samples from points throughout the facility on a continuous and sequential basis and analyzing the samples with gas chromatography. Once a leak has been detected, a portable analyzer will be used to locate the emission point, whereupon the facility is required to repair the leak as soon as possible, but no later than 15 calendar days after detection. A facility will have up to 6 months to complete repairs that require plant shutdown. Although NESHAPs do not address emergency planning either, they do formulate measures for the prevention of fugitive emissions throughout a facility and its boundaries.

USEPA's *recommended* final standards in 1997 for particulate matter and ozone (otherwise known as soot and smog) will be a major step forward in protecting the public from the health hazards of air pollution. These updated standards, the product of many years of intense scientific review, move the agency toward fulfilling the *Clean Air Act's* goal of ensuring Americans that their air is safe to breathe. The agency indicates that the new standards will provide new health protections to 125 million Americans, including 35 million children.

The standard for coarse particles remains essentially unchanged, while a new standard for fine particles will be set at an annual limit of 15 micrograms per cubic meter, with a 24-hour limit of 65 micrograms per cubic meter. Details on the fine particulate standard are:

1. This is the first time ever that the government has set a public health standard for fine particle pollution
2. Scientists say that fine particles – those measuring 2.5 micrometers in diameter and smaller – are some of the most damaging to human health because they penetrate and remain in the deepest passages of the lungs
3. This new standard, as revised from USEPA's original proposal, will provide new protections to nearly 70 million Americans, and will prevent approximately 15,000 premature deaths each year
4. By setting an annual limit of 15 micrograms per cubic meter, the new standard focuses on the most important issue – controlling the amount of pollution and exposure to which Americans are subjected – and therefore addresses the most significant health concerns
5. A more flexible 24-hour standard of 65 micrograms, revised from the original USEPA proposal, will give greater flexibility to individual sources of pollution, while still ensuring that the health of the American people is protected

For ozone, the recommended final standard will be updated from 0.12 parts per million of ozone measured over one hour to a standard of 0.08 parts per million measured over eight hours, with the average fourth highest concentration over a three-year period determining whether an area is out of compliance. Details are provided below.

1. This is the first time in 20 years that the ozone standard will be updated
2. The updated standard recognizes the current scientific view that exposure to ozone levels at and below the current standard causes significant adverse health effects in children and in healthy adults engaged in outdoor activities
3. The new 0.08 standard is much stronger and more protective than the old standard of 0.12. It will extend new health protections to 35 million people, bringing to 113 million the number of Americans protected by the air quality standard for ozone
4. For children, the new standard will reduce respiratory problems, such as asthma attacks. It will result in one million fewer incidences of decreased lung function in children each year
5. By moving from a one-hour to an eight-hour measurement, the standard will better reflect the real-world effects of ozone on human health
6. By focusing on concentration of ozone, the new standard will do more than merely designate high-pollution areas as out of compliance – for the first time, it will also respond to health concerns based on how much an area is out of compliance
7. Using the fourth maximum rather than the third (as originally proposed by USEPA) will provide greater stability in the designation of areas, consistent with providing strong public health protections[3]

2.4 WATER POLLUTION LEGISLATION

The *Clean Water Act* and its amendments were passed to eliminate the discharge of pollutants into the navigable waters of the United States. The purpose of the *Clean Water Act* was to restore and maintain the chemical, physical, and biological integrity of the nation's waters. It also requested the states to develop water quality standards that provide for the protection and propagation of fish, shellfish, and wildlife. Most important, contingency plans were to be developed to facilitate the establishment and designation of strike forces in the event of an emergency, and to set up a system of surveillance and notice. Contingency plans were not required of federal and state agencies alone. The act also requires certain companies to prepare *Spill Prevention Control and Countermeasures* (SPCC) plans, which initially were directed toward the discharge of oil. SPCC plans set out the procedures, methods, and equipment requirements for facilities subject to the legislation.

Any non-transportation-related onshore and offshore facility that has the potential to discharge oil into navigable waters is subject to SPCC plans. This includes not only facilities that produce oil, but also industrial, commercial, agricultural, and public facilities that use or store oil. An SPCC plan must have the full approval of management at a level with authority to commit the necessary resources. A complete SPCC plan would follow the sequence outlined below for the minimal prevention requirements.

1. A facility that has experienced one or more spill events within 12 months before the effective date must prepare a written description of each spill, corrective action taken, and plans for preventing recurrences
2. Where experience indicates a reasonable potential for equipment failure (such as a tank overflow, rupture, or leakage), the plan is to include a prediction of the direction, rate of flow, and total quantity of oil that could be discharged from the facility as a result of each major type of failure
3. Appropriate containment and/or diversionary structures or equipment to prevent discharged oil from reaching a navigable course should be provided
4. When it is determined that the installation of containment and/or diversionary structures or equipment to prevent discharged oil from reaching navigable waters is not practicable from any onshore or offshore facility, the owner/operator should clearly demonstrate such impracticability and provide the following:
   a. a strong spill contingency plan, and
   b. a written commitment of manpower, equipment, and materials required to expeditiously control and remove any harmful quantities of oil discharged
5. In addition to the minimal prevention standards, the plans should include a complete discussion of conformity to the following applicable guidelines

and their effective spill prevention plans and containment procedures (or if more stringent, to state rules, regulations and guidelines):

a.   facility drainage (onshore)
b.   bulk storage tanks (onshore)
c.   facility transfer operations, pumping and in-plant processes (onshore)
d.   facility tank car and tank truck loading/unloading rack (onshore)
e.   oil production facilities (onshore)
f.   oil drilling and workover facilities (onshore)
g.   oil drilling and workover facilities (offshore)
h.   inspection and records
i.   security (excluding oil production facilities)
j.   personnel training and spill prevention procedures

All SPCC plans must be certified by a registered professional engineer. In the event of a change in the facility that will affect the potential for discharge, the SPCC must be amended and recertified. SPCC plans must be evaluated at least once every 3 years and must be available on site during normal working hours for examination by the regional administrator of the USEPA. The establishment of SPCC plans is very important for the prevention of oil spills and, as discussed later in Section 2.6, these plans can be adapted to chemical spills as well.

The *Marine Protection, Research, and Sanctuaries Act* (MPRSA) of 1972, often referred to as the *Ocean Dumping Act*, has two basic objectives. These objectives are to regulate intentional ocean disposal of materials, and to authorize related research. Virtually all ocean dumping that occurs today is dredged material, i.e., sediments removed from the bottom of waterbodies in order to maintain navigation channels. The Corps of Engineers issues permits for ocean dumping of dredged material. The permits are to be based on the same criteria utilized by USEPA under the provisions of the *Ocean Dumping Act*, and to the extent possible, USEPA recommended dumping sites are used. Amendments enacted in 1992 expanded USEPA's role in permitting of dredged material by authorizing USEPA to impose permit conditions or even deny a permit, if necessary to prevent environmental problems. Permits issued under the *Ocean Dumping Act* specify the type of material to be disposed, the amount to be transported for dumping, the location of the dumpsite, the length of time the permit is valid, and special provisions for surveillance.[4]

The *Safe Drinking Water Act* (SDWA) is the key federal law for protecting public water supplies from harmful contaminants. First enacted in 1974 and substantively amended in 1986 and 1996, the act is administered through programs that establish standards and treatment requirements for public water supplies, control underground injection of wastes, finance infrastructure projects, and protect sources of drinking water. The 1974 law gave USEPA substantial discretionary authority to regulate drinking water contaminants and gave states the lead role in implementation and enforcement.

The *Safe Drinking Water Act Amendments of 1996* substantially revised the act. Among other changes, the 1996 law:

1. Added some flexibility to the act's standard setting provisions
2. Required USEPA to conduct cost-benefit analyses for most new standards
3. Added provisions to improve small system compliance and protect source waters
4. Expanded consumer information requirements
5. Authorized a *State Revolving Loan Fund* (SRF) program to help public water systems finance projects needed to meet SDWA requirements[4]

2.5 HAZARDOUS AND TOXIC WASTE LEGISLATION

Two pieces of legislation that address the growing concern about the accumulation of hazardous and toxic chemicals are the *Resource Conservation and Recovery Act* (RCRA) of 1976 and the *Comprehensive Environmental Response, Compensation, and Liability Act* (CERCLA).

RCRA

The *Resource Conservation and Recovery Act* addressed the development of economic and market incentives for fostering conservation and recovery through the newly created Resource Conservation Committee and through the procurement of products containing recycled materials. It formed technical assistance panels and called for the research, development, demonstration, and evaluation of technologies for recycling and resource recovery. RCRA also led to the development of state and regional programs for resource conservation and recovery through the planning and financial assistance programs of Subtitle C (Preparedness and Prevention) and Subtitle D (Contingency Plans and Emergency Procedures).

Under RCRA, each facility must contain a contingency plan designed to minimize hazards to human health or the environment from fires, explosions, or any unplanned sudden or nonsudden release of hazardous waste or hazardous waste constituent to air, soil, or surface water. The items that follow are applicable to each contingency plan.

1. The plan must describe the actions that facility personnel must take in response to fires, explosions, or any unplanned sudden or nonsudden release of hazardous wastes or hazardous waste constituents to air and/or surface water at the facility
2. An owner or operator who has already prepared *a Spill Prevention Control and Countermeasures* (SPCC) plan need only amend that plan to incorporate hazardous waste management provisions sufficient to comply with this section
3. The plan must describe arrangements agreed to by local police departments, fire departments, hospitals, contractors, and state and local emergency response teams to coordinate emergency services

4. The plan must list names, addresses, and phone numbers (office and home) of all persons qualified to act as emergency coordinator. When more than one person is listed, one must be named as primary emergency coordinator and others must be listed in the order in which they would assume responsibility as alternatives
5. The plan must include a list of all emergency equipment at the facility (such as fire extinguishing systems, spill control equipment, internal and external communications and alarm systems, and decontamination equipment). In addition, the plan must include for each item on the list a physical description, a brief outline of its capabilities, and its location
6. The contingency plan must include an evacuation plan for facility personnel if the possibility exists that evacuation might become necessary. The evacuation plan must describe signal(s) to be used to begin evacuation, evacuation routes, and alternate evacuation routes (in cases where primary routes could be blocked by releases of hazardous wastes or fires)[5]

Copies of the contingency plan and all its revisions must be maintained at the facility as well as submitted to all local police departments, fire departments, hospitals, and state and local emergency response teams.

RCRA has been amended nine times since its initial establishment in 1976. The most significant sets of amendments occurred in 1980, 1984, and 1992. As a result of the *Solid Waste Disposal Act Amendments of 1980*:

1. USEPA was provided tougher enforcement powers to deal with illegal dumpers
2. The Agency's authority to regulate certain high-volume, low-hazard wastes (known as "special wastes" was restricted
3. Funds were authorized to conduct an inventory of hazardous waste sites
4. RCRA authorizations for appropriations were extended

The most significant set of amendments to RCRA was the *Hazardous and Solid Waste Amendments of 1984* (HSWA). In addition to restrictions on land disposal, and expanding those required to follow hazardous waste regulations to include small-quantity hazardous waste generators (those producing between 100 and 1,000 kg of waste per month), HSWA created a new regulatory program for underground storage tanks. USEPA was directed to issue regulations governing those who produce, distribute, and use fuels produced from hazardous waste, including used oil. Under HSWA, hazardous waste facilities owned or operated by federal, state, or local government agencies must be inspected annually, and privately owned facilities must be inspected at least every two years. Each federal agency was requires to submit to USEPA an inventory of all the hazardous waste facilities it ever owned. In addition, the 1984 law:

1. Imposed on USEPA a timetable for issuing or denying permits for treatment, storage, and disposal facilities
2. Required permits to not exceed 10 years for fixed terms
3. Ended (in 1985) the "interim status" of land disposal facilities that existed prior to RCRA's enactment, unless they met certain requirements
4. Required permit applications to be accompanied by information regarding the potential for public exposure to hazardous substances in connection with the facility
5. Authorized USEPA to issue experimental permits for facilities demonstrating new technologies

The third major set of amendments was the *Federal Facility Compliance Act of 1992*. This act resolves the legal question of whether federal facilities are subject to enforcement actions under RCRA by waiving the government's sovereign immunity from prosecution. As a result, states, USEPA, and the Department of Justice can enforce the provisions of RCRA against federal facilities, and federal departments and agencies can be subjected to injunctions, administrative orders, and/or penalties for noncompliance.[6]

CERCLA

The *Comprehensive Environmental Response, Compensation, and Liability Act* (CERCLA) of 1980 was the first major response to the problem of abandoned hazardous waste sites throughout the nation. CERCLA was the beginning of the remediation of hazardous waste sites. This program was designed to:

1. Develop a comprehensive program to set priorities for cleaning up the worst existing hazardous waste sites
2. Make responsible parties pay for these cleanups wherever possible
3. Set up a $1.6 billion *Hazardous Waste Trust Fund*, properly known as "Superfund," for the twofold purpose of performing remedial cleanups when responsible parties could not be held accountable and responding to emergency situations involving hazardous substances
4. Advance scientific and technological capabilities in all aspects of hazardous waste management, treatment, and disposal

CERCLA requires the person in charge of a vessel or facility to notify *the National Response Center* (NRC) immediately when there is a release of a designated hazardous substance in an amount equal to or greater than the reportable quantity. CERCLA establishes the reportable quantity for releases of designated hazardous substances at one pound, unless otherwise specified. To ensure that the need for response can be evaluated and any response can be undertaken in a timely fashion, such releases require notification to government officials.

The development of the emergency planning and response actions under CERCLA is based primarily on a national contingency plan that was developed under *the Clean Water Act*. Although the actions of CERCLA have the capabilities to handle hazardous and toxic releases, the act was primarily directed toward the cleanup of abandoned hazardous waste sites.

Under Section 7003 of the RCRA legislation (1984), private citizens are authorized to bring legal action against companies, governmental entities, or individual citizens if past or present hazardous waste management practices are believed to pose an imminent danger. Section 7003 applies to past generators as well as to situations or sites where past acts or failures to act may have contributed to a present endangerment to human health and the environment. Citizen rights to sue are limited, however, (1) if USEPA or the state government is diligently bringing and prosecuting a related action under Section 7003 of RCRA or Section 106 of CERCLA, or (2) if USEPA or the state has settled a related action by entering into a consent decree. CERCLA was amended by the *Superfund Amendments and Reauthorization Act* (SARA) in 1986 (discussed in Section 2.6).

2.6 SUPERFUND AMENDMENTS AND REAUTHORIZATION ACT OF 1986 (SARA)

The *Superfund Amendments and Reauthorization Act of 1986* renewed the national commitment to correcting problems arising from previous mismanagement of hazardous wastes. While SARA was similar in many respects to the original law (i.e., CERCLA), it also contained new approaches to the program's operation. The 1986 Superfund legislation[7]:

1.   Reauthorized the program for 5 more years and increased the size of the cleanup fund from $1.6 billion to $8.5 billion
2.   Set specific cleanup goals and standards, and stressed the achievement of permanent remedies
3.   Expanded the involvement of states and citizens in decision making
4.   Provided for new enforcement authorities and responsibilities
5.   Increased the focus on human health problems caused by hazardous waste sites

The new law is more specific than the original statute with regard to remedies to be used at Superfund sites, public participation, and accomplishment of cleanup activities. The most important part of SARA with respect to public participation is Title III, which addresses the important issues of community awareness and participation in the event of a chemical release.

As mentioned earlier, Title III of SARA addresses hazardous materials releases; its subtitle is the *Emergency Planning and Community Right-to-Know Act of 1986*. Title III establishes requirements for emergency planning, hazardous emissions reporting, emergency notification, and "community right to know." The objectives of Title III are to improve local chemical emergency response capabilities, primarily through improved emergency planning and notification, and to provide citizens and local governments with access to information about chemicals in their localities. The major sections of Title III that aid in the development of contingency plans are as follows.

1. Emergency Planning (Sections 301-303)
2. Emergency Notification (Section 304)
3. Community Right-to-Know Reporting Requirements (Sections 311 and 312)
4. Toxic Chemicals Release Reporting-Emissions Inventory (Section 313)

Title III also developed time frames for the implementation of *the Emergency Planning and Community Right-to-Know Act of 1986.* Table 2.6.1 lists these initial dates (all of which have already passed).[8]

Sections 301-303 of Title III, which are responsible for emergency planning, are designed to develop state and local governments' emergency response and preparedness capabilities through better coordination and planning, especially within local communities. Section 301 required the governor of each state to establish *a State Emergency Response Commission* (SERC) by April 17, 1987. If a state commission had not been designated by that time, the governor was to operate as the commission until such a designation was made. The SERC was responsible for establishing emergency planning districts by July 17, 1987, and appointing, supervising, and coordinating local emergency planning committees (LEPCS) one month after the district was designated. The *State Emergency Response Commission* was also responsible for establishing procedures for receiving and processing public information requests as well as reviewing local emergency plans.

Each local emergency planning committee was responsible for developing an emergency response plan by October 17, 1987. An LEPC was to be composed of elected state and local officials; police, fire, civil defense, and public health professionals; representatives of community groups; environmental health and transport agencies; and the media. Title III also called for any facility subject to emergency planning requirements to designate a representative to participate in this planning process by September 17, 1987, as well as to act as facility emergency coordinator in the event of an accident. This was important because the facility emergency coordinator could provide technical assistance, an understanding of facility response procedure, information about chemicals and their potential effects on nearby persons and the environment, and response training opportunities. To assist local communities in preparing and reviewing emergency response plans, the *National Response Team* (NRT) was asked to publish the *Hazardous Materials Emergency Planning Guide.* This publication provided local planning committees with information on what should be included in an emergency response plan. In developing this plan, the local committees evaluate available

TABLE 2.6.1   Early Key Title III Dates[a]

Date	Action
November 17, 1986	USEPA publishes interim final List of Extremely Hazardous Substances and their Threshold Planning Quantities in Federal Register [Section 302(a)(2-3)]
November 17,1986	USEPA initiates comprehensive review of emergency systems [Section 305(b)]
January 27, 1987	USEPA publishes proposed formats for emergency inventory forms and reporting requirements in Federal Register [Section 311-312]
March 17, 1987	National Response Team publishes guidance for preparation and implementation of emergency plans [Section 303(f)]
April 17, 1987	State governors appoint SERCs [Section 301(a)]
May 17, 1987	Facilities subject to Section 302 planning requirements notify SERC [Section 302(c)]
June 1, 1987	USEPA publishes toxic chemicals release (i.e., emissions inventory) form [Section 302(c)]
July 17, 1987	SERC designates emergency planning districts [Section 301(b)]
August 17, 1987 (or 30 days after designation of districts whichever is earlier)	SERC appoints members of LEPCs [Section 301(c)]
September 17, 1987 (or 30 days after local committee is formed, whichever is sooner)	Facility notifies LEPC of selection of a facility representative to serve as facility emergency coordinator [Section 303(d)(1)]
October 17, 1987	Material safety data sheets or list of MSDS chemicals submitted to SERC, LEPC, and local fire department [Section 311(d))
March 1, 1988	Facilities submit their initial emergency inventory forms to SERC, LEPC, and local fire department [Section 312(a)(2)]
April 17, 1988	Final report on emergency systems study due to Congress [Section 305(b)]
July 1, 1988 (and annually hereafter)	Facilities to submit initial toxic chemical release forms to USEPA and designated state officials [Section 313(a)]
October 17, 1988	LEPCs complete preparation of an emergency plan [Section 303(a)]

[a] This table is a list of some key dates relative to the implementation of the Emergency Planning and Community Right-to-Know Act of 1986.

resources for preparing for and responding to a potential accident. An emergency response plan should include the following elements.

1. Identification of facilities at which an extremely hazardous substance is present in an amount in excess of the threshold planning quantity, whether it is being produced, used, or stored. Transportation routes of any extremely hazardous substance to or from the facility must be given, as well. Also to be identified were other facilities contributing or subjected to additional hazards because of their proximity to facilities such as hospitals or natural gas facilities

2. The methods and procedures to be followed by facility operators and local emergency and medical personnel in the event of a release of an extremely hazardous substance

3. The designation of community and facility emergency coordinators who shall make the necessary determinations in implementing an emergency plan

4. The procedures providing reliable, effective, and timely notification by the facility emergency coordinator and the community emergency coordinator to persons designated in the emergency plan and to the public in the event that a release has occurred

5. The methods for determining the occurrence of a release and the area or population likely to be affected

6. A description of emergency equipment and facilities in the community and at each facility in the community subject to Title III requirements, as well as the identification of the persons responsible for such equipment and facilities

7. Evacuation plans that include provisions for a precautionary evacuation and alternative traffic routes

8. Training programs that include schedules for training of local emergency response and medical personnel as well as facility personnel

9. The methods and schedules for exercising emergency response plans

These planning elements are presented in the same order that they appear in the *Hazardous Materials Emergency Planning Guide*. Table 2.6.2 is a sample outline of a *Hazardous Emergency Response Plan*. The emergency planning activities of the local emergency planning committee was to be initially based on, but not limited to, the *Extremely Hazardous Substance* list published on April 27, 1987.[9] A list of extremely hazardous substances and their threshold planning quantities is provided in Table 2.6.3. Emergency planning and response are discussed in detail in Chapter 3.

Section 304 of Title III is devoted to emergency notification and the procedures to be followed in the event of a release. The requirement for emergency comes into effect with the establishment of the *State Emergency Response Commission* and the local emergency planning committee. If the facility produces, uses, or stores one or more hazardous chemicals, and a release

TABLE 2.6.2 Sample Outline of a Hazardous Materials Emergency Plan[a,b]

A. Introduction
    1. Incident information summary
    2. Promulgation document
    3. Legal authority and responsibility for responding
    4. Table of contents
    5. Abbreviations and definitions
    6. Assumptions/planning factors
    7. Concept of operations
        a. Governing principles
        b. Organizational roles and responsibilities
        c. Relationship to other plans
    8. Instructions on plan use
        a. Purpose
        b. Plan distribution
    9. Record of amendments
B. Emergency assistance telephone roster
C. Response functions[c]
    1. Initial notification of response agencies
    2. Direction and control
    3. Communications (among responders)
    4. Warning systems and emergency public notification
    5. Public information/community relations
    6. Resource management
    7. Health and medical services
    8. Response personnel safety
    9. Personal protection of citizens
        a. Indoor protection
        b. Evacuation procedures
        c. Other public protection strategies
    10. Fire and rescue
    11. Law enforcement
    12. Ongoing incident assessment
    13. Human services
    14. Public works
    15. Others
D. Containment and cleanup
    1. Techniques for spill containment and cleanup
    2. Resources for cleanup and disposal
E. Documentation and investigative follow-up
F. Procedures for testing and updating plan
    1. Testing the plan
    2. Updating the plan

[a] "Hazardous Materials Emergency Planning Guide." National Response Team, March, 1987, Washington, DC.
[b] Depending on local circumstances, communities will develop some sections of the plan more extensively than other sections.
[c] These "response functions" are equivalent to the "functional annexes" of a multihazard emergency operations plan described in the "Guide for Development of State and Local Emergency Operations Plans," prepared by the Federal Emergency Management Association, March, 1987, Washington, DC.

of a listed hazardous substance occurs that exceeds the reportable quantity for that substance, the facility must notify immediately both the LEPC and the SERC. (Federally permitted releases are exempt from this requirement.) The initial notification can be by telephone, by radio or in person. Emergency notification requirements involving transportation incidents can be satisfied by dialing 911 or, in the absence of the 911 emergency number, by calling the operator. This emergency notification needs to include:

1. The identity of the substance
2. An indication of whether the substance is included as part of the list of extremely hazardous substances
3. An estimate of the quantity released into the environment
4. The time and duration of the release
5. The medium into which the release occurred (atmosphere, water, or land)
6. Any known or anticipated acute or chronic health risks associated with the emergency and, where appropriate, advise regarding medical attention necessary for exposed individuals
7. Proper precautions to take as a result of a release, including evacuation
8. The name and telephone number of the person reporting the incident

After a release, Section 304 requires a follow-up written emergency notice or notices to update information included in the initial notice, to tell more about actual response actions taken and to present any known or anticipated data on chronic health risks associated with the release, as well as advice regarding medical attention necessary for exposed individuals. Information included in such additional notifications (as well as additional information in the follow-up written notices) can be used by the local emergency planning committee to propose and/or revise the emergency plan. This information should be especially helpful in meeting the requirements to list methods for determining whether a

TABLE 2.6.3 List of Extremely Hazardous Substances and Their Threshold Planning Quantities

CAS No.	Chemical Name	Notes	Reportable quantity* (pounds)	Threshold planning quantity (pounds)
75-86-5	Acetone cyanohydrin		10	1,000
1752-30-3	Acetone thiosemicarbazide		1,000	1,000/10,000
107-02-8	Acrolein		1	500
79-06-1	Acrylamide		5,000	1,000/10,000
107-13-1	Acrylonitrile		100	10,000
814-68-6	Acrylyl chloride	h	100	100
111-69-3	Adiponitrile		1,000	1,000
116-06-3	Aldicarb	c	1	100/10,000
309-00-2	Aldrin		1	500/10,000
107-18-6	Allyl alcohol		100	1,000
107-11-9	Allylamine		500	500
20859-73-8	Aluminum phosphide	b	100	500
54-62-6	Aminopterin		500	500/10,000
373-497-2	Amiton oxalate		100	100/10,000
78-53-5	Amiton		500	500
7664-41-7	Ammonia		100	500
300-62-9	Amphetamine		1,000	1,000
62-53-3	Aniline		5,000	1,000
88-05-1	Aniline, 2,4,6-trimethyl-		500	500
7783-70-2	Antimony pentafluoride		500	500
1397-94-0	Antimycin A	c	1,000	1,000/10,000
86-88-4	Antu		100	500/10,000
1303-28-2	Arsenic pentoxide		1	100/10,000
1327-53-3	Arsenous oxide	h	1	100/10,000
7784-34-1	Arsenous trichloride		1	500
7784-42-1	Arsine		100	100
2642-71-9	Azinphos-ethyl		100	100/10,000
86-50-0	Azinphos-methyl		1	10/10,000
98-87-3	Benzal chloride		5,000	500
98-16-8	Benzenamine, 3-(trifluoromethyl)-		500	500
100-14-1	Benzene, 1-(chloromethyl)-4-nitro-		500	500/10,000
98-05-5	Benzenearsonic acid		10	10/10,000
3615-21-2	Benzimidazole, 4,5-dichloro-2-(trifluoromethyl)-	g	500	500/10,000
98-07-7	Benzotrichloride		10	100
100-44-7	Benzyl chloride		100	500

TABLE 2.6.3 (continued)

CAS No.	Chemical Name	Notes	Reportable quantity* (pounds)	Threshold planning quantity (pounds)
140-29-4	Benzyl cyanide	h	500	500
57-57-8	beta-Propiolactone		10	500
15271-41-7	Bicyclo[2.2.1]heptane-2-carbonitrile, 5-chloro-6-(((((methylamino)carbonyl)oxy)imino)-,(1-alpha,2-beta,4-alpha,5-alpha,6E))-		500	500/10,000
534-07-6	Bis(chloromethyl) ketone		10	10/10,000
4044-65-9	Bitoscanate		500	500/10,000
353-42-4	Boron trifluoride compound with methyl ether (1:1)		1,000	1,000
10294-34-5	Boron trichloride		500	500
7637-07-2	Boron trifluoride		500	500
28772-56-7	Bromadiolone		100	100/10,000
7726-95-6	Bromine		500	500
2223-93-0	Cadmium stearate	c	1,000	1,000/10,000
1306-19-0	Cadmium oxide		100	100/10,000
7778-44-1	Calcium arsenate		1	500/10,000
8001-35-2	Camphechlor		1	500/10,000
56-25-7	Cantharidin		100	100/10,000
51-83-2	Carbachol chloride		500	500/10,000
26419-73-8	Carbamic acid, methyl-, O-(((2,4-dimethyl-1,3-dithiolan-2-yl)methylene)amino)-	d	1	100/10,000
1563-66-2	Carbofuran		10	10/10,000
75-15-0	Carbon disulfide		100	10,000
786-19-6	Carbophenothion		500	500
57-74-9	Chlordane		1	1,000
470-90-6	Chlorfenvinfos		500	500
7782-50-5	Chlorine		10	100
24934-91-6	Chlormephos		500	500
999-81-5	Chlormequat chloride	h	100	100/10,000
79-11-8	Chloroacetic acid		100	100/10,000
107-07-3	Chloroethanol		500	500
627-11-2	Chloroethyl chloroformate		1,000	1,000
67-66-3	Chloroform		10	10,000
107-30-2	Chloromethyl methyl ether	c	10	100

TABLE 2.6.3 (continued)

CAS No.	Chemical Name	Notes	Reportable quantity* (pounds)	Threshold planning quantity (pounds)
542-88-1	Chloromethyl ether	h	10	100
3691-35-8	Chlorophacinone		100	100/10,000
1982-47-4	Chloroxuron		500	500/10,000
21923-23-9	Chlorthiophos	h	500	500
10025-73-7	Chromic chloride		1	1/10,000
10210-68-1	Cobalt carbonyl	h	10	10/10,000
62207-76-5	Cobalt, ((2,2'-(1,2-ethanediylbis (nitrilomethylidyne))bis (6-fluorophenylato))(2-)-N,N',O,O')-		100	100/10,000
64-86-8	Colchicine	h	10	10/10,000
56-72-4	Coumaphos		10	100/10,000
5836-29-3	Coumatetralyl		500	500/10,000
535-89-7	Crimidine		100	100/10,000
4170-30-3	Crotonaldehyde		100	1,000
123-73-9	Crotonaldehyde, (E)-		100	1,000
506-68-3	Cyanogen bromide		1,000	500/10,000
506-78-5	Cyanogen iodide		1,000	1,000/10,000
2636-26-2	Cyanophos		1,000	1,000
675-14-9	Cyanuric fluoride		100	100
66-81-9	Cycloheximide		100	100/10,000
108-91-8	Cyclohexylamine		10,000	10,000
17702-41-9	Decaborane(14)		500	500/10,000
8065-48-3	Demeton		500	500
919-86-8	Demeton-S-methyl		500	500
10311-84-9	Dialifor		100	100/10,000
19287-45-7	Diborane		100	100
111-44-4	Dichloroethyl ether		10	10,000
149-74-6	Dichloromethylphenylsilane	h	1,000	1,000
62-73-7	Dichlorvos		10	1,000
141-66-2	Dicrotophos		100	100
1464-53-5	Diepoxybutane		10	500
814-49-3	Diethyl chlorophosphate	h	500	500
71-63-6	Digitoxin	c	100	100/10,000
2238-07-5	Diglycidyl ether		1,000	1,000
20830-75-5	Digoxin	h	10	10/10,000
115-26-4	Dimefox		500	500
60-51-5	Dimethoate		10	500/10,000

TABLE 2.6.3 (continued)

CAS No.	Chemical Name	Notes	Reportable quantity* (pounds)	Threshold planning quantity (pounds)
2524-03-0	Dimethyl phosphorochloridothioate		500	500
77-78-1	Dimethyl sulfate		100	500
99-98-9	Dimethyl-p-phenylenediamine		10	10/10,000
75-78-5	Dimethyldichlorosilane	h	500	500
57-14-7	Dimethylhydrazine		10	1,000
644-64-4	Dimetilan	d	1	500/10,000
534-52-1	Dinitrocresol		10	10/10,000
88-85-7	Dinoseb		1,000	100/10,000
1420-07-1	Dinoterb		500	500/10,000
78-34-2	Dioxathion		500	500
82-66-6	Diphacinone		10	10/10,000
152-16-9	Diphosphoramide, octamethyl-		100	100
298-04-4	Disulfoton		1	500
514-73-8	Dithiazanine iodide		500	500/10,000
541-53-7	Dithiobiuret		100	100/10,000
316-42-7	Emetine, dihydrochloride	h	1	1/10,000
115-29-7	Endosulfan		1	10/10,000
2778-04-3	Endothion		500	500/10,000
72-20-8	Endrin		1	500/10,000
106-89-8	Epichlorohydrin		100	1,000
2104-64-5	EPN		100	100/10,000
50-14-6	Ergocalciferol	c	1,000	1,000/10,000
379-79-3	Ergotamine tartrate		500	500/10,000
1622-32-8	Ethanesulfonyl chloride, 2-chloro-		500	500
10140-87-1	Ethanol, 1,2-dichloro-, acetate		1,000	1,000
563-12-2	Ethion		10	1,000
13194-48-4	Ethoprophos		1,000	1,000
538-07-8	Ethylbis(2-chloroethyl)amine		500	500
371-62-0	Ethylene fluorohydrin	c, h	10	10
75-21-8	Ethylene oxide		10	1,000
107-15-3	Ethylenediamine		5,000	10,000
151-56-4	Ethyleneimine		1	500
542-90-5	Ethylthiocyanate		10,000	10,000
22224-92-6	Fenamiphos		10	10/10,000

TABLE 2.6.3 (continued)

CAS No.	Chemical Name	Notes	Reportable quantity* (pounds)	Threshold planning quantity (pounds)
115-90-2	Fensulfothion	h	500	500
4301-50-2	Fluenetil		100	100/10,000
7782-41-4	Fluorine	k	10	500
640-19-7	Fluoroacetamide	j	100	100/10,000
144-49-0	Fluoroacetic acid		10	10/10,000
359-06-8	Fluoroacetyl chloride	c	10	10
51-21-8	Fluorouracil		500	500/10,000
944-22-9	Fonofos		500	500
107-16-4	Formaldehyde cyanohydrin	h	1,000	1,000
50-00-0	Formaldehyde		100	500
23422-53-9	Formetanate hydrochloride	d, h	1	500/10,000
2540-82-1	Formothion		100	100
17702-57-7	Formparanate	d	1	100/10,000
21548-32-3	Fosthietan		500	500
3878-19-1	Fuberidazole		100	100/10,000
110-00-9	Furan		100	500
13450-90-3	Gallium trichloride		500	500/10,000
77-47-4	Hexachlorocyclopentadiene	h	10	100
4835-11-4	Hexamethylenediamine, N,N'-dibutyl-		500	500
302-01-2	Hydrazine		1	1,000
74-90-8	Hydrocyanic acid		10	100
7647-01-0	Hydrogen chloride (gas only)		5,000	500
7783-07-5	Hydrogen selenide		10	10
7664-39-3	Hydrogen fluoride		100	100
7722-84-1	Hydrogen peroxide (Conc.> 52%)		1,000	1,000
7783-06-4	Hydrogen sulfide		100	500
123-31-9	Hydroquinone		100	500/10,000
13463-40-6	Iron, pentacarbonyl-		100	100
297-78-9	Isobenzan		100	100/10,000
78-82-0	Isobutyronitrile	h	1,000	1,000
102-36-3	Isocyanic acid, 3,4-dichlorophenyl ester		500	500/10,000
465-73-6	Isodrin		1	100/10,000
55-91-4	Isofluorphate	c	100	100
4098-71-9	Isophorone diisocyanate		100	100
108-23-6	Isopropyl chloroformate		1,000	1,000
119-38-0	Isopropylmethylpyrazolyl dimethylcarbamate		1	500

TABLE 2.6.3 (continued)

CAS No.	Chemical Name	Notes	Reportable quantity* (pounds)	Threshold planning quantity (pounds)
78-97-7	Lactonitrile		1,000	1,000
21609-90-5	Leptophos		500	500/10,000
541-25-3	Lewisite	c, h	10	10
58-89-9	Lindane		1	1,000/10,000
7580-67-8	Lithium hydride	b	100	100
109-77-3	Malononitrile		1,000	500/10,000
12108-13-3	Manganese, tricarbonyl methylcyclopentadienyl	h	100	100
51-75-2	Mechlorethamine	c	10	10
950-10-7	Mephosfolan		500	500
1600-27-7	Mercuric acetate		500	500/10,000
21908-53-2	Mercuric oxide		500	500/10,000
7487-94-7	Mercuric chloride		500	500/10,000
10476-95-6	Methacrolein diacetate		1,000	1,000
760-93-0	Methacrylic anhydride		500	500
126-98-7	Methacrylonitrile	h	1,000	500
920-46-7	Methacryloyl chloride		100	100
30674-80-7	Methacryloyloxyethyl isocyanate	h	100	100
10265-92-6	Methamidophos		100	100/10,000
558-25-8	Methanesulfonyl fluoride		1,000	1,000
950-37-8	Methidathion		500	500/10,000
2032-65-7	Methiocarb		10	500/10,000
16752-77-5	Methomyl	h	100	500/10,000
151-38-2	Methoxyethylmercuric acetate		500	500/10,000
78-94-4	Methyl vinyl ketone		10	10
60-34-4	Methyl hydrazine		10	500
556-64-9	Methyl thiocyanate		10,000	10,000
556-61-6	Methyl isothiocyanate	h	500	500
79-22-1	Methyl chloroformate	h	1,000	500
3735-23-7	Methyl phenkapton		500	500
74-93-1	Methyl mercaptan		100	500
80-63-7	Methyl 2-chloroacrylate		500	500
676-97-1	Methyl phosphonic dichloride		100	100
74-83-9	Methyl bromide		1,000	1,000
624-83-9	Methyl isocyanate		10	500
502-39-6	Methylmercuric dicyanamide		500	500/10,000
75-79-6	Methyltrichlorosilane	h	500	500

TABLE 2.6.3 (continued)

CAS No.	Chemical Name	Notes	Reportable quantity* (pounds)	Threshold planning quantity (pounds)
1129-41-5	Metolcarb	d	1	100/10,000
7786-34-7	Mevinphos		10	500
315-18-4	Mexacarbate		1,000	500/10,000
50-07-7	Mitomycin C		10	500/10,000
6923-22-4	Monocrotophos		10	10/10,000
2763-96-4	Muscimol		1,000	500/10,000
505-60-2	Mustard gas	h	500	500
13463-39-3	Nickel carbonyl		10	1
65-30-5	Nicotine sulfate		100	100/10,000
54-11-5	Nicotine	c	100	100
7697-37-2	Nitric acid		1,000	1,000
10102-43-9	Nitric oxide	c	10	100
98-95-3	Nitrobenzene		1,000	10,000
1122-60-7	Nitrocyclohexane		500	500
10102-44-0	Nitrogen dioxide		10	100
62-75-9	Nitrosodimethylamine	h	10	1,000
991-42-4	Norbormide		100	100/10,000
95-48-7	o-Cresol		100	1,000/10,000
NONE	Organorhodium Complex (PMN-82-147)		10	10/10,000
630-60-4	Ouabain	c	100	100/10,000
23135-22-0	Oxamyl	d	1	100/10,000
78-71-7	Oxetane, 3,3-bis (chloromethyl)-		500	500
2497-07-6	Oxydisulfoton	h	500	500
10028-15-6	Ozone		100	100
2074-50-2	Paraquat methosulfate		10	10/10,000
1910-42-5	Paraquat dichloride		10	10/10,000
56-38-2	Parathion	c	10	100
298-00-0	Parathion-methyl	c	100	100/10,000
12002-03-8	Paris green		1	500/10,000
19624-22-7	Pentaborane		500	500
2570-26-5	Pentadecylamine		100	100/10,000
79-21-0	Peracetic acid		500	500
594-42-3	Perchloromethyl mercaptan		100	500
108-95-2	Phenol		1,000	500/10,000
64-00-6	Phenol, 3-(1-methylethyl)-, methylcarbamate	d	1	500/10,000

TABLE 2.6.3 (continued)

CAS No.	Chemical Name	Notes	Reportable quantity* (pounds)	Threshold planning quantity (pounds)
4418-66-0	Phenol, 2,2'-thiobis[4-chloro-6-methyl-		100	100/10,000
58-36-6	Phenoxarsine, 10,10'-oxydi-		500	500/10,000
696-28-6	Phenyl dichloroarsine	h	1	500
59-88-1	Phenylhydrazine hydrochloride		1,000	1,000/10,000
62-38-4	Phenylmercury acetate		100	500/10,000
2097-19-0	Phenylsilatrane	h	100	100/10,000
103-85-5	Phenylthiourea		100	100/10,000
298-02-2	Phorate		10	10
4104-14-7	Phosacetim		100	100/10,000
947-02-4	Phosfolan		100	100/10,000
75-44-5	Phosgene		10	10
732-11-6	Phosmet		10	10/10,000
13171-21-6	Phosphamidon		100	100
7803-51-2	Phosphine		100	500
2703-13-1	Phosphonothioic acid, methyl-, O-ethyl O-(4-(methylthio)phenyl) ester		500	500
50782-69-9	Phosphonothioic acid, methyl-, S-(2-(bis (1-methylethyl)amino)ethyl) O-ethyl ester		100	100
2665-30-7	Phosphonothioic acid, methyl-, O-(4-nitrophenyl) O-phenyl ester		500	500
3254-63-5	Phosphoric acid, dimethyl 4-(methylthio) phenyl ester		500	500
2587-90-8	Phosphorothioic acid, O, O-dimethyl-5-(2-(methylthio)ethyl)ester	c, g	500	500
10025-87-3	Phosphorus oxychloride		1,000	500
10026-13-8	Phosphorus pentachloride	b	500	500
7719-12-2	Phosphorus trichloride		1,000	1,000
7723-14-0	Phosphorus	b, h	1	100
57-47-6	Physostigmine	d	1	100/10,000
57-64-7	Physostigmine, salicylate (1:1)	d	1	100/10,000
124-87-8	Picrotoxin		500	500/10,000

TABLE 2.6.3 (continued)

CAS No.	Chemical Name	Notes	Reportable quantity* (pounds)	Threshold planning quantity (pounds)
110-89-4	Piperidine		1,000	1,000
23505-41-1	Pirimifos-ethyl		1,000	1,000
151-50-8	Potassium cyanide	b	10	100
1012-450-2	Potassium arsenite		1	500/10,000
506-61-6	Potassium silver cyanide	b	1	500
2631-37-0	Promecarb	d, h	1	500/10,000
106-96-7	Propargyl bromide		10	10
107-12-0	Propionitrile		10	500
542-76-7	Propionitrile, 3-chloro-		1,000	1,000
70-69-9	Propiophenone, 4'-amino	g	100	100/10,000
109-61-5	Propyl chloroformate		500	500
75-56-9	Propylene oxide		100	10,000
75-55-8	Propyleneimine		1	10,000
2275-18-5	Prothoate		100	100/10,000
129-00-0	Pyrene	c	5,000	1,000/10,000
504-24-5	Pyridine, 4-amino-	h	1,000	500/10,000
140-76-1	Pyridine, 2-methyl-5-vinyl-		500	500
1124-33-0	Pyridine, 4-nitro-, 1-oxide		500	500/10,000
53558-25-1	Pyriminil	h	100	100/10,000
14167-18-1	Salcomine		500	500/10,000
107-44-8	Sarin	h	10	10
7783-00-8	Selenious acid		10	1,000/10,000
7791-23-3	Selenium oxychloride		500	500
563-41-7	Semicarbazide hydrochloride		1,000	1,000/10,000
3037-72-7	Silane, (4-aminobutyl) diethoxymethyl-		1,000	1,000
13410-01-0	Sodium selenate		100	100/10,000
7784-46-5	Sodium arsenite		1	500/10,000
62-74-8	Sodium fluoroacetate		10	10/10,000
124-65-2	Sodium cacodylate		100	100/10,000
143-33-9	Sodium cyanide (Na(CN))	b	10	100
7631-89-2	Sodium arsenate		1	1,000/10,000
10102-18-8	Sodium selenite	h	100	100/10,000
26628-22-8	Sodium azide (Na(N3))	b	1,000	500
10102-20-2	Sodium tellurite		500	500/10,000
900-95-8	Stannane, acetoxytriphenyl-	g	500	500/10,000
57-24-9	Strychnine	g	10	100/10,000
60-41-3	Strychnine, sulfate		10	100/10,000

TABLE 2.6.3 (continued)

CAS No.	Chemical Name	Notes	Reportable quantity* (pounds)	Threshold planning quantity (pounds)
3689-24-5	Sulfote		100	500
3569-57-1	Sulfoxide, 3-chloropropyl octyl		500	500
7446-11-9	Sulfur trioxide	b	100	100
7446-09-5	Sulfur dioxide		500	500
7783-60-0	Sulfur tetrafluoride		100	100
7664-93-9	Sulfuric acid		1,000	1,000
77-81-6	Tabun	c, k	10	10
7783-80-4	Tellurium hexafluoride	k	100	100
107-49-3	Tepp		10	100
13071-79-9	Terbufos	h	100	100
78-00-2	Tetraethyl lead	c	10	100
597-64-8	Tetraethyltin	c	100	100
75-74-1	Tetramethyllead	c	100	100
509-14-8	Tetranitromethane		10	500
10031-59-1	Thallium sulfate	h	100	100/10,000
2757-18-8	Thallous malonate	c, h	100	100/10,000
6533-73-9	Thallous carbonate	c, h	100	100/10,000
7791-12-0	Thallous chloride	c, h	100	100/10,000
7446-18-6	Thallous sulfate		100	100/10,000
2231-57-4	Thiocarbazide		1,000	1,000/10,000
39196-18-4	Thiofanox		100	100/10,000
297-97-2	Thionazin		100	500
108-98-5	Thiophenol		100	500
79-19-6	Thiosemicarbazide		100	100/10,000
5344-82-1	Thiourea, (2-chlorophenyl)-		100	100/10,000
614-78-8	Thiourea, (2-methylphenyl)-		500	500/10,000
7550-45-0	Titanium tetrachloride		1,000	100
91-08-7	Toluene-2,6-diisocyanate		100	100
584-84-9	Toluene-2,4-diisocyanate		100	500
110-57-6	trans-1,4-Dichlorobutene		500	500
1031-47-6	Triamiphos		500	500/10,000
24017-47-8	Triazofos		500	500
1558-25-4	Trichloro(chloromethyl)silane		100	100
27137-85-5	Trichloro(dichlorophenyl) Silane		500	500
76-02-8	Trichloroacetyl chloride		500	500
115-21-9	Trichloroethylsilane	h	500	500
327-98-0	Trichloronate	k	500	500
98-13-5	Trichlorophenylsilane	h	500	500

TABLE 2.6.3 (continued)

CAS No.	Chemical Name	Notes	Reportable quantity* (pounds)	Threshold planning quantity (pounds)
998-30-1	Triethoxysilane		500	500
75-77-4	Trimethylchlorosilane		1,000	1,000
824-11-3	Trimethylolpropane phosphite		100	100/10,000
1066-45-1	Trimethyltin chloride		500	500/10,000
639-58-7	Triphenyltin chloride		500	500/10,000
555-77-1	Tris(2-chloroethyl)amine	h	100	100
2001-95-8	Valinomycin	c	1,000	1,000/10,000
1314-62-1	Vanadium pentoxide		1,000	100/10,000
108-05-4	Vinyl acetate monomer		5,000	1,000
129-06-6	Warfarin sodium	h	100	100/10,000
81-81-2	Warfarin		100	500/10,000
28347-13-9	Xylylene dichloride		100	100/10,000
1314-84-7	Zinc phosphide	b	100	500
58270-08-9	Zinc, dichloro(4,4-dimethyl-5((((methylamino)carbonyl)oxy)imino)pentanenitrile)-, (T-4)-		100	100/10,000

*Only the statutory or final RQ is shown. For more information, see 40 CFR Table 302.4.

Notes:

a.   This chemical does not meet acute toxicity criteria. Its TPQ is set at 10,000 pounds.

b.   This material is a reactive solid. The TPQ does not default to 10,000 pounds for non-powder, non-molten, non-solution form.

c.   The calculated TPQ changed after technical review as described in the technical support document.

d.   Indicates that the RQ is subject to change when the assessment of potential carcinogenicity and/or other toxicity is completed.

e.   Statutory reportable quantity for purposes of notification under SARA sect 304(a)(2).

f.   [Reserved]

g.   New chemicals added that were not part of the original list of 402 substances.

h.   Revised TPQ based on new or re-evaluated toxicity data.

j.   TPQ is revised to its calculated value and does not change due to technical review as in proposed rule.

k.   The TPQ was revised after proposal due to calculation error.

l.   Chemicals on the original list that do not meet toxicity criteria but because of their high production volume and recognized toxicity are considered chemicals of concern ("Other chemicals").

release has occurred and to identify the area and population most likely to be affected.

Information included in such additional notifications (as well as additional information in the follow-up written notices) can be used by the local emergency planning committee to propose and/or revise the emergency plan. This information should be especially helpful in meeting the requirements to list methods for determining whether a release has occurred and to identify the area and population most likely to be affected.

Section 303 gives local emergency planning committees access to information from facilities subject to Title III planning requirements. Sections 311 and 312 provide information about the nature, quantity, and location of chemicals at many facilities not subject to Section 303 requirements. For this reason, local emergency planning committees will find information from Sections 311 and 312 especially helpful when preparing a comprehensive plan for the entire planning district.

Two "community right-to-know" reporting requirements apply primarily to manufacturers and importers. Section 311 requires a facility that must prepare or have available *material safety data sheets* (MSDSs) under the *Occupational Safety and Health Administration* (OSHA) hazard communication regulations to submit either copies of MSDSs or a list of MSDS chemicals to the local emergency planning committee, the *State Emergency Response Commission*, and the local fire department (see Table 2.6.4 for MSDS details). Currently, only facilities in Standard Industrial Classification (SIC) Codes 20 to 39 (manufacturers and importers) are subject to these OSHA regulations, which deal primarily with health and safety in the workplace. Under OSHA, all facilities are required to establish accident prevention safety programs, especially those that use and/or store hazardous chemicals. These programs are instituted to make the working environment at a facility safe for its employees. Any facility that uses and/or stores a hazardous chemical must submit material safety data sheets to OSHA and all other required agencies. MSDSs were developed to keep track of the different kinds of chemicals a facility has on its premises and most important, what actions can be taken if a chemical release occurs. OSHA also requires facilities to provide for all employees training on accident prevention and response in the event of an accident. OSHA will be discussed further in Section 2.7.

In general, every material safety data sheet should provide the local emergency planning committee and the fire department in each community with the name of the chemical covered, as well as such general characteristics (see Table 2.6.4) as:

1. Toxicity, corrosivity, and reactivity
2. Known health effects, including chronic effects from exposure
3. Basic procedures in handling, storage, and use
4. Basic countermeasures to take in the event of a fire, explosion, or leak
5. Basic protective equipment to minimize exposure

More specific information or MSDSs is provided later in the book. Information on regulated chemicals is available in the literature.[10]

The above data should be useful for the planning to be accomplished by the local emergency planning committee and first responders, especially fire departments and HAZMAT teams. Both the hazards analysis (discussed in detail in Parts II and IV) and the development of emergency countermeasures should be facilitated by the availability of MSDS information. If significant new information regarding a chemical is discovered, revised material safety data sheets must be submitted.

The reporting requirements of Section 312 require facilities to submit an emergency and hazardous inventory form to the local emergency planning committee, *the State Emergency Planning Commission*, and the local fire department. The hazardous chemicals covered by Section 312 are the same chemicals for which facilities are required to submit MSDS forms or the list of MSDS chemicals for Section 311. The inventory form incorporates a two-tier approach. Under Tier I, facilities must submit the following aggregate information for each applicable OSHA category of health and physical hazard:

1.   An estimate (in ranges) of the maximum amount of chemicals for each category present at the facility at any time during the preceding year
2.   An estimate (in ranges) of the average daily amount of chemicals in each category
3.   The general location of hazardous chemicals in each category

Tier I information had to be submitted on or before March 1, 1988 and annually thereafter on March 1. The public may request additional information for a specific facility from the *State Emergency Response Commission* and the local emergency planning committee. Upon the request of the local emergency planning committee, the *State Emergency Response Commission*, or the local fire department, the facility must provide to the organization making the request the following Tier II information for each covered substance:

1.   The chemical name or the common name as indicated on the MSDS.
2.   An estimate (in ranges) of the maximum amount of the chemical present at any time during the preceding calendar year
3.   A brief description of the manner of storage of the chemical
4.   The location of the chemical at the facility
5.   An indication of whether the owner elects to withhold information from disclosure to the public

The information from Tier II and on-site inspections should help the local fire department in the development of prefire plans. The information submitted by facilities under Sections 311 and 312 must generally be made available to the public by state and local governments during normal working hours.

TABLE 2.6.4   MSDS Information[a]

MSDS sheets generally contain all or most of the following information:

1. Product or chemical identity used on the label
2. Manufacturer's name and address
3. Chemical and common names of each hazardous ingredient
4. Name, address, and phone number for hazard and emergency information
5. Preparation or revision date
6. The hazardous chemical's physical and chemical characteristics, such as vapor pressure and flashpoint
7. Physical hazards, including potential for fire, explosion, and reactivity
8. Known health hazards
9. Exposure limits
10. Emergency and first-aid procedures
11. Whether OSHA considers the ingredient as a carcinogen
12. Precautions for safe handling and use
13. Control measures such as engineering controls, work practices, hygienic practices or personal protective equipment required
14. Primary routes of entry
15. Procedures for spills, leaks, and clean-up

[a] Additional MSDS information is provided in Chapter 6

Section 313 of Title III required the USEPA to establish an inventory of toxic chemical emissions from certain facilities. Facilities subject to this reporting requirement were to complete a toxic chemical release form for specified chemicals. The form was to be submitted to USEPA and to the state officials designated by the governor on or before July 1, 1988, and annually thereafter on July 1, reflecting releases during each preceding calendar year. This information about chemical releases should assist in the research and development of regulations, guidelines, and standards.

Reports were to be filed by facilities that have 10 or more full-time employees, are in SIC Codes 20 through 39, and manufactured or processed any of the listed chemicals in quantities in excess of 75,000 pounds in 1987, 50,000 pounds in 1988, or 25,000 pounds in 1989 and thereafter, in quantities exceeding 10,000 pounds in any calendar year. The toxic chemical release form (see Figure 2.6.1), finalized on February 16, 1988, includes the following information for released chemicals.

1. The name, location, and type of business
2. Whether the chemical is manufactured, produced, or otherwise used, and the general categories of use of the chemical
3. An estimate (in ranges) of the maximum amounts of the toxic chemical present at the facility at any time during the preceding calendar year
4. Waste treatment and disposal methods and the efficiency of methods for each waste stream

(IMPORTANT: Type or print; read instructions before completing form)

EPA

FORM R

TOXIC CHEMICAL RELEASE INVENTORY REPORTING FORM

United States
Environmental Protection
Agency

Section 313 of the Emergency Planning and Community Right-to-Know Act of 1986,
also known as Title III of the Superfund Amendments and Reauthorization Act

| WHERE TO SEND COMPLETED FORMS: | 1. EPCRA Reporting Center
P.O. Box 3348
Merrifield, VA 22116-3348
ATTN: TOXIC CHEMICAL RELEASE INVENTORY | 2. APPROPRIATE STATE OFFICE
(See instructions in Appendix F) | Enter "X" here if this
is a revision

For EPA use only |

Important: See instructions to determine when "Not Applicable (NA)" boxes should be checked.

PART I. FACILITY IDENTIFICATION INFORMATION

SECTION 1. REPORTING YEAR _____

SECTION 2. TRADE SECRET INFORMATION

| 2.1 | Are you claiming the toxic chemical identified on page 2 trade secret?
☐ Yes (Answer question 2.2; Attach substantiation forms) ☐ No (Do not answer 2.2; Go to Section 3) | 2.2 | Is this copy ☐ Sanitized ☐ Unsanitized
(Answer only if "YES" in 2.1) |

SECTION 3. CERTIFICATION (Important: Read and sign after completing all form sections.)

I hereby certify that I have reviewed the attached documents and that, to the best of my knowledge and belief, the submitted information is true and complete and that the amounts and values in this report are accurate based on reasonable estimates using data available to the preparers of this report.

| Name and official title of owner/operator or senior management official | Signature: | Date Signed: |

SECTION 4. FACILITY IDENTIFICATION

4.1		TRI Facility ID Number
Facility or Establishment Name		Facility or Establishment Name or Mailing Address (if different from street address)
Street		Mailing Address
City/County/State/Zip Code		City/County/State/Zip Code

| 4.2 | This report contains information for
(Important: check a or b; check c if applicable) | a. ☐ An entire facility | b. ☐ Part of a facility | c. ☐ A Federal facility |

| 4.3 | Technical Contact Name | | Telephone Number (include area code) |

| 4.4 | Public Contact Name | | Telephone Number (include area code) |

| 4.5 | SIC Code (s) (4 digits) | Primary
a. | b. | c. | d. | e. | f. |

| 4.6 | Latitude | Degrees | Minutes | Seconds | Longitude | Degrees | Minutes | Seconds |

4.7	Dun & Bradstreet Number(s) (9 digits)	4.8	EPA Identification Number (RCRA I.D. No.) (12 characters)	4.9	Facility NPDES Permit Number(s) (9 characters)	4.10	Underground Injection Well Code (UIC) I.D. Number(s) (12 digits)
a.		a.		a.		a.	
b.		b.		b.		b.	

SECTION 5. PARENT COMPANY INFORMATION

| 5.1 | Name of Parent Company | NA ☐ | |
| 5.2 | Parent Company's Dun & Bradstreet Number | NA ☐ | |

Figure 2.6.1 USEPA's Toxic Chemical Release Inventory Reporting Form.

EPA FORM R PART II. CHEMICAL-SPECIFIC INFORMATION	TRI Facility ID Number
	Toxic Chemical, Category or Generic Name

SECTION 1. TOXIC CHEMICAL IDENTITY (Important: DO NOT complete this section if you completed Section 2 below.)

1.1	CAS Number (Important: Enter only one number exactly as it appears on the Section 313 list. Enter category code if reporting a chemical category.)
1.2	Toxic Chemical or Chemical Category Name (Important: Enter only one name exactly as it appears on the Section 313 list.)
1.3	Generic Chemical Name (Important: Complete only if Part 1, Section 2.1 is checked "yes". Generic Name must be structurally descriptive.)

SECTION 2. MIXTURE COMPONENT IDENTITY (Important: DO NOT complete this section if you completed Section 1 above.)

2.1	Generic Chemical Name Provided by Supplier (Important: Maximum of 70 characters, including numbers, letters, spaces, and punctuation.)

SECTION 3. ACTIVITIES AND USES OF THE TOXIC CHEMICAL AT THE FACILITY
(Important: Check all that apply.)

3.1 Manufacture the toxic chemical:	3.2 Process the toxic chemical:	3.3 Otherwise use the toxic chemical:
a. [ ] Produce b. [ ] Import	a. [ ] As a reactant	a. [ ] As a chemical processing aid
If produce or import c. [ ] For on-site use/processing	b. [ ] As a formulation component	b. [ ] As a manufacturing aid
d. [ ] For sale/distribution	c. [ ] As an article component	c. [ ] Ancillary or other use
e. [ ] As a byproduct	d. [ ] Repackaging	
f. [ ] As an impurity		

SECTION 4. MAXIMUM AMOUNT OF THE TOXIC CHEMICAL ONSITE AT ANY TIME DURING THE CALENDAR YEAR

4.1	[ ] (Enter two-digit code from instruction package.)

SECTION 5. QUANTITY OF THE TOXIC CHEMICAL ENTERING EACH ENVIRONMENTAL MEDIUM ONSITE

		A. Total Release (pounds/year) (Enter range code or estimate*)	B. Basis of Estimate (enter code)	C. % From Stormwater
5.1	Fugitive or non-point air emissions	NA [ ]		
5.2	Stack or point air emissions	NA [ ]		
5.3	Discharges to receiving streams or water bodies (enter one name per box)			
	Stream or Water Body Name			
5.3.1				
5.3.2				
5.3.3				
5.4.1	Underground injection onsite to Class I Wells	NA [ ]		
5.4.2	Underground injection onsite to Class II-V Wells	NA [ ]		

If additional pages of Part II, Section 5.3 are attached, indicate the total number of pages in this box [ ] and indicate the Part II, Section 5.3 page number in this box. [ ] (example: 1,2,3, etc.)

EPA form 9350-1(Rev. 04/97) - Previous editions are obsolete. * Range Codes: A= 1 - 10 pounds; B= 11- 499 pounds; C= 500 - 999 pounds.

Figure 2.6.1 (continued)

			TRI Facility ID Number	
EPA FORM R				
PART II. CHEMICAL - SPECIFIC INFORMATION (CONTINUED)			Toxic Chemical, Category or Generic Name	

SECTION 5. QUANTITY OF THE TOXIC CHEMICAL ENTERING EACH ENVIRONMENTAL MEDIUM ONSITE (Continued)

		NA	A. Total Release (pounds/year) (enter range code* or estimate)	B. Basis of Estimate (enter code)
5.5	Disposal to land onsite			
5.5.1A	RCRA Subtitle C landfills	☐		
5.5.1B	Other landfills	☐		
5.5.2	Land treatment/application farming	☐		
5.5.3	Surface impoundment	☐		
5.5.4	Other disposal	☐		

SECTION 6. TRANSFERS OF THE TOXIC CHEMICAL IN WASTES TO OFF-SITE LOCATIONS

6.1 DISCHARGES TO PUBLICLY OWNED TREATMENT WORKS (POTWs)

6.1.A Total Quantity Transferred to POTWs and Basis of Estimate

6.1.A.1. Total Transfers (pounds/year) (enter range code* or estimate)	6.1.A.2 Basis of Estimate (enter code)

6.1.B.___ POTW Name

POTW Address

City		State		County		Zip	

6.1.B.___ POTW Name

POTW Address

City		State		County		Zip	

If additional pages of Part II, Section 6.1 are attached, indicate the total number of pages in this box [   ] and indicate the Part II, Section 6.1 page number in this box [   ] (example: 1,2,3, etc.)

SECTION 6.2 TRANSFERS TO OTHER OFF-SITE LOCATIONS

6.2.___ Off-Site EPA Identification Number (RCRA ID No.)

Off-Site Location Name

Off-Site Address

City		State		County		Zip	

Is location under control of reporting facility or parent company?   ☐ Yes   ☐ No

EPA Form 9350-1 (Rev.04/97) - Previous editions are obsolete.   * Range Codes: A = 1 - 10 pounds, B = 11 - 499 pounds; C = 500 - 999 pounds.

Figure 2.6.1 (continued)

EPA FORM R	TRI Facility ID Number
PART II. CHEMICAL-SPECIFIC INFORMATION (CONTINUED)	Toxic Chemical, Category or Generic Name

SECTION 6.2 TRANSFERS TO OTHER OFF-SITE LOCATIONS (Continued)

A. Total Transfers (pounds/year) (enter range code* or estimate)	B. Basis of Estimate (enter code)	C. Type of Waste Treatment/Disposal/ Recycling/Energy Recovery (enter code)
1.	1.	1. M
2.	2.	2. M
3.	3.	3. M
4.	4.	4. M

6.2 ___ Off-Site EPA Identification Number (RCRA ID No.)

Off-Site Location Name

Off-Site Address

City		State	County		Zip

Is location under control of reporting facility or parent company? ☐ Yes ☐ No

A. Total Transfers (pounds/year) (enter range code* or estimate)	B. Basis of Estimate (enter code)	C. Type of Waste Treatment/Disposal/ Recycling/Energy Recovery (enter code)
1.	1.	1. M
2.	2.	2. M
3.	3.	3. M
4.	4.	4. M

SECTION 7A. ON-SITE WASTE TREATMENT METHODS AND EFFICIENCY

☐ Not Applicable (NA) – Check here if no on-site waste treatment is applied to any waste streams containing the toxic chemical or chemical category.

a. General Waste Stream (enter code)	b. Waste Treatment Method(s) Sequence [enter 3-character code(s)]			c. Range of Influent Concentration	d. Waste Treatment Efficiency Estimate	e. Based on Operating Data ?
7A.1a	7A.1b 1	2		7A.1c	7A.1d	7A.1e
	3	5			%	☐ Yes ☐ No
	6	7	8			
7A.2a	7A.2b 1	2		7A.2c	7A.2d	7A.2e
	3	5			%	☐ Yes ☐ No
	6	7	8			
7A.3a	7A.3b 1	2		7A.3c	7A.3d	7A.3e
	3	5			%	☐ Yes ☐ No
	6	7	8			
7A.4a	7A.4b 1	2		7A.4c	7A.4d	7A.4e
	3	5			%	☐ Yes ☐ No
	6	7	8			
7A.5a	7A.5b 1	2		7A.5c	7A.5d	7A.5e
	3	5			%	☐ Yes ☐ No
	6	7	8			

If additional pages of Part II, Section 6.2/7A are attached, indicate the total number of pages in this box and indicate the Part II, Section 6.2/7A page number in this box : ☐ (example: 1,2,3, etc)

EPA Form 9350-1 (Rev. 04/97) - Previous editions are obsolete   * Range Codes: A = 1 - 10 pounds; B = 11 - 499 pounds; C = 500 - 999 pounds.

Figure 2.6.1   (continued)

EPA FORM R

PART II. CHEMICAL-SPECIFIC INFORMATION (CONTINUED)

TRI Facility ID Number

Toxic Chemical, Category or Generic Name

SECTION 7B. ON-SITE ENERGY RECOVERY PROCESSES

☐ Not Applicable (NA) - Check here if no on-site energy recovery is applied to any waste stream containing the toxic chemical or chemical category.

Energy Recovery Methods [enter 3-character code(s)]

1. [____] 2. [____] 3. [____] 4. [____]

SECTION 7C. ON-SITE RECYCLING PROCESSES

☐ Not Applicable (NA) - Check here if no on-site recycling is applied to any waste stream containing the toxic chemical or chemical category.

Recycling Methods [enter 3-character code(s)]

1. [____] 2. [____] 3. [____] 4. [____] 5. [____]

6. [____] 7. [____] 8. [____] 9. [____] 10. [____]

SECTION 8. SOURCE REDUCTION AND RECYCLING ACTIVITIES

		Column A Prior Year (pounds/year)	Column B Current Reporting Year (pounds/year)	Column C Following Year (pounds/year)	Column D Second Following Year (pounds/year)
8.1	Quantity released [*]				
8.2	Quantity used for energy recovery onsite				
8.3	Quantity used for energy recovery offsite				
8.4	Quantity recycled onsite				
8.5	Quantity recycled offsite				
8.6	Quantity treated onsite				
8.7	Quantity treated offsite				
8.8	Quantity released to the environment as a result of remedial actions, catastrophic events, or one-time events not associated with production processes (pounds/year)				
8.9	Production ratio or activity index				
8.10	Did your facility engage in any source reduction activities for this chemical during the reporting year? If not, enter "NA" in Section 8.10.1 and answer Section 8.11.				

	Source Reduction Activities [enter code(s)]	Methods to Identify Activity (enter codes)		
8.10.1		a.	b.	c.
8.10.2		a.	b.	c.
8.10.3		a.	b.	c.
8.10.4		a.	b.	c.

8.11	Is additional information on source reduction, recycling, or pollution control activities included with this report? (Check one box)	YES ☐ NO ☐

[*] Report releases pursuant to EPCRA Section 329(8) including any spilling, leaking, pumping, pouring, emitting, emptying, discharging, injecting, escaping, leaching, dumping, or disposing into the environment. [**] Do not include any quantity treated onsite or offsite.

Figure 2.6.1 (continued)

5. The quantity of the chemical entering each environmental medium annually
6. A certification by a senior official that the report is complete and accurate

The USEPA must establish and maintain a national toxic chemical inventory based on all the release data submitted, which must be accessible by computer on a national database.

The list of toxic chemicals subject to the reporting requirements initially consisted of a combined list of the states of Maryland and New Jersey, whose reporting requirements were similar to those in this section. The USEPA can modify this combined list. In adding a chemical to the combined list, USEPA must ask the following questions.

1. Is the substance known to cause cancer or serious reproductive or neurological disorders, genetic mutations, or other chronic health effects?
2. Can the substance cause significant adverse acute health effects as a result of continuous or frequently recurring releases?
3. Can the substance cause an adverse effect on the environment because of its toxicity, persistence, or tendency to bioaccumulate?

Chemicals can be deleted if there is not sufficient evidence to establish positive responses to any of these questions.

In general, these Section 313 reports appear to be of limited value in emergency planning. Over time, however, they may contain information that can be used by local planners in developing a complete understanding of the total spectrum of hazards that a given facility may pose to the community.

2.7   OCCUPATIONAL SAFETY AND HEALTH ADMINISTRATION (OSHA)

The *Occupational Safety and Health Act* (OSH Act) was enacted by Congress in 1970 and established the *Occupational Safety and Health Administration* (OSHA), which addressed safety in the workplace. At the same time the USEPA was created. Both USEPA and OSHA are mandated to reduce the exposure of hazardous substances over land, sea, and air. The OSH Act is limited to conditions that exist in the workplace, where its jurisdiction covers both safety and health. Frequently, both agencies regulate the same substances but in a different manner as they are overlapping environmental organizations.

Congress intended that OSHA be enforced through specific standards in an effort to achieve a safe and healthful working environment. A "general duty clause" was added to attempt to cover those obvious situations that were admitted by all concerned but for which no specific standard existed. The OSHA standards are an extensive compilation of regulations, some that apply to all employers (such as eye and face protection) and some that apply to workers who engaged in a specific type of work (such as welding or crane operation).

Employers are obligated to familiarize themselves with the standards and comply with them at all times.

Health issues, most importantly, contaminants in the workplace, have become OSHA's primary concern. Health hazards are complex and difficult to define. Because of this, OSHA has been slow to implement health standards. To be complete, each standard requires medical surveillance, record keeping, monitoring, and physical reviews. On the other side of the ledger, safety hazards are aspects of the work environment that are expected to cause death or serious physical harm immediately or before the imminence of such danger can be eliminated.

Probably one of the most important safety and health standards ever adopted is the OSHA hazard communication standard, more properly known as the "right to know" laws. The hazard communication standard requires employers to communicate information to the employees on hazardous chemicals that exist within the workplace. The program requires employers to craft a written hazard communication program, keep *material safety data sheets* (MSDSs) for all hazardous chemicals at the workplace and provide employees with training on those hazardous chemicals, and assure that proper warning labels are in place.

The *Hazardous Waste Operations and Emergency Response Regulation* enacted in 1989 by OSHA addresses the safety and health of employees involved in cleanup operations at uncontrolled hazardous waste sites being cleaned up under government mandate, and in certain hazardous waste treatment, storage, and disposal operations conducted under RCRA. The standard provides for employee protection during initial site characterization and analysis, monitoring activities, training and emergency response. Four major areas are under the scope of the regulation:

1.   Cleanup operations at uncontrolled hazardous waste sites that have been identified for cleanup by a government health or environmental agency
2.   Routine operations at hazardous waste TSD (Transportation, Storage, and Disposal) facilities or those portions of any facility regulated by 40 CFR Parts 264 and 265
3.   Emergency response operations at sites where hazardous substances have or may be released
4.   Corrective actions at RCRA sites

The regulation addresses three specific populations of workers at the above operations. First, it regulates hazardous substance response operations under CERCLA, including initial investigations at CERCLA sites before the presence or absence of hazardous substance has been ascertained; corrective actions taken in cleanup operations under RCRA; and those hazardous waste operations at sites that have been designated for cleanup by state or local government authorities. The second worker population to be covered is those employees engaged in operations involving hazardous waste TSD facilities. The third

employee population to be covered is those employees engaged in emergency
response operations for release or substantial threat of releases of hazardous
substances, and post emergency response operations to such facilities (29 CFR,
March 6, 1989; 29 CFR February 24, 1992).[11]

2.8   USEPA'S RISK MANAGEMENT PROGRAM

Developed under the *Clean Air Acts* (CAA's) Section 112(r), the *Risk
Management Program* (RMP) rule is designed to reduce the risk of accidental
releases of acutely toxic, flammable and explosive substances. USEPA finalized
its list of regulated substances (138 chemicals) and defined threshold quantities
for these chemicals.

   In the RPM rule, USEPA requires a *Risk Management Plan* that
summarizes how your facility is complying with USEPA's RMP requirements.
It details methods and results of the hazard assessment, accident prevention, and
emergency response programs instituted at the facility. The hazard assessment
shows the area surrounding the facility and the population potentially affected
by "worst-case" and more likely releases. A "worst-case" scenario is a release,
over a 10-minute period, of the largest quantity of a regulated substance
resulting from a vessel or process piping failure. USEPA proposed a three-tiered
approach that would simplify requirements for most affected facilities. A facility
is affected if a process unit manufactures, processes, uses, stores, or otherwise
handles any of the listed chemicals at or above the threshold limit. The tiered
approach is summarized in Table 2.8.1. For example, USEPA proposed defining
Tier 1 facilities as those that can demonstrate that they do not pose an off-site
risk to their nearest neighbor or sensitive habitat. Tier 1 facilities would have to
prepare a brief RMP demonstrating and certifying that the source's worst-case
release would not reach any public or environmental receptors of concern. To
qualify for Tier 2, a streamlined risk management program including description
of the facility's five-year accident history, accident prevention steps, and an
emergency response plan would be required. Tier 3 facilities would be required
to comply fully with the RMP rule as originally proposed. Tier 3 facilities
belong to specific industrial categories identified by USEPA as historically
accounting for most industrial accidents resulting in off-site risk. These
categories include pulp, plastics and resins, chloralkalis, industrial organic and
inorganic chemicals, fertilizers, and refineries.[12]

2.9 REGULATORY PROBLEM AREAS AND POTENTIAL
   ENVIRONMENTAL VIOLATIONS

While some pollution is an unfortunate consequence of modern industrial life,
there are national, state, and local laws that limit the amount and kinds of
pollution allowed. In some cases these laws completely prohibit certain types of
pollution.

TABLE 2.8.1 RMP Tiered Approach

Tier	Description
1	Facilities will need only to register, certify that hazards will be contained on-site, post local caution signs, and report significant releases should they occur.
2	Facilities will need to implement a "streamlined" RMP. This category provides facilities with substantial flexibility to address the specific program elements.
3	These facilities, belonging to one of eight industrial classes, will be required to implement and document the full RMP rule. This will include hazard assessment, preventive steps, emergency preparation and response (see next chapter) and a risk management plan.

Several areas to be concerned about are:[13]

1. Illegal transport of wastes and/or chemicals
2. Storage or disposal of hazardous waste
3. Improper hazardous-waste inventory reporting (SARA 313)
4. Failure to report releases of hazardous substances
5. Polychlorinated biphenyl (PCB) releases
6. Mislabeling of solvents
7. Demolition or renovation involving asbestos
8. Dredging or filling of wetlands improperly and/or without a permit
9. Contamination of groundwater
10. Falsification of reporting required by a federal, state or local permit
11. Improper air-emission monitoring
12. Underground storage tank mismanagement
13. General process emissions (e.g., degreasers, asphalt patch, etc.)
14. Contaminating sewers and locks
15. Illegal filling of wetlands and marshes
16. Contaminating drinking water
17. Creating smoke and/or odor

Violations that may result from some of the above activities include:

1. Reckless, knowing, or willful disregard of environmental laws or regulations
2. Evidence of deceit or concealment of wrongdoing (e.g., false reporting, fraudulent certifications or failure to report)
3. Failure to comply with agency orders
4. Activities that result in significant harm to the environment
5. Economic benefit gained by the offender as a result of illegal activities
6. Complicity of a corporate manager or other responsible corporate official
7. Offender's history of noncompliance
8. Perceived deterrent value to the offender and others

If there is a high level of citizen involvement or there is significant media attention, the resulting penalty for any of the above violations will normally become exacerbated. In addition, Wilcox and Theodore have provided information on probable criminal prosecution for unethical conduct.[14]

2.10   THE POLLUTION PREVENTION ACT OF 1990

Another act that has had a major impact on the general area of health, safety and accident management is the *Pollution Prevention Act of 1990*. The major theme of the act was to provide an importance to reduce the generation of wastes/pollutants/chemicals that can create health, safety and accident management problems. Details of the act are provided later.[15]

Signed into law in November 1990, the *Pollution Prevention Act of 1990* is the most important regulation regarding pollution prevention. The act establishes pollution prevention as a "national objective" and notes that:

> There are significant opportunities for industry to reduce or prevent pollution at the source through cost-effective changes in production, operation and raw materials use. . . . The opportunities for source reduction are often not realized because existing regulations, and the industrial resources they require for compliance, focus upon treatment and disposal, rather than source reduction. . . . Source reduction is fundamentally different and more desirable than waste management and pollution control.

The 1990 act establishes the pollution prevention hierarchy as national policy, declaring that pollution should be prevented or reduced at the source wherever feasible, while pollution that cannot be prevented should be recycled in an environmentally safe manner. In the absence of feasible prevention or recycling opportunities, pollution should be treated; disposal or other release(s) into the environment should be used as a last resort. The *Pollution Prevention Act* also formalized the establishment of USEPA's *Office of Pollution Prevention*, independent of the single-medium programs, to carry out the functions required by the act and to develop and implement a strategy to promote source reduction. Among other provisions, the law directs USEPA to:

1. Facilitate the adoption of source reduction techniques by businesses and by other federal agencies
2. Establish standard methods of measurement for source reduction
3. Review regulations to determine their effect on source reduction
4. Investigate opportunities to use federal procurement to encourage source reduction
5. Develop improved methods for providing public access to data collected under federal environmental statutes

6.   Develop a training program on source reduction opportunities, model source reduction auditing procedures, a source reduction clearinghouse, and an annual award program

Under the act, facilities required to report releases to USEPA for the *Toxic Releases Inventory* (TRI) must now also provide information on pollution prevention and recycling for each facility and for each toxic chemical. The information includes the quantities for each toxic chemical entering the waste stream and the percentage change from the previous year, the quantities recycled and the percentage change from the previous year, source reduction practices, and changes in production from the previous year. Finally, the Act requires USEPA to report to congress within 18 months (and biennially afterward) on the actions needed to implement the strategy to promote source reduction.

States have been at the forefront of the pollution prevention movement, providing a direct link to industry, local governments and consumers. Through grants to states, USEPA has enhanced the capabilities of states to demonstrate innovative and results-oriented programs and has assisted states in implementing a multimedia prevention approach. Local governments can also play a significant role in promoting pollution prevention in the industrial, consumer, transportation, agricultural, and public sectors of the community. Many local governments have already taken the lead in putting successful recycling programs into place. A variety of tools are available to promote prevention. Local governments (including cities, counties, sewer and water agencies, planning departments, and other special districts) can provide:

1.   Educational programs to raise awareness in businesses and the community of the need to reduce waste and pollution and conserve resources
2.   Technical assistance programs that provide on-site help to companies and organizations in reducing pollution at the source
3.   Regulatory programs that promote prevention through mechanisms such as codes, licenses, and permits
4.   Procurement policies regarding government purchase of recycled products, reusable products, and products designed to be recycled

Many local governments have passed resolutions and ordinances relating to waste reduction, energy conservation, automobile use, procurement policies, and so on. Such resolutions and ordinances can be useful steps in signaling a public commitment to operate under environmentally sound principles: they also help to define goals and targets and to delineate the specific responsibilities of different local agencies.

Section 6607(c) of the *Pollution Prevention Act* provides enforcement authority under Title III of the *Superfund Amendments and Reauthorization Act* (also known as the *Emergency Planning and Community Right-to-Know Act*). Civil, administrative, and criminal penalties are authorized for non-compliance against a facility, USEPA, a Governor, or a SERC. The Act requires USEPA to file a report on implementation of its *Pollution Prevention Strategy* biannually. Authorization for appropriations under the *Pollution Prevention Act* expired September 30, 1993, but appropriations have continued.

2.11 ILLUSTRATIVE EXAMPLES

Example 2.1

Differentiate between the USEPA and OSHA.

Solution 2.1

The USEPA and OSHA are government agencies that have the authority to issue regulations to reduce the exposure of hazardous substances over land, sea, and air. OSHA is limited to conditions that exist in the workplace. OSHA is authorized by the *Occupational Safety and Health Act* to issue regulations that protect workers from the hazardous chemicals they use in manufacturing processes. If these hazardous chemicals are emitted by the plant and effect the surrounding community but do not expose the workers in the plant, OSHA is not authorized to issue an order to stop the practice. However, this issue would remain within the jurisdiction of the USEPA, which is not limited to conditions that exist in the workplace. Frequently, both agencies regulate the same substances but in a different manner. They are overlapping environmental organizations.

Example 2.2

Briefly describe the role of *State Implemented Plans* (SIPs).

Solution 2.2

SIPs are intended to prevent air pollution emergency episodes. The plans are directed toward preventing excessive buildup of air pollutants that are known to be harmful to the population and the environment when concentrations exceed certain limits. The compounds affected under the implementation plans are sulfur dioxide, particulate matter, carbon monoxide, nitrogen dioxide, and ozone. A contingency plan, which will outline the steps to be taken in the event that a particular pollutant concentration reaches the level at which it can be considered to be harmful, must be included in each implementation plan. The implementation plans are solely based on the continuous emission of the previously stated air pollutants. They do not mandate any actions to be taken in the event of an accidental toxic release.

Example 2.3

Review Table 2.6.4 and provide a one-sentence explanation of the need for each piece of information on a MSDS sheet.

Solution 2.3

1.   Product or chemical identity used on the label:

     This ensures that the correct chemical is being used and alerts the worker to the potential hazards of working with the chemical.

2.   Manufacturer's name and address:

     Contacting the manufacturer would help clarify any uncertainties concerning the chemicals being used and could also, in the event that it has been discovered that the manufacturer has made an error in the production or delivery of a certain chemical, prevent a potential catastrophe elsewhere caused by the use of the same chemical.

3.   Chemical and common names of each hazardous ingredient:

     This serves as a reference for those working with chemicals to check and see if the chemicals being used are hazardous.

4.   Name, address, and phone number for hazard and emergency information:

     Their assistance may be necessary if an accident should occur or if there is any uncertainty concerning a certain chemical.

5.   Preparation or revision date:

     The preparation and/or revision date of a certain material tells the worker whether the material is still fit to be used or if more of the material needs to be prepared.

6.   The hazardous chemical's physical and chemical characteristics, such as vapor pressure and flashpoint:

     This information can be used to control the environment that the hazardous chemical is going to be used in.

7.   Physical hazards, including potential for fire, explosion, and reactivity:

This aids in the analysis of a "worst-case" scenario that could result from a simple accident.

8. Known health hazards:

This alerts workers to use special caution when working with the hazardous materials.

9. Exposure limits:

This helps to protect those working with the material.

10. Emergency and first-aid procedures:

In the event of an accident, these procedures could save an afflicted worker's life.

11. Whether OSHA considers the ingredient as a carcinogen:

Alerts workers to the potential risk of developing cancer from working with the ingredient and encourages special caution to be taken in working with the ingredient.

12. Precautions for safe handling and use:

This helps to protect those working with the material.

13. Control measures such as engineering controls, work practices, hygienic practices or personal protective equipment required:

These measures are used in an attempt to minimize the risk involved in working with hazardous materials.

14. Primary routes of entry:

In the event that an accident should occur and help is called for, these are the routes by which help will most likely arrive.

15. Procedures for spills, leaks, and clean-up:

These procedures are used to minimize the damage caused by these accidents.

Example 2.4

List the advantages of applying the pollution prevention principle to an environmental management program.

Solution 2.4

The advantages of applying the pollution prevention principle to an environmental management program are:

1. The Pollution Prevention Act of 1990 calls on companies to disclose and report a great deal about their operations. Widespread inspections to determine compliance would be very expensive. It also would severely strain the government's manpower.
2. The law aims at creating a more cooperative relationship between the environmental agencies and industry. Strict enforcement provisions with penalties for incomplete compliance could do the opposite and actually create a disincentive to critical self-auditing, self-policing and voluntary disclosure.
3. Companies have an incentive to voluntarily comply with the law because having smaller quantities of chemicals to dispose of could actually save money while giving the company a public relations edge.

2.12 SUMMARY

1. The fear of accidents is probably the major reason for the promulgation of emergency planning and response legislation.
2. All the early water legislation addressed the issue of pollution, but the issue of emergency planning and response was overlooked.
3. States are required to propose implementation plans, SIPs, for the prevention of air pollution emergency episodes. Each implementation plan must include a contingency plan, which will outline the steps to be taken in the event that a particular pollutant concentration reaches the level at which it can be considered to be harmful.
4. As a result of the *Clean Water Act*, contingency plans were to be developed to facilitate the establishment and designation of strike forces in the event of an emergency, and to set up a system of surveillance and notice.
5. RCRA formed technical assistance panels and called for the research, development, demonstration, and evaluation of technologies for recycling and resource recovery. CERCLA was the beginning of the remediation of hazardous waste sites.
6. The *Superfund Amendments and Reauthorization Act of 1986* renewed the national commitment to correcting problems arising from previous mismanagement of hazardous wastes.

7. Probably one of the most important safety and health standards ever adopted is the OSHA hazard communication standard, more properly known as the "right to know" laws.
8. The *Risk Management Program* (RPM) rule is designed to reduce the risk of accidental releases of acutely toxic, flammable and explosive substances.
9. Certain areas of concern have come to the forefront of potential environmental violations.
10. The *Pollution Prevention Act of 1990* is the most important regulation regarding pollution prevention. The 1990 act establishes the pollution prevention hierarchy as national policy, declaring that pollution should be prevented or reduced at the source wherever feasible, while pollution that cannot be prevented should be recycled in an environmentally safe manner.

PROBLEMS

1. Discuss the original objectives of Title III of SARA and indicate the level of participation of community members.
2. Outline an emergency response plan under Title III.
3. List the major objectives of the *Resource Conservation and Recovery Act* (RCRA).
4. Discuss the *Occupational Safety and Health Administration* (OSHA) involvement with emergency planning and response.
5. Compare and note the differences between a spill prevention control and countermeasure plan and the emergency response plan of Title III.
6. Review the literature and obtain the various classifications of the *Standard Industrial Classification* (SIC) codes.
7. Describe USEPA's pollution prevention hierarchy.
8. Discuss some of the barriers to implementing a pollution prevention program.

REFERENCES

1. *Rivers and Harbors Act of 1899*, Public Law, Washington, DC, 1899.
2. *Solid Waste Disposal Act*, Public Law 80-272, 89th Congress, Washington, DC, Oct. 20, 1965.
3. "Fact Sheet – EPA's Recommended Final Ozone and Particulate Matter Standards," http://www.epa.gov/ttn/oarpg/naasqsfin/o3pm.html, June 25, 1997.
4. C. Copeland and M. Tiemann, "Environmental Laws: Summaries of Statutes Administered by the Environmental Protection Agency," January 12, 1999.
5. *Content of Contingency Plans*, 40 Code of Federal Regulations, *Federal Register*, p. 27480, Washington, DC, May 20, 1981.

6.   J. E. McCarthy and M. Tiemann, "Summaries of Environmental Laws Administered by the EPA: Solid Waste Disposal Act/ Resource Conservation and Recovery Act," July 15, 2000.
7.   J. W. Porter, "SARA: A First Year in Review," *Waste Age*, pp. 27-28, February 1988.
8.   *Title III Fact Sheet-Emergency Planning and Community Right to Know*, EPA Publication, Washington, DC, 1987.
9.   "Extremely Hazardous Substances List..." *Federal Register*, pp. 13378-13410, Washington, DC, Apr. 22, 1987.
10.   J. Spero, B. Devito, L. Theodore, *Regulatory Chemicals Handbook*, Marcel Dekker, New York City, 2000.
11.   M. K. Theodore and L. Theodore, *Major Environmental Issues Facing The 21st Century*, (originally published by Simon and Shuster), Theodore Tutorials, East Williston , NY, 1995.
12.   Adopted from: D. Heinold, "EPA's Risk Management Program – Don't Ignore It," *Environmental Protection*, 1995.
13.   J. W. Blattner and G. Bramble, "Environmental Manager," *Chemical Engineering Magazine*, Mc-Graw Hill, New York City, pp. 127-130, June, 1994.
14.   J. Wilcox and L. Theodore, *Engineering and Environmental Ethics*, John Wiley and Sons, New York City, 1998.
15.   R. Dupont, L. Theodore, and K. Ganesar, *Pollution Prevention*, Lewis Publications, Boca Raton, FL, 2000.

3

Emergency Planning and Response

3.1 INTRODUCTION

This chapter addresses planning for emergencies and responding appropriately when they occur. Although much of the material in this chapter may appear to be dated, the response procedures in place still apply. The reader should note that the presentation is geared primarily for local and state personnel. However the same basic approach is applied to planning and response activities for industrial applications.

Section 3.2 explains some reasons for planning ahead and discusses laws that require community groups to develop emergency response plans. Regardless of the existence of such laws, however, it makes good sense to plan ahead. Once an explosion has occurred, for example, it is probably too late for analysis. The topic of Section 3.3 is the planning committee; this group should be composed of people who can make it successful i.e. government leaders, industry specialists, police, firefighters, health specialists, and local residents. Section 3.4 describes the Hazards Survey. Before a plan can be developed, an inventory of the potential hazards in a community must be gathered; then the risks associated with each hazard must be assessed and prioritized. The assessment of risks is the main topic of Parts III and IV of this book.

The main section of this chapter, Section 3.5, details the items that should be included in an emergency plan, which will specify the actions to be taken during an emergency and identify the critical personnel and their responsibilities. A clear, concise, stepwise approach is the goal of the emergency plan. Section 3.6 discusses training. Service groups need to be trained for emergencies before such events occur. Public officials should be apprised of their roles in emergencies. Communications during an emergency

The authors gratefully acknowledge the assistance of Christopher Ruocco in researching, reviewing, and editing this chapter.

(Section 3.7) is critical. Notification of the proper government agencies is required by law; a clearly understood and well publicized notice is important to control the public's response. The manner in which information regarding an emergency is communicated can be just as crucial as the information communicated. Many injuries and deaths during a disaster are the result of panic, and panic is often triggered by misinformation or a lack of information. The implementation of the plan (Section 3.8) includes keeping it current. An occasional audit is imperative to keep the plan from becoming obsolete.[1]

The chapter concludes with a short section that examines the state initiatives on both environmental management and emergency response planning.

3.2 THE NEED FOR EMERGENCY RESPONSE PLANNING

Emergencies have occurred in the past and will continue to occur in the future. A few of the many common sense reasons to plan ahead are provided below.[2]

1. Emergencies will happen; it is only a question of time
2. When emergencies occur, the minimization of loss and the protection of people, property, and the environment can be achieved through the proper implementation of an appropriate emergency response plan
3. Minimizing the losses caused by an emergency requires planned procedures, understood responsibility, designated authority, accepted accountability, and trained, experienced people. With a fully implemented plan, these goals can be achieved
4. If an emergency occurs, it may be too late to plan. Lack of preplanning can turn an emergency into a disaster

A particularly timely reason to plan ahead is to ease the "chemophobia," or fear of chemicals, which is so prevalent in society today. So much of the recent attention to emergency planning and so many newly promulgated laws are a reaction to the tragedy at Bhopal. The probable causes of "chemophobia" are lack of information and misinformation. *Fire* is hazardous, and yet it is used regularly at home. Most adults have understood the hazard associated with fire since the time of the caveman. By the same token, hazardous chemicals, necessary and useful in our technological society, are not something to fear. Chemicals need to be carefully used and their hazards understood by the general public. A well-designed emergency plan that is understood by the individuals responsible for action, as well as by the public, can ease concerns over emergencies and reduce "chemophobia". People will react during an emergency; how they react can be somewhat controlled through education. When ignorance is pervasive, the likely behavior during an emergency is panic.

An emergency plan can minimize loss by helping to assure the proper responses. "Accidents become crises when subsequent events, and the actions of people and organizations with a stake in the outcome, combine in unpredictable ways to threaten the social structures involved."[3] The wrong response can turn an accident into a disaster as easily as no response. For example, if a chemical fire is doused with water, which causes the emission of toxic fumes, it would have been better to let the fire burn itself out. For another example, suppose people are evacuated from a building into the path of a toxic vapor cloud; they might well have been safer staying indoors with closed windows. Still another example is offered by members of a rescue team who become victims because they were not wearing proper breathing protection. The proper response to an emergency requires an understanding of the hazards. A plan can provide the right people with the information they need to respond properly during an emergency.

In addition to the above-mentioned commonsense reasons, there are legal reasons to plan. Recognizing the need for better preparation to deal with chemical emergencies, Congress enacted the *Superfund Amendments and Reauthorization Act of 1986* (SARA), discussed in detail in Chapter 2. One part of SARA is a free-standing act called *Title III* (the *Emergency Planning and Community Right-to-Know Act of 1986*). This act requires federal, state, and local governments, and industry to work together in developing emergency plans and "community right-to-know" reporting on hazardous chemicals. These new requirements build on EPA's Chemical Emergency Preparedness Program and numerous state and local programs, which are aimed at helping communities deal with potential chemical emergencies.[1]

Most larger industries have long had emergency plans designed for on-site personnel. The protection of people, property, and thus profits, has made emergency plans and prevention methods common in industry. On-site emergency plans are often required by insurance companies. One way to minimize the effort required for emergency planning is to expand existing industry plans to include all significant hazards and all people in a given community.

3.3 THE PLANNING COMMITTEE

Emergency planning should grow out of a team process coordinated by a leader. The team may be the best vehicle for gathering into the planning process people representing various areas of expertise, thus producing a more meaningful and complete plan. The team approach also encourages planning that will reflect a consensus of the entire community. Some individual communities and/or areas that include several communities had already formed advisory councils before the SARA requirements. These councils can serve as an excellent resource for the planning team.[4]

When selecting the members of a team that will bear overall responsibility for emergency planning, the following considerations are important.

1. Emergencies will happen; it is only a question of time
2. The group must possess, or have ready access to, a wide range of expertise relating to the community, its industrial facilities, its transportation systems, and the mechanics of emergency response and response planning
3. The members of the group must agree on their purpose and be able to work cooperatively
4. The group must be representative of all the elements of the community that have substantial interest in reducing the risks posed by emergencies

While many individuals have an interest in reducing the risks posed by hazards, their differing economic, political, and social perspectives may cause them to favor different means of promoting safety. For example, people who live near an industrial facility that manufactures, uses, or emits hazardous materials are likely to be greatly concerned about avoiding threats to their lives. They are likely to be less concerned about the costs of developing accident prevention and response measures than some of the other team members. Others in the community, i.e., those representing industry or the budgeting group, for example, are likely to be more sensitive to costs. They may be more anxious to avoid expenditures for unnecessarily elaborate prevention and response measures. Also, industry facility managers, although concerned with reducing risks posed by hazards, may be reluctant, for proprietary reasons, to disclose materials and process information beyond what is required by law. These differences can be balanced by a well-coordinated team that is responsive to the needs of its community.

Agencies and organizations bearing emergency response responsibilities may have differing views about the role they should play in case of an incident. The local fire department, an emergency management agency, and a public health agency are all likely to have some responsibilities during an emergency. However, each of these organizations might envision a very different set of actions at the emergency site. The plan will serve to detail the actions of each response group during an emergency.

In organizing the community to address the problems associated with emergency planning, it is important to bear in mind that all affected parties have a legitimate interest in the choices among planning alternatives. Therefore, strong efforts should be made to ensure that all such groups are included in the planning process. The need for unity of the committee during both the planning process and implementation increases for larger numbers of different community groups. Each group has a right to participate in the planning, and a well-structured, well-organized planning committee should serve the entire community.

By law, the planning committee should include:[5]

1. Elected and state officials
2. Civil defense personnel

3. First aid personnel
4. Local environmental personnel
5. Transportation personnel
6. Owners and operators of facilities subject to SARA
7. Law enforcement personnel
8. Fire-fighting personnel
9. Public personnel
10. Hospital personnel
11. Broadcast and print media
12. Community groups

Other individuals who could also serve the community well and should be a part of the committee include technical professionals, city planners, academic and university researchers, and local volunteer help organizations.[6]

The local government has a great share of the responsibility for emergency response within its community. The official who has the power to order evacuation, fund fire and emergency units, and educate the public is a key person to emergency planning and the response effort. For example, an entire plan might fail if a necessary evacuation were not ordered on time. Although politics should be disassociated from technical decisions, such linkage is inevitable in emergency planning. Distasteful options that require political courage are often necessary. In a given situation, for example, one may need to decide whether to evacuate a section of town where there is some doubt about the necessity of evacuation, but the worst-case consequence of not evacuating would be deadly. A public official can build support for future candidacy by using the issue of chemical safety as a bandwagon, but mistakes in handling emergencies are measured by a strong instrument i.e. the election and a failed emergency plan can be fatal to a political career. Politics is a social feedback device which, when used properly, can aid government leaders in making correct decisions. A political career can also be destroyed by an error in reading the social feedback. An effective plan can save elected officials hours of media criticism after a crisis because the details of the response were organized by someone on the team as events occurred. Because of the power elected officials have locally, they are likely to take the leadership places on such committees.

The independence of fire and police units from political control is somewhat traditional. Recognition of the freedom necessary to conduct public safety work gives police and fire units the option of rejecting outside control. However, to be successful, community emergency response planning must be universally developed and implemented. The fire and police departments, which are likely to be first on the emergency scene, will be required to act immediately. Their knowledge of the hazards and the plan is important to plan effectiveness as well as to their safety. The police are best suited for evacuation and crowd control or protection of evacuated areas. They will need to understand all such assigned roles. The fire service groups will likely be on the scene to control the effects of the accident. The fire service people must bring to the committee their expertise for treating emergencies, a real asset. The firefighters must also learn from the committee the special considerations to be given to emergencies other than structural fires.

The environmental agencies may be among the best suited for evaluating risk. Their expertise usually includes a sound knowledge of the particular features of the local environment, such as location of flood plains and water resources, and the hazards of certain chemicals. They should be used to support the risk evaluating effort. The local or state environmental agencies are also a source for inventory of hazards on industrial sites. This information will serve the committee.

The state or local health agencies will help the committee to understand adverse health effects. For example, the risks associated with different chemicals can be evaluated by health agency personnel or by their contacts at research institutions. These experts can assist in evaluating the risk of various hazards. The owners and operators of facilities handling hazardous chemicals can be an asset in emergency planning because of their knowledge of the safety features already in place. The representatives of industry also have access to information about the hazards of each chemical either from the supplier or from the company's research department. Knowledge of on-site prevention features at an industrial site can help to sort through the potential hazards listed and to focus on the significant ones.

The local planners in a city or community may also be equipped to assist in emergency planning. The agendas of these groups typically include developing the community, creating jobs, and establishing economic stability; thus the planners have their own reasons for wanting a community to be viewed as safe. The planners are also likely to have detailed information about the community: not only road maps and transportation routes, but locations of highly populated areas and industrial sections. The understanding of the locations of people and places in a community is important to planning and assessing the significant risks. The local planners can serve the committee because of their knowledge of these demographic and industry related features.

Toxicologists, meteorologists, chemists, and environmental and chemical engineers are among the technical professionals who have experience and knowledge about chemicals, hazards, and preventive designs. The committee needs individuals with such expertise to assist in the preparation of plans that are technically and scientifically sound as well as safe. The evaluation of hazards and the design of appropriate emergency responses reflect choices, not political options.

The management or control of the committee during planning, and especially during implementation, is essential. As suggested earlier, the emergency plan will be generated by different individuals with different priorities. The different groups will have their own legitimate interests, and each interest will have to be weighed against its value to the plan. The committee leader must demonstrate respect for the interests of each of the individuals, as well as for each member's contribution. The committee leader is likely to be chosen for several reasons; among these should be

1. The degree of respect held for the person by groups and individuals with an interest in the emergency plan
2. The time and resources the person will be able to devote to the work of the committee
3. The person's history of working relationships with concerned community agencies and organizations
4. The person's management skills and communication skills
5. The person's present responsibilities and background related to emergency planning, prevention, and response

Personal considerations, as well as institutional ones, should be weighed when selecting a committee leader. If one candidate has all the right resources to address the issues of emergency planning and implementation, but is unable to interact with local officials, someone else may be a better choice. Since the committee leader must coordinate this large group of people with different priorities and realms of expertise, the choice of the leader is critical to the success of the committee.[7]

3.4 HAZARDS SURVEY

To characterize potential disasters by type and extent, a survey of hazards or foreseeable threats in the community must be performed and evaluated. Without such information, an appropriate plan cannot be developed. An inventory of the community protection assets, hazard sources, and risks must be done before the actual plan is written. The procedures followed here is similar to that provided in Part IV of this book – Hazard Risk Assessment.

Although a plan for a city divided by a river may not be applicable to a desert city on a seismic fault, duplication can be an enemy of cost efficiency. Thus wherever possible, any emergency plans that already exist in the community should be used as a starting point. Community groups that may have developed such plans include civil defense organizations, fire departments, the Red Cross, public health agencies, and local industry councils. Existing plans should be studied and their applicability to the proposed community plan evaluated.

Local government departments such as transportation, water, power, and sewer, may have valuable resources. These should be listed and then compared to the needs of the plan. Some examples are provided below.

1. Trucks
2. Equipment (e.g., backhoes, flatbeds)
3. Laboratory services (e.g., water department)
4. Fire vehicles

5.   Police vehicles
6.   Emergency suits
7.   Breathing apparatus
8.   Gas masks
9.   Number of trained emergency people
10.   Number of volunteer personnel (e.g., Red Cross)
11.   Buses or cars
12.   Communication equipment (e.g., hand radios)
13.   Local TV and radio stations
14.   Ambulances
15.   Trained medical technicians and first aid personnel
16.   Stocks of medicines
17.   Burn treatment equipment
18.   Fall-out shelters

The potential sources of hazards should be listed for risk assessment. SARA requires certain industries to provide information to the planning committee. Information about small as well as large industries is necessary to permit the committee to evaluate the significant risks. The information required by SARA (some of which was provided in Chapter 2) includes:

1.   The chemical name
2.   The quantity stored over a period of time
3.   The type of chemical hazard (e.g., toxic, flammable, ignitable, corrosive)
4.   Chemical properties and characteristics (e.g., liquid at certain temperatures; gas at certain pressures; reacts violently with water)
5.   Storage description and storage location on the site
6.   Safeguards or prevention measures associated with the hazardous chemical storage or handling design, such as dikes, isolation of incompatible substances, fire resistant equipment
7.   Control features for prevention such as temperature and pressure controllers and fail-safe design devices, if included in the handling design
8.   Recycle control loops intended for accident prevention
9.   Emergency shutdown features

The planning committee should designate hazard sources on a community map. This information probably already exists and can be obtained locally from the transportation department, environmental protection agency, city planning department, community groups, and industry sources. Some of the data to be represented on the community map are:

1.   Industrial and other sites of possible chemical accidents
2.   Wastewater and water treatment plants in which chlorine is stored

3. Potable and surface water
4. Drainage and runoff
5. Population location and density in different areas
6. Transportation routes for children
7. Commuter routes
8. Truck transport roads
9. Railroad lines, yards, and crossings
10. Major highways, noting merges and downhill curves
11. Hospitals and nursing homes
12. Fall-out shelters

The potential for natural disaster, based on the history and knowledge of the region and earth structure, should be indicated in the plan. Items such as seismic fault zones and flood plains, and potentials for hurricanes and winter storms should be noted.

The risk inventory or risk evaluation is the next part of the hazard survey. It is not practical to expect the plan to cover every potential accident. When the hazards have been evaluated, the plan should be focused on the most significant ones. This risk assessment stage requires the technical expertise of many people to compare the pieces of data and determine the relevance of each. Among the important factors to be considered in performing the risk evaluation are the following:

1. The routes of transport of hazardous substances should be reviewed to determine where a release could occur
2. Industry sites are not the only sources of hazards; thus the proximity of hazards to people and other sensitive environmental receptors should be examined
3. The toxicology of different exposure levels should be reviewed

When the significant risks have been listed, the hazard survey is complete and the plan can be developed.

3.5 PLAN FOR EMERGENCIES

Successful emergency planning begins with a thorough understanding of the event or potential disaster being planned for. The impacts on public health and the environment must also be estimated. Some of the emergencies that should be included in the plan are[8]:

1. Earthquakes
2. Explosions
3. Fires
4. Floods
5. Hazardous chemical leaks, i.e., gas or liquid

6.   Power or utility features
7.   Radiation incidents
8.   Tornadoes or hurricanes
9.   Transportation accidents

To estimate the potential impact on the public or the environment of accidents of different types, the likely emergency zone must be studied. For example, a hazardous gas leak, fire, or explosion may cause a toxic cloud to spread over a great distance. The minimum affected area, and thus the area to be evacuated, should be estimated on the basis of an atmospheric dispersion model. Various models can be used; the more difficult models produce more realistic results, but the simpler and faster models may provide adequate data for planning purposes.[9]   A more thorough discussion of atmospheric dispersion is presented in Part III – Health Risk Assessment.

In formulating the plan, some general assumptions may be made.

1.   Organizations do a good job when they have specific assignments
2.   The various resources will need coordination
3.   Most of the necessary resources are likely to be already available in the community (in plants or city departments)
4.   People react more rationally when they have been apprised of a situation
5.   Coordination is basically a social process, not a legal one
6.   Disorganization and reorganization are common in a large group
7.   Flexibility and adaptability are basic requirements for a coordinated team

The objective of the plan should be a procedure that uses the combined resources of the community in a way that will:

1.   Safeguard people during emergencies
2.   Minimize damage to property and the environment
3.   Initially contain the incident and ultimately bring it under control
4.   Effect the rescue and treatment of casualties
5.   Provide authoritative information to the news media (for transmission to the public)
6.   Secure the safe rehabilitation of the affected area
7.   Preserve relevant records and equipment for subsequent inquiry into causes and circumstances

During the development of the plan, the assumptions and objectives should be kept in mind. Although prevention is an important goal in accident and emergency management, it is not really the objective of this plan. The plan should focus on minimizing damage when emergencies occur.[2] Key components of the emergency action plan include[1]:

1. Emergency actions other than evacuation
2. Escape procedures when necessary
3. Escape routes clearly marked on a site map, and perhaps also on the roads
4. A method of accounting for people after evacuation
5. Description and assignment of rescue and medical duties
6. A system for reporting emergencies to the proper regulatory agencies
7. A means of notification of the public by an alarm system
8. Contact and coordination person responsibilities

SARA called for each community group, as designated by the governor, to have a plan by the fall of 1988. Specific requirements include:

1. The identification of all facilities as well as transportation routes for extremely hazardous substances
2. The establishment of emergency response procedures, both on plant sites and off (facility owner and operator actions, as well as the actions of local emergency and medical personnel)
3. The establishment of methods of determining when releases occur and what areas and populations may be affected by them
4. A listing of community and industry emergency equipment and facilities, along with the names of those responsible for the equipment and its upkeep
5. The description and scheduling of a training program to teach methods for responding to chemical emergencies
6. The establishment of methods and schedules for exercises or drills to test emergency response plans
7. The designation of a community coordinator and a facility coordinator to implement the plan
8. The designation of facilities (e.g., hospitals, natural gas plants) that are subject to added risk and provision for their protection

A standard format that could be followed might incorporate

1. A statement promulgating the plan
2. A purpose for the plan
3. Assumptions made in developing the plan
4. A discussion of the plan's weaknesses and vulnerabilities
5. A clear statement of when the plan will be executed
6. A stepwise narrative explanation of how the plan works (for those who will direct or coordinate the plan)
7. A chart of the major disaster functional groups, including the departments and volunteers who are responsible for coordinating or supporting each function

8.   A description of the responsibilities of each functional group (e.g., duties and actions of police)
9.   A list of the necessary equipment, its location, and contact person for obtaining each item or unit
10. A method for communicating each type of emergency to the public, the functional groups, and the responsible agencies
11. A list of the emergency coordinator's tools or resources
12. Training details and schedules
13. The plan implementation schedule, including slots for routine audits and updates

Different emergencies are likely to require different response actions. Specific steps for coping with four types of emergency situations are outlined.

1.   Volatile toxic release
     a.   The release should be deluged with water
     b.   The people who will possibly be affected by the toxic cloud should be warned to close their windows or, if necessary, evacuated
     c.   Police with protective equipment should check the homes that have been affected
2.   Flammable chemical fire
     a.   Access to the area should be controlled
     b.   The fire should be prevented from spreading
     c.   The fire should be extinguished by professionals using proper personal protective gear and modern fire-fighting equipment
3.   Chemical spill
     a.   The spilled substance should be contained
     b.   Medical personnel with protective equipment should be available to administer to those affected
     c.   A rescue team with protective equipment should collect the spilled material in containers
4.   Tornado
     a.   Emergency warnings should be issued to people to move to shelters
     b.   Equipment in factories should be shut down
     c.   Squads of rescue teams should be rushed to the affected areas after the tornado has passed

The details of the plan will be different from community to community, and the appropriate responses will differ according to the event anticipated. Obviously, each community must develop a plan tailored to its own needs.

3.6 TRAINING OF PERSONNEL

The education of the public is critical to securing public support of the emergency plan; the real hazards in the community must be made known, as well as what to do in an emergency. Most people are not aware of the reality of hazards in their communities. The common perception is that hazards exist elsewhere, as do the resulting emergencies.[10] The education of the populace about the true hazards associated with routine discharges from plants in the neighborhood and preparing that populace for emergencies is a real challenge to the community committee. People must be taught how to react to an emergency-how to recognize and report an incident, how to react to alarms, and what other action to take. A possible initial result of SARA Title III may be a fear of industrial discharges on the part of the public.[11] News stories can be misleading if based on hazardous chemical inventories, accidental release data, or annual emissions reports of questionable accuracy or if taken out of context. Through training programs, it should be possible to put such information into perspective.

The personnel at an industrial plant who are trained in the operation of the facility are critical to proper emergency response. They must be taught to recognize abnormalities in operations and to report them immediately. Plant operators should be taught how to respond to various types of accidents. Internal emergency squads also can be trained to contain the emergency until outside help arrives, or, if possible, to terminate the emergency. It is especially important to train plant personnel in shutdown and evacuation procedures.

Training is important for the emergency teams to assure that their roles are clearly understood and that accidents can be reacted to safely and properly without delay. The emergency teams include the police, fire, and medical people, and the volunteers who will be required to take action during an emergency.[11] These people must be knowledgeable about the potential hazards. For example, specific antidotes for different health related conditions must be known by medical personnel. The whole emergency team must also be taught the use of personal protective equipment.

Local government officials also need training. Since these officials have the power to order an evacuation, they must be aware of the circumstances under which such action is necessary, and they must understand before an emergency occurs that the timing of an evacuation is critical. Local officials also control the use of city equipment and therefore must know what is needed for an appropriate response to a given emergency.

Media personnel, such as print and broadcast reporters, editors, etc., must also be involved in the training program, since it is important that the public receive accurate information. If incorrect or distorted information about an emergency is disseminated, panic can easily result. For this reason, it is important for print and broadcast journalists to be somewhat knowledgeable about the potential hazards and the details of emergency responses.

Training for emergencies should be done routinely:

1. When a new member is added to the group
2. When someone is assigned a new responsibility within the community
3. When new equipment or materials are acquired for use in emergency

response
4.   When emergency procedures are revised
5.   When a practice drill shows inadequacies in performance of duties
6.   At least once annually

Any training program should address five questions.

1.   How are potential hazards recognized? (This can be determined by periodic review of hazards and prevention measures.)
2.   What precautions (e.g., donning personal protective equipment) are to be taken when responding to an emergency?
3.   Where are the evacuation routes?
4.   To whom should a hazard be reported?
5.   What actions constitute proper responses to special alarms or signals?

It is important for emergency procedures to be performed as planned. This requires regular training to ensure that people understand and remember how to react. The best plan on paper is likely to fail if the persons involved are reading it for the first time as an emergency is occurring. People must be trained *before* an emergency happens.

3.7 NOTIFICATION OF PUBLIC AND REGULATORY OFFICIALS

Notifying the public of an emergency is a task that must be accomplished with caution. People will react in different ways. Many will simply not know what to do, some will not take the warning seriously, and others will panic. Proper training in each community, as discussed in Section 3.6, can help minimize panic and condition the public to make the right response in a time of stress.
Methods of communicating the emergency will differ from community to community, depending on size and resources. Some techniques for notifying the public are:

1.   The sounding of fire department alarms in different ways to indicate emergencies of certain kinds
2.   Chain phone calls (this method usually works well in small towns)
3.   Announcements made through loudspeakers from police cars or the vehicles of volunteer teams

Once the emergency has been communicated, an appropriate response by the public must be evoked. For this to occur, an accepted plan that people know and understand must be put into effect. Since an emergency can quickly become a disaster if panic ensues, the plan should include the appropriate countermeasures.
Information reported to the emergency coordinator must be carefully screened. A suspected "crank call" should be checked out before an alarm is

sounded. By taking no immediate action, however, the team runs the obvious risk that the plan will not be implemented in time. Therefore, if a call cannot be verified as bogus, a response must begin and local police should be dispatched quickly to the scene of the reported emergency to provide firsthand information.

The print and broadcast media can be a major resource for communication, and one job of the emergency coordinator is to prepare information for reporters. The emergency plan should include a procedure to pass along information to the media promptly and accurately.

Certain types of emergency must be reported to government agencies; it is not always sufficient to notify just the response team. For example, state and federal laws require the reporting of hazardous releases and nuclear power plant problems. There are also more specific requirements under SARA Title III for reporting chemical releases. Facilities that produce, store, or use a listed hazardous substance must immediately notify the local emergency planning committee and the *State Emergency Response Commission* if there is a release of one or more substances specifically listed in SARA. These substances include 402 extremely hazardous chemicals on the list prepared by the *Chemical Emergency Preparedness Program* and chemicals subject to the reportable quantities requirements of the original Superfund.[1] The initial notification can be made by telephone, radio, or in person. Emergency notification requirements involving transportation incidents can be satisfied by dialing 911 or calling the operator. The emergency planning committee should provide a means of reporting information on transportation accidents quickly to the coordinator.

SARA requires that the notification of an industrial emergency include:

1. The name of the chemical released
2. Whether it is known to be acutely toxic
3. An estimate of the quantity of the chemical released into the environment
4. The time and duration of the release
5. Where the chemical was released (e.g., air, water, land)
6. Known health risks and necessary medical attention
7. Proper precautions, such as evacuation
8. The name and telephone number of the contact person at the plant or facility at which the release occurred

As soon as is practical after the release, there must be a written follow-up emergency notice, updating the initial information and giving additional information on response actions already taken, known or anticipated health risks, and advice on medical attention.

Law has required the reporting and written notices since October 1986.

3.8 PLAN IMPLEMENTATION

Once an emergency plan has been developed, its successful implementation can be assured only through constant review and revision. Helpful ongoing procedures are:

1. Routine checks of equipment inventory, status of personnel, status of hazards, and population densities
2. Auditing of the emergency procedure
3. Routine training exercises
4. Practice drills

The coordinator must assure that the emergency equipment is always in readiness. Siting the control center and locating its equipment is also the coordinator's responsibility. There should be both a main control center and an alternate, in carefully chosen locations. The following items should be present at the control center:

1. Copies of the current emergency plan
2. Maps and diagrams of the area
3. Names and addresses of key functional personnel
4. Means to initiate alarm signals in the event of a power outage
5. Communication equipment (e.g., phones, radio, TV, and two-way radios)
6. Emergency generators and lights
7. Evacuation routes mapped out on the area map
8. Self-contained breathing equipment for possible use by the control center crew
9. Miscellaneous furniture, including cots

Inspection of emergency equipment such as fire trucks, police cars, medical vehicles, personal safety equipment, and alarms should be done routinely.
       The plan should be audited on a regular basis, at least annually; to assure that it is current. Items to be updated include the list of potential hazards and emergency procedures (adapted to any newly developed technology). A guideline for auditing the emergency response plan, adapted from literature published by the Chemical Manufacturers Association, is presented below in question format.[11]

General Questions

1. What types of emergency have occurred since the last audit?
2. Are all potential emergency types covered by the plan?
3. Who is responsible for maintaining the written plan?

4. Who is authorized to activate the plan?
5. When was the last revision?

Emergency Organization

1. Does the plan have an organization chart that defines responsibilities?
2. Who has overall responsibility?
3. Who is responsible for each of the emergency teams (e.g., fire, police, rescue)?
4. Who directs emergency activities in the field?
5. Are the key responsible positions covered during off hours?

Emergency Actions

1. Will the emergency action contain the incident (e.g., reduce a toxic cloud, contain a spill)?
2. Will the emergency action harm the environment?
3. Is evacuation the only alternative listed?
4. Could the emergency action be improved?

Alarm

1. How is the alarm activated?
2. What provisions are made for an alarm during a power failure?
3. Can all affected people hear the alarm or see the designated signal?
4. If the alarm is activated, what actions are taken?
5. How does the alarm differentiate among emergency types?
6. How are key coordinators notified during off hours?
7. What maintenance and testing is done on the alarm system?

Communications

1. Who handles communications with the media and public officials?
2. Who arranges and maintains communication equipment?
3. Who has access to special phone numbers?
4. What instructions do people have for use of phones in an emergency?
5. What radio channels are available?
6. What procedures exist for messengers?

Evacuation

1. Where are the people moved to?
2. Where are the alternate locations?
3. Are the evacuation signals clear?
4. Are the evacuation routes clearly marked and passable?

Accounting for Personnel

1.   Who maintains records of people?
2.   Who coordinates head counts?
3.   Who keeps track of injuries and fatalities?
4.   How are visitors provided for?

First Aid

1.   Where are the stations?
2.   Who is responsible for rescue?
3.   Who is responsible for treatment of the injured?
4.   What ambulance service is available?
5.   How is the emergency coordinator advised of the status of injured persons?
6.   How is training for emergency and special hazards accomplished?

Transportation

1.   What provisions have been made for moving the injured or disabled?
2.   What provisions have been made for transporting fire, police, and rescue squads?
3.   What provisions have been made for transporting medical supplies?
4.   Who is responsible for assembling vehicles at designated spots?

Security

1.   Who controls the emergency perimeter?
2.   Are these people protected?
3.   Who is responsible for training them?

Fire Fighting

1.   Who maintains the equipment?
2.   Who does the training?
3.   How are fire fighters notified when an emergency occurs?

Outside Agencies

1.   Who decides when to alert them?
2.   Who actually alerts them?

Training

1. Who handles training of all emergency personnel?
2. Who does public awareness training?
3. Who briefs the media?
4. Are local officials kept informed?
5. How is the training evaluated?
6. How often is training done?

Certain operational aspects of the plan should be practiced to assure that the proper response will be realized if and when an actual emergency occurs. The drill scenario should be prepared almost as carefully as the emergency plan itself. Both pre-announced and surprise drills should be held, observed, and evaluated to pinpoint deficiencies in the plan and to determine whether new training is required. The following questions should be used in evaluating drills.

1. What kinds of drills are performed?
2. What aspects are tested?
3. How often are the drills held?
4. Are there both announced and unannounced drills?
5. Comparing response times for announced and unannounced drills, are times for unannounced drills much longer?
6. What time of the day are the drills held?
7. Who is responsible for evaluating the drill?

Once deficiencies have been identified, the plan should be revised to correct them. Such testing and revision should be done regularly; the interval between tests and revisions should not exceed one year.

For the interested reader, further information on emergency planning and response is available in the literature.[12-21]

3.9 OTHER STATE REGULATORY INITIATIVES

In addition to the rules, regulations and procedures provided above for emergency planning and response, each state (in the United States) is actually involved in other environmental management activities. Recently, Matystik et al[28] have prepared a summary outline for interested users on these other state regulatory initiatives. The outline is a "baker's dozen" of information which include the following for each state:

1. Agency name (which varies widely)
2. Acronym
3. URL for the home environmental agency page

4. Date the website was last updated, if listed on the agency's home page
5. A general description of the agency
6. A contact of each agency with e-mail, phone, fax, information where available
7. Main office, mailing address, street address, e-mail, phone, and fax along with information on other offices. *[Note: Other contacts and e-mail addresses where listed in various tables are for the primary contact for that division (if available) or for the division or agency head (if unavailable). In some cases, the e-mail address may be general (non-personal) or it may be the e-mail address of another person designated by the agency as the contact. E-mail links were randomly tested, but in many cases one cannot determine whether a non-functional address may have been incorrectly listed by the agency or if there was a system problem when tested. In cases where any "mail to" links are found to be inoperable, the reader is encouraged to re-send the inquiry to the agency director's office requesting that it be forwarded appropriately.]*
8. Basic descriptions of three major environmental areas: Air, Water, and Solid/Hazardous Waste/Land; contacts; permits and file downloads (for each area when available)
9. Basic laws of the three major environmental areas
10. Online laws, rules, and regulations, i.e., links to specific state laws, rules, and regulations
11. The availability of downloadable forms, files, permits, and publications with the URL for the download area and information on the formats used to determine if one will require any special software, such as *Adobe Acrobat^{TM} Reader* for ".pdf" files or *Microsoft Word^{TM}* for ".doc" files, etc.
12. Specialized information on pollution prevention
13. Special features of a particular state's website, with positives and negatives

New Jersey Toxic Catastrophe Prevention Act (TCPA)

Perhaps the only state that truly addressed catastrophic accidents was New Jersey (as of this printing). Two important features of the above Act were:

1. Applicable (registration) to plants handling, using, manufacturing, storing or having the capability to generate any "extraordinarily hazardous substances" (defined by NJDEP) in at least "registration quantity"
2. Must have an acceptable "risk management program" in place or undertake an "extraordinarily hazardous substance accident risk assessment"

New Jersey also detailed what constituted as Acceptable Risk Management Program (RMP). It included:

1. At least one safety review report prepared in the last two years
2. At least one hazard analysis report with risk assessment, where required, prepared in the last four years
3. Hazard Analysis Report – Hazard and Operability Study (HAZOP), failure mode and effect analysis, quantitative fault tree analysis or what/if check list (see Part IV for details in theses subjects)

The TCPA defined a Risk Management Program through which a registered facility was to demonstrate to the Department how it handled the risks associated with using Extremely Hazardous Substances (ESPs). The RMP is the heart of the TCPA program. The following are the minimum eight elements of an RMP:

1. Safety Review – to ensure that the EHS facility is operated as designed, that no unauthorized modifications were performed, and that new facilities and modifications are designed according to state of the art technology
2. Standard Operating Procedures – to ensure that approved procedures covering all aspects of the handling of an EHS are in place for the appropriate operators
3. Preventive Maintenance Program – to ensure that EHS equipment is routinely tested and inspected and only authorized modifications are performed
4. Operator Training Program – to establish the critical initial training for anyone handling an EHS at the facility and provide annual refresher training
5. Accident Investigation Procedures – to evaluate any EHS accidents that occur, with the focus on prevention of a recurrence
6. Risk Assessment – to identify the risk of an EHS accident associated with a particular EHS operation and, if required, determine the likelihood and consequences of the accidental release in order to develop a risk reduction plan that focuses on accident prevention
7. Emergency Response Program – to develop a response plan to be implemented in the event of an EHS accident
8. Audit Procedures – to ensure that all RMP elements are being implemented

Civil administrative penalties for each violation were set at:

- $10,000 for the first offense
- $20,000 for the second offense
- $50,000 for the third offense

3.10 ILLUSTRATIVE EXAMPLES

Example 3.1

Although this chapter addresses emergency response planning from an industrial perspective, explain why it would be advantageous and explore emergency response planning at the home or office.

Solution 3.1

1. Keep stairs clear of debris. The same applies at a plant. This is a safety and accident concern
2. Install fire/smoke detectors. This is a safety and accident concern.
3. Keep the house well ventilated. The same applies to the enclosed area of a plant. This is a health concern.

Example 3.2

With reference to Matystik et al[28], provide output on "General Information" for a particular state.

Solution 3.2

The sample state selected is New York since both authors reside in New York. The output for "General Information" is provided below.

1. Regulate the disposal, transport, and treatment of hazardous and toxic wastes in an environmentally sound manner
2. Manage the state program for oil and chemical spills
3. Provide for the abatement of water, land, and air pollution, including pesticides
4. Monitor environmental conditions and test for contaminants
5. Encourage recycling, recovery, and reuse of all solid waste to conserve resources and reduce waste
6. Administer fish and wildlife laws, carry out sound fish and wildlife management practices, and conduct fish and wildlife research
7. Manage New York's marine and coastal resources
8. Conduct sound forestry management practices on state lands, provide assistance to private forest landowners, and provide fire prevention and control
9. Manage the Adirondack and Catskill forest preserves and recreational facilities, including campsites and the Belleayre Mountain ski center
10. Protect tidal and freshwater wetlands and flood plains
11. Promote the wise use of water resources

12. Administer the wild, scenic, and recreational rivers program
13. Regulate mining, including reclamation of mined lands, extraction of oil and gas, and underground storage of natural gas and liquefied petroleum gas
14. Inform the public about environmental conservation principles and encourage their participation in environmental affairs

Example 3.3

A chemical reactor at a plant site has exploded. Provide specific steps for coping with this accident/emergency.

Solution 3.3

1. Check if anyone is hurt
2. Leave the accident if necessary
3. Check if another explosion may occur
4. Call 911 as soon as possible
5. Notify as many people as possible

Example 3.4

Provide examples of the duties of "Air, Water, and Solid Waste" divisions.

Solution 3.4

Air Division

1. Conducts research on air pollutants in the atmosphere through modeling and data analysis
2. Involved with air permitting and air toxics assessments
3. Deals with new vehicle technology, programs to regulate fuels, diesel fuel testing and analysis, and research on fuels and vehicle controls
4. Provides air quality data analysis, source modeling and assessment, meteorological services, small business technical assistance, a source management system, and quality assurance and calibration standards for air monitoring
5. Air compliance and enforcement
6. Provide inspections and monitoring protocols

Water Division

1. Protects water quality in lakes, rivers, aquifers, and coastal areas by regulating wastewater discharges monitoring bodies of water, and controlling surface runoff

2. Manages availability of freshwater resources and helps communities prevent flood damage and beach erosion.
3. Promote water stewardship and education

Solid Waste Division

1. Control of hazardous waste from generation, through shipping and handling, to disposal
2. The regulation of the application of registered pesticides by certified applicators
3. Overseeing of clean-up of sites contaminated by radioactive material
4. The regulation of transport of discharges of low-level radioactive waste
5. Permitting of environmentally sound solid waste management facilities, including landfills and incinerators

Example 3.5

Describe the role of the "Pollution Prevention Unit".

Solution 3.5

The Pollution Prevention Unit coordinates pollution prevention efforts throughout New York State. It protects, air, water, and land. It also works with all types and sizes of businesses and facilities.

3.11 SUMMARY

1. There are many reasons to plan ahead (e.g., to minimize losses during an emergency). SARA Title III requires each community to prepare an emergency response plan.
2. The plan should be developed by a committee representing the community residents. Technical experts should be included to help make the plan appropriate.
3. The survey of hazards and the inventory of existing equipment will put the planning committee on the right track.
4. There is no cookbook approach to emergency response planning. A plan is as individual as the community it serves.
5. Training response teams, and orienting the public, the media, and the local officials, are critical to the success of the plan in an emergency. Education about the hazards associated with chemicals in the community can ease the fear the public has about industrial chemicals.
6. Communicating an emergency to all the people who need to know will require a carefully designed and properly executed scheme.
7. Implementation of the plan means that the community must understand

it, practice it, audit it, and rewrite it. The hope is that the plan is always ready but never needed.

8. Once an emergency plan has been developed, its successful implementation can be assured only through constant review and revision. Helpful ongoing procedures are: routine checks of equipment inventory, auditing of the emergency procedure, routine training exercises and practice drills.

9. In addition to the rules, regulations and procedures provided in Section 8 for emergency planning and response, each state (in the United States) is actually involved in other environmental management activities.

PROBLEMS

1. What are the reasons to plan before an emergency strikes?
2. Who should be on the planning committee if it is to be successful? Under SARA, who is required to be on the committee?
3. Why should a Hazards Survey be done before the emergency response plan is developed?
4. What are the objectives of the emergency plan?
5. Why teach the public about the hazardous chemicals in the community?
6. What purpose is served by training the emergency response team?
7. In an emergency, what methods can be used to notify the public? What method could be used if the emergency includes a power failure?
8. What four steps should be taken routinely to ensure that the plan is viable?
9. Discuss the advantages of implementing a pollution prevention procedure in an emergency response procedure.

REFERENCES

1. "Other Statutory Authorities: Title III: Emergency Planning and Community Right-to-Know," EPA J., 13 (1) (1987).
2. M. Krikorian, *Disaster and Emergency Planning*, Institute Press, Loganville, A, 1982.
3. P. Shrivastava, *Bhopal: Anatomy of a Crisis*, Ballinger, Cambridge, MA, 1987.
4. W. Beranek et al., "Getting Involved in Community Right-to-Know," *Chem. Eng. News*, p. 62, Oct. 26, 1987.
5. *Hazardous Materials Emergency Planning Guide*, National Response Team, March 1987, Washington, DC.
6. R. H. Schulze, *Superfund Amendments and Reauthorization Act of 1986 (SARA Title III)*, Trinity Consultants Incorporated, Richardson, TX, May 1987.
7. J. T. O'Reilly, *Emergency Response to Chemical Accidents. Planning and Coordinating Solutions*, McGraw-Hill, New York, 1987.

8. E. J. Michael, O. W. Bell, and J. W. Wilson, "Emergency Planning Considerations for Specialty Chemical Plants," Stone and Webster Engineering Corporation, Boston, August 1986.
9. "Title III Fact Sheet. Emergency Planning Community Right-to-Know," U.S. EPA, Washington, DC, 1987.
10. Chemical Manufacturers Associations, "Title III: The Right to Know, The Need to Plan," *Chemecology*, March 1987.
11. Chemical Manufacturers Association, "Community Awareness & Emergency Response," Program Handbook, CMA, Washington, DC, April 1985.
12. "San Francisco. Corporate Disaster Planning Guide," American Red Cross, San Francisco, January 1986.
13. R. Perry and D. Greer, "Perry's Chemical Engineering Handbook", 7th Ed., McGraw-Hill, New York City, 1997.
14. G. F. Bennett, F. S. Feates, and J. Wilder, *Hazardous Materials Spills Handibook*, McGraw-Hill, New York, 1982.
15. "Toxic Chemical Release Reporting," *Federal Register*, Feb. 16, 1988, p. 4500, Washington, DC.
16. E. J. Michael and R. E. Vanesse, "Planning and Implementation of Emergency Preparedness Exercises Including Scenario Preparation," Stone and Webster Engineering Corporation, Boston, June 1985.
17. J. Spero, B. Devito, and L. Theodore, "Regulating Chemicals Handbook", Marcel Decker, New York City, 2000
18. "Pre-Emergency Plan," Industrial Risk Insurers, Hartford, CT, February 1981.
19. C. L. Elkins and J. L. Makris, "Emergency Planning and Community Right-to-Know," J. *Air Pollut. Control Assoc.*, 38 (3), 243-247 (1988).
20. H. Beim, J. Spero, and L. Theodore, "Rapid Guide to Hazardous Air Pollutants", John Wiley and Sons, New York City, 1997.
21. D. J. McNaughton, G. G. Worley, and P. M. Bodner, "Evaluating Emergency Response Models for the Chemical Industry," *Chem. Eng. Prog.*, pp. 46-51, January 1987.
22. E. J. Michael, "Elements of Effective Contingency Planning," Stone and Webster Engineering Corporation, Boston, November 1985.
23. D. G. Smith, "Role of Real-Time Atmospheric Dispersion Assessment System," ERT, Inc., Concord, MA, March 1987.
24. Chemical Manufacturers Association, "Site Emergency Response Planning," CMA, Washington, DC, 1986.
25. G. Burke, B. Singh, and L. Theodore, "Handbook of Environmental Management and Technology", 2nd Ed., John Wiley and Sons, New York City, 2000.
26. R. Dupont, L. Theodore, and K. Ganesan, "Pollution Prevention", Lewis Publishers, Boca Raton, Fl., 2000.

27. M.K. Theodore and L. Theodore, "Major Environmental Issues Facing the 21st Century, (originally published by Simon & Shuster), Theodore Tutorials, East Williston, NY, 1995.
28. W. Matystik, L. Theodore and R. Diaz, "State Environmental Agencies on the Internet", Government Institutes, Rockville, MD, 1999.

Part II

Process and Plant Accidents

High numbers of casualties, severe material damage, and production losses due to fires, explosions, and other accidents in recent years have motivated all sectors of industry and responsible authorities to initiate scientific investigations of these deadly phenomena. The ultimate goal of the investigations is to develop safety measures that will prevent or limit the effects of such accidents.

Many of the accidents described in Chapter 1 were plant and/or process related. For this reason, Chapter 4 is devoted solely to process fundamentals. The material is presented in the traditional engineering format, but without specific referenced to accidents, and emergencies. Chapter 5 serves as a review of process equipment including plant layout and siting. Safeguards in design and the use of protective equipment to minimize industrial accidents are also discussed. Chapter 6 focuses on the classification of accidents, particularly as they apply to chemical processes since all process plants have the potential for accidents resulting from equipment breakdown, control system failure, utilities and ancillary system outages, human error, and other possibilities. Chapter 7 provides a basic overview of fires, explosions, hazardous spills, and toxic emissions. Part II concludes with Chapter 8, which addresses the process application of several chemicals that are considered to be highly toxic; these include chlorine, ammonia, hydrogen cyanide, hydrogen fluoride, and sulfuric acid. Physical and chemical properties, health effects, and methods of manufacture of these chemicals are discussed in conjunction with potential causes of release.

Note: Units used throughout Part II are those used in the references. No effort was made to provide consistent units.

Part II

Process and Plant Accidents

4

Process Fundamentals

4.1 INTRODUCTION

Since many industrial accidents are plant and/or process related, this chapter is devoted to process fundamentals. The material is presented in the traditional engineering format, but without specific references to accidents and emergencies. The last three chapters of Part II are primarily concerned with plant-related accidents.

Following a brief section on Units and Dimensions (Section 4.2), Sections 4.3 and 4.4 review some of the key physical and chemical properties, respectively. Three important conservation laws are presented in Section 4.5. Basic engineering principles are discussed in Section 4.6, to present a foundation for the theory underlying the proper design and operation of a chemical process.

Because the six main topics are somewhat unrelated, this chapter admittedly lacks a certain cohesiveness. However all the material presented here will find use in later chapters.

4.2 UNITS AND DIMENSIONS

Most of the units used in this book are consistent with those adopted by the engineering profession in the United States. For engineering work, International Systeme (SI) and English units are most often employed; in the United States, the English engineering units are generally used. These systems of units are shown in Table 4.2.1.[1]

The authors gratefully acknowledge the assistance of Tara E. Fleck in researching, reviewing, and editing this chapter.

Conversion of Units

Converting a measurement from one unit to another can conveniently be accomplished by using *unit conversion factors*; these factors are obtained from the simple equation that relates the two units numerically. The following is an example of a unit conversion factor

$$12 \text{ in}/1 \text{ ft} = 1 \tag{4.2.2}$$

Since this factor is equal to unity, multiplying some quantity (e.g., 18 ft) by the factor cannot alter its value. Hence

$$18 \text{ ft } (12 \text{ in}/1 \text{ ft}) = 216 \text{ in} \tag{4.2.3}$$

Note that in Equation (4.2.3), the old units of feet on the left-hand side cancel out leaving only the desired units of inches.

Physical equations must be dimensionally consistent. For the equality to hold, each term in the equation must have the same dimensions. This condition can be and should be checked when solving engineering problems. Throughout the text, and in particular in Part II, great care is exercised in maintaining the dimensional formulas of all terms and the dimensional homogeneity of each equation. Equations will generally be developed in terms of specific units rather than general dimensions (e.g., feet, rather than length). This approach should help the reader to more easily attach physical significance to the equations presented in these chapters.

Significant Figures and Scientific Notation

Significant figures provide an indication of the precision with which a quantity is measured or known. The last digit represents, in a quantitative sense, some degree of doubt. For example, a measurement of 8.12 inches implies that the actual quantity is somewhere between 8.315 and 8.325 inches. This applies to calculated and measured quantities; quantities that are known exactly (e.g., pure integers) have an infinite number of significant figures.

The significant digits of a number are the digits from the first nonzero digit on the left to either (a) the last digit (whether it is a nonzero or zero) on the right if there is a decimal point, or (b) the last nonzero digit of the number if there is no decimal point. For example:

370	has 2 significant figures
370.	has 3 significant figures
370.0	has 4 significant figures
28,070	has 4 significant figures

0.037   has 2 significant figures

0.0370 has 3 significant figures

0.02807 has 4 significant figures

Whenever quantities are combined by multiplication and/or division, the number of significant figures in the result should be equal to the lowest number of significant figures of any of the quantities. In long calculations, the final result should be rounded off to the correct number of significant figures. When quantities are combined by addition and/or subtraction, the final result cannot be more precise than any of the quantities added or subtracted. Therefore, the position (relative to the decimal point) of the last significant digit in the number that has the lowest degree of precision is the position of the last permissible significant digit in the result. For example, the sum of 3702, 370, 0.037, 4, and 37, should be reported as 4110 (without a decimal). The least precise of the five numbers is 370, which has its last significant digit in the tens position. The answer should also have its last significant digit in the tens position.

4.3 PHYSICAL PROPERTIES

Physical properties are important considerations in any study of accidents and emergencies. A substance may exhibit certain characteristics under one set of conditions of temperature, pressure, and composition. However, if the conditions are changed, a once-safe operation may become a hazard by virtue of vulnerability to fire, explosion, or rupturing. To promote a better understanding of these effects, many of which are covered in Chapter 7, a brief review of some key physical and chemical properties is provided in this and the next section.

Temperature

Whether in the gaseous, liquid, or solid state, all molecules possess some degree of kinetic energy; that is, they are in constant motion - vibrating, rotating, or translating. The kinetic energies of individual molecules cannot be measured, but the combined effect of these energies in a very large number of molecules can. This measurable quantity is known as *temperature*; it is a macroscopic concept only (i.e., it does not exist on the molecular level).

Temperature can be measured in many ways; the most common method makes use of the expansion of mercury with increasing temperature. (On process equipment, however, thermocouples or thermometers are more commonly employed). The two most commonly used temperature scales are the Celsius and Fahrenheit scales. The Celsius is based on the boiling and freezing points of water at 1 atm pressure; to the former a value of 100°C is assigned, and to the latter a value of 0°C. On the older Fahrenheit scale, the corresponding temperatures are 212 and 32°F, respectively. Equations (4.3.1) and (4.3.2) show the conversion from one scale to the other.

$$°F = 1.8(°C) + 32 \qquad\qquad (4.3.1)$$
$$°C = (°F - 32)/1.8 \qquad\qquad (4.3.2)$$

The volume of a gas would theoretically be zero at a temperature of approximately -273°C or -460°F. This temperature, which has become known as *absolute zero*, is the basis for the definition of two absolute temperature scales, the *Kelvin* (K) and *Rankine* (°R) scales. The former is defined by shifting the Celsius scale by 273-Celsius degrees so that 0 K is equal to −273°C. Equation (4.2.3) shows this relation.

$$K = °C + 273 \qquad\qquad (4.3.3)$$

The Rankine scale is defined by shifting the Fahrenheit scale 460 Fahrenheit degrees, so that

$$°R = °F + 460 \qquad\qquad (4.3.4)$$

Temperature is an important parameter for safety. At extreme temperatures, the risk of metal fatigue, stress corrosion cracking, and vessel rupture increases dramatically.

Pressure

In the gaseous state, molecules possess a high degree of translational kinetic energy, which means that they are able to move quite freely throughout the body of the gas. For example, when gas is in an enclosed container, the molecules are constantly bombarding the container walls. The macroscopic effect of this bombardment by a tremendous number of molecules - enough to make the effect measurable - is called *pressure*. The natural units of pressure are those of force per unit area.

A number of units are used to express a pressure measurement. Some are based on a force per unit area: for example, *pound (force) per square inch* (psi) or *dyne per square centimeter* (dyne/cm^2). Others are based on a fluid height, such as *inches of water* (in H_2O) or *millimeters of mercury* (mmHg); units such as these are convenient when the pressure is indicated by a difference between two levels of a liquid, as in a manometer or barometer. *Barometric pressure* is a measure of the ambient air pressure. *Standard* barometric pressure is 1 atm and is equivalent to 14.696 psi and 29.921 in Hg.

Measurements of pressure by most gauges indicate the difference in pressure either above or below that of the surrounding atmosphere. The system pressure is greater than atmospheric if the gauge (reading) is positive; if the gauge (reading) is negative, it is less than atmospheric pressure; the term *vacuum* designates a negative gauge pressure. Gauge pressures are usually identified by the letter *g* after the pressure unit (e.g., psig). Since gauge pressure is the pressure relative to the prevailing atmospheric pressure, the sum of the two gives the absolute pressure, as is often indicated by the letter *a* after the unit (e.g., psia):

$$P_a = P + P_g \qquad (4.3.5)$$

A pressure of *zero psia* is the lowest possible pressure theoretically achievable, i.e., perfect vacuum.

Pressure can have a significant impact on safety. Extreme pressures can cause severe metal stress, which can lead to a vessel rupture or explosion. High temperatures, plugged lines, and many other conditions can cause increased pressure.

Moles and Molecular Weights

An atom consists of protons and neutrons in a nucleus surrounded by electrons. An electron has such a small mass relative to that of the proton and neutron that the weight of the atom, the *atomic weight*, is approximately equal to the sum of the weights of the particles in its nucleus. Atomic weight may be expressed in *atomic mass units per atom* or in *grams per gram-atom*. The number of atoms in one gram-atom, called *Avogadro's number*, is 6.02×10^{23}.

The *molecular weight* (MW) of a compound is the sum of the atomic weights of the atoms that make up one molecule of the material. Units of *grams per gram-mole* (g/gmol) are used for molecular weight. One gram-mole contains 6.02×10^{23} molecules. For the English system, one pound-mole (lbmol) contains $454 \times 6.02 \times 10^{23}$ molecules.

For mixtures of substances, it is convenient to express compositions in *mole fractions* or *mass fractions*. A process stream seldom consists of a single component. It may also contain two or more phases, or a mixture of one or more solutes in a liquid solvent. The following definitions are often used to represent the composition of component A in a mixture of components.

$$w_A = \text{mass of A/total mass of stream} = \textit{mass fraction} \text{ of A} \qquad (4.3.6)$$
$$y_A = \text{moles of A/total moles of stream} = \textit{mole fraction} \text{ of A} \qquad (4.3.7)$$

Trace quantities of substances are very often expressed in *parts per million* (ppm) or, if the amount is even smaller, as *parts per billion* (ppb). These are usually represented on a volume basis for gases, and the following equations apply.

$$\text{ppm} = 10^6 \, y_A \qquad (4.3.8)$$
$$\text{ppb} = 10^3 \, \text{ppm} \qquad (4.3.9)$$

Density

The *density* (ρ) of a substance is the ratio of its mass to its volume and may be expressed in units of *pounds per cubic foot* (lb/ft^3) or *kilograms per cubic meter* (kg/m^3). For solids, density can be determined easily by placing a known mass of the substance in a liquid and measuring the displaced volume. The density of a liquid can be measured by weighing a known volume of the liquid in a graduated cylinder. For gases, the ideal gas law (to be discussed in Section 4.6) can be used to calculate the density from the temperature, pressure, and molecular weight of the gas.

The *specific gravity* (SG) is the ratio of the density of a substance to the density of a reference substance at a specific condition.

$$SG = \rho/\rho_{ref} \qquad (4.3.10)$$

The reference most commonly used for solids and liquids is water at its maximum density, which occurs at 4°C; this reference density is 1.000 g/cm^3, 1000 kg/m^3, or 62.43 lb/ft^3.

pH

An important chemical property of an aqueous solution is its *pH* value, which indicates the acidity or basicity of the substance. In a neutral solution such as pure water, the hydrogen (H^+) and hydroxyl (OH^-) ion concentrations (C) are equal. At ordinary temperatures, this concentration is

$$C_{H^+} = C_{OH^-} = 10^{-7} \text{ g-ion/L} \qquad (4.3.11)$$

The notation *g-ion* is the unit for *gram-ion*, which represents Avogadro's number (6.02×10^{23}) of ions. The pH is a direct measure of the hydrogen ion concentration and is defined by

$$pH = -\log C_{H^+} \qquad (4.3.12)$$

Thus, an acidic solution is characterized by a pH below 7 (the lower the pH, the higher the acidity), a basic solution by a pH above 7, and a neutral solution by a pH of 7.

Viscosity

Viscosity is a property associated with a fluid's resistance to flow; more precisely, this property accounts for the energy losses resulting from shear stresses that occur between different portions of the fluid, moving at different velocities. The *absolute viscosity* (μ) has units of mass per length-time; the fundamental unit is the *poise*, which is defined as 1 g/cm-s. Viscosities are frequently given in *centipoises* (0.01 poise), the abbreviation for which is *cP*. In English units, the absolute viscosity is expressed as pounds (mass) per foot-second (lb/ft-s). The absolute viscosity depends primarily on temperature and to a lesser degree on pressure. The *kinematic viscosity* (v), the absolute viscosity divided by the density of the fluid, is useful in certain fluid flow applications; the units for this quantity are length squared per time, e.g., ft^2/s. Because fluid viscosity changes rapidly with temperature, a numerical value of viscosity has no significance unless the temperature is specified.

Heat Capacity

The *heat capacity* of a substance is defined as the quantity of heat required to raise the temperature of that substance by 1°; the *specific heat capacity* is the heat capacity on a unit mass basis. The term *specific heat* is frequently used in place of specific heat capacity. This is not strictly correct because traditionally, specific heat has been defined as the ratio of the heat capacity of a substance to the heat capacity of water. However, since the specific heat of water is approximately 1 cal/g-°C or 1 Btu/lb-°F, the term *specific heat* has come to imply heat capacity per unit mass. For gases, the addition of heat to cause the 1° temperature rise may be accomplished either at constant pressure or at constant volume. Since the amounts of heat necessary are different for the two cases, subscripts are used to identify which heat capacity is being used - C_p for constant pressure or C_v for constant volume. This distinction does not have to be made for liquids and solids since there is little difference between the two. Values of heat capacity are available in the literature.[1,2]

Thermal Conductivity

Experience has shown that when a temperature difference exists across a solid body, energy in the form of heat will transfer from the high temperature region to the low temperature region until thermal equilibrium is reached. This mode of heat transfer in which vibrating molecules pass along kinetic energy through the solid is called *conduction*. Liquids and gases may also transport heat in this fashion. *Thermal conductivity* provides a measure of how fast heat flows through a substance. This property is defined as the amount of heat that flows in a unit of time through a unit surface area of unit thickness as a result of a unit difference in temperature. Typical units for conductivity are *Btu/h-ft-°F*.

Diffusivity

The *diffusion coefficient*, or *diffusivity* (D), is a measure of the rate of transfer of one substance through another. The diffusivity for component A migrating through a solute B, for example, is given by the ratio of the flux J_A (mass transferred per unit time through a unit area) to its concentration gradient dC_A/dz.

$$D = \frac{-J_A}{dC_A/dz} \qquad (4.3.13)$$

The negative sign indicates that diffusion occurs in the direction of decreasing concentration. Diffusivity is a function of temperature, pressure, concentration, phase, and the nature of the other components.

Vapor Pressure

Vapor pressure is an important property of liquids, and to a much lesser extent, of solids. If a liquid is allowed to evaporate in a confined space, the pressure of the vapor phase increases as the amount of vapor increases. If there is sufficient liquid present, the pressure in the vapor space eventually comes to equal exactly the pressure exerted by the liquid at its own surface. At this point, a dynamic equilibrium exists in which vaporization and condensation take place at equal rates and the pressure in the vapor space remains constant. The pressure exerted at equilibrium is called the vapor pressure of the liquid. Solids, like liquids, also exert a vapor pressure. Evaporation of solids (sublimation) is noticeable only for the few solids characterized by appreciable vapor pressures.

Boiling Point

The *boiling point* is the temperature at which the vapor pressure of a liquid is equal to its surrounding pressure. Since the vapor pressure remains constant during boiling, so does the temperature. The higher the system pressure, the higher the temperature must be to induce boiling. The boiling point is specific to each individual substance and is a strong function of pressure.

Freezing Point

The *freezing point* is the temperature at which the liquid and solid state can coexist in equilibrium at a given pressure. At this point the rate at which the substance leaves the solid state equals the rate at which it leaves the liquid state.

4.4   CHEMICAL PROPERTIES

Flash Point

The *flash point* of a flammable liquid is defined as the temperature at which the vapors can be ignited under conditions defined by the test apparatus and method employed. The three major methods of measuring the flash point are:

1. The Cleveland open cup method
2. The Pensky-Martens closed cup tester
3. The tag closed cup method

These are experimental tests that measure the lowest temperature at which application of a test flame causes the vapor overlying the sample to ignite. Tables of flash points for selected substances are available in the literature.[3]

Autoignition Temperature (AIT)

The *autoignition temperature* (AIT) or the *maximum spontaneous ignition temperature* is defined as the maximum temperature at which combustion occurs in a combustible bulk gas mixture when the temperature of a flammable gas-air mixture is raised in a uniformly heated apparatus. The AIT represents a threshold below which chemicals and combustibles can be handled safely. (The AITs of selected substances are available in the literature.[4]) The AIT is strongly independent on the nature of hot surfaces. The AIT may be reduced by as much as 100-200°C when the surfaces are contaminated by dust. When the temperature of a flammable mixture is raised to or above the autoignition temperature, ignition is not spontaneous. Most notably in liquids, there is a finite delay before ignition takes place, i.e., a lapse between the time there is a flammable mixture reaches its flame temperature and the first appearance of a flame. An equation that correlates with the ignition temperature is also available in the literature.[4]

Heat (Enthalpy) of Reaction

Many chemical reactions evolve or absorb heat. When applying energy balances (conservation law for energy) in technical calculations the heat (enthalpy) of reaction is often indicated in mole units so that they can be directly applied to demonstrate its chemical change. To simplify the presentation that follows, examine the equation:

$$aA + bB \rightarrow cC + dD \qquad (4.4.1)$$

The *heat (enthalpy) of reaction*, ΔH, is given by

$$\Delta H = c(\Delta H_f)_C + d(\Delta H_f)_D - a(\Delta H_f)_A - b(\Delta H_f)_B \qquad (4.4.2)$$

Thus, the heat of a reaction is obtained by taking the difference between the heat of formation (ΔH_f) of the products and reactants. If the heat of reaction is negative (exothermic), as is the case of most combustion reactions, then energy is liberated due to the chemical reaction. Energy is absorbed if ΔH is positive (endothermic).

The standard (approximately 16-30°C) heat of reaction, $\Delta H°$, is given by

$$\Delta H = c(\Delta H_f°)_C + d(\Delta H_f°)_D - a(\Delta H_f°)_A - b(\Delta H_f°)_B \qquad (4.4.3)$$

Thus, the standard heat of a reaction is obtained by taking the difference between the standard heat of formation of the products and reactants. Once again, if the standard heat of reaction of formation is negative, as is the case of most combustion reactions, then energy is liberated due to the chemical reaction. Energy is absorbed if $\Delta H°$ is positive.

Tables of standard enthalpies of formation, combustion and reaction are available in the literature for a wide variety of compounds.[1,2] It is important to note that these are valueless unless the stoichiometric equation and the standard state of reactants and products are included.

Enthalpy of reaction and standard enthalpy of reaction are not always employed in engineering reaction/combustion calculations. The two other terms that have been used are the gross (or higher) heating value and the net (or lower) heating value. These are discussed later in this Section.

The *heat (enthalpy) of combustion* at temperature T is defined as the enthalpy change during the chemical reaction where 1 mole of material is burned in oxygen, where all reactants and products are at temperature T. This quantity finds extensive application in calculating enthalpy changes for incineration reactions and is often given in the literature for a standard state temperature of 60°F (16°C). Although much of the literature data on standard heats of reaction are given for 25°C, there is little sensible enthalpy change between these two temperatures and the two sets of data may be considered compatible. Thus, the standard heat of combustion is obtained by taking the difference between the standard heat of formation $\Delta H°_f$ of the products and that of the reactants. Once again, if the standard heat of combustion is negative, as is the case of most combustion/incineration reactions, energy is liberated due to the chemical reaction. Energy is absorbed if enthalpy of combustion is positive. Extensive standard heat of formation and standard enthalpy of combustion data at 298K (25°C) are provided in the literature.[1,2]

Gross Heating Value

The heating value of a waste and/or fuel is one of its most important chemical properties. This value represents the amount of heat evolved in the complete reaction (usually combustion) of a given quantity of the waste and/or fuel.

The *gross heating value* or *higher heating value* (HHV) represents the enthalpy change or heat released when a compound is stoichiometrically combusted at 60°F with the final (flue) products at 60°F and any other water present in the liquid state. Stoichiometric combustion requires that no oxygen be present in the flue gas following complete combustion of the starting fuel.

Net Heating Value

Due to the varying amounts of water that may be produced in a combustion reaction, a second method for expressing heating value is also in use, the *net heating value*. The net heating value is similar to the gross heating value except that it is calculated with the produced water in the vapor state. The net heating value is also known as the *lower heating value* (LHV).

Theoretical Adiabatic Flame Temperature (TAFT)

Most combustion reactions do not operate at stoichiometric or zero percent excess air. Incomplete combustion and high carbon monoxide levels would result under this condition. If all the heat liberated by the reaction goes into heating up the products of combustion (the flue gas) the temperature achieved is defined as the *flame temperature*. If the reaction process is conducted adiabatically, with no heat transfer loss to the surroundings, the final temperature achieved in the flue gas is defined as the *adiabatic flame temperature*. If the combustion process is conducted with theoretical or stoichiometric air (0% excess air), the resulting temperature is defined as the *theoretical flame temperature*. (Theoretical or stoichiometric air is defined as that exact amount of air required the completely react with the compound to produce oxidized end products. Any air in excess of this stoichiometric amount is defined as excess air.)

     If the reaction is conducted both adiabatically and with stoichiometric air, the resulting temperature is defined as the *theoretical adiabatic flame temperature* (TAFT). It represents the maximum temperature that the products of combustion (flue) can achieve if the reaction is conducted both stoichiometrically and adiabatically. For this condition, all the energy liberated from combustion at or near standard conditions ($\Delta H°_C$ and/or $\Delta H°_{298}$) appears as sensible heat in raising the temperature of the flue products, ΔH_p, that is:

$$\Delta H°_C + \Delta H_p = 0 \qquad (4.4.4)$$

where $\Delta H°_C$ = standard heat of combustion at 25°C
          ΔH_p = enthalpy change of the products as the temperature increases from 25°C to the theoretical adiabatic flame temperature

4.5 CONSERVATION LAWS

Three key conservation laws – mass, energy, and momentum – are discussed in this section.

Conservation of Mass

The *conservation law for mass* can be applied to any process or system. The general form of this law is given by Eq. (4.5.1).

$$\text{mass in - mass out + mass generated = mass accumulated} \qquad (4.5.1)$$

Equation (4.5.1) may be applied to the total mass involved or to a particular species on either a mole or mass basis. The conservation law for mass can be applied to steady-state or unsteady-state processes and to batch or continuous systems. To isolate a system for study, it is separated from the surroundings by a boundary or envelope. This boundary may be real (e.g., the walls of a vessel) or imaginary. Mass crossing the boundary and entering the system is part of the *mass-in* term in Eq. (4.5.1), while that leaving the system is part of the *mass-out* term. Equation (4.5.1) may be written for any compound whose quantity is not changed by chemical reaction and for any chemical element, regardless of whether it has participated in a chemical reaction. It may be written for one piece of equipment, around several pieces of equipment, or around an entire process. It may be used to calculate an unknown quantity directly, to check the validity of experimental data, or to express one or more of the independent relationships among the unknown quantities in a particular problem.

A *steady-state* process is one in which there is no change in conditions (temperature, pressure, etc.) or rates of flow with time at any given point in the system. The *accumulation* term in Eq. (4.5.1) is then zero. If there is no chemical reaction, the *generation* term is also zero. All other processes are *unsteady state*.

In a *batch process*, a given quantity of reactants is placed in a container, and by chemical and/or physical means, a change is made to occur. At the end of the process, the container holds the product or products. In a *continuous process*, reactants are fed in an unending flow to a piece of equipment or to several pieces in series, and products are continuously removed from one or more points. A continuous process may or may not be steady state.

As indicated previously, Eq. (4.5.1) may be applied to the total mass of each stream (referred to as an *overall* or *total material balance*) or to the individual components of the streams (referred to as a *componential* or *component material balance*). Often the primary task in preparing a material balance is to develop the quantitative relationships among the streams.

Conservation of Energy

A presentation of the *conservation law for energy* would be incomplete without a brief review of some introductory thermodynamic principles. *Thermodynamics* is defined as the science that deals with the relationships among the various forms of energy. A system may possess energy due to its temperature, velocity, position, molecular structure, surface, and so on. The energies corresponding to these conditions are internal, kinetic, potential, chemical, and surface, and so on. Engineering thermodynamics is founded on three basic laws. Energy, like mass and momentum, is conserved. Application of the conservation law for energy gives rise to the *first law of thermodynamics*. This law, in steady-state form for batch flow processes, is presented here (potential, kinetic, and other energy effects have been neglected).

$$\Delta E = Q - W \tag{4.5.2}$$

For flow processes,

$$\Delta H = Q - W_s \tag{4.5.3}$$

where Q = heat energy transferred across the system boundary
        W = work energy transferred across the system boundary
        W_s = mechanical work energy transferred across the system boundary
        E = internal energy of the system
        H = enthalpy of the system
        $\Delta E, \Delta H$ = changes in internal energy and enthalpy during the process

The internal energy and enthalpy in Eqs. (4.5.2) and (4.5.3), as well as in the other equations in this discussion, may be on a mass or a mole basis, or they may represent the total internal energy and enthalpy of the entire system. Most industrial facilities operate in a steady-state flow mode. If no significant mechanical or shaft work is added or withdrawn from the system, Eq. (4.5.3) reduces to

$$Q = \Delta H \tag{4.5.4}$$

If a unit or system is operated adiabatically, where $Q = 0$, Eq. (4.5.4) then becomes

$$\Delta H = 0 \tag{4.5.5}$$

Although the topics of material and energy balances are covered separately in this section, it should be emphasized that this segregation or compartmentalization does not occur in reality; one must work with both energy and material balances simultaneously.

Perhaps the most important thermodynamic function the engineer works with is the enthalpy. The *enthalpy* is defined by

$$H = E + PV \tag{4.5.6}$$

where P and V are pressure and volume, respectively.

The terms E and H are *state* or *point functions*. By fixing a certain number of variables on which the function depends, the numerical value of the function is automatically fixed; i.e., it is single-valued. For example, fixing the temperature and pressure of a one-component single-phase system immediately specifies the enthalpy and internal energy. The change in enthalpy as the system undergoes a change in state from (T_1, P_1) to (T_2, P_2) is given by

$$\Delta H = H_2 - H_1 \qquad (4.5.7)$$

The correlations needed to calculate the values of enthalpy are not presented here; rather, the reader is directed to the literature.[2]

Conservation of Momentum

The equation of momentum transfer – more commonly called the equation of motion – can be derived from momentum considerations by applying a momentum balance on a rate basis. The total momentum within a system is unchanged by an exchange of momentum between two or more masses of the system. This is known as the principle or law of conservation of momentum. This differential equation describes the velocity distribution and pressure drop in a moving fluid.

The equations of momentum are readily applied to problems dealing with flowing fluids. Theoretical, semi-empirical, and empirical equations employed in nearly all engineering fluid flow applications have roots in the momentum equations. Although beyond the scope of this text, details are available in the literature.[2]

4.6 ENGINEERING PRINCIPLES

This section examines and reviews some of the basic principles that engineers and scientists employ in performing design calculations and predicting the performance of plant equipment. Topics include the thermochemistry, chemical reaction equilibrium, chemical kinetics, the ideal gas law, partial pressure, phase equilibrium, and the Reynolds Number. These basic principles will assist the reader in acquiring a better understanding of some of the material that appears later in the book.

Thermochemistry

Consider now the energy effects associated with a chemical reaction. To introduce this subject, the reader is reminded that engineers and applied scientists rarely are concerned with the *magnitude* or *amount* of energy in the system; their primary concern is with *changes* in the amount of energy. It has been found in measuring energy changes for systems that the enthalpy is the most convenient term to work with. There are many different types of enthalpy effects; these include sensible heat, latent heat, and heat of reaction.

The *heat of reaction* (see Section 4.4) is defined as the enthalpy change of a system undergoing chemical reaction. If the reactants and products are at the same temperature and in their *standard states*, the heat of reaction is termed the *standard* heat of reaction. For engineering purposes, the *standard state* of a chemical may be taken as the pure chemical at 1 atm pressure. Heat of reaction data for many reactions is available in the literature.[2]

Chemical Reaction Equilibrium

With regard to chemical reactions, two important questions are of concern to the engineer:

1. How far will the reaction go?
2. How fast will the reaction go?

Chemical thermodynamics provides the answer to the first question; however, it provides information about the second. Reaction rates fall within the domain of *chemical kinetics* and are treated later in this section. Both equilibrium and kinetic effects must be considered in an overall engineering analysis of a chemical reaction.

Chemical reaction equilibrium calculations are structured around another thermodynamic term called the *free energy*. This so-called free energy G is a property that also cannot be defined easily without some basic grounding in thermodynamics. However, no such attempt is made here, and the interested reader is directed to the literature.[1,2] Note that free energy has the same units as enthalpy and internal energy and may be on a mole or total mass basis. Some key equations and information is provided below.

Consider the equilibrium reaction

$$aA + bB = cC + dD \qquad (4.6.1)$$

where   A, B, C, D = chemical formulas of the reactants and products
        a, b, c, d   = stoichiometric coefficients

and the equals sign is a reminder that the reacting system is at equilibrium. For this reaction, the change in free energy is given by

$$\Delta G = cG_C + dG_D - aG_A - bG_B \qquad (4.6.2)$$

The following equation is used to calculate the chemical reaction equilibrium constant K at a temperature T.

$$\Delta G_T = RT (\ln K) \qquad (4.6.3)$$

The problem that remains is to relate K to understandable physical quantities; for gas phase reactions the term K in Eq. (4.6.3) may be approximately represented in terms of the partial pressures of the components involved. This relationship is given by Eqs. (4.6.4) and (4.6.5)

$$K = K_P \qquad (4.6.4)$$

where K is an equilibrium constant based on partial pressures and defined by

$$K_P = P_C^c P_D^d / P_A^a P_B^b \qquad (4.6.5)$$

where P_A = partial pressure of component A, etc.

The definition of K_P obviously applies to the reaction of Eq. (4.6.1). Assuming that a K value is available or calculable, this equation may be used to determine the partial pressures of the participating components at equilibrium. For liquid phase reactions, K is given approximately by

$$K = K_C \qquad (4.6.6)$$

where $K_C = C_C^c C_D^d / C_A^a C_B^b$
         C_C = concentration of component C (gmol/L), etc.

Chemical Kinetics

Chemical kinetics involves the study of reaction rates and the variables that affect these rates. It is a topic that is critical for the analysis of reacting systems. The objective in this sub-section is to develop a working understanding of this subject that will permit us to apply chemical kinetics principles in the area of safety. The topic is treated from an engineering point of view, that is, in terms of physically measurable quantities.

The rate of a chemical reaction can be described in any of several different ways. The most commonly used definition involves the time rate of change in the amount of one of the components participating in the reaction; this rate is usually based on some arbitrary factor related to the reacting system size or geometry, such as volume, mass, or interfacial area. The definition shown in Eq. (4.6.7), which applies to homogeneous reactions, is a convenient one from an engineering point of view.

$$R_A = (1/V)(dn_A/dt) \qquad (4.6.7)$$

where R_A = reaction rate based on component
         V = volume of reacting system
         n_A = number of moles of A at time t

$$t \quad = \text{time}$$

If the volume term is constant, one may write Eq. (4.6.7) as

$$R_A = dC_A/dt \qquad (4.6.8)$$

where C_A is the molar concentration of A at time t.

Based on Eq. (4.6.8), the reaction rate is positive if species A is being formed (since C_A increases with time), and negative if A is reacting (since C_A decreases with time). The rate is zero if the system is at chemical equilibrium.

An equation expressing the rate in terms of measurable and/or desirable quantities may now be developed. Based on experimental evidence, the rate of reaction is a function of the concentration of the components present in the reaction mixture, temperature, pressure, and catalyst variables. In equation form,

$$R_A = R_A \ (C_i, \ P, \ T, \ \text{catalyst variables}) \qquad (4.6.9)$$

Equation (4.6.9) may be condensed to

$$R_A = (\pm)k_A f(C_i) \qquad (4.6.10)$$

where k_A incorporates all the variables in Eq. (4.6.9) other than the concentration variable. The $(\pm)$ notation is included to indicate whether component A is being consumed (-) or produced (+). The term k_A may be regarded as a constant of proportionality and it is termed the *reaction velocity constant*. Although this "constant" is independent of concentration, it is a function of other variables. This term is very definitely influenced by temperature and catalyst activity. For most applications, it is assumed that k is solely a function of temperature. Thus,

$$R_A = (\pm)k_A(T)f(C_i) \qquad (4.6.11)$$

The effect of temperature on k is generally represented by the Arrhenius equation[5] and the values of k are obtained from experimental data.

Ideal Gas Law

The *ideal gas law* was derived from experiments in which the effects of pressure and temperature on gaseous volumes were measured over a moderate range of temperatures and pressures. As a general rule, this law works best when the molecules of the gas are far apart, that is, when the pressure is low and the temperature is high. Under these conditions, the gas is said to behave *ideally*. For engineering calculations the ideal gas law is almost always assumed to be valid, since it generally works well for the temperature and pressure ranges used in most applications.

The two precursors of the ideal gas law were *Boyle's law* and *Charles' law*. Boyle found that the volume of a given mass of gas is inversely proportional to the absolute pressure if the temperature is kept constant, that is,

$$P_1V_1 = P_2V_2 \qquad\qquad (4.6.12)$$

where V_1 and V_2 are the volumes of gas at absolute pressures P_1 and P_2, respectively. Charles found that the volume of a given mass of gas varies directly with the absolute temperature at constant pressure

$$V_1/T_1 = V_2/T_2 \qquad\qquad (4.6.13)$$

where V_1 and V_2 are the volumes of gas at absolute temperature T_1 and T_2, respectively. *Boyle's* and *Charles' laws* may be combined into a single equation in which neither temperature nor pressure need be held constant:

$$P_1V_1/T_1 = P_2V_2/T_2 \qquad\qquad (4.6.14)$$

This equation indicates that for a given mass of a specific gas, PV/T has a constant value. Since, at the same temperature and pressure, volume and mass must be directly proportional, this statement may be extended to

$$PV/nT = R \qquad\qquad (4.6.15)$$

where n is the number of moles and R is the universal gas constant. Equation (4.6.15) is called the *ideal gas law*. Numerically, the value of R depends on the units of P, V, T, and n.

Partial Pressure

In engineering practice, mixtures of gases are more often encountered than single or pure gases. The *ideal gas law* is based on the number of molecules present in the gas volume; the kind of molecules is not a significant factor. This law applies equally well to mixtures and to pure gases. Since pressure is caused by gas molecules colliding with the walls of the container, it seems reasonable that the total pressure of a gas mixture is made up of pressure contributions due to each of the component gases. These pressure contributions are called *partial pressures*. Dalton defined the partial pressure of a component as the pressure that would be exerted if the same mass of the component gas occupied the same total volume alone at the same temperature as the mixture. The sum of these partial pressures equals the total pressure:

$$P = P_1 + P_2 + P_3 + \dots + P_n = \sum_{i=1}^{n} P_i \qquad\qquad (4.6.16)$$

For ideal gases, the partial pressure of component i is defined by

$$P_i = y_i P \qquad (4.6.17)$$

Phase Equilibrium

The term "phase" for a pure substance indicates a state of matter - that is, solid, liquid, or gas. For mixtures, however, a more stringent connotation must be used, since a totally liquid or solid system may contain more than one phase. A *phase* is characterized by uniformity or homogeneity; the same composition and properties must exist throughout the phase region. At most temperatures and pressures, a pure substance normally exists as a single phase. At certain temperatures and pressures, two or perhaps even three phases can coexist in equilibrium.

The most important equilibrium phase relationship is that between liquid and vapor. *Raoult's* and *Henry's laws* theoretically describe liquid-vapor behavior and, under certain conditions, are applicable in practice. *Raoult's law*, sometimes useful for mixtures of components of similar structure, states that the partial pressure of any component in the vapor is equal to the product of the vapor pressure of the pure component and the mole fraction of that component in the liquid. It may be written in the following manner

$$P_i = P_i' x_i \qquad (4.6.18)$$

where P_i = partial pressure of component i in the vapor
        P_i' = vapor pressure of pure i
        x_i = mole fraction of component i in the liquid

This expression may be applied to all components. If the gas phase is ideal, Eq. (4.6.18) becomes

$$y_i = (P_i'/P) \, x_i \qquad (4.6.19)$$

where y_i = mole fraction of component i in the vapor
        P = total system pressure

Unfortunately, relatively few mixtures follow *Raoult's law*. A more empirical relationship used for representing data on many systems is *Henry's law* - also valid for low x_i:

$$P_i = H_i \, x_i \qquad (4.6.20)$$

where H_i is Henry's law constant for component i (with units of pressure). Values for Henry's law constant can be found in the literature.[2]

Many equilibrium calculations are accomplished using the phase equilibrium constant K_i. This constant has been referred to in industry as a *componential split factor*, since it provides the ratio of the mole fractions of a component in two equilibrium phases. The defining equation is

$$K_i = y_i/x_i \qquad\qquad (4.6.21)$$

This equilibrium constant is a function of the temperature, the pressure, the components in the system, and the mole fractions of these components. However, as a first approximation, K_i is generally treated as a function of temperature and pressure only. For this condition, K_i may be approximated by

$$K_i = P_i'/P \qquad\qquad (4.6.22)$$

The Reynolds Number

The Reynolds number, Re, is a dimensionless number that indicates whether a fluid flowing is in the laminar or turbulent mode. Laminar flow is characteristic of fluids flowing slowly enough so that there are no eddies (whirlpools) or macroscopic mixing of different portions of the fluid. (Note: In any fluid, there is always molecular mixing due to the thermal activity of the molecules; this is distinct from macroscopic mixing due to the swirling motion of different portions of the fluid.) In laminar flow, a fluid can be imagined to flow like a deck of cards, with adjacent layers sliding past one another. Turbulent flow is characterized by eddies and macroscopic currents. In practice, moving gases are generally in the turbulent region. For flow in a pipe, a Reynolds number above 2100 is an indication of turbulent flow.

The Reynolds number is dependent on the fluid velocity, density, viscosity, and some characteristic length of the system or conduit; for pipes, this characteristic length is the inside pipe diameter.

$$Re = vD\rho/\mu = vD/\nu \qquad\qquad (4.6.23)$$

where   Re = Reynolds number
          D = inside diameter of the pipe (ft)
          v   = fluid velocity (ft/s)
          ρ = fluid density (lb/ft^3)
          μ = fluid viscosity (lb/ft-s)
          ν = fluid kinematic viscosity (ft^2/s)

Any consistent set of units may be used with Eq. (4.6.23).

4.7   ILLUSTRATIVE EXAMPLES

Example 4.1

a.   What is the molecular weight of nitrobenzene ($C_6H_5O_2N$)?
b.   How many moles are there in 50.0g of nitrobenzene?
c.   If the specific gravity is 1.203, what is the density in g/cm^3?
d.   What is the volume occupied by 50.0g of nitrobenzene in cm^3, in ft^3, and in in^3?
e.   If the nitrobenzene is held in a cylindrical container with a base of 1inch in diameter, what is the pressure at the base? What is it in gauge pressure?
f.   How many molecules are contained in 50.0g of nitrobenzene?

Solution 4.1

a.
Component	Atomic Weight
C_6	72 g/gmol
H_5	5 g/gmol
O_2	32 g/gmol
N	14 g/gmol

   Total = 123 g/gmol = molecular weight

b.
Component	Moles
$C_6H_5O_2N$	50 g / (123 g/gmol) = 0.407 gmol

c.   Employ Equation (4.3.10)
   $SG = \rho/\rho_{ref} = 1.203$
   Rearranging the equation
   $\rho = \rho_{ref} (SG)$
   $= (1.000 \ g/cm^3)(1.203)$
   $= 1.203 \ g/cm^3$

d.   Volume = mass / ρ = 50.0g / 1.203 g/cm^3 = 41.6 cm^3
   Volume = 41.6 cm^3 (1 ft^3/28,317 cm^3) = 1.47 x 10^{-3} ft^3
   Volume = 1.5 x 10^{-3} ft^3 (1728 in^3/1 ft^3) = 2.54 in^3

e.   area = $\pi d^2/4 = \pi(1 \ in)^2/4 = 0.785 \ in^2$
   The definition of pressure is
   P_a = force/area = (50.0g)(1lb/453.593g)/0.79in^2 = 0.110 psi

   Employ Equation (4.3.5)
   $P_a = P + P_g$
   Rearranging the equation
   $P_g = P_a - P = 0.110$ psi - 14.696 psi
   $= -14.586$ psi

f. Note that Avogadro's number = 6.02×10^{23}. From Part (b), it is known that there is 0.407 gmol in 50.0 g. Thus,
0.407 gmol $(6.02 \times 10^{23}/1 \text{ gmol}) = 2.45 \times 10^{23}$ gmol nitrobenzene

Example 4.2

Convert the following temperature
a.   20°C to °F, K, and °R
b.   20°F to °C, K, and °R

Solution 4.2

a.   Equation (4.3.3): $K = °C + 273 = 20°C + 273 = 293K$
     Equation (4.3.1): $°F = 1.8(°C) + 32 = 1.8(20°C) + 32 = 68°F$
     Equation (4.3.4): $°R = °F + 460 = 68°F + 460 = 528°R$

b.   Equation (4.3.4): $°R = °F + 460 = 20°F + 460 = 480°R$
     Equation (4.3.2): $°C = (°F - 32)/1.8 = (20°F - 32)/1.8 = -6.7°C$
     Equation (4.3.3): $K = °C + 273 = -6.7°C + 273 = 266.3K$

Example 4.3

a.   What is the density of air at 115°F and 2 atm?
b.   What volume does 100g of air occupy at the condition of Part (a)?
c.   What is the new volume if the temperature is changed to 200°F and the pressure is 1 atm?

Solution 4.3

a.   First, the temperature must be converted.
     Equation (4.3.4): $°R = °F + 460 = 115°F + 460 = 575°R$
     The ideal gas law may be rearranged to give
     $\rho = (P)(MW)/RT$
        $= (2 \text{ atm})(29 \text{ lb/lbmol}) / (0.7302 \text{ ft}^3\text{-atm/lbmol-}°R)(575°R)$
        $= 0.138 \text{ lb/ft}^3$

b.   Rearranging the equation once again leads to
     $V = nRT / P$
        $= (1 \text{ mol}) (0.7302 \text{ ft}^3\text{-atm/lbmol-}°R) (575°R) / 2 \text{ atm}$
        $= 209.9 \text{ ft}^3$

c.   Equation (4.6.14): $P_1 V_1/T_1 = P_2 V_2/T_2$
     Solving for V_2 gives
     $V_2 = (2 \text{ atm}) (209.9 \text{ ft}^3/115°F) (200°F/1 \text{ atm})$
        $= 730.1 \text{ ft}^3$

Example 4.4

A liquid with a viscosity of 0.78 cP and a density of 1.50 g/cm^3 flows through a 1-inch diameter pipe at 20 cm/s. Calculate the Reynolds number. What region (laminar or turbulent) is it in?

Solution 4.4

Equation 4.6.23: $Re = vD\rho/\mu = vD/\nu$
First the viscosity and inside diameter must be converted to consistent units.
$\mu = 0.78$ cP $(1 \times 10^{-2}$ g/cm-s / 1 cP$) = 0.0078$ g/cm-s
$v = 1$ in$(2.54$ cm/1 in$) = 2.54$ cm
Thus,
$Re = vD\rho/\mu = (2.54$ cm$) (20$ cm/s$) (1.50$ g/cm$^3) / 0.0078$ g/cm-s
$= 9769.23$

For flow in a pipe, a Reynolds number above 2100 is an indication of turbulent flow. Thus, with a Reynolds number of 9769.23, the flow is in the turbulent region.

4.8 SUMMARY

1. This chapter primarily serves as a review of process fundamentals such as units, dimensions, chemical and physical properties, conservation laws, and engineering principles.
2. In general, engineering equations should be dimensionally consistent.
3. Physical properties are important considerations in any study of accidents and emergencies. A substance may exhibit certain characteristics under one set of physical conditions, but may become hazardous if the conditions are changed.
4. Chemical properties are also important considerations when studying any accident or emergency. A substance can also become hazardous when chemical conditions are changed.
5. The three basic conservation laws are: mass, energy, and momentum.
6. Several basic principles that engineers and scientists employ in performing design calculations and predicting the performance of plant equipment includes thermochemistry, chemical reaction equilibrium, chemical kinetics, the ideal gas law, partial pressure, phase equilibrium, and the Reynolds Number.

PROBLEMS

1. Calculate conversion factors for the following:
   a. feet to meters,
   b. pounds to kilograms,

c. Btu to joules, and
d. Pounds (force) to newtons.
2. Assuming an ideal gas mixture at atmospheric pressure, calculate the mole fraction and ppm of a component if its partial pressure is 19 mmHg.
3. The inlet gas to a spray tower is at 1600°F. It is piped through a 3.0 ft inside diameter duct at 25 ft/s to the spray tower. The scrubber cools the gas to 500°F. In order to maintain the velocity of 25 ft/s, what size duct would be required at the outlet of the unit? Neglect the pressure across the spray tower and any moisture considerations.
4. What is the density of air at 60°F and 14.7 psia? Calculate the volume (in ft^3) of 1.0 lbmol of any ideal gas at 77°F and 14.7psia. Also calculate the density of a gas (MW=32) in g/cm^3 at 20°C and 1.0 atm using the ideal gas law.
5. Data from an incinerator indicate a volumetric flow rate of 10,000 scfm (60°F, 1 atm). If the operating temperature and pressure of the unit are 1950°F and 1 atm, respectively, calculate the actual flow rate in cubic feet per minute.

REFERENCES

1. J. Santoleri, J. Reynolds, and L. Theodore, "Introduction to Hazardous Waste Incineration", 2nd ed., Wiley, New York, 2000.
2. R. H. Perry and D. Green, "Perry's Chemical Engineers' Handbook", 7th ed., McGraw-Hill, New York, 1996.
3. R. C. Weast, "CRC Handbook of Chemistry and Physics", 5th ed., CRC Press, Boca Raton, Florida, 1994.
4. R. Chang, "Chemistry", 4th ed., McGraw-Hill, New York, 1991.
5. L. Theodore, "Chemical Kinetics", A Theodore Tutorial, East Williston, New York, 1994.

5

Process Equipment

5.1 INTRODUCTION

This chapter provides details on a number of commonly used process units: reactors, heat exchangers, columns of various types (distillation, absorption, adsorption, evaporation, extraction), dryers, and grinders. The purpose of each unit or operation and the many configurations in which the units can be found are also discussed.

Some important factors regarding a safe plant can be better understood if the reader is familiar with such process equipment as reactors (Section 5.2), mass transfer units (Section 5.3), heat exchanges (Section 5.4), ancillary equipment (Section 5.5), environmental equipment (Section 5.6), and utilities (Section 5.7).   Protective equipment is reviewed in Section 5.8.   Process diagrams, which illustrate the various possible arrangements of plant equipment, valves, piping, and control systems, are presented in Section 5.9. Plant siting and layout are discussed in Section 5.10 - this last section illustrates the factors that can contribute to proper plant operation.

The reader is referred to the literature for more extensive details on the above topics.[1, 2]

5.2 REACTORS

The *reactor* is often the heart of a chemical process. It is the place in the process where raw materials are usually converted into products, and the reactor design is therefore a vital step in the overall design of the process. The treatment of reactors in this section is restricted to a discussion of the appropriate reactor types for a process and the design of an industrial chemical reactor must satisfy requirements in four fundamental areas.

1. *Chemical Factors.* These involve mainly the kinetics of the reaction. The design must provide sufficient residence time for the desired reaction to proceed to the required degree of conversion.
2. *Mass Transfer.* The reaction rate of heterogeneous reactions may be controlled by the rates of diffusion of the reacting species, rather than the chemical kinetics.
3. *Heat Transfer Factors.* These involve the removal, or addition of the heat of reaction.
4. *Safety Factors.* These involve the confinement of any hazardous reactants and products, as well as the control of the reaction and the process conditions.

The need to satisfy these interrelated and often contradictory factors makes reactor design a complex and difficult task. However, in many instances one of the factors predominates, hence determining the choice of reactor type and the design method.

Reactor Types

The characteristics normally used to classify reactor designs are:

1. Mode of operation - batch or continuous
2. Phases present - homogeneous or heterogeneous
3. Reactor geometry - flow pattern and manner of contacting the phases

The four major classes of reactors are:

1. Stirred tank reactor
2. Tubular reactor
3. Packed bed, fixed and moving
4. Fluidized bed

In a *batch* process, all the reagents are added at the beginning of the reaction. As the reaction proceeds, the compositions change with time. The reaction is stopped and the product is withdrawn when the required conversion has been reached. Batch processes are suitable for small-scale production and for processes that use the same equipment to make a range of different products or grades. Examples include pigments, dyestuffs, pharmaceuticals, and polymers.

In *continuous* processes, the reactants are fed to the reactor and the products withdrawn continuously; the reactor usually operates under steady state conditions. Continuous production normally gives lower production costs than batch production, but lacks the flexibility of batch production. Continuous reactors usually are selected for large-scale production. Processes that do not fit the definition of batch or continuous are often referred to as *semicontinuous* or *semibatch*. In a *semibatch* reactor, some of the reactants may be added to, or some of the products withdrawn from the batch as the reaction proceeds. A semicontinuous process is basically a continuous process that is interrupted periodically-for instance, for the regeneration of catalyst.[3]

Homogeneous reactions are those in which the reactants, products, and any catalysts used form one continuous phase (gaseous or liquid). Homogeneous gas phase reactors are almost always operated continuously, whereas liquid phase reactors may be batch or continuous. *Tubular* (pipeline) reactors are normally used for homogeneous gas phase reactions (e.g., in the thermal cracking of petroleum of dichloroethane to vinyl chloride). Both *tubular* and *stirred tank* reactors are used for homogeneous liquid phase reactions.

In a *heterogeneous* reaction two or more phases exist and the overriding problem in reactor design is to promote mass transfer between the phases. The possible combinations of phases are

1. *Liquid-liquid* - immiscible liquid phases
2. *Liquid-solid* - with one or more liquid phases in contact with a solid; the solid may be a reactant or catalyst
3. *Liquid-solid-gas* - where the solid is normally a catalyst
4. *Gas-solid* - where the solid may take part in the reaction or act as a catalyst

The reactors used for established processes are usually unique and complex designs that have been developed over a period of years to suit the requirements of the process. However, it is convenient to classify reactors into the broad categories discussed below.[3]

A *stirred tank (agitated) reactor* consists of a tank fitted with a mechanical agitator and a cooling jacket or coils. It may be operated as batch reactor or continuously. Several reactors may be used in series. The stirred tank reactor can be considered the basic chemical reactor, modeling on a large scale the conventional laboratory flask. Tank sizes range from a few liters to several thousand liters. This equipment is used for homogeneous and heterogeneous liquid-liquid and liquid-gas reactions and for reactions that involve finely suspended solids, which are held in suspension by the agitation. Since the degree of agitation is under the designer's control, stirred tank reactors are particularly suitable for reactions that require good mass transfer or heat transfer. When operated as a continuous process, the composition in the reactor is constant and equivalent to the composition of the product stream.

Except for very rapid reactions, the conversion that can be obtained in one stage of a continuous process is limited.[3]

Tubular reactors are generally used for gaseous reactions, but are also suitable for some liquid phase reactions. If high heat transfer rates are required, small-diameter tubes are used to increase the ratio of surface area to volume. Several tubes may be arranged in parallel, connected to a manifold, or fitted into a tube sheet in an arrangement similar to a shell and tube heat exchanger. For high temperature reactions, the tubes may be arranged in a furnace.[3]

There are two basic types of *packed-bed reactors*: those in which the solid is a reactant and those in which the solid is a catalyst. Many examples of the first type can be found in the extractive metallurgical industries. In the chemical process industries, the designer normally meets the second type, catalytic reactors. Industrial packed-bed catalytic reactors range in size from units with small tubes (a few centimeters in diameter) to large-diameter packed beds. Packed-bed reactors are used for gas and gas-liquid reactions. Heat transfer rates in large-diameter packed beds are poor and where high heat transfer rates are required, *fluidized beds* should be considered.[3]

The essential feature of a *fluidized-bed reactor* is that the solids are held in suspension by the upward flow of the reacting fluid; this promotes high mass and heat transfer rates and good mixing. Heat transfer coefficients in the order of 200 W/m-°C between jackets and internal coils are typically obtained. The solids may be a catalyst, a reactant (in some fluidized combustion processes), or an inert powder added to promote heat transfer.

Though the principal advantage of a fluidized bed over a fixed bed is the higher heat transfer rate, fluidized beds are also useful when large quantities of solids must be transported as part of the reaction processes.

5.3   HEAT EXCHANGERS

The transfer of heat to and from process fluids is an essential part of most chemical processes. The most commonly used type of heat transfer equipment is the shell and tube heat exchanger. The chemical process industries use four principal types of heat exchanger.

1. *Double-pipe exchangers* - the simplest type, used for cooling and heating.
2. *Shell and tube exchangers* - used for all applications.
3. *Plate and frame exchangers (plate heat exchangers)* - used for heating and cooling.
4. *Direct contact exchangers* - used for cooling and quenching.

The word "exchanger" applies to all types of equipment in which heat is exchanged, but often it is used specifically to denote equipment in which heat is transferred between two process streams. An exchanger in which a process fluid is heated or cooled by a plant service stream is referred to as a *heater* or *cooler*.

One of the simplest and cheapest types of heat exchanger is the concentric pipe arrangement known as the *double-pipe heat exchanger*. Such equipment can be made up from standard fittings and is useful where only a small heat transfer area is required. Several units can be connected in series to extend the capacity.[3]

The *shell and tube exchanger* is by far the most commonly used type of heat transfer equipment in the chemical and allied industries. The advantages of this type are

1. Large surface area in a small volume
2. Good mechanical layout - that is, a good shape for pressure operation
3. Reliance on well-established fabrication techniques
4. Wide range of construction materials available
5. Easily cleaned equipment
6. Well-established design procedures

Essentially, a shell and tube exchanger consists of a bundle of tubes enclosed in a cylindrical shell. The ends of the tubes are fitted into tube sheets which separate the shell-side and tube-side fluids. Baffles are provided in the shell to direct the fluid flow and increase heat transfer.[3]

In direct contact heat exchange, there is no wall to separate hot and cold streams, and high rates of heat transfer are achieved. Applications include reactor off-gas quenching, vacuum condensers, desuperheating, and humidification. Water-cooling towers are a particular example of a direct contact heat exchanger. In direct contact cooler-condensers, the condensed liquid is frequently used as the coolant.

Direct contact heat exchangers should be considered whenever the process stream and the coolant are compatible. The equipment used is basically simple and cheap and is suitable for use with heavily fouling fluids. For liquids containing solids, spray chambers, spray columns, and plate and packed columns are used.

5.4 MASS TRANSFER EQUIPMENT

Distillation

Distillation is probably the most widely used separation (mass transfer) process in the chemical and allied industries. Its applications range from the rectification of alcohol, which has been practiced since antiquity, to the fractionation of crude oil. The separation of liquid mixtures by distillation is based on differences in volatility between the components. The greater the relative volatilities, the easier the separation. Vapor flows up the column and liquid flows countercurrently down the column. The vapor and liquid are brought into contact on plates, or packing. Part of the condensate from the condenser is returned to the top of the column to provide liquid flow above the feed point (*reflux*). Part of the liquid from the base of the column is vaporized in the reboiler and returned to provide the vapor flow.

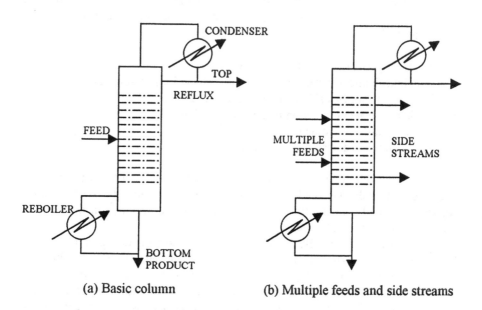

(a) Basic column   (b) Multiple feeds and side streams

Figure 5.4.1. Distillation column.

In the *stripping section*, which lies below the feed, the more volatile components are stripped from the liquid. Above the feed, in the *enrichment* or *rectifying section*, the concentration of the more volatile components is increased. Figure 5.4.1a shows a column producing two product streams, referred to as *tops* and *bottoms*, from a single feed. Columns are occasionally used with more than one feed, and with side streams withdrawn at points up the column (cf. Fig. 5.4.1b). This does not alter the basic operation but it does complicate the analysis of the process to some extent. If the process requirement is to strip a volatile component from a relatively nonvolatile solvent, the rectifying section may be omitted, and the column then is called a *stripping column*.

In some operations where the top product is required as a vapor, the liquid condensed is sufficient only to provide the reflux flow to the column. In this arrangement the condenser is referred to as a *partial condenser*. In a partial condenser, the reflux will be in equilibrium with the vapor leaving the condenser. When the liquid is totally condensed the condenser is referred to as a total condenser. The liquid returned to the column will have the same composition as the top product. Virtually pure top and bottom products can be achieved by using many stages or additional columns.

Adsorption

Adsorption is influenced by the surface area of the adsorbent, the nature of the solvent being adsorbed, the pH of the operating system, and the temperature of operation. These are important parameters to be aware of when designing or evaluating an adsorption process.

The adsorption process is normally performed in a column. The column is run as either a packed- or fluidized-bed operation. The adsorbent, after it has reached the end of its useful life, can either be discarded or regenerated. For further information, the reader is directed to the literature.[1,2]

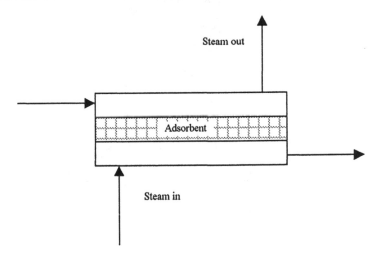

Figure 5.4.2. Adsorber.

Absorbers

The process of *absorption* conventionally refers to the intimate contacting of a mixture of gases with a liquid so that part of one or more of the constituents of the gas will dissolve in the liquid. The contact usually takes place in some type of packed column.

Packed columns are used for the continuous contact between liquid and gas. The countercurrent packed column is the most common type of unit encountered in gaseous pollutant control for the removal of the undesirable gas, vapor, or odor. This type of column has found widespread application in the chemical industries. The gas stream moves upward through the packed bed against an absorbing or reacting liquid that is injected at the top of the packing. This results in the highest possible efficiency. Since the concentration in the gas stream decreases as it rises through the column, there is constantly fresher liquid available for contact. This provides a maximum average driving force for the diffusion process throughout the bed.

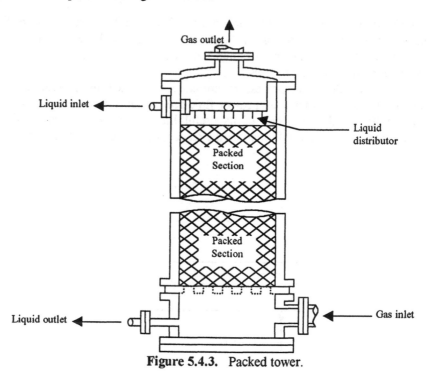

Figure 5.4.3. Packed tower.

Evaporation

The processing industry has given operations involving heat transfer to a boiling liquid the general name of evaporation. The most common application is the removal of water from a processing stream. Evaporation is used in the food, chemical, and petrochemical industries, and usually it results in an increase in the concentration of a certain species.

The factors that affect the evaporation process are concentration in the liquid, solubility, pressure, temperature, scaling, and materials of construction. An evaporator is a type of heat exchanger designed to induce boiling and evaporation of a liquid. The major types of evaporator are

1. Open kettle or pan
2. Horizontal-tube natural convection
3. Vertical-tube natural convection
4. Forced-convection evaporators

The efficiency of an evaporator can be increased by operating the equipment in single or multiple-effect modes.

Extraction

Extraction (sometimes called *leaching*) encompasses liquid-liquid as well as liquid-solid systems. *Liquid-liquid extraction* involves the transfer of solutes from one liquid phase into another liquid solvent; it is normally conducted in *mixer settlers, plate and agitated-tower contacting equipment,* or *packed or spray towers. Liquid-solid extraction,* in which a liquid solvent is passed over a solid phase to remove some solute, is carried out in fixed-bed, moving-bed, or agitated-solid columns.

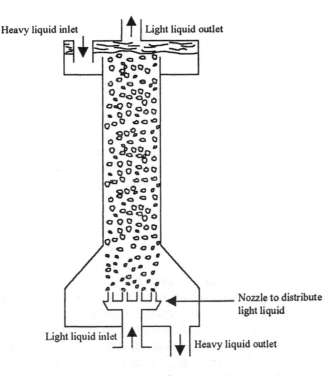

Figure 5.4.4. Spray tower.

Drying

Drying generally involves the removal of relatively small amounts of water or organic liquids from solids-whereas *evaporation* removes larger amounts. Drying removes the liquid as a vapor by warm gas (usually air) currents. Drying can be accomplished on a batch or continuous basis. The drying process is carried out in one of four basic dryer types. The first type is a *continuous tunnel dryer*. Trays with wet solids are moved through the system, in which warm air is blown over the trays. The second is a *rotary dryer*, which consists of an inclined hollow cylinder that rotates. The wet solids and hot air are fed into the cylinder in a countercurrent fashion. The dried solids are then removed from the cylinder. A *drum dryer* is a heated cylinder in which the wet solids are spread across the outside of the hot rotating drum. The wet solids are dried on this surface and then scraped off. The final type of dryer is a *spray dryer* where a liquid or slurry is sprayed into the dryer via a nozzle. The fine liquid droplets are dried by a hot gas. This operation may be run cocurrently, countercurrently, or in some combination of the two modes.

Grinding

The term *grinding*, or *crushing*, is used to describe the subdividing of larger solid particles into smaller particles. Solids are reduced in size by a number of methods. *Compression* is the reduction of hard solids to coarse sizes. *Impacting* reduces hard solids to coarse, medium, or fine products. *Attrition* yields only fine products. *Cutting* is also used to produce particles in specific size ranges. The most widely used pieces of equipment for this operation are jaw, gyratory, and roll crushers, as well as hammer and revolving grinder mills. Care should be exercised for fire and explosion hazards when grinding combustibles. In addition, equipment must be grounded and explosion venting must be provided if the material is combustible. The reader is referred to the literature for additional information.[4]

5.5   ANCILLARY EQUIPMENT

The discussion in this section begins with devices for transporting gases and liquids to, from, or between units of process equipment. Some of these devices are simply conduits for the moving material - *pipes, ducts, fittings, stacks*; other devices control the flow of material - *valves*; and others provide the mechanical driving force for the flow - *fans, pumps*, and *compressors*. This section also covers storage facilities, holding tanks, materials handling devices and techniques, and utilities (e.g., gas, steam, water), along with air, water, and solid waste control equipment.

Pipes

The most common conduits for fluids are pipes and tubing. Both have circular cross sections, but pipes tend to have larger diameters and thicker walls. Because of the heavier walls, pipes can be threaded, while tubing cannot. Process systems usually handle large flow rates that require the larger diameters associated with pipes.

Tubing and pipes are manufactured from many construction materials. The selection of the material depends on the corrosivity of the fluid and the flow (system) pressure. If special piping is required to accommodate corrosive liquids or high standards of purity, stainless steel, nickel alloys, or materials of high resistance to heat and mechanical damage are used. Steel pipe can be lined with tin, plastic, rubber, lead, or other coatings for special purposes. If problems of corrosion or contamination are the controlling factors, the use of a nonmetallic pipe such as glass, porcelain, thermosetting plastic, or hard rubber is often acceptable. There are several techniques used to join pipe sections. For small pipes, threaded connectors are the most common; for larger pipes, flanged fittings or welded connections are normally employed.

Ducts

Pipes and tubing are used as conduits for the transporting of liquids or gases; ducts are used only for gases. Pipes, with their thicker walls, can be used for flows at higher pressures; ducts are always thin walled and are generally used for gas flows close to ambient pressure. Pipes are usually circular in cross section. Ducts come in many shapes (circular, oval, rectangular, etc.). In general, ducts are larger in cross section than pipes because they carry fluids with low densities and high volumetric flow rates.

Ducts are most often constructed of field-fabricated galvanized sheet steel, although other materials such as fibrous glass board, factory-fabricated round fibrous glass, spiral sheet metal, and flexible duct materials are becoming increasingly popular. Other duct construction materials include black steel, aluminum, stainless steel, plastic and plastic-coated steel, cement, asbestos, and copper.

In safety applications, the corrosion resistance of the duct materials deserves special consideration. Since material costs generally increase along with corrosion resistance, the selection of material must be determined by the desired life span in the anticipated environment; this environment is a function of the characteristics of the chemical being processed and the operating conditions of the reactor. For maximum resistance to moisture or corrosive gases, stainless steel and copper are used where their cost can be justified. Aluminum sheet is used where lighter weight and superior resistance to moisture are needed.

Fittings

A fitting is a piece of equipment that has one or more of the following functions:

1. The joining of two pieces of straight pipe (e.g., couplings and unions)
2. The changing of pipeline direction (e.g., elbows and tees)
3. The changing of pipeline diameter (e.g., reducers and bushings)
4. The joining of two streams (e.g., tees and y's)

A *coupling* is a short piece of pipe threaded on the inside and used to connect straight sections of pipe. A *union* is also used to connect two straight sections but differs from a coupling in that it can be opened conveniently without disturbing the rest of the pipeline. When a coupling is opened, a considerable amount of piping usually must be dismantled. An *elbow* is an angle fitting used to change flow direction, usually by 90°. A *tee* (shaped like the letter T) can also be used to change flow direction but is more often used to combine two streams into one. A *reducer* is a coupling for two pipe sections of different diameter. A *bushing* is another connector for pipes of different diameter, but, unlike the reducer coupling, it is threaded on both inside and outside. The large pipe screws onto the outside of the bushing and the smaller pipe screws into the inside. A *Y* (shaped like the letter Y) is similar to the T and is used to combine two streams.

Valves

Valves have two main functions in a pipeline: to control the amount of flow, or to stop the flow completely. There are many different types but the most commonly used are the *gate valve* and the *globe valve*. The gate valve contains a disk that slides at right angles to the flow direction. This type of valve is used primarily for on-off control of a liquid flow. Because small lateral adjustments of the disk cause extreme changes in the flow cross-sectional area, this type of valve is not suitable for adjusting flow rates.

Unlike the gate valve, the globe valve is designed for flow control. Liquid passes through a globe valve in a somewhat circuitous route - in one form, the seal is a horizontal ring into which a plug with a slightly beveled edge is inserted when the stem is closed. Good control of flow is achieved with this type of valve, but pressure loss is higher than with a gate valve.

Some other types of valves are the *check valve*, which permits flow in one direction only; the *butterfly valve*, which operates in damperlike fashion by rotating a flat plate to either a parallel or a perpendicular position relative to the flow; the *plug valve*, in which a rotating tapered plug provides on-off service; the *needle valve*, which is a variation of the globe valve that gives improved flow control; and the *diaphragm valve*, a valve specially designed to handle fluids such as very viscous liquids, slurries, or corrosive liquids that might clog the moving parts of other valves.

Fans

The terms *fans* and *blowers* are often used interchangeably, and no distinction is made between them in the discussion that follows. Whatever is stated about fans equally applies to blowers. Strictly speaking, however, *fans* are used for low pressure (drop) operation, generally below 2 psi. *Blowers* are generally employed when pressure heads in the range of 2.0 to 14.7 psi are generated Operations at higher pressures require *compressors*.

Fans are usually classified as the *centrifugal* or the *axial-flow* type. In centrifugal fans, the gas is introduced into the center of the revolving wheel (the eye) and is discharged at right angles to the rotating blades. In axial-flow fans, the gas moves directly (forward) through the axis of rotation of the fan blades. Both types are used in industry.

Pumps

Pumps may be classified as *reciprocating, rotary*, or *centrifugal*. The first two are referred to as *positive-displacement* pumps because, unlike the centrifugal type, the liquid or semiliquid flow is broken into small portions as it passes through the pump.

Reciprocating pumps operate by the direct action of a piston on the liquid contained in a cylinder. As the liquid is compressed by the piston, the higher pressure forces it through discharge valves to the pump outlet. As the piston retracts, the next batch of low pressure liquid is drawn into the cylinder and the cycle is repeated.

The *rotary* pump combines rotation of the liquid with positive displacement. The rotating elements mesh with elements of the stationary casing in much the same way that two gears mesh. As the rotating elements come together, a pocket is created that first enlarges and then draws in liquid from the inlet or suction line. As rotation continues, the pocket of liquid is trapped, reduced in volume, and then forced into the discharge line at a higher pressure. The flow rate of liquid from a rotary pump is a function of size and speed of rotation and is slightly dependent on the discharge pressure. Unlike reciprocating pumps, rotary pumps deliver nearly constant flow rates. Rotary pumps are used on liquids of almost any viscosity as long as the liquids do not contain abrasive solids.

Centrifugal pumps are widely used in the process industry because of simplicity of design, low initial cost, low maintenance, and flexibility of application. Centrifugal pumps have been built to move as little as a few gallons per minute against a pressure of several hundred pounds per square inch. In its simplest form, this type of pump consists of an impeller rotating within a casing. Fluid enters the pump near the center of the rotating impeller and is thrown outward by centrifugal force. The kinetic energy of the fluid increases from the center of the impeller to the tips of the impeller vanes. This high velocity is converted to a high pressure as the fast-moving fluid leaves the impeller and is driven into slower moving fluid in the volute or diffuser.

Compressors

Compressors operate in a similar fashion to pumps and have the same classifications: rotary, reciprocating, and centrifugal. An obvious difference between the two operations is the large decrease in volume resulting from the compression of a gaseous stream compared to the negligible change in volume caused by the pumping of a liquid stream.

Centrifugal compressors are employed when large volumes of gases are to be handled at low to moderate pressure increases (0.5-50 psi). Rotary compressors have smaller capacities and can achieve discharge pressures up to 100 psi. Reciprocating compressors are the most common type used in industry and are capable of compressing small gas flows to as much as 3500 psig. With specially designed compressors, discharge pressures as high as 25,000 psig can be reached. However, these devices are capable of handling only very small volumes and do not work well for all gases.

Other Prime Movers

Other prime movers/drivers include:

1. Steam turbines
2. Steam engines
3. Gas turbines
4. Internal combustion engines

These are rarely used, although gas turbines may be the selection of choice if gas is inexpensive.

Stacks

Gases are discharged into the ambient atmosphere by means of stacks (referred to as *chimneys* by some in industry) of several types. *Stub* or *short stacks* are usually fabricated of steel and extend a minimum distance up from the discharge of an induced draft fan. These are constructed of steel plate, either unlined or refractory-lined, or entirely of refractory and structural brick. *Tall stacks*, which are constructed of the same materials as short stacks, provide a driving force (draft) of greater pressure difference than that resulting from the shorter stacks. In addition, tall stacks ensure more effective dispersion of the gaseous effluent into the atmosphere. Some chemical and utility applications use metal stacks that are made of a double wall with an air space between the metal sheets. The insulating air pocket created by the double wall prevents condensation on the inside of the stack, thus avoiding corrosion of the metal. The reader should note that the authors consider stacks a process unit that could also be considered an air pollution device (see next section).

Transportation and Storage of Materials

This subsection deals with the handling, storage, and transportation of solids, liquids, and gases. Each form is considered individually.

Gases

The type of equipment best suited for the pumping of gases in pipelines depends on the flow rate, the differential pressure required, and the operating pressure. In general, fans are used where the pressure drop is small. Axial-flow compressors are employed for high flow rates and moderate differential pressures, and centrifugal compressors are chosen for applications with high flow rates and, by staging, high differential pressures. Reciprocating compressors can be used over a wide range of pressures and capacities but normally are specified in preference to centrifugal compressors only where high pressures are required at relatively low flow rates.

Gases are stored at high pressures in order to fulfill a process requirement or to reduce the storage volume. For some gases, the volume can be further reduced by liquefying the gas using pressure or refrigeration. Cylindrical and spherical vessels (Horton spheres) are used.

Liquids

The transportation of liquids is usually accomplished with pumps. The type of pump used-centrifugal, reciprocating, diaphragm, or rotary (gear or sliding vane) depends on the operating pressures and capacity range needed. Liquids are usually stored in bulk in vertical cylindrical steel tanks. Fixed- and floating-roof tanks are used. In a floating-roof tank, a movable piston floats on the surface of the liquid and is sealed to the tank walls. Floating-roof tanks are used to eliminate evaporation losses and, for flammable liquids, to obviate the need for inert gas blanketing in order to prevent the formation of an explosive mixture above the liquid (as would occur with a fixed-roof tank). Horizontal cylindrical tanks and rectangular tanks are also used for storing liquids, usually for relatively small quantities.

Solids

Solids usually are more expensive to move and store than liquids and gases. The best equipment to use will depend on a number of factors, including throughput, length of travel, change in elevation, and nature of the solids (size, bulk density, angle of repose, abrasiveness, corrosiveness, wet or dry, etc.).

Belt conveyors are the most commonly used type of equipment for the continuous transport of solids. They can economically carry a wide range of materials over long and short distances, either horizontally or at an appreciable angle, depending on the angle of repose of the solids. *Screw conveyors*, also called *worm conveyors*, are used for free-flowing materials. The modern conveyor consists of helical screw rotating in a U-shaped trough. This type of device can be used horizontally or, with some loss of capacity, at an incline to lift materials.

Where a vertical lift is required, the most widely used equipment is the *bucket elevator*, consisting of buckets fitted to a chain or belt which passes over a driven roller or sprocket at the top end. Bucket elevators can handle a wide range of solids from heavy lumps to fine powders and are suitable for use with wet solids and slurries.

The simplest way to store solids is to pile them on the ground in the open air. This is satisfactory for the long-term storage of materials that do not deteriorate on exposure to the elements - for example, coal, which is seasonally stockpiled at utilities. For large stockpiles, permanent facilities are usually installed for distributing and reclaiming the material; traveling cranes, grabs, and drag scrapers feeding belt conveyors are used. For small, temporary storages, mechanical shovels and trunks will suffice. Where the cost of recovery from the stockpile is large compared with the value of the stock held, storage in silos or bunkers should be considered.

Overhead bunkers, also called *bins* or *hoppers*, are normally used for the short-term storage of materials that must be readily available for the process. The units are arranged so that the material can be withdrawn at a steady rate from the base of the bunker on to a suitable conveyor.[3]

5.6   ENVIRONMENTAL CONTROL EQUIPMENT

Air Pollution Equipment

Electrostatic Precipitators

Electrostatic precipitators are satisfactory devices for removing small particles from moving gas streams at high collection efficiencies. They have been used almost universally in power plants for removing fly ash from the gases prior to discharge. Electrostatic precipitators have the capability of fine particulate control. Resistivity plays an important role in determining whether a particle can be readily collected in this device.

There are three classes of electrostatic precipitators: *dry electrostatic precipitators* (ESP), *wet electrostatic precipitators* (WEP), and *ionizing wet scrubbers* (IWS).

The dry unit is by far the most popular as it has been used successfully to collect both solid and liquid particulate matter from many operations including smelters, steel furnaces, petroleum refineries, and utility boilers.

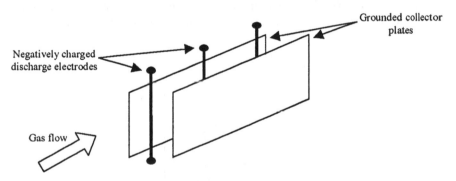

Figure 5.6.1.   Flat-surface-type electrostatic precipitator.

Wet Scrubbers

Wet scrubbers have found widespread use in cleaning contaminated gas streams because of their ability to effectively remove both particulate and gaseous pollutants. Specifically, wet scrubbing describes the technique of bringing a contaminated gas stream into intimate contact with a liquid. The types most widely used for particulate control are spray towers, packed-bed units, ionizing wet scrubbers, and venturi scrubbers.

Baghouses

One of the oldest, simplest, and most efficient methods for removing solid particulate contaminants from gas streams is by filtration through fabric media. The fabric filter is capable of providing high collection efficiencies for particles as small as 0.5 µm and will remove a substantial quantity of particles as small as 0.01 µm. In its simplest form, the industrial fabric filter consists of a woven or felted fabric through which dust-laden gases are forced. A combination of factors results in the collection of particles on the fabric filters. When woven fabrics are used, a dust cake eventually forms. This, in turn, acts predominantly as a sieving mechanism. When felted fabrics are used, the dust cake is minimal or nonexistent.

As particles are collected, the pressure drop across the fabric-filtering medium increases. Partly because of fan limitations, the filter must be cleaned at predetermined intervals. Dust is removed from the fabric by gravity and/or mechanical means. The fabric filters or bags are usually tubular or flat. The structure in which the bags hang is referred to as a *baghouse*. The number of bags in a baghouse may vary from a couple to several thousand. Quite often, when great numbers of bags are involved, the baghouse is compartmentalized to permit the cleaning of one compartment while others are still in service.

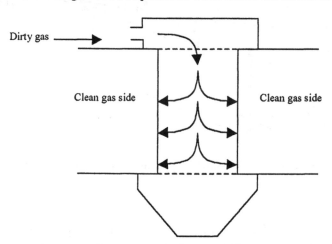

Figure 5.6.2. Baghouse filtration with top feed.

Water Pollution Equipment

To control water pollution, a waste stream can be subjected to at least one, or perhaps a combination, of chemical, biological, and physical treatments. Some of these processes are discussed below.

Oxidation

Oxidation is a process in which one or more electrons are transferred from the chemical being oxidized to the chemical initiating the transfer. The main purpose of treating wastes by oxidation is detoxification. Oxidation can also aid in the precipitation of ions in cases of oxidized ions that have a solubility lower than that of the original ions.

Reduction

In chemical *reduction*, one or more electrons are transferred from the reducing agent to the chemical being reduced. This process can reduce the toxicity of a solution, making it safer for disposal.

Precipitation

Precipitation involves the alteration of the ionic equilibrium to produce insoluble precipitates. To remove the sediment, chemical precipitation is allied with solids separation processes such as filtration. Undesirable metal ions and anions are commonly removed from waste streams by converting them to an insoluble form. The process is sometimes preceded by chemical reduction of the metal ions to a form that can be precipitated more easily. Chemical equilibrium can be affected by a variety of means to change the solubility of certain compounds. For example, precipitation can be induced by alkaline agents, sulfides, sulfates, and carbonates. Precipitation with chemicals is a common waste stream treatment process and is effective and reliable. The treatment of sludges is covered next.

Activated Sludge

The *activated sludge* process uses microorganisms to decompose organic materials in aqueous waste streams. The microorganisms absorb the organics into the cell, through the cell wall, and into the cytoplasm, where the organic substances are broken down by enzyme oxidation and hydrolysis to produce energy, as well as other cellular material. Besides taking in the organics as food, the biomass acts as a filter to collect colloidal matter and suspended solids. Volatile organics can be driven off somewhat by the aeration process. Some metals are collected in the organisms and in the sludge.

The activated sludge process usually must be preceded by neutralization and some metal removal, and possibly by solids removal. The process is normally followed up by a separation step, usually sedimentation, to remove the biological sludge from the waste liquid stream.

Anaerobic Digestion

Anaerobic digestion relies on microorganisms that do not require oxygen for respiration. This process is useful for the degradation of simple organics. The microbiology is not well understood, but essentially some parts of the organic compounds are used by the cells for growth and other parts are converted to methane and carbon dioxide. A delicate equilibrium is required, and this makes the process less suitable for industrial waste streams.

Anaerobic digestion takes place in a closed vessel with no agitation mechanisms. The gases rise to the top and are collected for possible use as a heating fuel. The digested sludge settles to the bottom. When the sludge is stable and inert, it can be disposed of by landfilling.

Waste Stabilization Ponds

Waste stabilization ponds are shallow basins into which wastes are fed for biological decomposition. The chemical reactions involved are the same as those that occur in the other biological processes. Aeration is provided by the wind, and anaerobic digestion may also occur near the bottom of deeper ponds. The ponds are very commonly used for sewage treatment and dilute industrial wastes. Waste stabilization ponds are normally used as the final treatment step for effluents because they are not efficient enough to be used on their own.

Centrifugation

Centrifugation is a well-established liquid-solid separation process popular in commercial and municipal waste treatment facilities. It is usually used to reduce slurry and sludge volumes and to increase the solids concentration in these waste streams. It is a technically and economically competitive process and is commonly used on waste sludges produced from water pollution control systems and on biological sludges produced in industry and municipal treatment facilities.

Centrifugation is performed in a closed system and is therefore an excellent choice for treating volatile fluids. The liquid and solid are mechanically separated by centrifugal force. The removal of most of the liquid increases the solid concentration in, and reduces the volume of, the waste stream. The collected solid waste may then be treated and disposed of or recovered. Three types of units are available for centrifugation: the *solid bowl*, the *disk*, and the *basket*. The first two are used in large plants, the third in smaller plants.

Filtration

Filtration is a popular liquid-solid separation process commonly used in treating wastewater and sludges. In wastewater treatment, it is used to purify the liquid by removing suspended solids. This is usually followed by *flocculation* or *sedimentation* (discussed below) for further solids removal. In sludge treatment, it is used to remove the liquid (sludge dewatering) and concentrate the solid waste, thereby lowering the sludge volume. This method is highly competitive with other sludge-dewatering processes. Filtration may also be used in treating nonaqueous liquid waste streams.

In the filtration process, a liquid containing suspended solids is passed through a porous medium. The solids are trapped against the medium, and the separation of solids from liquids results. For large solid particles, a thick barrier such as sand may be used; for smaller particles, a fine filter such as a filter cloth is preferable. Fluid passage may be induced by gravity, positive pressure, or a vacuum. A few of the more popular filter types are the *plate and frame filter press*, and *shell and leaf* and *cartridge filters*.

Flocculation and Sedimentation

Flocculation and *sedimentation* are two processes used to separate waste streams that contain both a liquid and a solid phase. Both are well-developed, highly competitive processes, which are often used in the complete treatment of waste streams. They may also be used instead of, or in addition to, filtration. Some applications include the removal of suspended solid particles and soluble heavy metals from aqueous streams. Many industries use both processes in the removal of pollutants from their wastewaters. These processes work best when the waste stream contains a low concentration of the contaminating solids. Although they are applicable to a wide variety of aqueous waste streams, these processes are not generally used to treat nonaqueous or semisolid waste streams such as sludges and slurries.

Flocculation is a physiochemical process in which fine suspended particles, which are difficult to settle out of the liquid, are brought together by flocculating chemicals to form larger, more easily collected particles. *Sedimentation* is a physical process in which suspended solids are settled out through the use of gravitational and inertial forces acting on the solids. The solids removed by these processes form a sludge that may be disposed of or treated to recover any valuable material.

Solid Waste Equipment

Solid wastes are disposed of by two basic methods. The first is by some type of dumping or landfill procedure; the second is by incinerating (burning) the waste. This section focuses on incinerators, namely the *rotary kiln, liquid injection, fluidized-bed,* and *multiple-hearth* devices, which are the four types of incinerator widely used for the disposal of solid wastes. Extensive information, including calculation details, is available in the literature.[1]

Rotary Kiln

Rotary kiln process incinerators were originally designed for lime processing. The rotary kiln is a cylindrical refractory-lined shell that is mounted at a slight incline from the horizontal plane to facilitate mixing the waste materials with circulating air. The kiln accepts solid and liquid waste materials of all types, with heating values between 1000 and 15,000 Btu/lb, and even higher. Solid wastes and drummed wastes are usually fed by a pack-and- drum feed system, which may consist of a bucket elevator for loose solids and a conveyor system for drummed wastes. Pumpable sludges and slurries are injected into the kiln through a nozzle. Temperatures for burning vary from 1500 to 3000°F. The kiln may be equipped with a lime or other caustic injection system to neutralize acid gas and combustion end products. Although liquid wastes may be incinerated in rotary kilns, these devices are designed primarily for combustion of solid wastes. They are exceedingly versatile in this regard, being capable of handling slurries, sludges, bulk solids of varying sizes, and containerized wastes.

Rotary kiln systems usually have a secondary combustion chamber after the kiln to ensure complete combustion of the wastes. Airtight seals close off the high end of the kiln while the lower end is connected to the secondary combustion chamber or mixing chamber. In some cases, liquid waste is injected into the secondary combustion chamber. The kiln acts as the primary chamber to volatilize and oxidize combustibles in the wastes. Inert ash is then removed from the lower end of the kiln. The volatilized combustibles exit the kiln and enter the secondary chamber where additional oxygen is available and ignitable liquid wastes or fuel can be introduced. Complete combustion of the waste and fuel occurs in the secondary chamber.

Liquid Injection

Liquid injection incinerators are currently the most commonly used type of incinerator for hazardous waste disposal. A wide variety of units are marketed today, mainly horizontally and vertically fired types; a less common unit is the tangentially fired vortex combustor.

As the name implies, the liquid injection incinerator is confined to hazardous liquids, slurries, and sludges with a viscosity value of 10,000 SUS (Saybold universal seconds) or less. This limitation reflects the requirement that a liquid waste be converted to a gas before combustion. This change is brought about in the combustion chamber and is generally expedited by increasing the waste surface area through atomization. An ideal droplet size is in the range of 40 to 100 µm, which is attainable mechanically using rotary cup or pressure atomization or via gas-fluid nozzles and high pressure air or steam.

The key to efficient destruction of liquid hazardous wastes lies in minimizing unevaporated droplets and unreacted vapors. Just as for the rotary kiln, temperature, residence time, and turbulence may be optimized to increase destruction efficiencies. Typical combustion chamber residence time and temperature ranges are 0.5-2 s and 1300-3000°F. Liquid injection incinerators vary in dimensions and have feed rates up to 1500 gal/h of organic wastes and 4000 gal/h of aqueous waste.

Fluidized Bed

Fluidized-bed process incinerators have been used mostly in the petroleum and paper industries, and for processing nuclear wastes, spent cook liquor, wood chips, and sewage sludge disposal. Wastes in any physical state can be applied to a fluidized-bed process incinerator. Auxiliary equipment includes a fuel burner system, an air supply system, and feed systems for liquid and solid wastes. The two basic bed design modes, bubbling bed and circulating bed, are distinguished by the extent to which solids are entrained from the bed into the gas stream.

The *bubbling fluidized-bed* design takes its name from the behavior of a granular bed of nonreactive sand, stirred by passing a gaseous oxidizer (air, oxygen, or nitrous oxide) through the bed at a rate high enough to cause the bed to expand and act as a fluid. Preheating of the bed to a start-up temperature may be accomplished by a burner located above and impinging down on the bed. The waste is passed directly into the sand. After combustion, the exhaust and almost all of the ash pass out of the top of the unit.

Multiple Hearth

The *multiple-hearth* incinerator is a flexible unit that has been used to dispose of sewage sludges, tars, solids, gases, and liquid combustible wastes. A typical multiple-hearth furnace includes a refractory-lined steel shell, a central shaft that rotates, a series of solid flat hearths, a series of rabble arms with teeth for each hearth, an air blower, fuel burners mounted on the walls, an ash removal system, and a waste feeding system. Side ports for tar injection, liquid waste burners, and an afterburner may also be included. Sludge and/or granulated solid combustible waste is fed through the furnace roof by a screw feeder or belt and flapgate. The rotating, air-cooled central shaft with air-cooled rabble arms and teeth distributes the waste material across the top hearth to drop holes. The waste falls to the next hearth and then the next until discharged as ash at the bottom. The waste is agitated as it moves across the hearths to make sure that fresh surface is exposed to hot gases. When combustion is complete, the remaining ash is screw-conveyed out of the unit.

5.7 UTILITIES

Today the word "utilities" generally designates the ancillary services needed in the operation of any production process. These services normally are supplied from a central site facility and usually include:

1. Electricity
2. Steam for process heating
3. Cooling water
5. Water for general use
6. Demineralized water
7. Compressed air
8. Inert gas supplies
9. Refrigeration

Electricity

The power required for processes (motor drives, lighting, and general use) may be generated on site, but more often it is purchased from the local utility.

Steam

The steam for process heating is generated in either fire or water-tube boilers, using the most economical fuel available. The process temperatures required usually can be obtained with low pressure steam (typically 25 psig), and steam is distributed at a relatively low pressure (typically 100 psig). Higher steam pressures are needed for high process temperatures.

Cooling Water

Natural and forced-draft cooling towers are generally used to provide the cooling water required on a site, unless water can be drawn from a convenient river or lake in sufficient quantity. Seawater or brackish water can be used at coastal sites, but if used directly will necessitate more expensive materials of construction for heat exchangers.

Water for General Use

The water required for general purposes on a site is usually taken from the local supply, unless a cheaper source of suitable quality water (e.g., a river, lake, or well) is available.

Demineralized Water

Demineralized water is water from which all the minerals have been removed by ion-exchange. It is used where pure water is needed for a process and as boiler feed water. Mixed and multiple-bed ion-exchange units are used for this purpose, one resin converting the cations to hydrogen and the other removing the acid radicals. Water can be produced that has less than 1 ppm of dissolved solids.

Refrigeration

Refrigeration is needed for processes that require temperatures below those that can be economically obtained with cooling water. For temperatures down to around 10°C, chilled water can be used. For lower temperatures, down to -30°C, salt brines (NaCl and CaCl$_2$) are used to distribute the "refrigeration" around the site from a central refrigeration unit. Vapor compression machines are normally used.

Compressed Air

Compressed air is needed for general use and for the pneumatic controllers that usually serve for chemical process plant control. Air is often distributed at a pressure of 100 psig. Rotary and reciprocating single-stage or two-stage compressors are used. Instrument air must be dry and clean (free from oil).

Inert Gases

Large quantities of inert gas are required for the inert blanketing of tanks and for purging. This gas usually is supplied from a central facility. Nitrogen is normally used and can be manufactured on site in an air liquefaction plant or purchased as liquid in tankers.

5.8 PROTECTIVE AND SAFETY SYSTEMS

Protective systems are employed to reduce/eliminate hazards and risks, and may be viewed as a special category of process equipment. Ten examples of protective systems are provided below. The management of protective systems information should occur within the management of systems for process and equipment. However, with protective systems equipment, it is particularly important to record and disseminate information on promising new approaches, and on problems with older approaches.

Provision for protection and safety equipment should be incorporated in the original plant design. The size of the plant, nature of the hazards, and the exposure will determine the amount, kind, and location of this equipment.

Regarding fires, water is the primary extinguishing agent, and it should be available in adequate supply and pressure at all of the locations in the plant. The layout for various types of installations and the appropriate recommendations are found in the standards of the National Fire Protection Association. Fire hydrants, hose lines, automatic sprinkler and water spray systems should all be a part of the permanent equipment facilities of the plant.

Fire extinguishing systems can include foam, carbon dioxide and dry chemical. Wetting agents and high expansion foam have been used in some plant protection systems. All fire extinguishing systems should be evaluated for the potential health risks as well as overall effectiveness before incorporation into a plant.

Since many chemical plants have severe health hazards, it is essential to provide medical facilities and first aid stations. In addition, showers and eye wash stations are necessary in certain hazardous areas. Also, guards and covers should be provided for all moving equipment. Ten of the key protective equipment are listed below.

1. Pressure relief/vent collection
2. Release devices (flares, scrubbers, etc.)
3. Plant equipment isolation
4. Critical alarms/interlocks
5. Fire detection/protection
6. Gas detection
7. Flame arrestors
8. Emergency system services
9. Appropriate grounding and bonding
10. Personal protection equipment

Perhaps the most important of the above equipment services are the pressure relief systems. Pressure relief systems are incorporated in process units to protect the equipment and ancillary piping from failure due to pressures above the design operating limits. The relief system, usually a pressure relief valve, must discharge to a safe place. In a process plant, such systems commonly vent to a flare if flammable, toxic, or carcinogenic substances are involved. Occasionally, the relief system can discharge to a closed system. In an emergency, the discharge may be from the source at an elevated level. Calculation details are beyond the scope of this text, but are available in the literature.[2]

Personal protective equipment can include the following:

1. Hearing – earmuffs and earplugs
2. Eyes – goggles, shatterproof safety glasses, and face shields
3. Feet – puncture proof shoes, chemical resistance and insulated rubbers/boots with metal toe protection
4. Hands – metal mesh gloves, fire resistant gloves, and insulted gloves
5. Head – hard hat helmets
6. Breathing – respirators that provide and supply fresh air
7. Body – special clothing, including flame and chemical resistant material and impact softener (leather) coverings
8. Skin – barrier creams

It should also be noted that some protective systems are occasionally shared within a site, thus, providing protection to more than one unit or process.

5.9 PROCESS DIAGRAMS

To the practicing engineer, particularly the chemical engineer, the process flow sheet is the key instrument for defining, refining, and documenting a chemical process. The process flow diagram is the authorized process blueprint, the framework for specifications used in equipment designation and design; it is the single, authoritative document employed to define, construct, and operate the chemical process.

Flow Sheets

Beyond equipment symbols and process stream flow lines, there are several essential constituents contributing to a process flow sheet. These include equipment identification numbers and names, temperature and pressure designations, utility designations, volumetric or molar flow rates for each process stream, and a material balance table pertaining to process flow lines. The process flow sheet may show additional information such as energy requirements, major instrumentation, and physical properties of the process streams. When properly assembled and employed, this type of process schematic provides a coherent picture of the overall process. It can point out some deficiencies in the process that may have been overlooked earlier in the study-for example, by-products (undesirable or otherwise) and recycle needs. Basically, the flow sheet symbolically and pictorially represents the interrelations among the various flow streams and equipment and permits easy calculations of material and energy balances. A number of symbols are universally employed to represent equipment, equipment parts, valves, piping, and so on. These symbols obviously reduce, and in some instances replace, detailed written descriptions of the process. Note that many of the symbols are pictorial, which improves the descriptive power of the document.

A flow sheet usually changes over time with respect to the degree of sophistication and details. A crude flow sheet may initially consist of a simple, freehand block diagram offering information about the equipment only; a later version (described in the next subsection on Process Flow Diagrams) may include line drawings with pertinent process data such as overall and componential flow rates, utility and energy requirements, and instrumentation. During the later stages of a design project, the flow sheet will consist of a highly detailed *piping and instrumentation diagram* (P&ID), which is covered in a later subsection. For information on aspects of the design procedure, which is beyond the scope of this text, the reader is referred to the literature.

In a sense, flow sheets are the international language of the engineer, particularly the chemical engineer. Chemical engineers conceptually view a (chemical) plant as consisting of a series of interrelated building blocks that are defined as units or unit operations. The plant essentially ties together the various pieces of equipment that make up the process. Flow schematics follow the successive steps of a process by indicating where the pieces of equipment are located and when the material streams enter and leave each unit.

Process Flow Diagram (PFD)

The *process flow diagram*, or *PFD*, is a pictorial description of the process. It gives the basic processing scheme, the basic control concept, and the process information from which equipment can be specified and designed. It provides the basis for the development of the P&I diagram and also serves as a guide for the plant operator. The process flow diagram usually includes

1. Material balance data (may be on separate sheets)
2. Flow scheme equipment and interconnecting streams
3. Basic control instrumentation
4. Temperature and pressure at various points
5. Any other important parameters unique to each process

Data on spare and parallel equipment are often omitted. Valving is also generally omitted. A valve is shown only where its specification can aid in understanding intermittent or alternate flows. Instrumentation is indicated to show the location of variables being controlled and the location of the actuating device, usually a control valve. To help the reader better understand the process flow sheet, a list of commonly used symbols is presented in Fig. 5.9.1.

Process Piping and Instrument Flow Diagram

The *P&ID*, which provides the basis for detailed design, offers a precise description of piping, instrumentation, and equipment. This key drawing defines the plant system, describes equipment, and shows all instrumentation, piping, and valving. It is used to train personnel and aids in trouble-shooting during start-up and operation. The P&ID assigns item numbers to all equipment (e.g., towers, reactors and tanks), gives dimensions of equipment and vessel elevations, and shows all piping, including line numbers, sizes, and specifications, and all valves. All instrumentation is covered, with relevant numbers, function, and types (an indication of electronic or pneumatic control).

A general knowledge of the symbols for flow, level, pressure, and temperature controllers, as given in Fig. 5.9.2, is needed to comprehend flow diagrams like the simple example presented in Fig. 5.9.3. In this vessel, with an inlet feed on top of the tank equipped with a flow controller, the level in the tank is maintained by a level controlling device. When the level rises above the high level point, the level controller sends a signal to a valve actuator and the valve is opened to drop the level. When the level approaches a specified value, the valve is closed. Complicated systems can be analyzed in the same manner used in this basically simple example.

5.10 PLANT SITING AND LAYOUT

The proper location of a plant is as important to its success as the selection of a process. Not only must many tangible factors such as labor supply and raw material sources be carefully considered, but also a number of intangible factors, which are more difficult to evaluate. The selection of a plant site must be based on a very detailed study in which all factors are weighed as carefully as possible. Such a study often requires a substantial outlay of capital.

For many processes, one or more predominant factors effectively minimize the number of possibilities for plant location. Raw material and transportation costs may be such that a plant must be located near a source. Thus, only the sites near to sources of raw material need be studied and these may be few in number. Similarly, labor requirements may be heavy enough to eliminate cities below a certain size. These and other factors serve as effective screening agents that save both time and money.

Important factors to be considered in the study of areas and sites for plant location include raw materials, transportation, process water, waste disposal, fuel and power, and weather.[6] These are discussed individually.

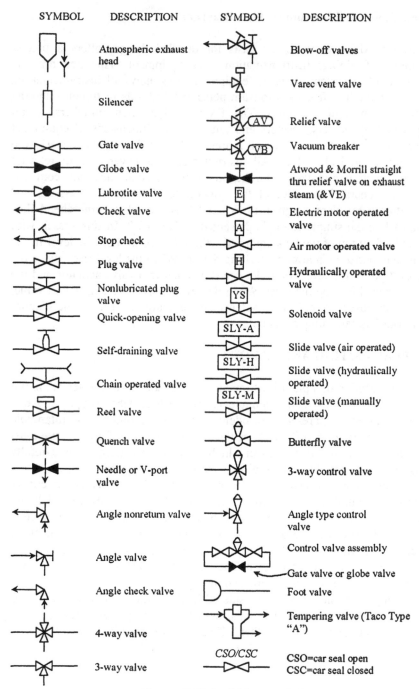

Figure 5.10.1. Valve symbols.

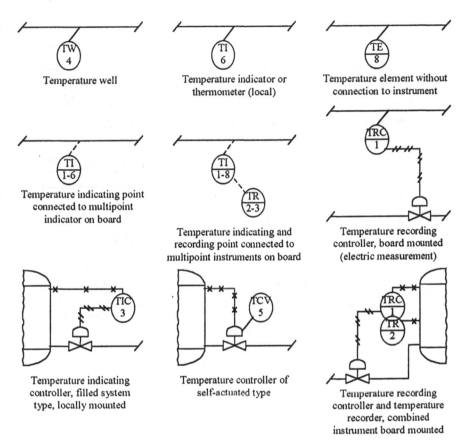

Temperature well

Temperature indicator or
thermometer (local)

Temperature element without
connection to instrument

Temperature indicating point
connected to multipoint
indicator on board

Temperature indicating and
recording point connected to
multipoint instruments on board

Temperature recording
controller, board mounted
(electric measurement)

Temperature indicating
controller, filled system
type, locally mounted

Temperature controller of
self-actuated type

Temperature recording
controller and temperature
recorder, combined
instrument board mounted

Figure 5.10.2a. Typical instrumentation symbols for temperature.

Raw Materials

Although the source of raw materials may not be at the plant site, it is an extremely important factor in the ultimate location of the plant. Process development work and economic studies will indicate the minimum standards for raw materials selection. When these standards have been determined, all possible sources of acceptable raw materials can be located and a detailed analysis can proceed.

The size of each raw material source must be determined in the light of existing and estimated future requirements. An attempt must be made to estimate the life of the raw material source based on future requirements. Alternate sources or substitutes in the area should also be located and evaluated. The cost of delivering raw material to the plant site can then be determined for all sources that meet the process quality and quantity specifications.

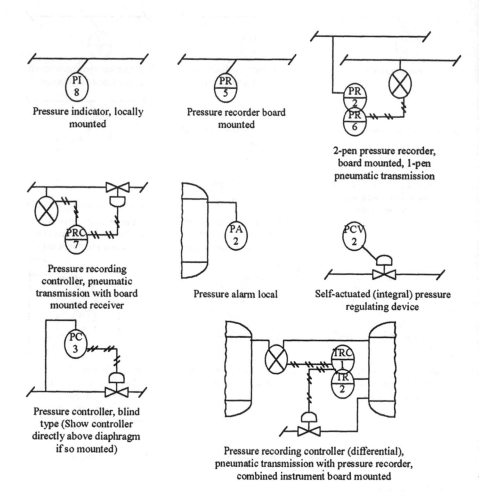

Pressure indicator, locally mounted

Pressure recorder board mounted

2-pen pressure recorder, board mounted, 1-pen pneumatic transmission

Pressure recording controller, pneumatic transmission with board mounted receiver

Pressure alarm local

Self-actuated (integral) pressure regulating device

Pressure controller, blind type (Show controller directly above diaphragm if so mounted)

Pressure recording controller (differential), pneumatic transmission with pressure recorder, combined instrument board mounted

Figure 5.9.2b. Typical instrumentation symbols for pressure.

Transportation

It is not possible to present a complete discussion of freight rates in this subsection. The engineer in charge of obtaining information related to plant location need only realize that experts must be consulted in establishing freight

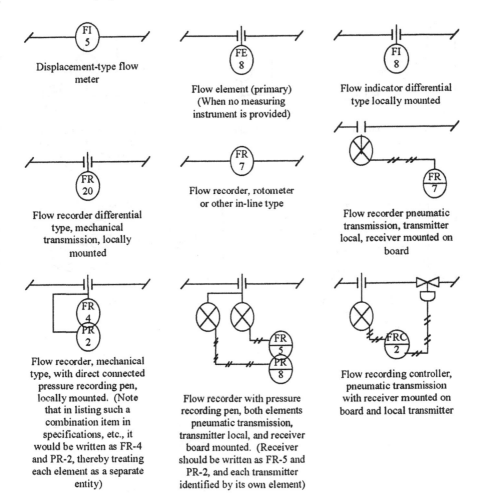

Figure 5.10.2c. Typical instrumentation symbols for flow.

charges and optimum location with respect to transportation. Such advice is available from freight agents and traffic experts from the railroads and other transportation facilities. In addition, the traffic manager of the company can be of great assistance in obtaining the necessary information and aiding in its interpretation.

The effect of transportation facilities and rates on plant location can be a controlling factor in plant siting. Industries such as the plastic industry, which must deliver many small shipments to various users in the minimum amount of time, find setting the location near the majority of users mandatory.

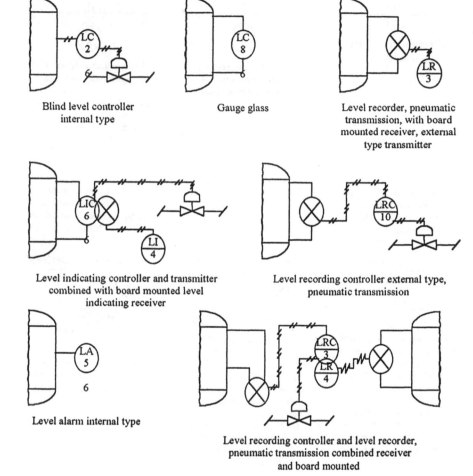

Figure 5.10.2d. Typical instrumentation symbols for level.

Less-than-carload-lot rates are very high. Therefore, the distance that the material must be shipped should be kept to a minimum. Each of the four major transportation methods in use today (railroad freight, trucking, water transport, and air travel) has its benefits and drawbacks.

Process Water

The process industries rank above all others as users of water. No process plant could operate without water as a cooling medium and as a direct raw material in certain phases of a process. The local water supply, therefore, must be

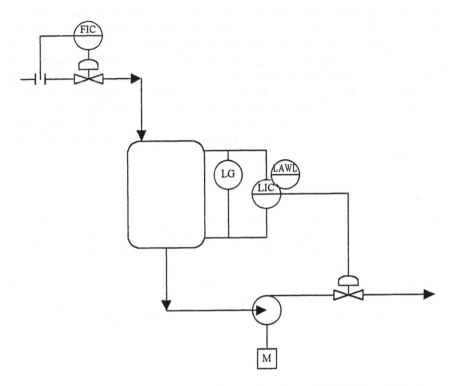

Figure 5.10.3. Vessels---level control on outlet. Legend: FIC, flow indicator
controller; LG, level gauge; LIC, level indicator controller;
LAWL, level alarm for water level, M, motor.

studied before an area can even be considered as a possible site. A detailed
estimate of present and future water requirements must precede the study.
Then the availability of water in the region being considered should be
carefully investigated. If well water is to be used, a complete study of the past
history of the underground water table is necessary. If underground water
supplies are adequate, they are preferred because of their lower temperature.

Surface water from streams or lakes also require careful consideration,
since they are often affected severely by seasonal variations. Freshwater streams
discharging into the ocean during times of low runoff can become salty as a
result of seawater backup. Under such conditions, plant design may need to
include large storage facilities for fresh water to be collected during the period
of high runoff and used during the season when the stream becomes salty.

Companies moving into relatively nonindustrialized areas often fail to
consider the possibility of other plants following suit. The size of the water
supply should be adequate not only for the future needs of the proposed plant

but also for the anticipated needs of other industries that might move into the area. It is also desirable to consider alternate sources of supply that may be required as the preferred water source becomes depleted.

Water quality must be studied as well as quantity. Chemical and bacteriological examination will indicate the extent of treatment required and will aid in the development of water cost estimates for comparison with other locations. The possible contamination of the water source by other industries in the area should be anticipated. Note that "contamination" may consist of raising the temperature of the water to a level that renders its use as a cooling medium impossible.

Waste Disposal and Noise Abatement

The forward-looking engineer will consider waste disposal and noise abatement to be just as important in thinly populated areas as in metropolitan regions having special ordinances addressing these problems. Nothing is so unprincipled or injudicious as dumping waste into the atmosphere of nearby streams. In addition to violating moral and ethical considerations, it is not good economics. Eventually the legislature, responding to demands of regulatory agencies and the public, will impose laws that may be burdensome enough to make profitable operation impossible.

Fuel and Power

All process plants require both steam and electric power in their operations. Power is either purchased from local utility companies or generated at the plant site. Even if power is generated by the process plant, arrangements for standby power from the local utility must be made for emergency purposes. Steam is rarely purchased but is generated at the plant for use in the process and as a driving medium for pumps and compressors.

A detailed knowledge of the quantity of power and steam required for the operation of the projected plant must be obtained before the study can proceed. The costs of all fuels available in the area should be carefully analyzed.

Labor

A large portion of the cost of any manufactured item is represented by labor. Although labor rates are more and more becoming similar in most parts of the country, factors such as skill, labor relations, and the general welfare of the work force affect labor productivity and efficiency. Each region being considered for plant location must be surveyed to determine the availability and the skills of the labor market. The skills need not exactly match with those required by the process plant. A stable labor force is valuable in successful plant operation.

Weather

Weather data for a number of years should be assembled for each community being studied. Particular attention should be given to such natural disasters as hurricanes, earthquakes, and floods, which often can be predicted from meteorological data. In certain locations these catastrophic events must be assumed to be probable and this increases construction costs. Extremely cold weather often hampers process plant operation and requires special construction features to protect equipment from freezing. Predominantly warm weather permits cheaper construction but may also reduce the efficiency of the labor force.

Factors in Planning Layouts

Rational design must include the arrangement of processing areas, storage areas, and handling areas with regard to such factors as

1. New site development or addition to a previously developed site
2. Future expansion
3. Economic distribution of services (water, process steam, power, and gas)
4. Weather conditions
5. Building code requirements
6. Waste disposal problems
7. Sensible use of floor and elevation-space
8. Safety considerations (possible hazards)

Methods of Layout Planning

To start a detailed planning study, space requirements must be known for various products, by-products, and raw materials, as well as for process equipment. A starting or reference point, together with a directional schematic flow pattern, will enable the design engineers to make a trial plot plan, as explained below. A number of such studies will be required before a suitable plot and elevation plan can be chosen.

The basic blocks with which to build an arrangement for plot plans are often used in the unit area concept. This method of planning is particularly well adapted to large plant layouts. Unit areas are often delineated by means of distinct process phases and operational procedures by the presence or absence of contamination and by safety requirements. Thus, the determination of the shape and extent of a unit area and the interrelationships of each area in a

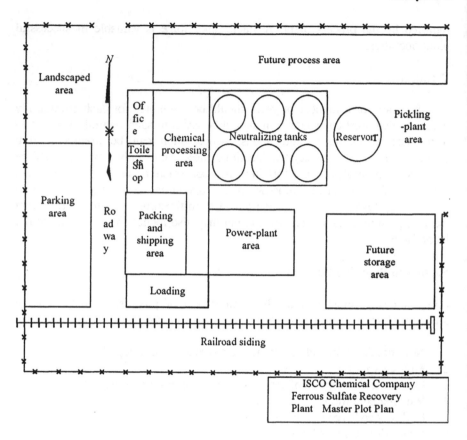

Figure 5.10.4. Typical master plot plan.

master plot plan is one of the first tasks of layout planning. Figure 5.10.1 demonstrates an example of this type of planning.

Modeling System Layouts

Years ago plastic scale models were fabricated for each plant under construction, providing excellent three-dimensional representations of the actual facilities. Today, in the age of the microcomputer, it is quicker, easier, and much cheaper to generate models by means of computer graphics.

Principles of Plant Layout

Storage facilities for raw materials and intermediate and finished products may be located in isolated or adjoining areas. Hazardous materials should be isolated because they menace life and property when stored in large quantities.

Storage in adjoining areas to reduce materials handling now may introduce an obstacle to future expansion of the plant.

Arranging storage of materials to facilitate or simplify handling is also a point to be considered in design. Where it is possible to pump a single material to an elevation so that subsequent handling can be accomplished by gravity into intermediate reaction and storage units, costs may be reduced. In making a layout, ample space should be assigned to each piece of equipment; accessibility is an important factor for maintenance.

It is extremely poor economic practice to fit the equipment layout too closely into a building. A slightly larger building than appears necessary will cost little more than one that is overcrowded, yet the extra cost will be small in comparison to the penalties that will result if, to iron out the kinks, the building must be expanded.

The operations that constitute a process are essentially a series of unit operations that may be carried on simultaneously. Since these operations are repeated several times in the flow of materials, it should be possible to arrange the equipment into groups of like pieces. Such a layout will make it possible for one or two operators to tend all equipment in a given group.

Access for initial construction and maintenance is a necessary part of planning. Space should be provided for repair and replacement equipment such as cranes and forked trucks, and specialized items such as snow removal equipment, as well as access ways around doors and underground hatches.

Expansion must always be kept in mind. The question of multiplying the number of units or increasing the size of the prevailing unit or units merits more study than can be given here. It suffices to say that one must exercise engineering judgment.

Whether floor space is a major factor in the design of a particular plant depends on the value of land near the proposed site. The engineer should, however, allot floor space economically, consistent with good housekeeping in the plant and should be sure to consider line flow of materials, access to equipment, space to permit working on parts of equipment that need frequent servicing, and the safety and comfort of the operators.

The distribution of gas, air, water, steam, power, and electricity is not always a major item since the flexibility of distribution of these services permits designing to meet almost any condition. However, proper placement of each of these services helps to insure ease of operation and orderliness and aids in reducing maintenance costs.

After a complete study of quantitative factors, the selection of the building or buildings must be considered. Standard factory buildings are to be desired but if none can be found satisfactory to handle the space and process requirements of the chemical engineer, an architect specializing in this area should be consulted to design a building around the process - as opposed to a beautiful structure into which a process must fit.

Consideration of equipment for materials handling is only a minor factor in most cases of arrangement, owing to the multiplicity of devices available for this purpose. Where materials handling is paramount in a process, however, serious thought must be given to it. Again, engineering judgment must be exercised. Whenever possible, one should take advantage of the topography of the site location.

Existing or possible future railroads and highways adjacent to the plant must be known in order to be able to plan rail sidings and access roads within the plant. Railroad spurs, roadways of the correct capacity and location, and an overall master track and road plan of the plant area should be provided for in a traffic study.

It must be recognized that there is no single solution to the problem of layout of the equipment. There are many rational designs. A plan must be decided on after exercise of engineering judgment, striking a balance between the advantages and disadvantages of each possible choice. A great deal of planning is governed by local and national safety and fire code requirements (see Chapter 3). Fire protection, consisting of reservoirs, mains, hydrants, hose houses, fire pumps, sprinklers in buildings, explosion barriers and directional routing of explosion forces to clear areas, and dikes for combustible product storage tanks, etc. (see earlier section) must be incorporated to comply with such codes in order to to protect costly plant investment and to reduce insurance rates.

Many of the safety considerations discussed above can be better understood by performing a Hazard and Operability Study (HAZOP). HAZOP is covered in detail in Part IV of this book.

5.11 ILLUSTRATIVE EXAMPLES

Example 5.1

Discuss the various classes of heat exchangers.

Solution 5.1

The chemical process industries use four principal types of heat exchangers.

1. Double-pipe exchangers-the simplest type with concentric pipe arrangement used for cooling and heating; several units can be connected in series to extend their capacity.[3]
2. Shell and tube exchangers-most commonly used for all applications in the chemical and allied industries; there are several advantages to this type of heat exchanger: large surface area in a small volume, good mechanical layout, reliance on well-established fabrication techniques, wide range of

construction materials available, easily cleaned equipment, and well-established design procedures.

3. Plate and frame exchangers (plate heat exchangers)-used for heating and cooling in reactor off-gas quenching, vacuum condensers, desuperheating, and humidification.

4. Direct contact exchangers - used for cooling and quenching whenever the process stream and the coolant are compatible; the equipment used is basically simple and cheap.

The reader is referred to the literature for the description of other units.

Example 5.2

What are potential consequences if a pinhole leak develops in a tube in the reboiler of a distillation column?

Solution 5.2

Potential consequences of a pinhole leak are:

1. Changes in pressure
2. Changes in temperature
3. Chemical reaction, with accompanying over-pressure, over-temperature, and formation of other phases
4. Leakage of toxics/flammables to an undesirable location
5. Corrosion, embrittlement, or similar effect

Example 5.3

List safety concerns (in the form of questions) associated with plant site location and layout.

Solution 5.3

1. Is there any exposure to or from other nearby plants from fire, noise, air pollution, stream pollution, or explosion hazards?
2. Is there adequate access for emergency vehicles?
3. Are major highways, airports or congested areas near the plant site capable of creating a blockage of access roads at any time of the day?
4. Are access roads well engineered to avoid sharp curves? Are traffic signs provided?
5. Do adequate fences and gates enclose the plant?

6. Will toxic fumes or fires, explosions or other accidents at the plant affect the surrounding community?
7. Is the plant well situated with regard to topography and adequate drainage?
8. Are utilities (water, gas, electricity, etc.) adequate?
9. Are waste disposal systems downwind from plant personnel and the surrounding community?
10. Are loading areas on the periphery of the plant away from any sources of ignition?
11. Will the climate/weather (earthquake, flood, fog, hurricane, lightning, smog, snow, tornado, and very low temperature) materially affect plant operations?
12. Is there a safe distance from the boundary to the fence?

Example 5.4

Discuss various concerns a plant manager might have regarding plant buildings and structures.

Solution 5.4

1. What standards are being followed in the design of stairways, platforms, ramps, and fixed ladders?
2. Are sufficient general exit and escape routes available for employees? Is there an alternate means of escape from the roof? Is protection provided to individuals along the line of the escape routes?
3. Is adequate lighting provided?
4. Are the doors and windows hung to avoid projecting into or blocking walkways and exits?
5. Are the buildings and structures properly grounded?
6. Are equipment, steam, water, air and electronic outlets arranged to keep aisles and operating floor areas clear of hoses and cables?
7. Is ventilation furnished for hazardous fumes, vapors, dust and excessive heat?
8. Is temporary storage provided for raw materials at process points and for finished products?

Example 5.5

List and briefly discuss the various types of fire system equipment.

Solution 5.5

Chemical plants require a well-engineered system for fire protection. Many of the basic fire protection facilities must be incorporated into the original plant design in order to achieve maximum effectiveness. Fire extinguishing facilities and equipment should include:

1. Water supply – It is the most important of all extinguishing agents for most chemical plant fires. The water supply should be sufficient to fulfill the demand for automatic protection and hose streams for at least a four-hour period. Allowance should be made for explosion damage to the system and protection against freezing.
2. Distribution system–This system should cover all facets of outside fire protection water demands in order to provide adequate water distribution for existing conditions and the possibility of plant expansion.
3. Monitors and deluge sets–Many chemical plants use monitors for general use and for high hazard locations in order to provide maximum water supply with a minimum of manpower exposure. Deluge sets supported by ample hose streams are preferred for some protection uses.
4. Sprinkler and water spray installations–Many process and storage area buildings should be protected by automatic sprinkler systems. The size and arrangement for water supply are dependent upon the nature of the hazard and the degree of protection desired. Water spray installations are particularly adapted for cooling uninsulated steel structures, elevated pipe lines, vessels, spheres, and similar plant installations.
5. Foam extinguishing system–Available in automatic and manual systems, foam has the ability to adhere to surfaces, and thus provide a blanketing as well as cooling effect.
6. Carbon dioxide systems–This system can be used where there is a handling and storage of gaseous and flammable materials, electrical equipment, and hazardous solids by introducing an inert gas (such as carbon dioxide) into the area in order to reduce the concentration of oxygen to the point where the fire will be extinguished.
7. Dry chemical extinguishing systems–This is used primarily for flammable liquid fires since they provide a rapid flame knockdown and extinguishment.
8. High expansion foam systems–This is a concentration of air-filled bubbles resulting from the mechanical expansion of a foam solution to smother surface fires involving flammable liquids.
9. Halogenated extinguishing systems–This system can be used to protect flammable liquid storage and processing units. It is used with caution since, in the appropriate concentration, the products of decomposition are

toxic and can result in injury or death of exposed personnel. These systems have become less popular in recent years.

10. Combination extinguishing systems–Combining dry chemicals and foam agents unites the fast flame control of the dry chemical with the cooling and sealing ability of foam to provide an efficient portable extinguishing system.

11. Portable extinguishers–These are used in small fires since they can easily be transported and operated. Ideal placement would be in laboratories and pilot plant installations where fires of limited size can be anticipated.

5.12 SUMMARY

1. There is a nearly infinite variety of process equipment that can be found at a plant site.

2. One of the major pieces of process equipment is a reactor. It is the place in the process where raw materials are usually converted into products. Thus, its design is a vital step in the overall design of the process.

3. A second major piece of process equipment is a heat exchanger. It transfers heat to and from process fluids in a chemical process.

4. Unit operations in the mass transfer category that may be found at a plant are evaporation, extraction, adsorption, and absorption.

5. Ancillary equipment consists of the support equipment, other than process equipment, needed for a plant to function properly. Pipes, valves, and primary movers (e.g., pumps) are included in this category.

6. There are a variety of pollution control equipment options that can be incorporated into a plant to reduce and control any waste or pollution that is generated.

7. Utilities are critical to the proper functioning of a plant.

8. Protection and safety systems should be used to reduce/eliminate hazards in a plant.

9. Process diagrams are a necessary part of a plant. Flow and P&I diagrams aid the process engineer in understanding, controlling, and troubleshooting the process. They also provide valuable information about the process in the event of an emergency.

10. Plant siting and layout are important for the safety of the workers in the plant as well as the surrounding communities. A proper plant layout can save both lives and dollars.

PROBLEMS

1. Discuss the advantages and disadvantages of the four classes of reactors.
2. List the various types of
   a. Air pollution control equipment

   b.   Water pollution control equipment

   c.   Solid waste control equipment

3.   Discuss the differences between the various mass transfer operations.

4.   Discuss the main differences between a process flow diagram (PFD) and a process piping and instrument diagram (P&ID).

5.   List the important factors to be considered in plant siting and layout.

6.   Discuss the advantages and disadvantages of physical versus computer modeling of plant systems.

7.   List some of the precautions that should be taken when wearing personal protective equipment.

8.   Describe safety requirements for pressure vessels.

REFERENCES

1. J. Santoleri, J. Reynolds, and L. Theodore, "Introduction to Hazardous Waste Incineration", 2nd ed., Wiley, New York, 2000.

2. R. H. Perry and D. Green, "Perry's Chemical Engineers' Handbook", 7th ed., McGraw-Hill, New York, 1996.

3. J. M. Coulson, J. F. Richardson, and R. K. Skinnott, "An Introduction to Chemical Engineering Design", Pergamon Press, Elmsford, New York, 1983.

4. C. J. Geankpolis, "Transport Processes and Unit Operations", Allyn & Bacon, Boston, 1978.

5. F. N. Nazaro, "Rupture Disks", Chemical Engineering Magazine, McGraw-Hill, New York, June 20, 1988.

6. F. C. Vilbrandt and C. E. Dryden, "Chemical Engineering Plant Design", 4th ed., McGraw-Hill, New York, 1959.

6

Classification of Accidents

6.1 INTRODUCTION

Classes of accidents that can result in financial and personnel losses include more than just natural disasters such as earthquakes and tornadoes, or occupational personnel mishaps such as tripping and slipping. These, and several others, are reviewed in this chapter. The section breakdown with respect to content is provided below:

Section 6.2:   Equipment Failure
Section 6.3:   Human Error and Occupational Mishaps
Section 6.4:   Transport Accidents
Section 6.5:   Electrical Failures and Computer Failures
Section 6.6:   Nuclear Accidents
Section 6.7:   Natural Disasters

Although the terms "emergency" and "accident" are discussed from a risk analysis point of view in Parts IV and V, they are now defined in relatively general terms, to help the reader differentiate between the two and to understand their application to the general subject of process "occurrences".

An *accident* is defined as an unexpected event, or alternately, an event that is not likely to occur. Accidents are usually uncontrolled events. They are the results of unforeseen circumstances and changes in otherwise controlled conditions.

An *emergency* is defined as a crisis or a sudden occasion that requires immediate remedial action to restore the previous (i.e., controlled) conditions. Therefore, an emergency precedes an accident, and also encompasses it. Before

---

The authors gratefully acknowledge the assistance of Dean Won in researching, reviewing, and editing this chapter.

an accident, there is an emergency or a pressing need to revert to or restore the previous conditions, which would alleviate the potential for an accident. During an accident there is an even more urgent necessity to alter the existing conditions in a manner that will not only alleviate the hazardous or critical condition(s) that are causing the accident but also will minimize the effects of the occurring accident.

Accidents such as fires, explosions, toxic emissions, and hazardous spills, are dealt with in the next chapter. The decision to include these in a separate chapter was not as easy as one would expect since the topics of both chapters could be classified as accidents. However, in order to treat the accidents in a cause and effect manner, it was felt that fires, explosions, etc., could well be considered an effect arising from the generic accidents discussed in this chapter.

6.2 EQUIPMENT ACCIDENTS

The academic training of engineers with respect to plant equipment has traditionally focused on design and predicting performance. Little to no effort was expended in attempting to answer the question:

WHAT CAN GO WRONG?

This question requires answers in today's high-tech environment. Failure to properly and realistically address and answer this question can result/lead to an:

ACCIDENT!

Industry now knows that equipment sometimes fail partially and sometimes fail catastrophically. In any event, it behooves the engineer to carefully and thoroughly examine the various pieces of plant equipment that can fail. A partial list of some of the more common equipment that can be expected to fail with some regularity in a plant is provided below. Details are available in the literature.[1]

1.   Foundations
2.   Structural steel
3.   Vessels and tanks
4.   Pumps
5.   Compressors
6.   Fans
7.   Heat exchangers
8.   Turbines
9.   Electrical systems
10.   Instrumentation and controls

11. Piping systems
12. Valves and joints
13. Mass transfer unit operations
14. Heaters and furnaces
15. Chemical reactors
16. Fire protection
17. Safety equipment

Deviations, i.e., abnormal conditions that can lead to failure and/or an accident, generally occur because of:

1. Abnormal temperatures
2. Abnormal pressures
3. Material flow stoppage
4. Equipment leaks
5. Equipment spills
6. Materials failure due to wear
7. Materials failure due to imperfections
8. Materials failure due to poor maintenance
9. Materials failure due to corrosion

Some specific operational failures include:[1]

1. Blocked outlet
2. Opening/closing valve
3. Cooling water failure
4. Power failure
5. Instrument air failure
6. Thermal expansion
7. Vacuum problems

Specific comments/recommendations/or procedures for the potential failure of equipment that can be found in a process plant are beyond the scope this text. However, a short list of the units of concern follows:

1. Heat exchangers should be constructed in accordance with applicable codes and regulations. Materials of construction should be carefully selected to resist corrosion and fouling. These units should be equipped with relief valves, bypassing piping (or the equivalent), and adequate drainage facilities. Only seamless tubes should be used for exchangers where tube leakage or rupture may result in an adverse reaction or excessive pressure. Heat exchanger tubes and piping should also allow for thermal expansion and contraction without causing excessive stress on connections. A proactive operation, maintenance and inspection (OM&I) program is a must.

2. Heaters and furnaces should also be designed in accordance with standards and codes. Boilers and heating units must be inspected periodically in accordance with codes, insurance requirements and state regulations. Proper controls, interlocks and fail-safe instrumentation must be provided. The heaters should also be provided with sight glasses for flame observation, monitoring devices for flame-out detection, and temperature alarms.

3. Pumps and compressors are equipment of choice for moving both liquid and gas. In general, pump selection and performance details are designed to meet specific operating conditions. Equipment utilized for flammables should be located outdoors whenever possible. Provision for inspection and maintenance is also an important safety condition of high speed equipment. Serious fires have been reported due to the failure of the pump packing glands on pumps used to transport flammable liquids. In such instances the pump may become inaccessible during the fire; for this condition, remotely controlled power switches and shut-off valves are desirable to control the flow of fuel during an emergency.

4. All reactors should meet or exceed ASME Boiler and Pressure Vessel Codes with respect to appropriate materials of construction. The unit should also be sized properly and equipped with safety relief valves and/or rupture discs and contain appropriate piping and controls.

5. The functions and limitations of each instrument should be defined and its reliability demonstrated. Certain basic requirements should be specified:

    - All instruments and controls must be of the "fail-safe" type
    - Instruments should be made of material capable of withstanding the corrosive or erosive conditions to which they are subjected
    - Instruments should be constructed and installed so that they can be easily inspected and maintained
    - The use of separate indicators for each critical hazard point of the operation is recommended
    - Audible and visual alarms should be provided so that operators will know what type of emergency is occurring

6.3 HUMAN ERRORS AND OCCUPATIONAL MISHAPS

Occupational mishaps are usually a function of three aspects of the work place: the human element, task variables (i.e., the job itself), and the environmental element.

Human Element

When evaluating human elements in relation to occupational mishaps, five factors are taken into consideration.

1. *Sex.* Within a given job setting, a worker's sex may increase the individual's propensity for accidents and injuries. For example, if a job originally designed for male workers is opened to female workers, the possibility of increased proneness to injuries of certain types (e.g., back injuries) should be considered.[2]
2. *Age.* Data seem to indicate that younger workers have a higher potential to become involved in accidents. This may be due to several factors, such as inexperience, often resulting in bad judgment, a tendency toward aggressiveness, and a willingness to take high risks. In addition, at the other end of the spectrum, older workers may begin to lose their eyesight, hearing, dexterity, and strength, which may also lead to higher accident and injury potential.[2]
3. *Personality.* Personality factors such as tendency to anger, discontent, excitability, and hostility, as well as low order of adjustment and high impulsiveness can also contribute to maladjustments and accidents.[2]
4. *Physical-Physiological Status.* A worker's physical and physiological capacity for work may have an impact on his or her accident and injury potential. For example, many accidents occur during the handling, lifting, and carrying of goods; frequently the result is permanent injury (such as back trouble or hernia) and long periods of absence from work.[2]
5. *Accident Proneness.* The old notion that certain people are accident-prone has been difficult to establish as fact. A more acceptable concept may be "accident liability," which can be related to factors that often are temporary and do not depend primarily on personality traits, such as work situation or stress.

Task Variables

Hazards inherent in the job itself contribute to the overall accident potential of occupational mishaps. Within the category of task variables, there arise hazards that can be described as mechanical, electrical, and thermal.

1. *Mechanical Hazards.* Injuries can arise from the improper use of hand tools, the use of defective tools, or the use of wrong tools. Moving machinery also can cause accidents. Many accidents occur because isolation of the moving parts of mechanical equipment is incomplete, exposing workers to the possibility of catching a hand or piece of clothing. In addition, many pieces of moving machinery present the hazard of running nips. A running nip occurs when material runs onto or over a rotating cylinder (e.g., a belt over a pulley wheel, a conveyor over a roller) or where two rollers come together. It is quite possible for a worker to catch clothing in a nip or even to fall into one while trying to make adjustments to the material running onto a cylinder.
2. *Electrical Hazards.* If grounding and bonding systems are absent or in-adequately maintained, or if electrical tool and equipment maintenance is

inadequate, serious accidents can result. For example, cables and plugs are vulnerable and require periodic attention. Wear occurs particularly at the point where a cable enters an electrical tool or plug.

3. *Thermal Hazards.* These hazards include hot surfaces, fire produced by reacting substances, and flame contacts from heat-producing equipment, welding operations, and so on.

Environmental Element

Injuries and occupational illnesses can be functions of the following environmental factors.

1. *Atmospheric Hazards.* Examples are the presence of toxic airborne chemical substances and particulate matter, biological agents, noise, vibration, radiation, extremes of temperature and humidity, and lack of illumination.
2. *Workplace Characteristics.* Examples include flammable and other hazardous materials, unsafe walking and working surfaces, and workplace layout and design that can cause excessive strain on the worker. These environmental variables have a direct influence on the amount of danger associated with a particular task. These variables, depending on their level, may add to the danger inherent in a task.[2]
3. *Social Factors.* The personalities of co-workers and supervisors are also factors to be considered when evaluating the workplace. The higher the employee morale, the lower the potential for accidents. Another factor is the relationship of one job to another, and whether the job requires the coordinating of information, materials, and human effort.

A program to minimize occupational accidents should incorporate measures in four areas.

1. *Training.* Operators should be trained in the use of equipment and made aware of why the operation must be performed in the prescribed manner.
2. *Design.* Equipment should be evaluated to determine whether error might be inherent in the equipment design. Equipment redesign may be necessary.
3. *Human Performance.* Minimizing the stress factor on employees is an important consideration. Stress can be caused by poor environmental conditions, bad equipment design, lack of time, fatigue, and anxiety.
4. *Monitoring.* Monitoring of the working atmosphere and monitoring of personnel is also important. Under certain conditions, for example, periodic medical checks on breath, urine, or blood might be considered.

In addition to natural disasters and occupational mishaps, the process industry is subject to "accidents" such as theft, vandalism, and power failures. Very few events of these types occur. However, steps and precautions should always be taken to minimize their potential for occurrence and to cope with any damage.

Precautions against theft and/or vandalism may include:

1. Securing all major equipment, entrances, and storage with alarm systems
2. Attaching, securing, and locking anti-vandalism and antitheft devices on items not protected by alarms
3. Keeping the plant site well lit
4. Posting warning notices
5. Employing security guards

Additional general details on human error can be found in the literature.[1] Specific details are available in a classic work by Kletz, an internationally recognized authority and a pioneer in the field.[3]

Precautions against power failures may include:

1. Supplying a backup power source such as a generator
2. Providing for manual safety and operational controls in addition to electric ones
3. Developing an emergency procedure plan

6.4 TRANSPORT ACCIDENTS

This section focuses on industrial accidents that are not plant related and reviewing accidents that occur during the transport and storage of hazardous substances. Frequently, such accidents result in fires, explosions, and toxic releases. This topic will be discussed in the next chapter.

Fire is more likely than an explosion where there is a loss of containment of a flammable material from a railroad car, barge, ship tank, or from a pipeline. However, both unconfined vapor cloud explosions (UVCES) and boiling liquid-expanding vapor explosions (BLEVES) can occur as a result of transport accidents. (see Section 7.5)

In addition to fires and explosions, the loss of containment of conventional toxic substances from a tank or pipeline may give rise to a large toxic gas cloud or may pollute water supplies. There is little published information on the extent to which ultratoxic substances are transported. In view of the danger of the contamination that may arise from even a small spillage, transport of a toxic substance is usually done by special arrangement. The initiating cause of transport accidents may lie with the cargo, the operations, or the transporter.

Cargo

The cargo may catch fire, explode, or corrode the tank. The use of the same container to carry different chemicals is quite normal; it is fairly common practice with road tankers, barges, ships, and pipelines. This not only creates potential problems of incompatibility between substances carried but also means that carriers are less familiar with the substances being transported. Thus it is essential that all personnel involved in the transport of chemicals be thoroughly trained in the properties of the chemicals, their possible hazards, and the emergency procedures associated with each hazard.[5]

Operations

Accidents can occur when operations such as charging and discharging are incorrectly executed or when tanks are accidentally overfilled. The field of transportation provides many examples.

Transporter

Assessments of the comparative safety of the different modes of transport should take into account both the risk to the public and the risk to terminal operators and transport crew. The transporter may be involved in a crash or derailment; drivers may be injured or killed. Thus, the events that can give rise to hazards include container failure, loading and unloading operations, and accident impact. Hazardous materials are moved by:

1. Roadway
2. Railway
3. Waterway (barges)
4. Pipeline
5. Sea
6. Aircraft

The remainder of this section focuses on accidents related to these specific modes of transport.

Road Transport

Special measures are required for certain chemicals that are transported by road. The following classes of substances, which have been prohibited for transport by road in substantial quantities in West Germany since 1971, indicate some of the chemicals regarded as particularly hazardous, which therefore require special measures for transport.[5]

Class 2: Gases such as chlorine, hydrogen bromide, hydrogen chloride, hydrogen fluoride, hydrogen sulfide, phosgene, sulfur dioxide.

Class 3: Flammable liquids such as carbon disulfide.

Class 6: Poisons such as acetone cyanohydrin, acetonitrile, acrylonitrile, allyl alcohol, allyl chloride, aniline, epichlorohydrin, lead alkyls, organophosphorus compounds.

Class 8: Corrosives such as bromine, fluoroboric acid, hydrazine, liquid chlorides.

Most road accidents leading to the release of hazardous materials involve tankers crashing or overturning. To help prevent accidents during the road transport of hazardous substances, proper training and operation should be implemented and appropriate safety equipment procured. All vehicles should carry emergency equipment such as tool kit, emergency lighting, first aid supplies, and protective clothing and breathing equipment. There should be a fire extinguisher adequate for dealing with a fire in the cargo in addition to the cab fire extinguisher normally provided.

Rail Transport

Rail accidents entailing escapes of hazardous material happen mostly in the wake of rail crashes or derailments. Mechanical failures of rolling stock and faults in the rail track are frequent causes of these. Another hazard of railways is fires along the track, caused by engine or brake shoe sparks.

Waterway Transport (Barge Transport)

In the United States, the responsibilities of the Coast Guard extend to inland waterways. The U.S. Army Corps of Engineers is also involved through its responsibility for waterfront structures, embankments, canals, bridges, and dams.

A principal hazard of barge transport is collision. It is important to minimize the risk of this hazard, not only when barges are on the move but also during loading operations, when barges are particularly vulnerable. Another hazard of barge transport arises when barges are used to carry a variety of chemicals. While skilled personnel are generally available to assist at the loading point, the barge operator may well be on his own at the unloading point. Barge operators, therefore, should be aware of the hazards of the various substances handled and of the necessary precautions. The Coast Guard encourages barge crews to receive training as chemical tankermen.[4]

Pipeline Transport

In the United States, pipeline transport is regulated by the Department of Transportation (DOT) through the *Natural Gas Pipeline Safety Act of 1968* and related legislation for other hazardous materials. A significant fraction of hazardous materials is transported by pipeline, in particular, natural gas, LPG, ethylene, ammonia, and chlorine. Lees reports that the Chlorine Institute has estimated that 20% of the chlorine transported in 1965 and 1966 traveled by pipeline.[4]

Some causes of pipeline failures are damage by outside forces, corrosion, and construction defects and materials failure. The main cause of pipeline failure is interference, particularly by earthmoving and excavating equipment. Most pipelines, for example, can be punctured by a tooth of a mechanical excavator, as may occur during ditching and land drain work in rural areas and from work on services such as water, gas, electricity, and sewage in urban areas. It is essential, therefore, for the pipeline operator to keep a close watch on activities that may give rise to this hazard and to keep in contact with planning and utility authorities. [4]

Sea Transport

Shipping differs from the other modes of transport in that the quantities of hazardous materials generally tend to be much larger. Note that this applies to both the quantities on the vessel itself and those in storage at terminals.

Hazards to populated areas on land may arise from accidents to ships in port or in coastal waters close to land. Principal hazards associated with ships are fire or explosion on the vessel itself, or spillage of a flammable or toxic material.

Spillage in a harbor is likely to be the result of collision or of loading/unloading accidents. The spillage may give rise to a cloud of flammable or toxic material. If spillage of flammable or toxic material occurs well out to sea, the effects are limited to the ship itself or occasionally to other ships nearby. But if the spillage takes place near the coast, particularly in port, people on land may be endangered.

The explosion hazard on ships is not confined to cargoes of explosives, however. Explosions can also occur from flammable gases and liquids, particularly during transfer and cleaning operations.

A further source of explosion is the engine room. Shipboard explosions may cause casualties and damage in themselves. They may also cause damage to land-based plants and thereby initiate more severe accidents.[4]

Air Transport

The carriage of hazardous materials by air is the concern of the Restricted Articles Board of the International Air Transport Association (IATA). The IATA Restricted Articles Regulations place limitations on the types and quantity of chemicals that are carried and specify requirements for packaging and labeling.

The regulations list some 2300 substances requiring special packing and handling. These include flammable liquids and solids, combustible liquids, explosives, oxidizing materials, organic peroxides, corrosive materials, poisons, noxious, or irritating substances, and radioactive materials. The regulations also forbid the carriage of certain other materials.[4]

In general, the carriage of hazardous materials does not appear to be a significant cause of, or aggravating factor in, aircraft accidents. However, improperly packed and loaded nitric acid was declared the probable cause of a cargo jet crash at Boston in 1973, in which three crewmen died.[4]

6.5 ELECTRICAL FAILURES

All wiring and electrical equipment in chemical plants should be installed in accordance with the National Electrical Code. Electrical equipment for use in hazardous locations should be recognized by Underwriters Laboratories (or other testing organizations recognized by the authority having jurisdiction) for the conditions to be encountered.

The proper installation of electrical equipment is very important from the standpoint of safety. Electrical equipment should be grounded for protection of both personnel and equipment. Precautions should be taken to prevent flammable gases or vapors from traveling through conduits to a point of ignition. Adequate clearance or insulation should be provided between conduits and hot surfaces to prevent damage to the wiring insulation. Lighting should be adequate for the purposes of good operational practices and conformance with standards.[7]

Static electricity presents a dangerous source of ignition in hazardous chemical processing. Equipment must be properly bonded and grounded to minimize this danger. Lightning, a natural occurrence, is a hazard to electric power lines, structures and hazardous chemical storage or process units. Sharp, high points on equipment should be avoided. Equipment should be shielded from lightning by protective grounded wires, rods, or masts.

From a basic physics standpoint, a circuit is a complete loop. Electric current can flow only if it returns to its source, i.e., completes the circuit. The path through which the current returns to its source is called the "return" or ground. The reason for the term "ground" is that the earth is literally used to provide the return path no matter what distance separates the equipment from the power source. Ground connections can be made to a cold water system as its components provide a reliable, low-resistance path for contact with the earth.

Shock occurs when the body becomes a part of the circuit, i.e. the current enters the body at one point and leaves at another.[8] Shock normally occurs in one of three ways. The person must come in contact with both wires of the electrical circuit; with one wire of an energized circuit and the ground; or with a metallic part that has become "hot" by being in contact with an energized wire while the person is also in contact with the ground.

The severity of the shock received when a person becomes a part of an electrical circuit is affected by three primary factors: the amount of current flowing through the body (measured in amperes); the path of the current through the body; and, the length of time the body is in the circuit. Other factors that may affect the severity of shock are the frequency of the circuit, the phase of the heart cycle when shock occurs, and the general health of the person before the shock. The effects from electric shock depend upon the type of circuit, its voltage, resistance, amperage, pathway through the body, and duration of the contact. Effects can range from a barely perceptible tingle to immediate cardiac arrest.

Electrical accidents appear to be caused by any one or any combination of three possible factors: unsafe equipment, unsafe installation, or both; workplaces made unsafe by the environment; and unsafe work practices. Possible ways to protect people from the hazards caused by electricity include insulation, guarding, grounding, mechanical devices, and safe work practices. These are detailed below.

1. *Insulation.* One way to safeguard individuals from electrically energized wires and parts is through insulation. An insulator is any material with high resistance to electrical current. Insulators, such as glass, mica, rubber, and plastic, are placed on conductors to prevent shock, fires, and short circuits. When preparing to work with electrical equipment, the insulation should be checked before making a connection to a power source to be sure there are no exposed wires. The insulation of flexible cords, such as extension cords, is particularly vulnerable to damage.

2. *Guarding.* Live parts of electric equipment must be guarded to prevent accidental contact. Live parts may be guarded by:

    - Locating them in a room, vault, or similar enclosure accessible only to qualified individuals
    - Using permanent, substantial partitions or screens to exclude unqualified individuals
    - Locating them on a suitable balcony, gallery, or platform elevated and arranged to exclude unqualified individuals
    - Elevating them 8 feet or more above the floor

    Entrances to rooms and other guarded locations containing exposed live parts must be locked and marked with conspicuous warning signs forbidding unqualified persons to enter.

3. *Grounding.* Grounding is another way to provide protection from electric shock; however, it is normally a secondary protective measure. By grounding a tool or electrical system, a low-resistance path to the earth through a ground connection or connections is intentionally created. When properly done, this path offers sufficiently low resistance and has sufficient current-carrying capacity to prevent the buildup of voltages that may result

in a personnel hazard. This precaution does not guarantee that no one will receive a shock, be injured, or be killed. It does, however, substantially reduced the possibilities of such accidents, especially when used in combination with other safety measures.

4. *Mechanical Devices.* A ground-fault circuit interrupter (GFCI) is a fast-acting circuit breaker that senses small imbalances in the circuit caused by leakages to ground.[9] When the leakage exceeds 5 ±1 milliamperes, the GFCI interrupts the circuit quickly enough to prevent electrocution, in some cases in as little as 1/40 second. This protection is required in addition to, not as a substitute for, normal grounding requirements. Circuits for systems involving or using wet processes should be equipped with GFCI's. That practice is also required for extension cords even if plugged into permanent wiring.

5. *Safe Work Practices.* Individuals working with electrical equipment must use safe work practices. All must know how to shut off power to a piece of equipment by using either the power switch on the equipment or the circuit breaker. Circuit breakers or switch boxes must be secure but readily accessible. Each circuit must be clearly labeled as to the equipment or area that it activates. Individuals need to know how to use the master electrical switch under emergency conditions. (The circuits should be checked to ensure that room lighting is on a separate box.) Individuals should be aware of the "left-hand rule." Anytime they prepare to turn the master switch back on, especially after changing a fuse, they should stand to the side, face the wall instead of the box, and use their left hand to push the switch back on. In this way, if the box explodes when power is restored, they are less likely to suffer severe burns to the face or even death.

The accidental or unexpected sudden starting of electrical equipment can cause severe injury or death. Before ANY inspections or repairs are made (even on the so-called low-voltage circuits) the current should be turned off at the switch box and the switch padlocked in the OFF position. At the same time, the switch or controls of the machine or other equipment being locked out of service should be securely tagged to show which equipment or circuits are being worked on. Lockouts and tagouts should be removed only by the individual(s) who installed them.

All electrical components and equipment must be inspected before use. Frayed or worn wiring and extension cords; cracked plugs, switches, or receptacle and switch-plate covers missing grounds; and, two-wire to three-wire plug adapters with missing "pigtails" or ground prong all render the device or system unfit for use until repairs have been made. Only qualified personnel should make repairs.

Individuals should immediately report any unsafe electrical conditions to prevent an accidental shock. Damaged or unauthorized extension cords must be taken out of service immediately. Individuals must shut off power at the breaker before disconnecting a damaged cord. To ensure that someone else does not use equipment with a damaged power cord, the plug can be clipped off with a pair of wire cutters. Because most damage occurs at the plug, the electrician will repair it anyway. As noted above, all electrical work must be done by experienced personnel. Equipment to be repaired should be tagged out and repairs verified before reuse.

All vessels containing flammable and/or explosive materials (liquids and dusts) and all transfer devices (pumps, conveyors, etc.) must be grounded to minimize the possibility of disaster as a result of stray currents, differences in electrical potential between different parts of the system, and/or static electricity. Grounds consist of effective, large-diameter conductors between the vessel or apparatus and the earth. A multi-strand braided conductor or a single large-diameter conductor may be used. Either should have sufficiently low resistance so that stray currents will more readily flow through the ground loop than through the equipment. Since cold water systems are buried in the soil, cold water pipes make excellent grounds.

When flammable and/or explosive materials are transferred from a fixed storage container to another container for transportation, an electrical bond must be established between the two pieces of equipment. The reverse is also true, e.g., in pouring or pumping a flammable solvent from a 5-gallon can into a feed tank. Bonds or "bond-clips" are made from heavy-gauge conductors with heavy-duty alligator, bear-paw, or post connectors at each end. Metal-to-metal contact must be made at each end of the bond, even if that necessitates scraping away some paint on the experimental apparatus. No material transfer may be made unless the grounds and bonds are in place.

Finally, when using electrically powered or controlled machines, the equipment as well as the electrical system itself must be properly grounded. Replacing frayed, exposed, or old wiring will also help protect the operator and others from electrical shocks or electrocution. Just as all power sources for machinery are potential sources of danger (and must be checked before use), high pressure systems also need careful inspection and maintenance to prevent possible failure from pulsation, vibration, or leaks. Such a failure could cause explosions or flying objects.

6.6 NUCLEAR ACCIDENTS

Nuclear or radioactive materials are used in many applications throughout today's society. Radioactive materials are used to generate power in nuclear power stations, and are used to treat patients in hospitals. The generators of radioactive waste in today's society are primarily the federal government, electrical utilities, private industry, hospitals and universities. Although each of these generators uses radioactive materials, the waste that is generated by each of them may be very different and must be handled accordingly. Any material that contains radioactive isotopes in measurable quantities is considered nuclear or radioactive waste. For the purposes of this chapter, the terms nuclear waste and radioactive waste will be considered synonymous.

Waste management is a field that involves the reduction, stabilization, and ultimate disposal of waste. Waste reduction is the practice of minimizing the amount of material that requires disposal. Some of the common ways in which waste reduction is accomplished are incineration, compaction, and de-watering. The object of waste disposal is to isolate the material from the biosphere, and in the case of radioactive waste, allow it time to decay to sufficiently safe levels.

The federal government has mandated that individual states or interstate compacts, which are formed and dissolved by Congress, be responsible for the disposal of the LLW (low level waste) generated within their boundaries. Originally, these states were to bring the disposal capacity online by 1993. Although access to the few remaining facilities is drawing to an end, none of the states or compacts have a facility available to accept waste. Some states are making progress, but none of the proposed facilities are currently in the construction phase.

Both the high level waste (HLW) and the transuranic (TRU) waste programs have sites defined for their respective facilities at Yucca Mountain, and at the Waste Isolation Pilot Plant (WIPP) in Carlsbad, New Mexico. The WIPP facility is a Department of Energy (DOE) research and development facility that has been designed to accept 6 million ft^3 of remote-handled TRU waste. The facility will accept defense-generated waste and place it into a retrievable geologic repository. A geologic repository is in this instance the salt formations located near Carlsbad. The facility has a design-based lifetime of twenty-five years.

As described in Chapter 1, the three largest radiological accidents of the last twenty years are the explosion at Chernobyl, the partial core meltdown at Three Mile Island Unit #2, and the mishandling of a radioactive source in Brazil. The least publicized, but perhaps the most appropriate of these accidents, with respect to waste management, was the situation in Brazil.

Although much still remains to be learned about the interaction between ionizing radiation and living matter, more is known about the mechanism of radiation damage on the molecular, cellular, and organ system level than most other environmental hazards. The radioactive materials warning sign is shown in Figure 6.6.1. A vast amount of quantitative dose-response data has been accumulated throughout years of studying the different applications of radionuclides. The following paragraphs will provide a brief description of the different types of ionizing radiation and the effects that may occur upon overexposure to radioactive materials.

Several different mechanisms, most importantly alpha particle, beta particle, and gamma ray emissions accomplish radioactive transformations. Each of these mechanisms is a spontaneous nuclear transformation. The result of these transformations is the formation of different stable elements. The kind of transformation that will take place for any given radioactive element is a function of the type of nuclear instability as well as the mass/energy relationship. The nuclear instability is dependent on the ratio of neutrons to protons; a different type of decay will occur to allow for a more stable daughter product. The mass/energy relationship states that for any radioactive transformation, the laws of conservation of mass and conservation of energy must be followed.

Figure 6.6.1 Radioactive materials warning sign.

An alpha particle is an energetic helium nucleus. The alpha particle is released from a radioactive element with a neutron to proton ratio that is too low. The helium nucleus consists of two protons and two neutrons. The alpha particle differs from a helium atom in that it is emitted without any electrons. The resulting daughter product from this type of transformation has an atomic number that is two less than its parent and an atomic mass number that is four less. Below is an example of alpha decay using polonium (Po); polonium has an atomic mass number of 210 (protons and neutrons) and atomic number of 84.

$$^{210}_{84}\text{Po} \rightarrow {}^{4}_{2}\text{He} + {}^{206}_{82}\text{Pb} \tag{6.6.1}$$

The terms He and Pb represent helium and lead, respectively.

The above is a useful example because the lead daughter product is stable and will not decay further. The neutron to proton ratio change from 1.5 to 1.51 is just enough to result in a stable element. Alpha particles are known as having a high LET (linear energy transfer). The alphas will only travel a short distance while releasing energy. A piece of paper or the top layer of skin will stop an alpha particle. Although, alpha particles are not external hazards, they can be extremely hazardous if inhaled or ingested.

Beta particle emission occurs when an ordinary electron is ejected from the nucleus of an atom.

$$^{1}_{0}\text{n} \rightarrow {}^{1}_{1}\text{H} + {}^{0}_{(-1)}\text{e} \tag{6.6.2}$$

Similar to beta decay is positron emission, where the parent emits a positively charged electron. Positron emission is commonly called betapositive decay. This decay scheme occurs when the neutron to proton ratio is too low and alpha emission is not energetically possible. The positively charged electron, or positron, will travel at high speeds until it interacts with an electron. Upon contact, each of the particles will disappear and two gamma rays will result. When two gamma rays are formed in this manner it is called annihilation radiation.

Unlike alpha and beta radiation, gamma radiation is an electromagnetic wave with a specified range of wavelengths. Gamma rays cannot be completely shielded against, but can only be reduced in intensity with increased shielding. Gamma rays typically interact with matter through the photoelectric effect, Compton scattering, pair production, or direct interactions with the nucleus.

The response of humans to varying doses of radiation is a field that has been widely studied. The observed radiation effects can be categorized as stochastic or nonstochastic effects, depending upon the dose received and the time period over which such dose was received. Contrary to most biological effects, effects from radiation usually fall under the category of stochastic effects. The nonstochastic effects can be noted as having three qualities: a minimum dose or threshold dose must be received before the particular effect is observed; the magnitude of the effect increases as the size of the dose increases; and a clear, casual relationship can be determined between the dose and the subsequent effects.

Stochastic effects, on the other hand, occur by chance. Stochastic effects will be present in a fraction of the exposed population as well as in a fraction of the unexposed population. Therefore, stochastic effects are not unequivocally related to a noxious agent as the above example implies. Stochastic effects have no threshold. Any exposure will increase the risk of an effect, but will not wholly determine if any effect will arise. Cancer and genetic effects are the two most common effects linked with exposure to radiation. Cancer can be caused by the damaging of a somatic (non-reproductive) cell, while genetic effects are caused when damage occurs to a germ cell that results in a pregnancy.

6.7 NATURAL DISASTERS

An event may be labeled an accident through the assessment of the following factors:

1. Degree of expectedness,
2. Degree of misjudgment
3. Degree of intention
4. Degree of warning
5. Degree of negligence
6. Degree of avoidability

Items 3-6 above do not apply to natural accidents. The natural accidents are often termed "acts of God," and can include:

1. Floods
2. Lightning

3.   Windstorms
4.   Landslides
5.   Earthquakes
6.   Tornadoes
7.   Hurricanes
8.   Volcanic eruptions
9.   Avalanches

Although accidents of these types occur infrequently, they may present a greater potential for loss than fires, explosions, or spills. Since natural disasters are difficult to predict and prevent, one is obliged to rely more heavily on precautions designed to minimize the impact of an occurrence of a natural disaster, such as emergency planning.

    The objective of emergency planning is to reduce the probability of serious loss due to a particular hazardous accident. The probability of an occurrence of a hazardous accident is first evaluated. It is then it is assumed that, if the accident occurs, the worst consequences will follow (the so-called worst-case scenario). Procedures for handling a particular accident are then developed and practiced, both to minimize the exposure of personnel and to prevent escalation of the original incident.

    Plant personnel should always be aware of the special climactic, geological, and topographic conditions that exist in the area, since specific accident control programs will vary according to these conditions. In addition, plant personnel should be cognizant of accidents that can follow a natural disaster. For example, floods can cause pipelines to fail, lightning can start storage tank fires, hurricanes can cause power outages, and prevailing winds can cause the rapid spread of fires and toxic releases.

    Other natural disasters and "external events" are listed below in Table 6.7.1., along with an abbreviated comment.

6.8 ILLUSTRATIVE EXAMPLES

Example 6.1

Explain why a plant accident is more likely to happen during startup of a new plant or a retro-fit process. Refer to Chapter 20 and careful review the presentation or the "bathtub" curve that is represented by the Weibull distribution.

Solution 6.1

A plant accident is more likely to happen during the startup of a new plant or a retro fit process because "new equipment usually experiences a high failure rate during the early or break-in period. The overall process is best described by the Weibull distribution. The Weibull Distribution or the "bathtub" curve is a three-

TABLE 6.7.1. Examples of Natural Disasters and "External Events"

Disaster/Event	Comments
Aircraft Impact	Sites less than 3 miles from airport have higher frequencies
Drought	May impact the availability of cooling water for pant site
Earthquake (listed above)	Damage to foundations of structures and equipment
External Flooding	May impact storm water drainage impacts
Extreme Winds	Can create large numbers of "missiles"
Extreme Summer Temperature	Will impact on vapor pressure of chemicals in storage systems
Fog	May increase frequency of accidents
Forest Fire	Review location of plant relative to large areas of standing trees
Frost	Frost heave may damage foundations of plant structures
Ice Cover	Ice blockage of rivers, loss of cooling, and mechanical damage due to falling ice are possible
Landslide	Can be excluded for most sites in the United States
Lighting (listed above)	Should be considered during design since computer control systems are vulnerable; may also damage plant power grid
Low Lake or River Level	May halt raw material and product shipping; alternative truck or rail shipping may be used
Extreme Winter Temperature	Thermal stresses and embrittlement may occur in storage tanks
Meteorite Impact	All sites have approximately same frequency of occurrence
Missile Impact	Shrapnel and large pieces of pressure vessels are possible from explosions; rocks, bolts, and lumber may become missiles as a result of extreme winds
Intense precipitation	Flooding affects need to be considered
Sabotage	Disgruntled employee may deliberately damage or destroy vital plant systems
Sandstorm	May damage equipment and block air intakes
Ship Wreck	May halt raw material and production shipping; alternative truck or rail shipping may be used
Snow	Review design load of roofs; may increase frequency of in-plant accidents
Terrorist Attack	High explosives and weapons may be used against selected targets; essential personnel may be held for ransom or killed

TABLE 6.7.1. (Cont'd)

Disaster/Event	Comments
Transportation Accidents	Site specific; accident on major highway may cause evacuation of plant site
Tsunami	Site specific; series of damaging waves in coastal areas can cause flooding
Volcanic Activity (listed above)	May cause extensive downstream flooding; volcanic ash may damage equipment and plug air intakes
War	Damage caused by high intensity combat will probably be greater that that caused by worst credible case from plant site

stage curve. The first stage is the aforementioned break-in stage with a declining failure rate, a useful life stage characterized by a near constant failure rate, and a wear-out period characterized by an increasing failure rate.

Example 6.2

Describe radioactive transformations as they apply to alpha particles.

Solution 6.2

Alpha particle radioactive transformations are best described by the following example:

$$^{210}_{84}\text{Po} \rightarrow {}^{4}_{2}\text{He} + {}^{206}_{82}\text{Pb}$$

The above is a useful example because the lead daughter product is stable and will not decay further. The neutron to proton ratio change from 1.5 to 1.51 is just enough to result in a stable element. Alpha particles are known as having a high LET (linear energy transfer). The alphas will only travel a short distance while releasing energy. A piece of paper or the top layer of skin will stop an alpha particle. Although alpha particles are not external hazards, they can be extremely hazardous if inhaled or ingested.

Example 6.3

Hot equipment surfaces can lead to serious health related problems. Describe a procedure that could be used to reduced this health hazard.

Solution 6.3

Hot surfaces can produce serious burns to personnel in chemical plant operations. Surfaces that are within reach of personnel should be insulated or the areas shielded from accidental contact. Many critical areas in chemical plant operations should be restricted to all but experienced workers.

Example 6.4

Briefly describe factors that should be considered in an accident arising from an airplane crash.

Solution 6.4

The incidence of aircraft impacts may be significantly higher in certain areas (e.g., in the vicinity or airports). The aircraft crash hazard is site specific and the failure is strongly dependent on the kinetic energy of the aircraft. Two types of data are needed to analyze for aircraft impact: the aircraft crash rate in the site vicinity (per unit area per year) and the effective target area of the vulnerable item. Crash rates for different categories of aircraft can be obtained from state and national authorities (e.g., FAA). The proximity of the site to airfields must be taken into account because crashes are much more frequent within a radius of approximately 3 miles.

Example 6.5

List six safeguards that can be used to protect against mechanical hazards.

Solution 6.5

1. Use personal protective equipment-wear rubber insulated gloves if you handle electrical equipment. You may also need insulated clothing, such as rubber-soled shoes or boots-especially if it's wet. Check with your supervisor about what PPE you need for your specific job. Never wear metal jewelry that could turn you into a conductor.
2. Turn off power to electrical equipment before tests and repairs; then lock and tag it out so it does not turn on by accident. Follow your company's lockout/tagout procedures.
3. Inspect tools regularly. If a tool shocks, smokes, smells, or sparks, do not use it.
4. Keep your distance from power lines-notify the company before working near power lines. Never work closer than 10 feet on a ladder near power lines. Use insulated equipment (nothing metal). Never touch fallen power lines.

5. Take special precautions near flammables–make sure your electrical equipment is identified as safe for use around these materials. Also be sure that it doesn't spark or get hot enough to ignite flammable substances.

Example 6.6

List several procedures that can be instituted at a plant to reduce the possibility of a personal electrical accident.

Solution 6.6

1. Individuals should immediately report any unsafe electrical conditions to prevent an accidental shock.
2. Damaged or unauthorized extension cords must be taken out of service immediately.
3. Individuals must shut off power at the breaker before disconnecting a damaged cord.
4. To ensure that someone else does not use equipment with a damaged power cord, the plug can be clipped off with a pair of wire cutters. Because most damage occurs at the plug, the electrician will repair it anyway.
5. Experienced personnel must do all electrical work.
6. Equipment to be repaired should be tagged out and repairs verified before reuse.

6.9 SUMMARY

1. Classes of accidents that can result in financial and personnel losses include more than just natural disasters such as earthquakes and tornadoes, or occupational personnel mishaps such as tripping and slipping.
2. The academic training of engineers with respect to plant equipment has traditionally focused on design and predicting performance. Little to no effort has been expended on attempting to answer the question: What can go wrong?   This question requires answers in today's high-tech environment. Failure to properly address and answer this question can result/lead to an accident!
3. Occupational mishaps are usually a function of three aspects of the work-place: the human element, task variables (i.e., the job itself), and the environmental element.
4. Fire is more likely than explosion where there is a loss of containment of a flammable material from a railroad car, barge, ship tank, or from a pipeline.
5. The proper installation of electrical equipment is very important from the standpoint of safety. Electrical equipment should be grounded for protection of both personnel and equipment.
6. The response of humans to varying doses of radiation is a field that has been widely studied. The observed radiation effects can be categorized as

stochastic or nonstochastic effects, depending upon the dose received and the time period over which such dose was received. Contrary to most biological effects, effects from radiation usually fall under the category of stochastic effects. The nonstochastic effects can be noted as having three qualities: a minimum dose or threshold dose must be received before the particular effect is observed; the magnitude of the effect increases as the size of the dose increases; and a clear, casual relationship can be determined between the dose and the subsequent effects.

7. The objective of emergency planning is to reduce the probability of serious loss due to a particular hazardous accident. The probability of an occurrence of a hazardous accident is first evaluated; then it is assumed that, if the accident occurs, the worst consequences will follow (the so-called worst-case scenario). Procedures for handling a particular accident are then developed and practiced, both to minimize the exposure of personnel and to prevent escalation of the original incident.

PROBLEMS

1. Why do plant equipment fail?
2. Why are naturally occurring accidents often referred to as "acts of god"?
3. Why are engineers duty-bound to report what they view as potential accidents?
4. With respect to electrical problems, why is the training of personnel important?
5. Describe the basic differences between electrical grounding and electrical bonding.
6. Briefly describe factors that should be considered in plant accidents resulting from an earthquake.
7. Describe radioactive waste treatment and disposal.

REFERENCES

1. R. Perry and D. Green, "Perry's Chemical Engineering Handbook," McGraw-Hill, New York, 1997.
2. T. A. Kletz, "What Went Wrong? Case Histories of Process Plant Disasters," Gulf Publishing, Houston, TX, 1985.
3. T. A. Kletz, "An Engineer, View of Human Error," 2nd ed, Institution of Chemical Engineers, Rugby, Warwickshire, UK, 1991.
4. Adopted from: R. Perry and D. Green, "Perry's Chemical Engineering Handbook," McGraw-Hill, New York, 1997.
5. U.S. Department of Labor: Hand and Power Tools, Publ. No. 3080 (revised), Washington, DC (1980).
6. Adopted from; R. Bertea, "Incorporation of Occupational Safety and Health Into Unit Operations Laboratory Courses," NIOSH, Cincinnati, 1991

7. Hazard Survey of the Chemical and Allied Industries, American Insurance Association, New York City, 1975.
8. Hill, V.H.:   Control of Noise Exposure, Ch. 37 in The Industrial Environment-Its Evaluation and Control, U.S. Dept. of Health & Human Services, NIOSH Publ. No.74-117, U.S. Govt. Printing Office, Washington DC.
9. U.S. Department of Labor: Ground-Fault Protection on Construction Sites, Publ. No. OSHA 3007 (revised), Washington, DC (1987).

7

Fires, Explosions, Toxic Emissions, and Hazardous Spills

7.1 INTRODUCTION

Some of the preceding chapters have dealt with the history and legislation of emergency and accidents; this chapter addresses specifically the fundamentals of plant fires, explosions, and certain other plant- and non-plant-related accidents.

The remainder of this chapter provides information on relative physical properties of materials (flash points, upper and lower explosive limits, threshold limit values, etc.) and methods to calculate the conditions that approach or are conducive to hazardous levels. Fire hazards in industrial plants are covered in Sections 7.2 and 7.3, and Sections 7.4 and 7.5 focus on accidental explosions. Sections 7.6 and 7.7 address toxic emissions and hazardous spills respectively.; these latter types of accident frequently result in fires and explosions; they can cause deaths, serious injuries and financial losses.

7.2 FIRE FUNDAMENTALS

By definition, fire is the combining of oxygen and fuel in proper proportions and at the proper temperature to sustain combustion. Combustion is the process in which a substance chemically reacts with fuel and oxygen at a rapid rate, producing light and heat. To produce combustion, four conditions must coexist: (1) presence of fuel, (2) presence of oxygen, (3) heat, and (4) mixing. If one of these conditions is missing, combustion may not take place, and if one condition is altered during combustion, the fire may become extinguished.

The authors gratefully acknowledge the assistance of Rebecca Hobden in researching, reviewing, and editing this chapter.

Conditions of Combustion

To better understand the principles underlying fires, certain fire characteristics must be defined.

1. *Fuel.* Wood, paper, coal, and gas are just a few of the products commonly thought of as fuels. However, from a chemical standpoint, the common fuel elements are carbon (C) and hydrogen (H). Carbon is found in coal, coke, lignite, and peat. Other carbon fuels include fat, petroleum, and natural gas. Hydrogen is commonly found in conjunction with these carbon compounds.

2. *Oxygen.* Quantities of air must be available for complete combustion to occur; otherwise a fire will smolder. The amount of oxygen required to sustain a fire may depend on the form and characteristics of the substance that burns. A liquid or solid as it is heated evolves vapor. As the concentration of vapor increases, it forms a flammable mixture with the oxygen of the air. As a result, it may not be necessary to remove all the oxygen to extinguish such a fire. Liquid fires can generally be put out by reducing the oxygen concentration below 12 to 16% (by volume). Solid fires may require a greater reduction of oxygen concentration-below 5% for surface smoldering and as low as 2% for deep-seated smoldering.[1]

3. *Heat.* Fuel will not burn until it reaches a certain temperature, which depends on the type of fuel and on factors such as the exposed surface, the vapor present, and the presence or absence of other fuels.

4. *Mixing.* The chemical union of fuel and oxygen requires the proper mixing of these components. For example, if the ratio of oxygen to fuel is either too high or too low, a fire will be extinguished. The proper oxygen-to-fuel mix must be maintained to sustain a fire.

Transmission of Heat

The growth and spread of fires occurs through heat transfer or the migration of burning materials. There are three main modes of heat transfer: conduction, convection, and radiation.

Conduction is important particularly in allowing heat to pass through a solid barrier, causing ignition of material on the other side. However, most of the heat transfer from fires is by convection and radiation. *Convection* is the process by which heat is transferred by the motion of the heated matter, such as the flow of water through a hot pipe or a current of hot air. In *radiation*, heat travels in the same fashion as light is propagated-as a wave moving in all directions in straight lines until it is absorbed or reflected by another object.[2]

Flammability Characteristics

Determining the fire potential of a fuel is best accomplished through an evaluation of the fuel's flammability characteristics. No single factor, however, defines a substance's flammability. When a flammability comparison is to be made between different substances, the following factors should be considered:

1. Flammability limits
2. Flash point
3. Burning velocity
4. Ignition energy
5. Autoignition temperature

Each of these factors is discussed below.

Flammability limits (or explosion limits) for a flammable gas define the concentration range of a gas-air mixture within which an ignition source can start a self-propagating reaction. The minimum and maximum fuel concentrations in air that will produce a self-sustaining reaction under given conditions are called the *lower flammability limit* (LFL) and the *upper flammability limit* (UFL). (The abbreviations LEL and UEL, for *lower* and *upper explosivity limits*, are sometimes used.) The flammability limits are functions of

1. Ignition energy
2. Ignition pressure
3. Ignition temperature of the mixture or substance
4. Inert gas concentration
5. Relative humidity of the mixture or substance

The range of the flammability limits becomes greater when the ignition energy is higher, moving the UFL to a higher concentration. The flammability limits also increase when the initial pressure and/or initial temperature at which the ignition source is activated increases. In addition, the increase in pressure and temperature will increase the rate at which a flame propagates through a gas.[3] Flammability limits can be significantly altered by changing the oxygen content or adding an inert gas to the gas-air mixture. The heat capacity of the dilutent, or inert gas, plays a role in flammability because the dilutent will act as a heat sink. Thus, carbon dioxide is a better dilutent than nitrogen because it has a higher heat capacity. The flammability limits also increase for drier mixtures. Table 7.2.1 shows the flammability limits of some flammable gases. The flammability of a gas mixture can be calculated by using Le Chatelier's law given the flammability of the gas components.

$$LFL(mix) = \frac{1}{f_1/LFL_1 + f_2/LFL_2 + ... + f_n/LFL_n} \qquad (7.2.1)$$

$$UFL(mix) = \frac{1}{f_1/UFL_1 + f_2/UFL_2 + ... + f_n/UFL_n} \qquad (7.2.2)$$

Where $f_1, f_2, \ldots, f_n$ = volume or mole fraction of each of the n components

LFL(mix), UFL(mix) = mixture lower and upper flammability limits in volume or mole fraction

LFL_1, LFL_2, ..., LFL_n = component lower flammability limits in volume or mole fraction

UFL_1, UFL_2, ..., UFL_n = component upper flammability limits in volume or mole fraction

If data are not available for a particular gas mixture, it is possible to estimate a flammability limit by taking data for a similar material and applying Eq. (7.2.3)

$$LFL_A = (M_B/M_A)\, LFL_B \qquad (7.2.3)$$

where M_A and M_B are the relevant molecular weights of components A and B, respectively.

The *flash point* of a flammable liquid is defined as the temperature at which the vapor pressure of the liquid is the same as the vapor pressure corresponding to the lower flammability limit concentration. The three major methods of measuring the flash point are

1. The Cleveland open cup method
2. The Penskey-Martens closed cup tester
3. The tag closed cup method

These are empirical tests that measure the lowest temperature at which application of a test flame causes the vapor overlying the sample to ignite. Table 7.2.2 shows the flash points of selected substances.

Since liquids are condensed and fuel vapor (not the liquid fuel) is the ignition participant, the ignition concepts developed for gases often apply to liquids as well. In general, liquid reactants are inclined to be more sluggish and to contain longer time lags than gaseous reactions. Time is required for the liquid to vaporize so that it can mix with air and become ignited.[5]

The physical properties of a flammable solid, such as hardness, texture, waxiness, particle size, melting point, plastic flow, thermal conductivity, and heat capacity, impart a wide range of characteristics to the flammability of solids. A solid ignites by first melting and then producing sufficient vapor, which in turn mixes with air to form a flammable composition.

TABLE 7.2.1 Flammability Limits of Flammable Compounds Under Normal Pressure, Room Temperature

Compound	Limits of Flammability	
	Lower (% v/v)	Upper (% v/v)
Acetone	2.6	13.0
Acetylene	2.5	100.0
Ammonia	15.0	28.0
Amylene	1.8	8.7
Benzene	1.4	8.0
n-Butane	1.8	8.4
i-Butane	1.8	8.4
1-Butene	2.0	10.0
2-Butene	1.7	9.7
Carbon disulfide	1.3	50.0
Carbon monoxide	12.5	74.0
Cyclohexane	1.3	7.8
Decane	0.8	5.4
Ethane	3.0	12.4
Ethyl alcohol	3.3	19.0
Ethylene	2.7	36.0
Ethylene dichloride	6.2	15.9
Ethylene oxide	3.0	100.0
Heptane	1.2	6.7
Hexane	1.4	7.4
Hydrogen	4.0	75.0
Kerosene	0.7	5.0
Methane	5.0	15.0
Methanol	6.4	37.0
Methyl ethyl ketone	1.7	11.4
Naphthalene	0.9	7.8
n-Pentane	1.8	7.8
Petroleum ether	1.1	5.9
Propane	2.1	9.5
Propylene	2.4	11.0
Isopropyl alcohol	2.0	12.0
n-Propyl alcohol	2.1	13.7
Styrene	1.1	6.1
Toluene	1.3	7.0
Turpentine	0.8	
o-Xylene	1.0	6.0
Vinyl Chloride	4.0	22.0
2,2-Dimethylpropane	1.3	7.5
2,3-Dimethylpentane	1.1	6.8

[a] From *Manual of Industrial Hazard Assessment Techniques.*[4]

TABLE 7.2.2 Flash Points of Selected Substances in Air at Atmospheric Pressure and Room Temperature[a]

Compound	Flash Point (°C)	
	Closed Cup	Open Cup
Acetone	-18	-9
Benzene	-11	
n-Butane	-60	
Carbon disulfide	-30	
Cyclohexane	-20	
Ethane	-135	
Ethylene	-121	
Ethylene dichloride	13	18
Ethylene oxide		-20
Propane	-104	
Propylene	-108	
Styrene	32	38
Toluene	4	7
Vinyl chloride		-78

[a] From Lees.[3]

When ignited, a flammable gas mixture of fixed pressure, temperature, and composition will propagate a combustion wave at a constant rate. The propagation mechanism is quite complex, but the interaction of these mechanisms is indicated by an observable burning velocity. *Burning velocity* is the velocity of the flame relative to the motion of the unburned gas or gas mixture. This measurable quantity is an excellent index to the energetic state of the reaction. The burning velocity is a function of gas mixture composition, pressure, and temperature. The maximum velocity is observed at the stoichiometric concentration of the mixture. The effect of pressure on the burning velocity is different for different pressure ranges. For example, the stoichiometric burning velocity of a methane-air mixture decreases with increasing pressure in the pressure range of 0.5-20 atm, but it increases in the pressure range of 0.2-0.5 atm. The effect of temperature can be expressed by [3]

$$S_b = A + BT^n \tag{7.2.4}$$

where S_b = burning velocity
  T = absolute temperature
  A, B = constants
  n = index (e.g., $n = 2$ for paraffinic hydrocarbons)

The maximum fundamental burning velocities of selected substances are given in Table 7.2.3.

TABLE 7.2.3   Maximum Fundamental Burning Velocities of Selected Substances[a]

Compound	Maximum Fundamental Burning Velocity S_b[b]	
	cm/s	ft/s
In Air		
Methane	36.4[c]	1.2
n-Butane	45.0[c]	1.5
n-Hexane	38.5	1.3
Cyclohexane	20	
Ethane	40.1	
Ethylene	68.8[c]	2.3
Town gas[d]		3.7
Acetylene	173	5.8
Propane	45.0[c]	1.5
Hydrogen	320	11
Benzene	40.7[c]	
In Oxygen		
Methane	393[e]	
Propane	390	
Ethylene	550	
Acetylene	1140	
Hydrogen	1175	

[a] From Lees.[3]
[b] The values given in the first column are those of Flock (1955) and in the second column those of the HSE (1965 HSW). The values quoted from Flock are for initial pressure (atmospheric) and initial temperature (room temperature) and in most cases dry gas and generally are selected from several values listed.
[c] Some higher values are also listed by Flock.
[d] For town gas containing 63% H_2.
[e] For stoichiometric mixture.

A flammable gas-air mixture can be ignited by a local source of ignition (flame, spark, hot gas, compression, shock wave, adiabatic heating, etc.), provided the local source possesses the *minimum ignition energy*-that is, the ignition energy required to raise the temperature of the gas-air mixture above the threshold temperature and to initiate a reaction. Ignition energy is a function of a flammable mixture's composition. The minimum ignition energy is usually measured at the stoichiometric composition of the mixture. Table 7.2.4 provides minimum ignition energies for selected materials in air.

The *maximum spontaneous ignition temperature* (SIT) or *autoignition temperature* (AIT) is defined as the minimum temperature at which combustion occurs in a bulk gas mixture when the temperature of a flammable gas-air mixture is raised in a uniformly heated apparatus. The autoignition temperature represents a threshold below which chemicals and combustibles can be handled safely. The AITs of selected substances are given in Table 7.2.5. The AIT is strongly dependent on the nature of hot surfaces. When the surfaces are contaminated by dust, the autoignition temperature may be reduced by as much as 100-200°C. When the temperature of a flammable mixture is raised to or above the autoignition temperature, ignition is not instantaneous. Most notably in liquids, there is a finite delay before ignition takes place, that is, a lapse between the time a flammable mixture reaches its flame temperature and the first appearance of a flame. Equation (7.2.5) was developed by Semenov[6] to correlate the time delay and ignition temperature. The effect of temperature on the AIT, plus the LFL and UFL, is provided in Figure 7.2.1.

$$\ln t = k_1 (E/T) + k_2 \qquad\qquad (7.2.5)$$

where E   = apparent activation energy
     T   = absolute temperature
    k_1, k_2 = constants
    t   = time delay before ignition

The time delay decreases as the ignition temperature increases. This means that when a flammable fluid in a flow system comes in contact with a hot surface for a time (shorter than the time delay), ignition may not occur.

An important factor in assessing the causes and effects of fires is the behavior of a fire's flame. Knowledge of a flame's spreading rate and heat intensity can reduce fire hazard potentials and fire damage. The classifications of flame behavior are :

1. Orifice flames
2. Pool flames
3. Jet flames
4. Fireball flames
5. Flash fire flames

Orifice or Pipe Flames

Orifice flames can be characterized as either premixed or diffusion flames. In a premixed flame, the air for combustion is already mixed with the fuel gas before it leaves the orifice or pipe. In a diffusion flame, fuel exiting the orifice is pure and the air needed for combustion diffuses into the fuel gas from the surroundings. Orifice flames can also be characterized by the *flame Reynolds number*. The flame length of a diffusion flame can be calculated by Jost's equation.[7]

$$L = d^2 U_f / 4D \qquad\qquad (7.2.6)$$

TABLE 7.2.4 Minimum Ignition Energies for Selected Substances in Air[a]

Compound	Minimum Ignition Energy (mJ)
Ammonia	up to 100
Acetylene	0.02
Benzene	0.22
n-Butane	0.25
Carbon disulfide	0.01–0.02
Ethane	0.24
Ethylene	0.12
n-Hexane	0.25
Hydrogen	0.019
Methane	0.29
Propane	0.25

[a] From Lees.[3]

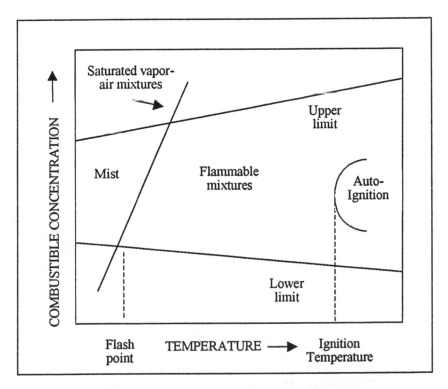

Figure 7.2.1. **Effect of temperature on the autoignition temperature and the flammibility for a combustible vapor in air.**

The equation of Hawthorne, Wenddell, and Hottel[8] can be used to predict the flame length in the turbulent regime.

$$L = (5.3\,d/C_f)(T_a/T_f)[C_f + (1 - C_f)(M_a/M_f)]^{1/2}$$ (7.2.7)

where C_f = fuel concentration in stoichiometric mixture (mole fraction)
  d = diameter of orifice or pipe (m)
  M_f = molecular weight of fuel
  M_a = molecular weight of air
  T_a = adiabatic flame temperature (K)
  T_f = temperature of fuel in the pipe (K)

The diameter of the flame as a function of distance along the flame length is calculated using Baron's equation.[9]

$$D_f = (0.29)[\ln(L/x)]^{1/2}$$ (7.2.8)

Where D_f = diameter of the flame (m)
  x = axial distance (m)
  L = flame length (m)

The maximum flame diameter, which occurs at $x/L = 0.61$, is given by

$$D_{max} = 0.12L$$ (7.2.9)

where D_{max} is the maximum flame diameter in meters.

Pool Flames

Examples of pool flames are flames on ground pools and flames on tanks. The rate of evaporation and dispersion of a spreading pool of liquid affects the burning rate of a liquid. A typical liquid burning rate ranges from 0.75 to 1.0 cm/min. Burges and Zabetakis developed Eqs. (7.2.10) and (7.2.11) to predict the liquid burning rate on ground pools in windless conditions.[10]

$$U_{lb} = U(1 - e^{-k_1 d})$$ (7.2.10)

$$U = 0.0076(-H_c)/H_v$$ (7.2.11)

Where U_{lb} = liquid burning rate (cm/min)
  D = diameter of the pool (cm)
  U = liquid burning rate for a pool of infinite diameter (cm/min)
  U = 0.79 for butane and 0.17 for methanol

k_1 = constant (cm^{-1})
k_1 = 0.027 for butane and 0.046 for methanol
H_c = net heat of combustion (kJ/gmol)
H_v = total heat of vaporization (kJ/gmol)

Pool burning time is defined as the time for the pool fire to burn itself out. It is calculated by

$$t_b = V_p/A_p U_{lb} \tag{7.2.12}$$

where t_b = burning time (s)
     U_{lb} = liquid burning rate (m/s)
     A_p = surface area of the pool (m^2)
     V_p = volume of the pool (m^3)

The heat flux is given by

$$q = U_{lb} H_c X_e \tag{7.2.13}$$

where q   = heat flux (kJ-m/s)
     U_{lb} = liquid burning rate (m/s)
     H_c = heat of combustion (kJ)
     X_e = fraction of the heat produced as radiation; X_e normally takes a value in the range of 0.13 to 0.35 (to be conservative, use 0.35)

The flame radiation at a distance r from the pool center is given by

$$I = q/(4\pi r^3) \tag{7.2.14}$$

where I   = flame radiation $(kJ/s-m^2)$
     r   = distance from pool center (m)

Fireballs

Fireballs radiate intense heat, which can cause fatal burns and can quickly ignite other materials. The diameter and persistence time of the fireballs can be calculated by the equations developed by Gayle and Bransford.[11]

$$D_f = 9.56 \, W^{0.325} \tag{7.2.15}$$

$$t_p = 0.196 \, W^{0.349} \tag{7.2.16}$$

where D_f = fireball diameter (ft)
     W   = total mass of combustion reactants including oxygen (lb)
     t_p = persistence time of fireball (s)

Jet Flame

Tanks, cylinders, and pipelines that contain gases under pressure (e.g., liquefied gases) may discharge gases at a high speed if they are somehow punctured or broken during an accident. Subject to an ignition source, a flammable gas can produce a flame jet of considerable length (possibly hundreds of feet from a hole less than a foot in diameter). The radiative heat flux of a jet flame or flare can be determined through Eq. (7.2.17).

$$I = X_g q/(4\pi r^2)$$ (7.2.17)

where I   = radiative heat flux
      X_g = emissivity factor dependent on the nature of the combustible material
            involved (a value of 0.2 is suggested for jet flames).
      r   = distance from the particular point in reference to the flame
      q   = heat flux

Flash Fires

In a flash fire, a gas cloud or plume forms and moves in a downwind direction. Subject to contact with an ignition source, a wall of flame will flash back to the vapor source, sometimes with explosive force. The behavior of flash fire flames is not well documented. It is generally assumed that a flash fire will spread throughout the vapor cloud emitted and that it can be calculated by gas outflow dispersion. It is generally assumed that the fire is lethal to anyone within the contours of the cloud who is not wearing special protective gear. [5]

Fire Types and Sources

In addition to assessing a fire's flame characteristics to determine the cause and effect of a fire, knowledge of the fire accident type will provide insight into its ignition source and possible fire protection and prevention methods. Electrical, chemical, and metal fires can occur in a solid, liquid, or gaseous phase. Section 7.3 presents a detailed discussion of these fires and their ignition sources.

7.3 PLANT FIRES

Having discussed the fundamental characteristics of fires in general and the different types of fire, we now examine more closely fire accidents that occur in process plants. Specifically, we review plant fire classifications, sources, causes, damage potentials, and detection and protection systems.

Classes of Fires

The National Fire Protection Association recognizes four general classification of fires.[2]

1. *Class A Fires.* Class A fires, which occur in highly flammable solid materials such as wood, cellulose, paper, and excelsior, are extinguished by bringing the burning materials below their ignition temperatures with the quenching and cooling effects of water. Under certain circumstances, these fires may be extinguished by the blanketing or smothering effects of dry chemical and carbon dioxide fire extinguishers.
2. *Class B Fires.* Class B fires are those that occur in the vapor-air mixture overlying the surface of flammable liquids such as oil, greases, alcohol, kerosene, and gasoline. Class B fires are most successfully extinguished by limiting the air that supports combustion. Fire extinguishers, dispersing of dry chemicals, carbon dioxide, foam, halogenated hydrocarbon agents, and fog streams of water are recommended for Class B fires.
3. *Class C Fires.* Class C fires involve electrical equipment. The extinguishing agents recommended are dry chemicals, carbon dioxide, compressed gas, and vaporizing liquid.
4. *Class D Fires.* The last classification is reserved for fires occurring in combustible metals such as magnesium, lithium, sodium, and aluminum. Class D fires require special extinguishing methods and agents, such as the graphite-based type.

Ignition Sources

Industrial plants contain a great number of possible ignition sources. A study made by the Factory Mutual Engineering Corporation of almost 25,000 industrial fires reported over a decade indicates that, for the majority of fires, the origins can be traced to the following general sources:[12]

1. Electrical (23%)
2. Smoking (18%)
3. Friction (10%)
4. Overheated materials (8%)
5. Hot surfaces (7%)
6. Burner flames (7%)
7. Combustion sparks (5%)
8. Spontaneous ignition (4%)
9. Cutting and welding (4%)
10. Exposure (3%)
11. Incendiarism (3%)
12. Mechanical sparks (2%)
13. Molten substances (2%)

14. Chemical action (1%)
15. Static electricity (1%)
16. Lightning (1%)
17. Miscellaneous (1%)

Some of these ignition sources are briefly discussed.[2]

Electrical

Electrical accidents, including electrical arcing, short circuits, and overheated electrical equipment, are the leading cause of industrial fires.

Smoking

Many fires are started by careless or neglectful workers who do not abide by rules that prohibit smoking in areas containing flammable liquids, gases, vapors, dusts, fibers, and so on.[12]

Friction

These fires are caused by heat produced by, for example, inadequate lubrication, poorly controlled cutting or grinding operations, of maladjustment of power drives and conveyors. Friction can be created by two moving parts, or by one moving and one stationary part, as can be found in a conveyor system.

Overheated Materials

Processes or operations that require heating of flammable materials, liquids, and ordinary combustibles are subject to fires in this category.

Hot Surfaces

Conduction, convection, and radiation of heat (from boilers, furnaces, forges, etc.) cause the ignition of flammable liquids and combustibles.

Burner Flames

Improper use or poor maintenance of portable torches, boilers, driers, and portable heating equipment comprise this source of ignition.

Combustion Sparks

This ignition source includes sparks released from foundry cupolas, furnaces, and incinerators.

Spontaneous Ignition

When combustibles and oxygen in the air are heated sufficiently, reaction begins and continues until the combustible material reaches a temperature at which the reaction becomes self-sustaining. This temperature level is the *autoignition temperature* (AIT) defined in Section 7.2. Examples of materials that may undergo spontaneous combustion are oil rags left on steam pipes, dust coatings on hot surfaces, and coal stored in piles on the ground. For convenience, substances subject to spontaneous ignition have been divided into

1. Substances that are not combustible but may cause ignition if combustibles (e.g., Wet calcium oxide and wet unslaked lime) are present
2. Substances having ignition points below ordinary temperatures (e.g., Sodium and potassium in the presence of water)
3. Combustible substances (e.g., Easily oxidized vegetable oils, which generate heat sufficient to cause ignition) that may undergo sufficient oxidation at ordinary temperatures to reach their AIT
4. Organic combustible substances subject to microbial thermogenesis (e.g., Agricultural products such as hay and grain)

Cutting and Welding

Fires of this origin are most commonly caused by sparks and hot metal from cutting and welding operations, by defective gauges, or deteriorated gas lines on welding equipment.

Exposure

In this category, ignition results from converted or radiated heat from adjoining or nearby facilities.

Incendiarism

This source includes fires started maliciously (e.g., by employees, intruders, or professional arsonists).

Mechanical Sparks

Fires caused by molten metal released from a ruptured furnace or by molten metal spilled during handling are included in this ignition source category.

Chemical Action

This category includes chemicals reacting with other chemicals or materials and the decomposition of unstable chemicals (e.g., the reaction of some substances with water or moist air or with strong oxidizing and reducing agents).

Static Electricity

Static electricity is generated when two dissimilar bodies come in contact, then separate, and the developed charge in one or both bodies is retained as a result of poor conductivity of the body. A low conductivity liquid flowing through a pipeline can generate a charge at a rate of 10^{-9} to 10^{-6} A. A powder coming out of a grinding mill can carry a charge of 10^{-8} to 10^{-4} A. At a charging rate of 10^{-6} A, the potential of an insulated container can rise at a rate of 1000 V/s^2. The most common sources of static electricity in the process industry are solids handling operations (e.g., grinding, sieving, and pneumatic conveying).

The ignition sources described above cannot be eliminated, but they can be controlled by careful layout of the plant, proper design and maintenance of equipment, and the use of trip systems.

Causes of Plant Fires

The most common cause of fire accidents in process plants is equipment failure. This is primarily a result of poor equipment maintenance or poor equipment layout and design. Maintenance performed according to a detailed and well structured schedule will significantly reduce the occurrence of fire accidents. The second largest cause of fire accidents is ignorance of the properties of a specific chemical or chemical process. Proper training of employees will increase their knowledge of the properties of a specific chemical and chemical process and can prevent many of these chemical fire accidents.

Leaks and spills resulting from an equipment failure will frequently cause fires. Large leaks can occur when a vessel, pipe, or pump fails; failure of a pump seal, pipe flange, or bore connection will cause a somewhat smaller leak. Since fire at pumps and/or at flanges are highly possible, it is essential to assess the effect of such fires on the equipment above or near pump or flange. Incidence of pump fires can be minimized by using double mechanical seals. Pipe flanges also can be protected against fires by placing an insulating pad between the joint ring and pipe bore and by specifying more massive flanges to minimize the temperature gradient.

Thermal insulation (or lagging) on plant equipment may become soaked or impregnated with oils and other flammable liquids. When the lagging gets hot, spontaneous combustion can occur. Lagging fires are affected by oil leaks, insulation material, and temperature. Spontaneous combustion occurs only when the oil is nonvolatile, since volatile oil evaporates more easily, thus delaying the accumulation of oil that reaches its AIT. A continuous leak may actually inhibit lagging fires, since the lagging fire can be over-saturated. A good insulating material with a low thermal conductivity is favored by the lagging fire. A low density, porous structure provides surface area for the air to diffuse in. A thick lagging gives good insulation; therefore spontaneous combustion is more likely on a thicker lagging. To prevent lagging fires, the following precautions are recommended:

1. Prevent leaks
2. Use appropriate sealings (cement finish or aluminum foil)
3. Use special insulation materials such as foam glass or crimped aluminum sheeting

Fire Hazards and Damage

The potential for fire hazards is rather high in the chemical industry. However, this potential is generally judged to be less than that of an explosion or toxic release, as discussed later. The scale of a fire hazard can be determined by assessing the following factors.[3]

1. *Inventory.* The larger the inventory of material, the greater is the loss potential (see discussion on Flixborough accident in Chapter 1).
2. *Energy.* The more energy available for a release, such as the stored energy in a material state or chemical reaction, the greater the potential.
3. *Time.* A higher rate of release of a hazard and a long warning time (i.e., period before emergency countermeasures can be taken) increase the loss potential.
4. *Exposure.* The intensity of the hazard and the distance over which it may cause injury or damage also directly affect the potential.

Most fire accidents involve a large loss of life and extensive property damage. Frequently, a fire is preceded by an explosion. Explosions are usually more lethal than fires (see Section 7.4). Some major fires in industrial plants have been described in detail in the literature.[3]

The thermal radiation intensity and the time duration of fires often are used to estimate injury and damage due to a fire. Various tables have been compiled to set up criteria for fire damage to people and property. Table 7.3.1 shows a relationship between heat radiation intensity and burn injury.[13]

Fire Protection and Prevention

The methods of fire protection and prevention are dealt with elsewhere. However, to conclude this section we present a brief discussion on plant fire fighting methods and equipment.

TABLE 7.3.1 Damage Caused at Various Incident Levels of Thermal Radiation[a]

Incident Flux (kW/m²)	Type of Damage Caused[b]
37.5	Sufficient to cause damage to process equipment: 100% lethality
25	Minimum energy required to ignite wood at infinitely long exposures (nonpiloted): 100% lethality
12.5	Minimum energy required for piloted ignition of wood, melting plastic tubing: 100% lethality
4	Sufficient to cause pain to personnel if unable to reach cover within 20 s; blistering of skin (first degree burns are likely): 0% lethality
1.6	Will cause no discomfort for long exposure

[a]From Bertknecht.

[b]At the lower levels, where time is required to cause serious injury to people, there is often the possibility to escape or take shelter.

The accuracy of the incident flux damage relationship is considered to be adequate for initial hazard assessments and within the estimation of hazardous incidents.

The correlation of thermally induced damage or injury may be applied to hazard assessment.

The best way to fight a fire is to remove any one of three essential conditions required to sustain the fire:

1. The fuel source (e.g., by eliminating the leaks in a process plant)
2. Heat (e.g., by dousing the fire with water)
3. The supply of oxygen (e.g., by applying foams or inert gases)

Various fire extinguishing systems can be used. These include:

1. Water systems
    a. Automatic sprinklers
    b. Fire hoses
2. Foam systems
    a. Chemical foams: formed by the chemical reaction in which bubbles of CO_2 gas and a foaming agent combine to produce and expand froth; often used on Class B fires.
    b. Mechanical foams: bubbles of air, which are produced when air and water are mechanically agitated with a foam-making agent; often used on Class B fires.
    c. High expansion foams: tiny foam bubbles filled with air, created by a fan blowing the air through a mesh over which a detergent based solution is flowing. High expansion foam is used where water damage is a problem or access to an area is not feasible. For example, such a foam

would be suitable if a flammable liquid held in an atmospheric storage tank were to spill as a result of overfilling, causing the formation of an ignitable vapor cloud.

3. Carbon dioxide systems: often used on Class B fires.
4. Dry chemical systems (e.g., sodium bicarbonate). These are not toxic and do not conduct electricity or freeze.
5. Water spray systems. These are used for exposure protection of buildings, tanks, and control of Class B flammable liquid fires.
6. Steam jet systems. These are used to smother some fires in closed containers or in confined spaces.

7.4 EXPLOSION FUNDAMENTALS

An *explosion* is defined by Strehlow and Baker[14] as an event in which energy is released over a sufficiently small period of time and in a sufficiently small volume to generate a pressure wave of finite amplitude traveling away from the source. This energy may have been originally stored in the system as chemical, nuclear, electrical, or pressure energy. However, the release is not considered to be explosive unless it is rapid and concentrated enough to produce a pressure wave that can be heard.

Many types of processes can lead to explosions in the atmosphere. They may cause accidental explosions such as condensed phase explosions with or without confinement (liquids), combustion explosions in enclosures (gases and dusts), explosions in pressure vessels, boiling liquid-expanding vapor explosions (BLEVEs), and unconfined vapor cloud explosions (UVCEs). There also are intentional explosions such as nuclear weapon explosions, condensed phase high explosions, vapor phase high explosions, gunpowder propellant explosions, and explosions due to natural phenomena such as lightning, volcanic eruptions, and meteor landings. This section focuses mainly on the fundamentals of accidental explosions.

To understand the basic or fundamental principles and characteristics of explosions, certain terms must be defined. The discussion begins by reviewing explosion (or flammability) limits, as covered in the preceding section. An explosion limit is the concentration range of a flammable or explosive gas-air mixture within which an ignition source can start a self-propagating reaction. The two ends or limits of this concentration range are known as the *lower explosion limit* (LEL or LFL) and the *upper explosion limit* (UEL or UFL). For dust-air mixtures, the lower and upper explosion limits can be determined by calculations if the chemical nature and heat of combustion of the dust are known. Typical ranges for industrial dust are from 20 to 60 g/m^3 for the lower explosion limit and between 2000 and 6000 g/m^3 for the upper explosion limit.

An *explosion pressure*, P_{ex}, is the pressure in excess of the initial pressure at which the explosive mixture is ignited. The rate of pressure rise is represented by dP/dt, a pressure change with respect to time. This is a measure of the speed of the flame propagation, hence of the violence of the explosion. Typical values of maximum explosion pressures in a closed vessel range from 7 to 8 bar. The rate of pressure rise can vary considerably with the flammable gas. The influence of vessel volume on the maximum rate of pressure rise for a given flammable gas is characterized by the cubic law

$$(dP/dt)_{max} \, V^{1/3} = K_G = \text{constant} \qquad (7.4.1)$$

Where V $\qquad$ = vessel volume (m^3)
$\quad$ K_G $\qquad$ = constant (bar-m/s)
$\quad$ $(dP/dt)_{max}$ = maximum pressure rise (bar/s)

The cubic law may be applied only to systems that are similar with respect to vessel shape, degree of turbulence, ignition source, and concentration of the gas-air mixture.[13] Table 7.4.1 provides K_G values of some flammable gases obtained by using spark gap ignition with an ignition energy of 10 J. A similar cubic law equation applies for dust explosions in closed vessels. Table 7.4.2 gives K_{St} values for fine dusts.

$$(dP/dt)_{max} \, V^{1/3} = K_{St} \qquad \text{for } V > 0.04 \, m^3 \qquad (7.4.2)$$

Where K_{St} = constant (bar-m/s)

Table 7.4.3 shows the effect of ignition source type and ignition source energy on the maximum explosion energy and the rate of pressure rise for dusts. Other factors influencing the maximum explosion pressure and the rate of pressure rise for dust explosions are vessel shape and turbulence. The particle size, the particle size distribution of the dust, and the available surface area of the dust affect the violence of the explosion. The smaller the median value of dust's particle size, the higher the maximum explosion pressure and the maximum rate of pressure rise. The higher the specific particle. surface area, the higher the maximum rate of pressure rise. The maximum explosion pressure is not affected by the particle surface area.

The explosion of a flammable mixture in a vessel or enclosure can be either a *deflagration* or a *detonation*. A *deflagration* is an explosion that occurs when (1) the concentration of the flammable mixture is within the flammability range, (2) there is a source of ignition, and (3) the mixture is above its autoignition temperature. In other words, deflagrations are the result of a combustion reaction with a self-propagating flame front. These reactions occur with a low rate of pressure rise and a low peak pressure fall. A *detonation* occurs at similar conditions except that the mixture composition is limited to a narrower range than that for an explosion.

TABLE 7.4.1 Average K_G Value of Gases, Ignited at Zero Turbulence[a]

Flammable Gas	K_G (bar-m/s)
Methane	55
Propane	75
Hydrogen	550

[a] From Bartknecht.[1]

TABLE 7.4.2 K_{St} Values of Technical Fine Dusts: High Ignition Energy[a]

Type of Dust	P_{max} (bar)	K_{St} (bar-m/s)
PVC	6.7 - 8.5	27 – 98
Milk powder	8.1 - 9.7	58 – 130
Polyethylene	7.4 - 8.8	54 – 131
Sugar	8.2 - 9.4	59 – 131
Resin dust	7.8 - 8.9	108 – 174
Brown coal	8.1 - 10.0	93 – 176
Wood dusts	7.7 - 10.5	83 – 211
Cellulose	8.0 - 9.8	56 – 229
Pigments	6.5 - 10.7	28 – 344
Aluminum	5.4 - 12.9	16 – 750

[a] From Bartknecht.[1]

TABLE 7.4.3 Influence of the Type of Ignition Source and of the Ignition Energy on the Explosion Data of Combustible Dusts (1-m^3 explosion chamber)[a]

Type of Dust	Ignition Sourde[b]	Ignition Energy (J)	P_{max} (bar)	K_{St} (bar-m/s)
Lycopodium	D	10,000	8.2	186
	C	0.080	8.3	199
	S	10	8.4	153
Cellulose	D	10,000	9.7	150
	C	0.040	9.2	147
	S	10	8.2	63
2-Naphthol	D	10,000	8.0	100
	C	0.005	7.7	90
	S	10	7.9	90
2-Nitro-4-propionyl-	D	10,000	8.3	84
aminoanisol	C	16	8.0	95
	S	10	7.3	52
	D	10,000	6.4	47
Dibutyltin dioxide	C	8	6.6	55
	S	10	0	0

[a] From Bartknecht.[1]
[b] D = chemical detonator, C = condenser discharge, S = permanent spark gap.

initiated directly. However, it is also possible for a deflagration to undergo transition to a detonation.

The maximum explosion pressure is a function of the initial pressure, P. If the initial pressure is increased by a factor of 2, the maximum explosion pressure and the maximum pressure rise will also increase by a factor of about 2 for both flammable gas and dust mixtures. When the initial pressure is less than 10 mbar, it is usually no longer possible to have an explosion.

The preceding discussion applies only to nonturbulent or laminar mixtures in vessels of cylindrical shape, having a diameter-to-length ratio of 1:1.5. The higher the value of this ratio, the more severe the violence of the explosion, due to the axial propagation of the flame front. At high turbulence, an unburned mixture can be reached by the flame much faster than in laminar mixtures and the violence of the explosion can be increased greatly. For example, the violence of a methane-air mixture explosion can be increased by a factor of 9 at high turbulence.'

Another method of estimating the effects of explosions in process plants is represented by

$$P_1^0 / P_1^0 = (W_1^{1/3} / W_2^{1/3})^n (r_2 / r_1)^n \qquad (7.4.3)$$

Where P_1^0, P_2^0 = peak overpressure
W_1, W_2 = mass of explosion
r_1, r_2 = distance
n = 1.6 for overpressure of 1-10 psi
n = 2.3 for overpressure of 10-100 psi

Equation (7.4.3) was developed from data on the high explosive, trinitrotoluene (TNT).

Fluid expansion energy in an explosion W_{ex} is determined by the equation

$$W_{ex} = RT\ln(P_1/P_2) \quad \text{(isothermal expansion)} \qquad (7.4.4)$$

or

$$W_{ex} = \frac{P_1 V_1 - P_2 V_2}{k-1} \qquad (7.4.5)$$

where k is the expansion index.

The energy released in an explosive expansion of a liquid is given by

$$W_{ex} = \left(\frac{1}{2}\beta\right)P^2V \qquad (7.4.6)$$

where β = bulk modulus of liquid
V = volume of vessel

Vessel strain energy W_S, in a cylindrical vessel, neglecting end effects, is given by[12]

$$W_s = \frac{P^2V}{2Y}\left[3(1-2\gamma)+\frac{2K^2(1+\gamma)}{K^2-1}\right] \qquad (7.4.7)$$

where V = vessel volume
Y = Young's modulus
γ = Poisson's ratio
K = diameter ratio

Blast Waves

An explosion in air is a process by which a pressure wave of finite amplitude is generated in air by a rapid release of energy and is accompanied by a sudden or instantaneous rise in pressure and by the formation of shock waves or blast waves. As a blast wave passes through the air, rapid changes in density, temperature, and particle velocity are also encountered. The blast wave then generates overpressure which apply stresses to any nearby structures. This load results in such adverse effects as damage to buildings and/or equipment and injury to people.

The shape of the pressure profile near the center of the explosion depends on the type of explosion. At some distance from the explosion center, the region of positive pressure (or overpressure) in the shock wave is followed by a region of negative pressure (or underpressure). The underpressure is usually very weak, seldom exceeding 4 psi.[12]

In Fig. 7.4.1, an idealized representation of the blast wave, the pressure pulse is shown as a function of the distance from the explosion center. The shape of the curve at A is not shown; the curves at B through D and times 2 through 4 show the decrease in peak overpressure as the wave moves outward. Both positive and negative pressures are observed in the curve at point D and time 4. Figure 7.4.2 illustrates the variation of overpressure P^0 with time at point D; t_s and t_d are the peak overpressure arrival time and duration time, respectively. The modified Friedlander equation is most commonly used to describe the overpressure curve.

$$P = P^0\left[1-(t/t_d)\right]e^{-t/t_d} \qquad (7.4.8)$$

where P = overpressure
       P^0 = peak overpressure
       t   = time
       t_d = duration time

        Other properties of the blast wave are the *shock velocity*, which is the rate or speed of the blast wave as it travels through the air, the *particle velocity* (or *peak wind velocity*), the *peak dynamic pressure*, and the *peak rejected overpressure*.

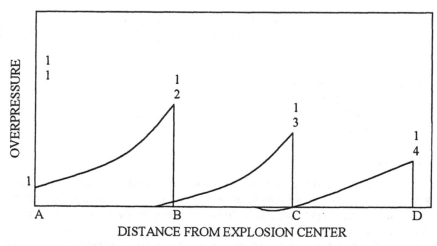

Figure 7.4.1. Idealized representation of a blast wave.

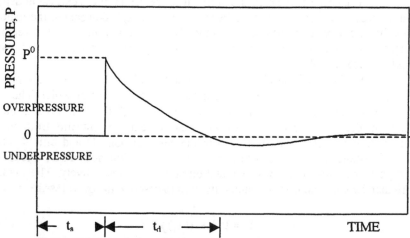

Figure 7.4.2. Variation with time of overpressure from a blast wave.

The *shock velocity* in air is given by

$$U = \frac{C_o}{1 + 6\left[P^o/7P\right]^{1/2}}$$ (7.4.9)

where U = shock velocity
C₀ = velocity of sound in air
P^0 = peak overpressure
P = absolute ambient pressure

Confined Explosions

Confined combustion explosions (gaseous or liquid) usually occur when a flammable vapor leaks into an enclosure and mixes with air to form a flammable mixture, whereupon this mixture contacts an ignition source that was present before the leak occurred. This type of explosion can also occur in storage tanks or ships where the vapor space above the stored flammable liquid (fuel) is in the explosivity range. In this case, an ignition source accidentally introduced will cause an explosion.

Confined dust explosions occur primarily in coal mines, grain elevators, feed and hour silos, boilers, and chemical plants. Coal mine, explosions have been occurring since the dawn of the Industrial Revolution. Explosions in mines are normally started by the ignition of a pocket of methane. Larger dusts explosions occur as the layered dust is picked up by a primary explosion and triggers a disastrous secondary explosion. Dust explosions in the chemical and pharmaceutical industries are generally confined to process equipment, since dust control techniques are sophisticated.

If the enclosure or vessel has a length-to-diameter ratio of approximately 1, the enclosure usually goes through a simple overpressure explosion. The rate of pressure rise is relatively slow and the explosion is a low density blast. If the enclosure has a large length-to-diameter ratio or contains a large number of obstacles such as internal partitions or pieces of equipment, flame propagation causes gas motion ahead of the flame, generating large-scale eddies (turbulence) at points where the flow separates from the obstacles. This, in turn, causes rapid increases in effective flame area, which cause a rapid rise of pressure and additional turbulence flame interactions. These processes may lead to gas phase detonations, which usually occur as far from the ignition source as is possible within the enclosure. These explosions produce strong blast waves and high velocity fragments, and therefore can cause more damage to the surroundings than simple overpressure explosions.[3]

The violence of the confined explosions (flammable vapor, gas, and dust) can be calculated by the explosion pressure and energy equations introduced earlier. The behavior of flammability (or explosivity) limits of flammable mixtures and dusts is also covered in preceding sections.

Unconfined Vapor Cloud Explosions

An unconfined vapor cloud explosion (UVCE) is one of the most serious hazards in the process industry. Not only is an unconfined vapor cloud explosion large and destructive, but it may occur some distance from the vapor release source. Thus, its zone of influence can span a large area. Vapor clouds are usually caused by vessel failures, failures of piping, valves, or fittings, and releases from venting facilities. The material escaping may be a gas, a volatile liquid, a superheated liquid, or a refrigerated liquid. Flashing superheated liquids tend to cause the largest vapor clouds.

The distance traveled by a cloud of flammable vapor is site specific and relies on several dispersion factors, which are discussed in Part III of this book. It is not likely that a vapor cloud would travel far in any industrial or urban area. In open areas with few sources of ignition, a vapor cloud may drift several miles. The time before ignition can range from 10 seconds to 15 minutes.

The factors that affect unconfined vapor cloud explosions are not well understood. In a model developed by William, it is assumed that ignition occurs at a point source, that the flame front travels out from the core at a flame speed S, and that the pressure waves produced by the flame generate a weak shock wave that travels ahead of the flame with a time-dependent velocity.[15] The equation for the flame speed for spherical systems is

$$S = \frac{a_0 (\rho_0/\rho_1)(T_0/T_2)(P_1/P_0)(P_1/P_0 - 1)^{1/2}}{(1 - \rho_0 T_0 P_1/\rho_1 T_1 P_0)^{1/3}(1 - \rho_0/\rho_1)^{1/6}} \qquad (7.4.10)$$

where S = flame speed
a_0 = speed of sound
ρ_0 = density in undisturbed gas
ρ_1 = density in shocked gas
T_0 = absolute temperature in undisturbed gas
T_1 = absolute temperature in shocked gas
T_2 = absolute temperature in burnt gas (core)
P_0 = absolute pressure in undisturbed gas
P_1 = absolute pressure in shocked gas

Several common substances amenable to unconfined vapor cloud explosions are hydrogen, methane, and liquefied natural gas (LNG).

Explosion Types and Sources

Generally, explosions have an identifiable accidental, natural, or intentional cause. Table 7.4.4 lists a number of explosion sources according to these three categories. Extensive calculational details on a host of explosion types is available in the literauture.[16]

Accidental explosions are potentially the most dangerous; they are a major concern for any industrial plant that deals with either pressurized or flammable gases. An accidental explosion occurs not by design and therefore is not similar to an intentional explosion, where the conditions are planned and can be controlled. Accidental explosions usually are the result of equipment failure or operator error. Although accidental explosions are by definition unforeseen events, the procedures discussed in the next chapter may be implemented either to minimize their effects or to prevent them entirely.

A *natural* explosion is also an uncontrolled and unexpected event. However, unlike an accidental explosion it cannot be prevented. Some examples of natural explosions include lightning and volcanic eruptions. Under the right critical conditions natural explosions also trigger and cause accidental explosions.

An *intentional* explosion usually has no undesired effects because it is controlled and limited to an isolated area. Examples include explosions from nuclear weapons, blasting lasers, and gun powder blasts, and contained explosions (see note *b*, Table 7.4.4).

7.5 PLANT EXPLOSIONS

Section 7.4 presented three major types of explosion. This section more closely examines plant-related explosions, focusing on causes, results, and damage potentials. The means for explosion protection and prevention are the subject of a subsequent chapter.

Many factors may lead or contribute to a plant explosion. However, plant explosions most often encountered in industry are caused either by faulty operational procedures or by faulty equipment. We now discuss several of the more common types of plant explosion: chemical, nuclear, expanding vapors, and pressurized gas explosions.

Chemical Explosions

A *chemical explosion* tends to have a slower pressure buildup compared with accidents of other types; its most likely cause is a runaway chemical reaction. Such an unstable and potentially dangerous situation can result from the combination of two or more incompatible chemicals, or the improper storage of potentially explosive or highly reactive chemicals. A runaway, or uncontrolled, reaction usually is exothermic and due to system upset (e.g., presence of excess catalyst and/or reactants or inadequate cooling or mixing), which permits the pressure to build up to high enough levels to cause vessel rupture. This type of explosion tends to occur when the contents of the reactor are gaseous. However, if the contents are liquid and the temperature is above the flash evaporation temperature, the result is called a *boiling liquid-expanding vapor explosion* (BLEVE), referred to earlier and described in more detail below.

TABLE 7.4.4 Types of Explosions

Natural Explosions	Intentional Explosions	Accidental Explosions

Lightning
Volcanoes
Meteors
	Nuclear weapon
	explosions
	Condensed phase high
	explosives
	  Blasting
	  Military
	  Pyrotechnic separators
	Vapor phase high
	explosives
	Gun powders/propellants
	  Muzzle blast
	  Recoilless rifle blast
	Exploding sparks
	Exploding wires
	Laser sparks
	Contained explosions[b]
		Condensed phase
		explosions
		  Light or no confinement
		  Heavy confinement
		Combustion explosions
		in enclosures (no
		pressure)
		  Gases and Vapors
		  Dusts
		Pressure Vessel (gaseous
		content)
		  Simple failure (inert
		  contents)
		  Combustion-generated
		  failure
		  Failure followed by
		  immediate combustion
		  Runaway chemical
		  reaction before failure
		  Runaway nuclear
		  reaction before failure
		BLEVEs (boiling liquid-
		expanding vapor
		explosion: pressure
		vessel containing a flash-
		evaporating liquid)
		  External Heating
		    Immediate combustion
		    after release
		    No combustion after
		    release
		Unconfined vapor cloud
		explosions (UCVEs)
		Physical vapor
		explosions

[a] From Baker et al.[15]

[b] Contained vessel explosions such as those used in gas and dust explosion research, and explosions in internal combustion engine cylinders are examples.

Nuclear Explosions

The most serious accident that can occur in a nuclear plant is a *reactor core meltdown*. In a core meltdown, the enclosed gases physically melt through the reactor vessel, and once contacting with cooler liquids or vapors either in a cooling jacket or in the outer environment, cause a physical explosion to occur. However, the hazard caused by the explosion itself is minimal and more localized compared with the release of radioactive material that accompanies such an accident.

Boiling Liquid-Expanding Vapor Explosions (BLEVE)

A *boiling liquid-expanding vapor explosion* occurs when a pressure vessel containing a liquid is heated to a temperature high enough to cause the metal to lose strength and rupture. The source of the heat is normally another fire near the vessel. The effects of a BLEVE depend on whether the liquid in the vessel is flammable. If the liquid is flammable, it may either cause a fire, which radiates heat, or form a vapor cloud, which could result in a second explosion.

Compressed Air Line Explosions

Compressed air lines are very susceptible to a combustion generation explosion, fueled by oil or char on the pipe walls. Explosions in pipelines can cause considerable damage. Pipelines within which gas, vapor, or dust explosions can occur must be designed to have sufficient mechanical strength to withstand pressure or stress beyond that required by the application.

Explosion Prevention and Containment

About two thirds of the major chemical plant losses have involved explosions. Explosions can be prevented or controlled. The release of energy is so rapid that the usual concepts of extinguishment are practically impossible to apply. It is essential during the risk evaluation procedure to identify those processes, operations and plant areas which are exposed to explosion hazards. Anticipating the type and degree of hazard well in advance is essential if the proper explosion prevention steps are to be taken.

The identification process referred to above may be based on a number of factors including:

1. Past history of similar operations
2. Laboratory and pilot plant experience
3. Technical information on the explosion tendencies of materials involved
4. Technical information on the explosion tendencies of the processes or operations involved
5. Information resulting from an investigation of the plant equipment, operations and environmental conditions

All environmental conditions should be considered. All possible situations that could lead to a hazardous situation should be identified and explored.

A system for the identification of explosion potential by areas and process may be obtained from the definitions of hazardous locations in *the National Electrical Code (NFPA No. 70)*. This system is based on the classification of various atmospheric mixtures of hazardous gases, vapors and dusts under specific environmental conditions. Other "explosion potential" identification systems suggested in the literature are determined from a consideration of the materials handled, type of process involved, and the process conditions. Each system is essentially an organized means of accumulating pertinent technical information and then applying engineering judgement to determine the explosion potential.

Explosions are likely to occur whenever circumstances are favorable as a result of a variety of basic conditions:

1. Rapid release of energy through the ignition of atmospheric mixtures of flammable gases, vapors or combustible dusts within the explosive range
2. Rapid release of energy through deflagration or detonation of unstable chemicals after exposure to an initiating force or energy
3. Rapid release of energy through decomposition or exothermic chemical reactions
4. Rapid release of energy through the mechanical failure of a pressure container as a result of mechanical defect or the generation of excessive pressure

The magnitude of the explosion not only depends on the quantity and nature of the materials involved, but also on process or operating conditions, degree of confinement, temperature, pressure, type of equipment, nature of operation (outdoor or closed construction), and prevention or control measures applied.

Prevention and control essentially deals with the elimination of those conditions which make the explosion possible. This is accomplished through the application of basic safety standards, acceptable safe practices and good engineering judgement. Primary standards and safe practices applicable under the various explosion categories are available in the literature.[17]

Explosion protection (control) may be accomplished with one or a combination of the following:

1. Containment
2. Flame barriers
3. Explosion suppression
4. Relief venting
5. Quenching

7.6 TOXIC EMISSIONS

The next two sections discuss accidents that result in the release of a toxic emission or a hazardous spill. In general, a toxic emission can be considered to be either continuous or instantaneous. In this section only the atmospheric effects of toxic emissions are considered. However, hazardous spills (next section) usually denote a liquid contamination of either soil or water systems; in addition a hazardous chemical spill may lead to the release of toxic emissions.

The Federal *Clean Air* and *Clean Water* acts define as toxic "those pollutants or combinations of pollutants, including disease-causing agents, which after discharge and upon exposure, ingestion, inhalation, or assimilation into any organisms either directly from the environment or indirectly by ingestion through food chains will, on the basis of information available to the [EPA] administrator, cause death, disease, behavioral abnormalities, cancer, genetic mutations, physiological malfunctions, or physical deformations in such organisms or their offspring."

Threshold Limit Values

The general subject of toxicology is discussed in detail in Part III – Chapter 11. However, three toxicology terms are introduced here in order for the reader to better grasp the problems encountered with a toxic release. Threshold limit values (TLVs) can be categorized in three ways.[18,19]

1. TLV-TWA is defined as the time-weighted average concentration for a normal 8-hour workday and a 40-hour workweek, to which nearly all workers may be repeatedly exposed, day after day, without any long-term adverse effect.
2. TLV-STEL is defined as the concentration to which workers can be exposed continuously for a short period of time without suffering from irritation, chronic or irreversible tissue damage, or mucosis of sufficient degree to increase the likelihood of accidental injury. TLV-STEL is not an independent exposure limit but rather supplements the TWA limit, which cites recognized acute effects from a substance whose toxic effects are primarily chronic. A STEL (short-time exposure limit) is a 15-minute time-weighted average exposure, which should not be exceeded at any time during a work day.
3. TLV-C is defined as the threshold limit concentration value ceiling that should not be exceeded during any part of the working exposure. This ceiling limit places a definitive boundary on concentrations of toxic or otherwise hazardous substances that should not be exceeded.

Typical TLVs for toxic substances are provided in Table 7.6.1.

TABLE 7.6.1   Threshold Limit Values for Selected Toxic Substances[a]

Substance	TWA ppm	TWA mg/m^3	STEL ppm	STEL mg/m^3
Arsine	0.05	0.2		
Ammonia	25	18	35	27
Benzene	10	30[b]		
Bromine	0.1	0.7	0.3	2
Carbon dioxide	5000	9000	30,000	54,000
Carbon monoxide	50	55	400	440
Carbon tetrachloride	5[b]	30[b]		
Chlorine	1	3	3	9
Chlorine dioxide	0.1	0.3	0.3	0.9
Chlorobenzene	75	350		
Chloroform	10[b]	50[b]		
Hydrogen sulfide	10	14	15	21
Mercury	0.001	0.01		
Nitrogen dioxide	3	6	5	10
Sulfur dioxide	2	5	5	10
Toluene	0.005	0.04	0.02	0.15
Xylene	100	435	150	655

[a] From American Conference of Governmental Industrial Hygienists
[b] Suspected Human Carcinogen

Continuous versus Instantaneous Releases

As mentioned earlier, toxic releases may consist of continuous releases or instantaneous emissions. Continuous releases usually involve low levels of toxic emissions, which are regularly monitored and/or controlled. Such releases include continuous stack emissions and open or aerated chemical processes in which certain volatile compounds are allowed to be stripped off into the atmosphere through aeration or agitation. Mathematical models for these releases to the environment are covered in detail in Part III.

Greater concern is warranted for the case of an instantaneous release, which is usually the result of an uncontrolled process. Most of these incidents are the result of a highway or railway accident or a fire, windstorm, or other natural accident. However, the cause can sometimes be linked to the breakdown of normal safeguards in plants, factories, mines, or chemical storage facilities. Whatever the cause, the result is often a significant potential threat to life, property, and/or the environment. The accident at Bhopal is an excellent example of an instantaneous release, in which a chemical process malfunction seems to have led to an uncontrolled toxic release. A release is said to be "instantaneous" if a significant amount of hazardous or toxic material is emitted over a short period of time. Since the disaster at Bhopal (described in Chapter 1), the chemical process industries have implemented certain risk-reducing measures that have cost billions of dollars. These measures are aimed at a tighter, safer control for chemical processes that produce or use hazardous substances. The measures include improving the storage and processing of hazardous chemicals, substituting benign compounds for hazardous ones, and producing certain hazardous substances under more controlled conditions.

The release/emission/discharge from a pipe or a vessel can be described mathematically through the appropriate application of the conservation laws for momentum, mass and energy (see Chapter 4). Most of the models employed by industry are generally derived from the traditional "fluid flow" equation available in the literature. *Perry's Chemical Engineers' Handbook*[20] has done an excellent job in presenting key predictive equations for either a gas, a liquid, or a mixed class of releases. Both gas and liquid discharges are included here since the emission usually consists of a (volatile) flashing liquid and vapor plus noncondensable gases and (possibly) some solid particulate. In addition the discharge flow may be sonic and/or transient (varies with time). Note that some of these equations apply to the next section, which is concerned with hazardous spills. The following detailed information is provided in *Perry's Chemical Engineers' Handbook*[20]

Discharge flow regime
General two-phase flow relationship
Two-phase orifice discharge
Subcooled liquid orifice discharge
Compressed fluid orifice discharge
Choked two-phase flow
Full-bore and punctured pipe discharge
Two-phase pipe discharge
Subcooled liquid inclined pipe discharge
Inclined pipe discharge
Horizontal pipe discharge
Discharge coefficients
Blowdown modeling

The CCPS of the AIChE has done a commendable job of extracting some of the key equations applicable to a toxic gas emission[16]. This material follows (with minor editing).

There are two flow regimes corresponding to sonic (or choked) flow for higher pressure drops and subsonic flow for lower pressure drops. The transition between the two flow regimes occurs at the dimensionless critical pressure ratio, r_{crit}, which is related to the gas heat capacity ratio γ via

$$r_{crit} = \left(\frac{p}{p_a}\right)_{crit} = \left(\frac{\gamma+1}{2}\right)^{\gamma/(\gamma-1)} \qquad (7.6.1)$$

Where p = absolute upstream pressure (N/m^2)

p_a = absolute downstream pressure (N/m^2)
γ = gas specific heat ratio (C$_p$/C$_v$, dimensionless)

Typical values of γ range from 1.1 to 1.67, which give r_{crit} values of 1.71 to 2.05. Thus, for releases of most diatomic gases (γ = 1.4) to the atmosphere, upstream pressures over 1.9 bar absolute will result in sonic flow. Note that the inverse of r_{crit} is occasionally used by industry.

Gas flow through an orifice is given by:

$$G_v = C_d \frac{Ap}{a_o} \Psi \qquad (7.6.2)$$

Where G$_v$ = gas discharge rate (kg/s)
C$_d$ = discharge coefficient (dimensionless $\leq$ 1)
A = hole area (m^2)
P = absolute upstream pressure (N/m^2)
a$_o$ = sonic velocity of gas at T=(γRT/M)$^{1/2}$
M = gas molecular weight (kg-mol)
R = gas constant (8310 J/kg-mol/K)
T = upstream temperature (K)
Ψ = flow factor, dimensionless

The flow factor for equation 7.6.2, ψ, is dependant on the flow regime as follows:
For subsonic flows

$$\Psi\Psi = \left\{ \frac{2\gamma^2}{\gamma-1} \left(\frac{p_a}{p}\right)^{2/\gamma} \left[1 - \left(\frac{p_a}{p}\right)^{(\gamma-1)/\gamma}\right]\right\}^{1/2} \quad \text{for } \frac{p}{p_a} \leq r_{crit} \qquad (7.6.3)$$

For sonic (choked) flows

$$\Psi = \gamma \left(\frac{2\gamma}{\gamma+1}\right)^{(\gamma+1)/2(\gamma-1)} \quad \text{for } \frac{p}{p_a} \geq r_{crit} \qquad (7.6.4a)$$

$$\Psi = \gamma \left(\frac{1}{r_{crit}}\right)^{(\gamma+1)/2\gamma} \qquad (7.6.4b)$$

An important case of gas discharge is the flow from pressure relief valves and rupture disks. When relief is required due to fire exposure in a nonreacting system, a long established empirical method for estimating relief rates is available. It is based on the following equations for predicting heat flux:

$$Q = 34,500FA^{0.82}$$

(7.6.5)

where Q = heat input through vessel wall from fire (Btu/hr)
  A = total surface area (ft^2)
  F = environmental factor, dimensionless

The area, A, in this equation is the entire surface area of the vessel, not the wetted surface area which maybe used in related equations. However, the error introduced by this difference in the calculation for a full tank is small.

For water spray protection over the entire surface area of the tank designed with a density of 0.25 gpm/ft^2 or more F=0.3. For an approved fire resistant installation, F=0.3. For an underground or buried tank, F=0.3. For water spray with good drainage F=0.15.

The gas discharge rate from the relief valve, G_{rv}, can then be calculated using the following formula:

$$G_{rv} = Q/h_{fg}$$

(7.6.6)

where G_{rv} = gas discharge rate from relief valve (kg/s)
  h_{fg} = latent heat of vaporization at relief pressure (kJ/kg)

A detailed discussion of the formulas used in NFPA Codes can be found in Appendix B of the Flammable and Combustible Liquids Code Handbook[22]. API RP520[23] recommends a similar formula applicable to pressurized storage of liquids at or near their boiling point where the liquids have a higher molecular weight than that of butane.

All of the recommended heat flux equations in API 520 and NFPA Codes that are used to design relief valve assume that the liquids are not self-reactive or subject to runaway reaction. If this situation arises, it is necessary to include the heat of reaction and the rate of the reaction into account in sizing the relief device.

Liquid and two phase discharge/emissions are presented in the next section.

7.7 HAZARDOUS SPILLS

Hazardous wastes are legally defined as those that may cause adverse or chronic effects on human health or the environment when not properly controlled. Hazardous wastes are generated either because processes have converted harmless materials into hazardous substances or because natural materials that are hazardous to begin with have been concentrated and released into the environment These substances may be ignitable, reactive, corrosive, radioactive, infectious, or toxic. They may exist as solids, liquids, sludges, powders, or slurries. About 90% of them are liquid or semiliquid. Some of these wastes are nondegradable and may persist in nature indefinately.

Hazardous chemical spills may have adverse effects on natural water systems, the land environment, and whole ecosystems, as well as the atmosphere. Major spills evolve from accidents (see Chapter 6) that somehow damage or rupture vessels, tank cars, or piping used to store, ship, or transport hazardous materials. In such cases, the spills must be contained, cleaned up, and removed as quickly and effectively as possible.

Spills of hazardous materials from transporting vehicles pose one of the most significant problems in accidental contamination of the water ecosystem, with its associated danger and threat to public welfare and health. In accidents involving land transportation, once a leak or spill has been contained, immediate cleanup must ensue and the spilled material must be disposed of properly. A quick response time is necessary to collect spilled materials before they are able to breach dikes, be absorbed into the ground, or otherwise flood the environment.

Hazardous wastes can be classified under five categories:

1. Waste oils and chlorinated oils
2. Flammable wastes and synthetic organics
3. Toxic metals, etchants, pickling, and plating wastes
4. Explosive, reactive metals, and compounds
5. Salts, acids, and bases

A particular waste may overlap into any number of these five categories. Flammable wastes are comprised mainly of contaminated solvents; this category also includes many oils, pesticides, plasticizers, complex organic sludges, and off-specification chemicals. Synthetic organic compounds include halogenated hydrocarbon pesticides, polychlorinated biphenyls, and phenols.

Describing Equations

As indicated in the previous section, information on liquid emissions for a variety of conditions is available in the literature, including equations for two phase flow[16,20]. Key equations for liquid and two-phase discharges have been adopted from CCPS and provided below[16].

Discharge of pure (i.e., nonflashing) liquids through a sharp-edged orifice can be estimated from the following equation.

$$G_L = C_d A \rho \left(\frac{2(p - p_a)}{\rho} + 2gh \right)^{1/2} \tag{7.7.1}$$

Where G_L = liquid mass emission rate (kg/s)
 C_d = discharge coefficient (dimensionless)
 A = discharge hole area (m^2)
 ρ = liquid density (kg/m^3)
 p = liquid storage pressure (N/m^2 absolute)
 p_a = downstream (ambient) pressure (N/m^2 absolute)
 g = acceleration of gravity (9.81 m/s^2)
 h = height of liquid above hole (m)

The discharge coefficient for fully turbulent discharges from small sharp edged orifices has traditionally been assumed to be 0.61. Crane Co. provides values for smooth nozzles and gives a good description of how to account for pipe fittings and other obstructions when calculating discharge rates [24].

Regarding two phase flow, pressurized liquid above its normal boiling point will start to flash when released to atmospheric pressure, and two phase flow will result. Two-phase flow is also likely to occur from depressurization of the vapor space above a volatile liquid, especially if the liquid is viscous (e.g., greater than 500 cP) or has a tendency to foam. Fauske and Epstein have provided the following practical calculation guidelines for two-phase flashing flows. The discharge of subcooled or saturated liquids is described by

$$G_{2p} = C_d \sqrt{G_{sub}^2 + G_{ERM}^2} \Big/ N \qquad (7.7.2)$$

where G_{2p} is two phase mass flow rate (kg/m^2/s). Discussion of each term will follow. The effect of subcooling is accounted for by

$$G_{sub} = \sqrt{2(p-p_v)\rho_f} \qquad (7.7.3)$$

Where p = storage pressure (N/m^2)
 p_v = vapor pressure at storage temperature (N/m^2)
 ρ_f = liquid density (kg/m^3)

For saturated liquids, equilibrium is reached if the discharge pipe size is greater than 0.1 m (length greater than 10 diameters) and discharge rate is predicted by

$$G_{ERM} = \frac{h_{fg}}{\upsilon_{fg}(TC_P)^{1/2}} \qquad (7.7.4)$$

Where h_{fg} = latent heat of vaporization (kJ/kg)
 υ_{fg} = change in specific volume liquid to vapor (m^3/kg)
 T = storage temperature (K)

C_P = liquid specific heat (kJ/kg-K)

For discharge pipes less than 0.1 m, the flashing flow increases strongly with decreasing length, approaching all liquid flow as the discharge pipe length approaches zero. The nonequilibrium effect is estimated by the parameter N (dimensionless) in Equation (7.7.2) given by

$$N = \frac{h_{fg}^2}{2\Delta p \rho_f C_d^2 \upsilon_{fg} T C_p} + \frac{L}{L_e} \qquad (7.7.5)$$

Where L = pipe length in opening (m)
       L_e = 0.1 m

For L = 0, Equations (7.7.2) and (7.7.5) reduce to (7.7.1) (neglecting the head of liquid). Equivalently, the discharge rate of flashing liquids from sharp-edged orifices at vessels can be estimated as though there were no flashing.

It is worth noting that design calculations for the sizing of relief systems (relief valves, headers, scrubbers and knock-out drums, etc.) are conservative in order to protect the integrity of vessels and relief systems. The calculations used for risk assessments are those which most accurately describe the discharge rate from the hazardous incident being modeled.

7.8 ILLUSTRATIVE EXAMPLES

Example 7.1

Calculate the upper and lower flammability limits of a gas mixture that consists of 30% methane, 50% ethane, and 20% pentane by volume.

Solution 7.1

Using Equations (7.2.1) and (7.2.2) and Table 7.2.1:

$$LFL(mix) = \frac{1}{(f_m/LFL_m) + (f_e/LFL_e) + (f_p/LFL_p)}$$

$$LFL(mix) = \frac{1}{(0.3/0.046) + (0.5/0.35) + (0.2/0.014)}$$

$$LFL(mix) = 0.0285 = 2.85\% \text{ by volume}$$

$$UFL(mix) = \frac{1}{(f_m /UFL_m) + (f_e /UFL_e) + (f_p /UFL_p)}$$

$$UFL(mix) = \frac{1}{(0.3/0.142) + (0.5/0.151) + (0.2/0.078)}$$

$$UFL(mix) = 0.125 = 12.5\% \text{ by volume}$$

Example 7.2

A flammable vapor flows through a 2-inch-diameter insulated pipe at a flow rate of 4.5 acfm. A lagging fire started and heated a 4-foot length of the pipe to 150 °F, which is above the ignition temperature of the vapor. The ignition delay time of the vapor is expressed by

$$\ln (t) = (250/T) + 0.5$$

where t = ignition time delay (s)
    T = absolute temperature (R)

Will ignition take place?

Solution 7.2

The ignition time delay can be calculated from the equation in the problem statement:

$$\ln (t) = [250 / (150+460)] + 0.5$$

$$t = 2.5 \text{ s}$$

The average residence time of the vapor in the four foot section of pipe is:

$$t_r = V/vq$$

$$t_r = (\pi / 4) (2 / 12)^2 (4) \Big/ 4.5$$

$$t_r = 0.019 \text{ min} = 1.16 \text{ s}$$

The residence time of the vapor in the pipe section is less than the ignition delay time. Therefore, ignition will not take place.

Example 7.3

Calculate the liquid burning rate of an octane ground pool. The pool has a diameter of 4.0 m and a depth of 4 cm.

Solution 7.3

Applying Equation (7.2.10) the pool burning time is calculated as follow:

$$U_{lb} = U(1 - e^{-k_1 d})$$

$$U_{lb} = (0.79)\left[1 - e^{-(0.027)(400)}\right]$$

$$U_{lb} = 0.790 \, cm/min$$

Example 7.4

A round vessel filled with hydrogen is ignited. The volume of the vessel is 0.5 m³. Calculate the maximum pressure rise. Also calculate the maximum pressure rise for the same vessel filled with methane. KG for hydrogen and methane is 550 and 55, respectively.

Solution 7.4

Equation (7.4.1) is employed to calculate the maximum pressure rise:

$$(dP / dt)_{max} \, V^{1/3} = K_G$$

Rearranging gives

$$(dP / dt)_{max} = K_G / V^{1/3}$$

Substituting for hydrogen

$$K_G = 550 \, bar \cdot m/s, \quad V = 0.5 \, m^3$$

$$(dP / dt)_{max} = (550 \, bar \cdot m/s) / (0.5 \, m^3)^{1/3}$$

$$(dP/dt)_{max} = 693 \, bar/s$$

Maximum pressure rise for methane is:

$$K_G = 55 \, bar\text{-}m/s, \quad V = 0.5 \, m^3$$

$$(dP/dt)_{max} = (55 \, bar\text{-}m/s)/(0.5 \, m^3)^{1/3}$$

$$(dP/dt)_{max} = 69.3 \, bar/s$$

Example 7.5

A baghouse has been used to clean a particulate gas stream for nearly 30 years. There are 600, 8 inch diameter bags in the unit. 50,000 acfm of dirty gas at 250 °F enters the baghouse with a loading of 5.0 grains/ft³. The outlet loading is 0.3 grains/ft³. Local EPA regulations state that the maximum allowable outlet loading is 0.4 grains/ft³. If the system operates at a pressure drop of 6 inches of water, how many bags can fail before the unit is out of compliance? The Theodore-Reynolds equation (see below) applies and all the contaminated gas emitted through the broken bags may be assumed the same as that passing through the tube sheet thimble.

The effect of bag failure on baghouse fractional penetration (or efficiency) can be described by the following equation:

$$P_t^* = P_t + P_{tc}$$

$$P_{tc} = 0.582 \, (\Delta P)^{1/2} / \phi$$

$$\phi = \frac{q}{\left[LD^2(T+460)^{1/2}\right]}$$

where P_t^* = penetration after bag failure
P_t = penetration before bag failure
P_{tc} = penetration correction term; contribution of broken bags to P_t^*
ΔP = pressure drop, in. H₂O
ϕ = dimensionless parameter
q = volumetric flow rate of contaminated gas, acfm
L = number of broken bags

D   = bag diameter, inches
T   = temperature, °F

For a detailed development of the above equation, refer to "Effect of Bag Failure on Baghouse Outlet Loading," L. Theodore and J. Reynolds, JAPCA, August 1979, 870-2.
Solution 7.5

Calculate the effciency, E, and penetration, P, before the bag failure(S):

$$E = (\text{inlet loading-outlet loading}) / (\text{inlet loading})$$
$$E = (5.0 - 0.03) / (5.0)$$
$$E = 0.9940 = 99.40\%$$

$$P_t = 1 - 0.9940$$
$$P_t = 0.0060 = 0.60\%$$

Calculate the efficiency and penetration, P_t^*, based on regulatory conditions:

$$E = (5.0 - 0.4) / (5.0)$$
$$E = 0.9200 = 92.00\,\%$$

$$P_t^* = 1 - 0.9200$$
$$P_t^* = 0.0800 = 8.00\%$$

Calculate the penetration term, P_{tc}, associated with the failed bags:

$$P_{tc} = 0.0800 - 0.0060$$
$$P_{tc} = 0.0740$$

Write the equation(s) for P_{tc} in terms of the number of failed bags, L:

$$P_{tc} = 0.582\,(\Delta P)^{1/2}\big/\phi$$

where

$$\phi = q\Big/\Big[LD^2(T+460)^{1/2}\Big]$$

Calculate the number of bag failures that the system can tolerate and still remain in compliance:

$$L = qP_{tc} / (0.582)\,(\Delta P)^{0.5}\,(D)^2\,(T+460)^{0.5}$$

$$L = (50,000) \ (0.074) \ / \ (0.582) \ (6)^{0.5} \ (8)^2 \ (250+460)^{0.5}$$

$$L = 1.52$$

So, if two bags fail, the baghouse is out of compliance.

The importance of when to correct/replace a broken bag will depend on the type of collector and the resultant effect on outlet emissions. In "inside bag collection" types of collectors, it is very important that dust leaks be stopped as quickly as possible to prevent adjacent bags from being abraded by jet streams of dust emitting from the broken bag. This is called the "domino effect" of bag failure. "Outside bag collection" systems do not have this problem, and the speed of repair is determined by whether the outlet opacity has exceeded its limits. Often, it will take several broken bags to create an opacity problem and a convenient maintenance schedule can be employed instead of emergency maintenance.[21]

     In either type of collector, the location of the broken bag or bags has to be determined and corrective action taken. In a non-compartmentalized unit, this requires system shutdown and visual inspection. In inside collectors, bags often fail close to the bottoms, near the tube sheet. Accumulation of dust on the tube sheets, the holes themselves, or unusual dust patterns on the outside of the bags often occurs. Other probable bag failure locations in reverse-air bags are near anticollapse rings or below the top cuff. In shaker bags, one should inspect the area below the top attachment. Improper tensioning can also cause early failure[21].

     In outside collectors, which are normally top-access systems, inspection of the bag itself is difficult; however, location of the broken bag or bags can normally be found by looking for dust accumulation on top of the tube sheet, on the underside of the top-access door, or on a blow pipe[21].

Example 7.6

Estimate the ethane discharge relief through a pressure relief valve for an uninsulated tank with a 7.5 m^2 surface area that is exposed to fire. The following data is provided.

Environment factor $F = 1.0$
Latent heat of vaporization $h_{fg} = 154.8$ kJ/kg

Solution 7.6

First use Equation (7.6.5) to estimate the heat flux from the fire:

$$A = 7.5 \ m^2$$

The heat transfer area is then

$$A = 81\,ft^2$$

$$Q = 34500FA^{0.82}$$

$$Q = 34500\,(1)\,(81^{0.82})$$

$$Q = 1.3 * 10^6\,Btu/hr$$

$$Q = 22,279\,kJ/s$$

Applying Equation (7.6.6) yields the release rate:

$$G = Q/h_{fg}$$
$$G = 22,279/154.8$$

$$G = 143.9\,kg/s$$

7.9 SUMMARY

1. Fires, explosions, toxic emissions, and hazardous spills are major plant concerns.
2. Fire is the combining of oxygen and fuel in the proper proportions and at the proper temperature to sustain combustion. The fire potential of a fuel is determined by evaluating its flammability characteristics.
3. The flame behavior of a fire is important in determining the causes and effects of fires. There are several classifications of flames: orifice flames, pool flames, fireballs, jet flames, and flash fires. Orifice or pipe flames are characterized as either premixed flame or diffusion flames. Pool flames are flames on ground pools and flames on tanks. Fireballs radiate intense heat, which can cause fatal burns and can quickly ignite other materials. Jet flame or flares also radiate intense heat.
4. An explosion is an event in which energy is released over a time sufficiently short and in a volume sufficiently small to permit the generation of a pressure wave of finite amplitude traveling away from the source.
5. The maximum explosion pressure is a function of and is directly proportional to the initial pressure. Blast waves are pressure waves of finite amplitude that are generated in air by a rapid release in energy and an instantaneous rise in pressure. The most common plant explosion types encountered in industry are chemical, nuclear, expanding vapors, and pressurized gas.

6. Threshold limit values (TLVs) can be categorized by a time-weighted average taken for a normal 8-hour workday and a normal 40-hour workweek. A short term exposure limit that is taken for a period of time not exceeding 15 minutes. A ceiling limit value that is not to be exceeded at any time.

PROBLEMS

1. Calculate the upper and lower flammability limits of a gas mixture that consists of 50% methane, 10% ethane, and 40% pentane by volume.
2. Calculate the burning velocity of a paraffin hydrocarbon gas-air mixture at 150°C if the burning velocity of the mixture is 45 cm/s at 25°C and 80 cm/s at 38°C.
3. Calculate the total burning time of the octane pool in Illustrative Example 3.
4. Calculate the peak overpressure of a 50-pound TNT explosion at a distance 200 feet from the ignition point, if the peak overpressure at 1000 feet is 0.10 psi when 150 pounds of TNT is detonated.
5. Calculate the fluid expansion energy for an isothermal expansion for a cylindrical vessel at 550°C with initial and final pressures of 147 and 450 psi, respectively.
6. Calculate the discharge of butane through a 50-mm diameter hole at 10 barg, 25°C with 10 m liquid head. The following data is provided.

    Butane density = 773 kg/m^3
    Butane vapor pressure @ 25°C = 8.3 barg

7. With reference to problem 6, recalculate the butane emission rate in kg/s if the discharge occurs through 20-mm by 100-mm rectangular hole.

REFERENCES

1. W. Bartknecht, *Explosions*, Springer-Veriag, New York, 1981.
2. R. J. Firenze, *The Process of Hazard Control*, Kendall/Hunt Publishing Company, New York, 1979
3. F. P. Lees, *Loss Prevention in the Process Industries*, Vol. 1, Butterworths, Boston, 1980.
4. *Manual of Industrial Hazard Assessment Techniques*, Office of Environmental and Scientific Affairs, The World Bank, Oct., 1985, London.
5. D. R. Stuwl, *Fundamentals of Fires and Explosions*, Vol. 73, American Institute of Chemical Engineers-Dow Chemical Company, Midland MI, 1977.
6. N. N. Semenov, *Some Problems in Chemical Kinetics and Reactivity*, Pergamon Press, London, 1959.
7. W. Jost, *Explosion and Combustion Processes in Gases*, McGraw-Hill, New York, 1946.

8. W. R. Hawthorne, D. S. Wenddell, and M. C. Hottel, "Mixing and Combustion in Turbulent Gas Jets," in *Third Symposium on Combustion*, Williams & Wil-kins, Baltimore, 1949.
9. T. Baron, "Reactions in Turbulent Free Jets-The Turbulent Diffusion Flame," *Chem. Eng. Prog.*, 50, 73 (1954).
10. D. Burgess and M. G. Zabetakis, *Fire and Explosion Hazards of LNG*, Vervalin, C. H., 1964.
11. J. B. Gayle and J. W. Bransford, "Size and Duration of Fireballs from Propellant Explosions," NASA report TMX-53314, George C. Marshall Space Flight Center, Huntsville, AL, 1965.
12. W. R. D. Manning and S. Labrow, *High .Pressure Engineering*, Leonard Hill, London, 1971.
13. W. Bartknecht, "Brenngas und Staubexplosionen," Forschungebericht F 45 des Bundesinstitutes fur Arbeitsschutz, Koblenz, 1971.
14. R. A. Strehlow and W. E. Baker, "The Characterization and Evaluation of Accidental Explosions," *Prog. Energy Combust. Sci.*, 2 (1), 27-60 (1976).
15. W. E. Baker, P. A. Cox, P. S. Westine, J. J. Kulesz, and R. A. Strehlow, *Explosion Hazard and Evaluation*, Elsevier, New York, 1983.
16. "Chemical Process Quantitative Risk Analysis", Center for Chemical Process Safety, AIChE, New York City, 1989.
17. "Hazard Survey of the Chemical and Allied Industries", American Insurance Association, New York City, 1979.
18. Threshhold Limit Values and Biological Exposure Indeces for 1987-88, The American Conference of Governmental Industrial Hygienists, Cincinnati, OH, 1987.
19. L. Theodore, J. Reynolds, and K. Morris, "Health, Safety and Accident Prevention: Industrial Applications," *Theodore Tutorials*, East Westchester, NY, 1997.
20. R. Perry and D. Greeen, "Perry's Chemical Engineers' Handbook", McGraw-Hill, New York City, 1997.
21. L. Theodore and J. Reynolds, "Effect of Baghouse Failure and Baghouse Outlet Loading," J. Air
22. "Flammable and Combustible Liquids Code Handbook", Fire Protection Association, Quincy, MA, 1987.
23. "Design and Installation of Pressure-Release Systems in References; Part I – Design", API Recommended Practice 520, 4th ed., American Petroleum Institute, Washington DC, 1976.
24. "Flow of Fluids Through Valves, Fittings, and Pipes", Technical Paper 410M. Crane Co., New York, 1981.

8

Process Applications

8.1 INTRODUCTION

In 1988 producers of basic industrial chemicals, plastics, and fibers in the United States increased their sales at least 10% to about $90 billion (exclusive of foreign subsidiaries) primarily as a result of increased demand at home and abroad.[1] Along with the increase of chemical production; safety and accident prevention have become more critical and essential. Such dramatic releases of toxic chemicals as those that occurred in Bhopal and at Three Mile Island have heightened public concern for the integrity of process facilities that handle hazardous materials.

Toxic chemicals that could potentially cause a major problem if accidentally released into the atmosphere include chlorine, hydrogen fluoride, hydrogen chloride, ammonia, chloropicrin, gasoline lead additives, vinyl chloride, and benzene. This chapter addresses the process application of some chemicals from the foregoing list, as well as some others that are considered to be highly toxic: hydrogen cyanide, sulfuric acid, and ethylene. Process considerations, physical and chemical properties, health effects, and methods of manufacture of these chemicals are discussed in conjunction with potential causes of release.

It is virtually impossible to design a fail-safe operation of a chemical process. However, many companies have attempted to minimize hazardous conditions by developing a systematic approach to process design. Implemention of these actions hoped to achieve maximum protection to personnel, equipment, and the public.

8.2 CHLORINE

Chlorine is an elemental chemical that exists as a gas at ambient conditions but liquefies at moderate pressures. Some of its common physical properties are listed in Table 8.1.1. Chlorine is slightly water soluble, is yellow-green in the gaseous state, and has a strong characteristic odor. Because chlorine gas is about 2.5 times denser than air, it tends to stay close to the ground when released into the atmosphere. Liquid chlorine has a clear amber color; one volume of liquid can vaporize to about 460 volumes of gas. In addition, liquid chlorine has a large coefficient of thermal expansion.

Chlorine is considered to be neither explosive nor flammable in the normal sense. However, chlorine is an oxidizing agent that will, like oxygen, support the burning of most combustible materials. Because chlorine is a strong oxidizer, it reacts readily with reducing agents. Chlorine will react with metals and other elements as well as inorganic and organic compounds. The most significant chemical properties contributing to the potential for accidental releases are as follows.

1. As a result of hydrolysis, moist chlorine contains hydrochloric and hypochlorous acids, which are very corrosive to most common metals. Chlorine also can react explosively with powdered metals.
2. Under certain conditions, chlorine will react rapidly with most of the elements. It will, for example, react violently with hydrogen to form hydrogen chloride. This explosive reaction can take place if either component is present in a mixture at a concentration greater than approximately 15%.[8]
3. Because of its great affinity for hydrogen, chlorine tends to remove hydrogen atoms from other compounds. Chlorine reacts with ammonia or ammonium compounds to form various mixtures of chloramines, depending on the conditions. One of these chloramines is nitrogen trichloride, which becomes highly explosive, even at relatively low concentrations (in the range of a few percent).
4. Chlorine dissolves rapidly in strong alkali solutions to produce hypochlorite solutions. Under certain conditions these solutions are prone to decompose with explosive force.
5. The strong oxidizing ability of chlorine allows it to react vigorously with organics to form chlorinated derivatives and hydrogen chloride.

Chlorine is a highly toxic, severe skin and lung irritant.[9,10] Exposure to low concentrations causes a stinging or burning sensation in the eyes, nose, and throat; headaches due to irritation of the accessory nasal sinuses may also develop. There may be redness of the face, tearing, sneezing, coughing, and huskiness or loss of voice. Bleeding of the nose may also occur, and sputum

TABLE 8.2.1   Physical Properties of Chlorine (Cl₂): CAS Registry Number 07782-50-5

Property	Value	Ref.
Molecular Weight	70.914	
Normal Boiling Point	-29.3°F at 14.7 psia	3
Melting Point	-149.8°F	3
Liquid specific gravity (H_2O = 1)	1.41 at 68°F	2
Vapor specific gravity (air = 1)	2.5 at 68°F	2
Vapor pressure	93 psia at 68°F	4
Vapor pressure equation		

$$\log P' = A - \frac{B}{T + C}$$

   where P' = vapor pressure (mmHg)
       T = Temperature (°C)
       A = 6.93790, a constant
       B = 861.34, a constant
       C = 246.33, a constant

Liquid Viscosity	0.345 centipoise	5
Solubility in water	6.08 lb/100 gal at 68°F and 14.7 psia	2
Specific heat of vapor at constant volume	0.085 Btu/(lb-°F) at 59°F	2
Specific heat of vapor at constant pressure	0.115 Btu/(lb-°F)	2
Specific heat of liquid at constant pressure	0.226 Btu/(lb-°F)	2
Latent heat of vaporization	123.8 Btu/lb at −29.3°F	2
Liquid surface tension	25.4 dynes/cm at -22°F	6
Average coefficient of thermal expansion, 0-60°F	0.00110/°F	4

Additional properties useful in determining other properties from physical property correlations:

Critical temperature	291.2°F	2
Critical pressure	1118.36 psia	2
Critical density	35.77 lb/ft³	6
Energy of molecular interaction	357 kJ	7
Effective molecular diameter	4.115 Å	7

from the larynx and trachea may be blood-tinged. Inhalation of chlorine in higher concentrations affects both the upper and lower respiratory tract and also produces pulmonary edema. The most pronounced symptoms are suffocation, constriction in the chest, and tightness in the throat. A concentration of 833 parts per million (ppm) inhaled for 30-60 minutes has caused death.[9] Skin

TABLE 8.2.2   Exposure Limits for Chlorine

Limit	Concentration (ppm)	Description	Ref.
IDLH	25	This concentration poses an immediate danger to life and health (i.e., causes of irreversible toxic effects for a 30-minute exposure).	12
PEL	1	A time-weighted 8-hour exposure to this concentration, as set by OSHA, should result in no adverse effects for the average worker.	10
TC_{10}	15	This concentration is the lowest published concentration causing toxic effects (irritation) for a 1-minute exposure.	10
LC_{10}	430	This concentration is the lowest published lethal concentration for a human over a 30-minute exposure.	10

TABLE 8.2.3   Predicted Human Health Effects of Exposure to Various Concentrations of Chlorine[a]

Concentration (ppm)	Predicted Effect
3.5	Odor threshold
4	Maximum concentration tolerated without serious effects for a 1-hour exposure
30	Minimum concentration known to cause coughing
40-60	May be dangerous in 30 minutes
1000	Likely to be fatal after only a few deep breaths

[a] From *The Handbook of Chlorinating*, 2nd ed.[2]

contact with the liquid or vapor may result in ulceration and necrosis. Tables 8.2.2 and 8.2.3 summarize some of the relevant exposure limits for chlorine and predicted effects on human health of exposure to various concentrations of chlorine.

Chlorine is manufactured primarily by electrolysis of brine in diaphragm, mercury, or membrane cells. Approximately 95% of the chlorine in the United States is produced in such cells, with diaphragm cells being the predominant method.[11] Figure 8.2.1 is a flow diagram of a typical chlorine manufacturing process using a diaphragm cell.

In a typical *diaphragm cell*, sodium or potassium chloride brine is electrolyzed to chlorine gas at a graphite anode. Anodes usually are constructed of impregnated carbon, titanium, or tantalum. Sodium or potassium ions from the brine migrate through an asbestos diaphragm to the cathode, where sodium or potassium hydroxide is formed and hydrogen gas is liberated.

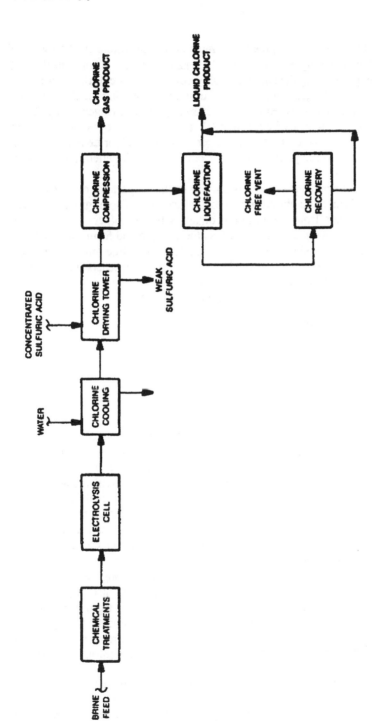

Figure 8.2.1. Conceptual process diagram of typical chlorine manufacturing process.

The diaphragm cell typically operates at a temperature in the range of 176-210°F.[13] The electrolyte is heated by the passage of current through the cell resistance. The gas leaving the anode is highly corrosive wet chlorine containing oxygen, nitrogen, hydrogen, and/or carbon dioxide, depending on the cell type used. The wet chlorine is cooled either in titanium heat exchangers or by direct contact with water in a packed tower; then it is dried by countercurrent scrubbing with sulfuric acid in a contact tower. The dried chlorine is transported as a gas by pipeline to the point of use or is compressed, liquefied, and pumped to storage tanks.

The *membrane cell* process is similar to the diaphragm except that an ion-exchange polymer membrane replaces the asbestos diaphragm to prevent chlorine and hydrogen from coming into contact with each other. The *mercury cell* process uses a moving bed of mercury as the cathode and forms an amalgam with the sodium or potassium ions as they migrate. The dilute amalgam is then fed to a decomposer, which is a packed bed reactor, where it reacts with water to form sodium hydroxide, hydrogen, and mercury.

High hazard areas in chlorine manufacture include:

1. The electrolytes cell
2. Chlorine cooling
3. The acid scrubber
4. The chlorine compressor
5. Chlorine liquefaction
6. Storage and transfer

The electrolysis cell is a critical area of the process, since a potentially hazardous situation exists as a result of chlorine-hydrogen and hydrogen-oxygen combinations present in the same manufacturing system. If the gases were allowed to come into contact, a highly explosive mixture could result. When both gases are present in concentrations greater than approximately 15%, the mixture can explode when initiated thermally or by ultraviolet radiations.[8] It should be noted that hydrogen levels above 4% in air can also lead to fire and/or explosions. The cooling section of the process is subject to the corrosive attack of wet chlorine, which can lead to equipment failure and release of chlorine. Loss of cooling could result in overpressure and release through a relief valve. The acid scrubber is subject to attack from wet chlorine and also sulfuric acid. In addition, if packed towers are used, plugging could result from a buildup of sodium sulfate, leading to overpressure in the tower or process piping. The chlorine compression section presents hazards of overpressure of a pressurized gas system and possible compressor failure from corrosion caused by insufficient water removal from the chlorine gas.

A potentially hazardous by-product of chlorine manufacture is nitrogen trichloride, which is unstable and highly explosive. It can be formed from a combination of chlorine with nitrogen compounds in the brine feed, ammonia in the water used in direct-contact cooling, or nitrogen compounds in sulfuric acid used in chlorine drying. An additional consideration in chlorine manufacture is the potential buildup of noncondensable gases in the raw chlorine from the electrolysis cell, which collect in the liquefaction equipment.

The most important industrial uses of chlorine are based on its general reactivity and its properties as an oxidizer. Aliphatic and aromatic organic compounds are chlorinated through addition and substitution reactions. Chlorine is also used to produce a wide variety of inorganic chemicals such as sodium hypochlorite for bleach. Another major use is in the area of water treatment i.e. chlorinating drinking water, wastewater and sewage treatment, and cooling tower water. Table 8.2.4 lists some of the end uses of chlorine.

A large number of accidents have occurred in systems used for the delivery, transportation, and storage of chlorine. Flexible metal hoses are generally used for loading and unloading barges with liquid chlorine, and several accidents have been attributed to hose failure. In most of these cases, the transfer was stopped with the transfer hose full of liquid; this resulted in over pressuring the line due to warming and thermal expansion of the chlorine, leading to failure. In one such case the pressure was estimated to be 1500 psig at the time of rupture. An employee working in the immediate area inhaled large quantities of chlorine gas and died of massive pulmonary edema and lung hemorrhaging. In another accident, a chlorine vaporizer equipped with a rupture disk in series with a safety valve failed to work properly. The vaporizer was being preheated prior to use. However, after 5 minutes a gasket failed and a large release of chlorine filled the area. The gasket failure occurred because of excessive pressure as a direct result of leakage from the rupture that caused the safety valve to corrode. Another case involved a structural failure in a reactor column, causing a section of the shell, complete with brick lining, to collapse. This allowed chlorine gas to escape to the atmosphere, contaminating the area. An employee working in the area was permanently disabled as a result of exposure.

8.3 AMMONIA

At atmospheric temperatures and pressures, anhydrous ammonia is a pungent, colorless gas, which may easily be compressed or cooled to a colorless liquid. Its more important physical and chemical properties are presented in Table 8.3.1. Pure liquid ammonia is lighter than water, and pure gaseous ammonia is lighter than air; therefore, a cloud of pure ammonia gas will rise into the atmosphere. However, depending on the pressure and temperature, denser-than-air mixtures of air and ammonia may also be formed. Water vapor may condense out of an air-ammonia mixture because of the cooling effect of evaporating ammonia, causing fog. Because of the higher specific gravity of the cooled air, this fog could spread laterally over the ground.[15]

TABLE 8.2.4   Typical Uses of Chlorine[a]

Organic Chemical
Manufacture

Allyl chloride	Chloroform	Methyl chloride
Amyl chloride	Chlorophenols	Methylene chloride
Benzene hexachloride	Chloroprene	Perchloroethylene
		Perchloromethyl
Carbon tetrachloride	Chlorosulfonic acid	mercaptan
Chloral	Chlorotoluenes	Phosgene
Chlorinated		Polychlorinated
naphthalenes	Dichlorobenzenes	biphenyls
	2,4- Dichlorophenoxyacetic	
Chlorinated paraffins	acid	Tetrachlorobenzene
		Tetrachlorophthalic
Chlorinated waxes	Dichloropropane	anhydride
Chloroacetic acid	Dichloropropenes	Trichlorobenzene
Chloroacetyl chloride	Ethyl chloride	1,1,1-Trichloroethane
Chloroanilines	Ethylene dichloride	1,1,2-Trichloroethane
Chloroanthraquinone	Hexachlorocyclopentadiene	Trichloroethylene
		2,4,5-
		Trichlorohenoxyacetic
Chlorobenzene	Hexachloroethane	acid
Chlorofluoro-		
hydrocarbons	Methallyl chloride	

Inorganic Chemical
Manufacture

Anhydrous aluminum		
chloride	Iodine monochloride	Silicon tetrachloride
Antimony		
pentachloride	Iodine trichloride	Sulfur dichloride
Antimony trichloride	Mercuric chloride	Sulfur monochloride
Arsenic chloride	Mercurous chloride	Sulfuryl chloride
Bismuth trichloride	Molybdenum pentachloride	Stannous chloride
Chlorinated		
isocyanurates	Phosphorous oxychloride	Titanium tetrachloride
Chlorine trifluoride	Phosphorous pentachloride	Titanium trichloride
Ferric chloride	Phosphorous trichloride	Zinc chloride
Hydrochloric acid		

Bleach Manufacture
Sanitizing and Disinfecting (e.g., for municipal water supplies, swimming pools)
Waste and Sewage Treatment
Slimicide

[a] From *Chemical Origins Markets*, 5th ed.[14]

TABLE 8.3.1 Physical Properties of Anhydrous (NH₃): CAS Registry Number 07664-41-7

Property	Value	Ref.
Molecular Weight	17.03	
Normal Boiling Point	-28.17°F at 14.7 psia	17
Melting Point	-107.93°F	17
Liquid specific gravity ($H_2O = 1$)	0.6815 at –27.7°F	17
Vapor specific gravity (air = 1)	0.5970 at 32°F	17
Vapor pressure	128.8 psia at 70°F	17
Vapor pressure equation		

$$\log P' = A - \frac{B}{T + C}$$

where P' = vapor pressure (mmHg)

T = Temperature (°C)

A = 7.36050, a constant

B = 926.132, a constant

C = 240.17, a constant

Liquid Viscosity	0.255 centipoise at –33.5°C	5
Solubility in water at various temperatures and 1 atm (wt%)	32°F 42.8	16
	50°F 33.1	
	68°F 23.4	
	86°F 14.1	
Specific heat of vapor at constant volume	0.38 Btu/(lb-°F) at 32°F	3
Specific heat of vapor at constant pressure	0.5 Btu/(lb-°F) at 32°F	3
Specific heat of liquid at constant pressure	1.10 Btu/(lb-°F) at 32°F	3
Latent heat of vaporization	588.2 Btu/lb at –27.7°F	17
Liquid surface tension	23.4 dynes/cm at 52°F	6

Additional properties useful in determining other properties from physical property correlations:

Critical temperature	270.32°F	17
Critical pressure	1639.1 psia	17
Critical density	14.66 lb/ft³	3

Liquid anhydrous ammonia has a large coefficient of expansion, which could lead to serious problems during storage and transfer due to the hydrostatic pressure exerted by the liquid. The flammability range of ammonia in air at atmospheric pressure is from 16 to 25% ammonia by volume. Increasing the temperature and pressure of the ammonia broadens the flammability range. Ammonia is readily absorbed in water to make ammonia liquor (ammonium hydroxide or aqua ammonia).   The dissolution of ammonia in water is accompanied by relatively large heats of solution.   Approximately 938 Btu of heat is evolved when 2.2 pounds of ammonia gas is dissolved in water.[16]

Pure ammonia is very stable under normal conditions; however, it is a highly reactive chemical, forming ammonium salts with inorganic and organic acids. Ammonia reacts with chlorine in dilute solution to give chloramines, an important reaction in water purification.[16] Because of the alkaline characteristics of ammonia, it is used as a neutralizing agent in a number of processes. Explosive materials that can be formed include metal hydrazines, which are produced from the reaction of alkali metals and liquid ammonia.   Acetylides, which are highly explosive in the dry state, are formed in the presence of ammonia solutions of copper, mercury, or silver salts.[17]

Depending on the concentration, the effects of exposure to ammonia gas range from mild irritation to severe corrosion of sensitive membranes of the nose, eyes, throat, and lungs.[16] Because of the high solubility of ammonia in water, it is particularly irritating to most skin surfaces. A concentration of 500 ppm has been designated as the IDLH (*immediately dangerous to life and health)* concentration, which is based on a 30-minute exposure.   Table 8.3.2 summarizes some of the relevant exposure limits for ammonia gas.

The predicted effects on human health of increasing concentrations of ammonia (Table 8.3.3) are immediately recognizable at low concentrations; it is highly unlikely that anyone would become overexposed unknowingly. Ammonia is not a cumulative metabolic poison; ammonium ions are actually important constituents of living systems. However, inhalation of high levels of ammonia gas may have fatal consequences as a result of the spasm, inflammation, and edema of the larynx and bronchi, chemical pneumonitis, and pulmonary edema. Exposure of the eyes to high concentrations may result in ulceration of the conjunctiva and cornea and destruction of all ocular tissues.[19] Contact of the skin with liquid ammonia may result in severe injury by freezing the tissue, since liquid ammonia vaporizes rapidly when released to the atmosphere and will absorb heat from any substance it contacts. If the skin is moist, it may also cause severe burns from the caustic action of the ammonium hydroxide produced.

Anhydrous ammonia is prepared by the reaction of hydrogen and nitrogen ("synthesis" or "syn" gas) in the presence of a catalyst at elevated temperatures and pressures. The manufacturing process consists of three basic steps: synthesis gas preparation, purification, and ammonia synthesis. The first step involves the production of hydrogen and the introduction of the stoichiometric amount of nitrogen. In the second step, catalyst poisons (carbon dioxide, carbon monoxide, and water) are removed from the synthesis gas. The third step includes the catalytic fixation of nitrogen at high temperatures and

TABLE 8.3.2 Exposure Limits for Anhydrous Ammonia[a]

Limit	Concentration (ppm)	Description
IDLH	500	This concentration poses an immediate danger to life and health (i.e., causes of irreversible toxic effects for a 30-minute exposure).
PEL	50	This concentration was determined by OSHA to be the time-weighted 8-hour exposure limit that should result in no adverse effects for the average worker, healthy, male worker.
LC_{10}	30,000	This concentration is the lowest published lethal concentration for a human over a 5-minute exposure.
TC_{10}	20	This concentration is the lowest published lethal concentration causing toxic effects (irritation).

[a] Data from Registry of Toxic Effects of Chemical Substances, 1981-1982 ed.[18]

TABLE 8.3.3 Predicted Effects on Human Health of Exposure to Various Concentrations of Anhydrous Ammonia[a]

Concentration (ppm)	Effect
5	Least perceptible odor
20-50	Readily detectable odor
40	A few individuals may suffer slight eye irritation
100	Noticeable irritation of eyes and nasal passages after a few minutes exposure
150-200	General discomfort and eye tearing; no lasting effect from short exposure
400	Severe irritation of the throat, nasal passages, and upper respiratory tract
700	Severe eye irritation; no permanent effect if the exposure is limited to less than 30 minutes
1700	Serious coughing, bronchial spasms, burning, and blistering of the skin; less than 30 minutes
5000-10,000	Serious edema, strangulation, asphyxia; rapidly fatal
10,000	Immediately fatal

[a] Data from Kirk-Othmer Encyclopedia of Chemical Technology, 3rd ed.,[16] and Anhydrous Ammonia, 7th ed.[17]

pressures and the recovery of ammonia. The specific processes used by the numerous producers of ammonia differ primarily in the source of hydrogen for the synthesis gas and in the temperature and pressure of the ammonia synthesis loop.

The main sources of hydrogen in modern ammonia plants are coal, petroleum fractions, and natural gas, with the latter being the principal source in

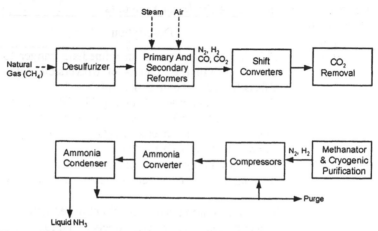

Figure 8.3.1. Conceptual diagram of typical ammonia production process.

commercial practice. In general, the most economic feedstock has the highest ratio of hydrogen in carbon. The two hydrogen generation techniques used for processing these raw materials are partial oxidation (reaction with oxygen) and steam reforming (reaction with steam). Of these, steam reforming is the more widely used process; partial oxidation processes are employed where steam-reformable feeds are not available or where favorable economic conditions exist, in certain special situations.

Figure 8.3.1 is a typical process diagram for the production of ammonia by steam reforming. The first step in the preparation of the synthesis gas is desulfurization of the hydrocarbon feed. This is necessary because sulfur poisons the nickel catalyst (albeit reversibly) in the reformers, even at very low concentrations. Steam reforming of hydrocarbon feedstock is carried out in the primary and secondary reformers.

The primary reformer is a refractory-lined furnace that contains vertically suspended tubes filled with a nickel-based catalyst. In the primary reformer, the feed reacts with steam to produce hydrogen gas and carbon monoxide. Some of the carbon monoxide also reacts with steam in the "shift conversion" reaction to produce carbon dioxide and hydrogen. The secondary reformer is a refractory-lined pressure vessel that contains additional reforming catalyst. Primary reformer effluent gas is mixed with air before entering the secondary reformer, thus serving the twofold purpose of supplying both the stiochiometric amount of nitrogen and oxygen for combustion, which furnishes the heat required for the reforming reaction.

Regardless of the hydrogen generation technique used, the unpurified syn gas contains the oxides of carbon, which deactivate the ammonia synthesis catalyst and must be removed. In the shift converters, carbon monoxide is catalytically converted to carbon dioxide, which is removed more easily than carbon monoxide and hydrogen gas. The next purification step is the removal of carbon dioxide. In commercial processes the gas generally is absorbed into a solvent under pressure and the solvent is recovered in a stripping column. The final purification step involves removal of the residual carbon monoxide and carbon dioxide in a methanator and, in some plants, cryogenic purification. In the methanator, carbon monoxide and carbon dioxide are catalytically converted to methane, which passes through the ammonia synthesis loop as an inert. The cryogenic purification removes the excess nitrogen that was added to the secondary reformer. The purpose of the nitrogen removal is to avoid excessive loss of hydrogen and excessive compression costs in the ammonia synthesis loop.

The purified syn gas next flows to the final production stage, which is ammonia synthesis and recovery. The first step is compression of the syn gas. Synthesis pressures range from 2000 to 10,000 psi depending on the quality of the syn gas and certain other conditions, such as production requirements per converter.[20]

Basically, there are two classes of ammonia converters, tubular and multiple bed. The tubular bed reactor is limited in capacity to a maximum of about 500 tons/day.[16] In most reactor designs, the cold inlet synthesis gas flows through an annular space between the converter shell and the catalyst cartridge. This maintains the shell at a low temperature, minimizing the possibility of hydrogen embrittlement, which can occur at normal synthesis pressures. The inlet gas is then preheated to synthesis temperature by the exit gas in an internal heat exchanger, after which it enters the interior of the ammonia converter, which contains the promoted iron catalyst.

Gas recirculation in the ammonia synthesis section is necessary because only 9-30% conversion is obtained per pass over the catalyst.[20] There are two types of synthesis loops. One type recovers ammonia product before recycle compression. Inerts entering with the makeup gas are removed with a purge stream. The ammonia is recovered by condensation, which requires refrigeration. Since anhydrous ammonia is readily available, it is normally used as the refrigerant.

The applications of anhydrous ammonia in the United States are very diverse. Some of the minor uses of ammonia include the manufacture of rubber, water purification, food and beverage treatment, the production of pulp and paper, the preparation of cleaners and detergents, and leather and textile treatment. Ammonia is also used in the manufacture of many important industrial chemicals, which are listed in Table 8.3.4. The largest consumer of anhydrous ammonia in the United States, accounting for 70-80% of all ammonia produced in the past decade, is the fertilizer industry. Ammonia is used as a fertilizer by direct application; however, it is also used to manufacture ammonium nitrate, ammonium sulfate, monoammonium phosphate, diammonium phosphate, and urea. Other major uses include the manufacture of nitric acid by oxidation of ammonia, the manufacture of hydrazine which is used

TABLE 8.3.4   Anhydrous and Aqueous Ammonia Products[a]

Ammonium acetate	Ammonium paratungstate
Ammonium adipate	Urea
Ammonium benzoate	Monoammonium phosphate
Ammonium bicarbonate	Diammonium phosphate
Ammonium bifluoride	Nitric oxide
Ammonium binoxalate	Acrylonitrile
Ammonium bisulfate	Caprolactam
Ammonium bitartrate	Monomethylamine
Ammonium tetraborate	Dimethylamine
Ammonium bromide	Hexamethylenetetramine
Ammonium carbonate	Trimethylamine
Ammonium chloride	Monoethanolamine
Ammonium citrate	Diethanolamine
Ammonium dichromate	Triethanolamine
Ammonium fluoride	Hydrogen Cyanide
Ammonium fluorosilicate	Fatty nitrogen compounds (nitriles,
Ammonium gluconate	amines, quaternary ammonium
Ammonium iodide	compounds)
Ammonium molybdate	Boron nitride
Ammonium nitrate	Calcium carbonate (precipitated from
Ammonium oxalate	calcium chloride)
Ammonium perchlorate	Hydrazine
Ammonium picrate	Hydrogen (high purity)
Ammonium polysulfide	Lead hydroxide
Ammonium salicylate	Lithium amide
Ammonium stearate	Methyl ethyl pyridine
Ammonium sulfate	Sodamide
Ammonium sulfide (hydrosulfide)	Sodium cyanide
Ammonium tartrate	Nitrogen dioxide
Ammonium thiocyanate	Nitric acid
Ammonium thiosulfate	

[a] From *Chemical Origins Markets,* 5th ed.[14]

as a rocket fuel, the production of a variety of resins and fibers, and as a refrigerant in low temperature applications.

Numerous accidents have been reported relating injuries to rupture of liquid-full transfer lined used to unload trucks and tank cars of anhydrous ammonia. As described previously, liquid anhydrous ammonia has a large coefficient of thermal expansion, which can cause serious problems when transfer operations are interrupted. Many of the accidents reported could have been prevented by more careful engineering. Part IV presents methods of hazard and risk assessment that should be applied to all areas of chemical processing, including transportation and storage.

As described earlier, the synthesis gas is scrubbed in a contact tower at approximately 2000 psig, usually with a weak aqua ammonia solution. At one such plant the solution was removed from the base of the tower by a level controller and was returned to a 2000-gallon vertical hold-up tank. When the level controller failed in the open position, the liquid and gas content of the scrubber was released into the hold-up tank, which was equipped with a 2-inch atmospheric vent. The surge of gas into the hold-up tank caused a rupture of the vessel. The tank head was blown upward, shearing the vent line and then striking and damaging an overhead conduit rack before coming to rest on the ground near the base of the tank. The supporting steel for the tank head was blown upward about 50 feet. The escaping gas, containing approximately 75% hydrogen, immediately ignited and caused a costly fire.

In July 1985 an explosion at another ammonia plant resulted in two fatalities and serious injuries to a third operator. A water feed for the contact spray tower used for purification of the synthesis gas was due for changeover during normal operation. Pump A with turbine was to be put into operation and pump B taken out. Three operators performed the change. However, the suction valve for the pump that was going into operation (pump A) was not opened. The water supply failed when the low level trip on the absorption tower switched off pump B, shutting down the plant. Because the failure of the water supply was not discovered, the shift supervisor and the operators decided to restart the plant with water pump A. When pump A was activated, however, the operator in the control room reported via radio to the operators in the pump hall that the water flow was varying between 400 and 1700 m^3/h. The operators then found the suction valve that was closed and tried to open it. Because of the high differential pressure against the gate valve, however, this effort proved fruitless. A small (50-mm) bypass valve that was already open had been giving the reported flow. At this moment, it was noticed that the axial bearing on the pump was becoming red hot. Soon there was water leakage on the suction side of the pump. Because the suction valve was still closed, the pump started cavitating, and the axial thrust force-balancing system lost its water flow. Consequently, the thrust collar of the bearing came into contact with the bearing cover, the temperature of which rapidly rose to an estimated 750°C. The result was a backflow of water from the absorption tower followed by hydrogen gas, which entered the pump hall via the leaking flange. The leaking hydrogen exploded and caused a shock wave that initiated several more detonations. The walls were blown out and the ceiling was lifted about 3 feet, breaking the main gas inlet line to the absorption tower. The explosion force is estimated to have been equivalent to 200-500 pounds of TNT.

8.4 HYDROGEN CYANIDE

Anhydrous hydrogen cyanide is a colorless or pale yellow liquid with a mild odor similar to that of bitter almonds. The liquid boils at 78.3°F and 1.0 atm and forms a colorless, flammable, toxic gas. Hydrogen cyanide is completely soluble in water and is slightly less dense than air, although a mixture of hydrogen cyanide in moist air may stay near ground level. The physical and chemical properties of anhydrous cyanide are listed in Table 8.4.1.

Three chemical properties of hydrogen cyanide contribute to the potential for an accidental release of the chemical. [21,22]

1. Hydrogen cyanide is flammable in air at concentrations from 6 to 41% hydrogen cyanide.
2. The addition of alkaline chemicals, water, and/or heat may promote self-polymerization and decomposition of hydrogen cyanide. The self-polymerization reaction is exothermic, and the heat released will promote further polymerization. The heat generation will also result in the decomposition of hydrogen cyanide into ammonia and formate. The pressure rise from polymerization or decomposition reactions can become explosive. Small amounts of acid, such as sulfuric or phosphoric, will help to stabilize the hydrogen cyanide against polymerization.
3. The addition of large quantities of acid (>15wt% of concentrated sulfuric acid) can cause rapid, and highly exothermic decomposition of hydrogen cyanide. When sulfuric acid is involved, the decomposition by-products will be sulfur dioxide and carbon dioxide.

Hydrogen cyanide is highly toxic if ingested, inhaled, or absorbed into the skin. It is a true non-cumulative protoplasmic poison (i.e., it can be detoxified readily). Hydrogen cyanide combines at the blood-tissue interfaces with the enzymes that regulate oxygen transfer to the cellular tissues. Unless the cyanide is removed, death results through asphyxia. The warning signs of hydrogen cyanide poisoning include dizziness, headache, rapid pulse, nausea, reddening skin, and bloodshot eyes. More prolonged exposure can cause vomiting and labored breathing, followed by unconsciousness, cessation of breathing, rapid and weak heart beat, and death. Severe exposure (by inhalation) can cause immediate unconsciousness; this rapid knockdown power without any irritation or odor detectable to some people makes hydrogen cyanide more dangerous than other materials of comparable toxicity (e.g., hydrogen sulfide). The high toxicity of hydrogen cyanide is of concern because even small leaks in a piping system can be dangerous to operating personnel. Tables 8.4.1 and 8.4.2 summarize some of the relevant exposure limits for hydrogen cyanide and predicted effects on human health of exposure to various concentrations of this poison.

Two processes for manufacturing hydrogen cyanide account for most of the total production in this country. The most widely used process produces hydrogen cyanide by reacting natural gas (methane), ammonia, and air. A second widely used process, actually a variation of the first, is called the BMA process; it produces hydrogen cyanide by reacting methane with ammonia. In the manufacture of hydrogen cyanide (Fig. 8.4.3), ammonia, methane or natural gas, and air are preheated to about 750-900°C, mixed, and sent to a packed bed

TABLE 8.4.1 Exposure Limits for Hydrogen Cyanide

Limit	Concentration (ppm)	Description	Ref.
IDLH	50	This concentration poses an immediate danger to life and health (i.e., causes of irreversible toxic effects for a 30-minute exposure).	23
PEL	10	A time-weighted, 8-hour exposure to this concentration, as set by OSHA, should result in no adverse effects for the average, healthy male worker	10
LC_{10}	178	This concentration is the lowest published lethal concentration for a human over a 10-minute exposure.	10

TABLE 8.4.2 Predicted Effects to Human Health of Exposure to Various Concentrations of Hydrogen Cyanide[a]

Concentration (ppm)	Predicted Effect
2-5	Odor threshold
20	Causes slight symptoms, including headache and dizziness after several hours
50	Causes disturbances within an hour
100	Dangerous for exposures of 30-60 minutes
300	Rapidly fatal unless prompt, effective first aid is administered

[a] From *Kirk-Othmer Encyclopedia of Chemical Technology*, 3rd ed.[21]

reactor. The reactor is typically packed with a catalytic wire gauze composed of platinum or a platinum-rhodium composite.[24] The exit gas from the reactor contains a mixture of hydrogen cyanide, ammonia, and water vapor (a by-product of the reaction). As illustrated in Fig. 8.4.1, this crude product mixture is sent to an ammonia absorption column, where the ammonia is absorbed in an ammonium phosphate solution.[25] Most of the hydrogen cyanide exists in the gas phase, where it is absorbed, washed, and treated with sulfur dioxide as an inhibitor to prevent polymerization. The ammonium phosphate solution is sent through a series of processing operations, where the ammonia is recovered and recycled back to the reactor.

As described earlier, hydrogen cyanide is a highly toxic substance and as such warrants a high hazard assessment from the point of chemical formation of HCN through the remainder of the process. Four pieces of process equipment, however, are generally considered to be in the high hazard category: (1) the cyanide reactor, (2) the hydrogen cyanide absorber, (3) the hydrogen cyanide stripper, and (4) the hydrogen cyanide fractionator.

TABLE 8.4.3 **Physical Properties of Hydrogen Cyanide (HCN): CAS**
Registry Number 74-90-8

Property	Value	Ref.
Molecular Weight	27.03	
Normal Boiling Point	78.3°F at 1 atm	21
Melting Point	8.17°F	22
Liquid specific gravity (H_2O = 1)	0.6884 at 68°F	21
Vapor specific gravity (air = 1)	0.947 at 68°F	21
Vapor pressure	0.348 atm at 32°F	21
Vapor pressure equation		

$$\log P' = A - \frac{B}{T+C}$$

where P' = vapor pressure (mmHg)
T = Temperature (°C)
A = 7.5282, a constant
B = 1329.5, a constant
C = 260.4, a constant

Liquid Viscosity	0.2014 centipoise at 68°F	21
Solubility in water	Complete	
Specific heat of vapor at constant pressure	16.94 Btu/(lbmol-°F) at 62.4°F	22
Latent heat of vaporization	10,834 Btu/lbmol at 77°F	22
Liquid surface tension	17.2 dynes/cm at 77°F	22
Heat of Combustion	287,000 Btu/lbmol	22
Autoignition temperature	1000°F	22
Explosive range (min-max) in air at 1 atm and 68°F	6-41 vol%	22
Flash Point, TCC (ASTM D-56)	0°F	22

Additional properties useful in determining other properties from physical property correlations:

Critical temperature	362.2°F	22
Critical pressure	53.2 atm	22
Critical density	12.2 lb/ft^3	22

The *cyanide reactor* is critical because of the high temperatures that are involved. Overheating the reactor could result in uncontrollable combustion reactions or explosions.[27] These uncontrollable combustion reactions or explosions could result in the physical breakdown of the reactor vessel by

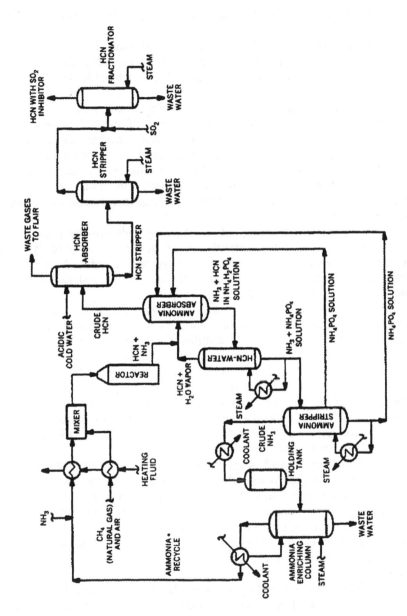

Figure 8.4.1. Conceptual process diagram of typical hydrogen cyanide manufacturing process.

thermal fatigue or overpressure. The three possible causes of overheating are:

1. Poor heat distribution within the reactor bed, resulting in hot spots
2. Overheating raw materials before they enter the cyanide reactor
3. Loss of composition or quantity control of raw material feeds

Hot spot formation within the reactor can result in catalyst breakdown or physical deterioration of the reactor vessel.[27] If the endothermic cyanide reaction has ceased (e.g., because of poor catalyst performance), the reactor is likely to overheat. Iron is a decomposition catalyst for hydrogen cyanide and ammonia under the conditions present in the cyanide reactor, and exposed iron surfaces in the reactor or reactor feed system can result in uncontrolled decomposition, which could in turn lead to an accidental release by overheating and overpressure.

Only a small inventory of hydrogen cyanide will be present in the cyanide reactor. Therefore, catastrophic failure of this unit is not likely to result directly in the release of large quantities of hydrogen cyanide. However, such failure could result in damage to other portions of the system where larger quantities of the gas are present.

The *hydrogen cyanide absorber, stripper*, and *fractionator* are high hazard areas because they contain inventories of concentrated hydrogen cyanide. All the associated pumps, piping, and fittings for these units are also high hazard areas. Controlling the pH of these systems is important because the vapor pressure of hydrogen cyanide is dependent on pH. Reliable pH control is particularly important at the hydrogen cyanide fractionator, where acid is intentionally added as a stabilizer to the feed stream. Although only a very small quantity of acid is added at this point, the potential still exists for a loss of acid flow control. Large excesses of acid can result in a violent hydrogen cyanide decomposition reaction.

The primary hazard associated with the storage of hydrogen cyanide is the potential for self-polymerization. If this occurs, rapid temperature and pressure increases in a storage system can result in a breach of containment. Additional hazards associated with hydrogen cyanide storage include the potential for overfilling, corrosion, and contamination caused by the backflow of process materials into the storage tank. The prevention of accidental releases relies on a combination of technological, administrative, and operational practices.

The major use of hydrogen cyanide in the United States today is for the production of adiponitrile, which is used primarily as an intermediate for hexamethylenediamine. This intermediate is the primary ingredient for the production of nylon 6,6.[28] Another major use of hydrogen cyanide is for the

TABLE 8.4.4 Typical Uses of Hydrogen Cyanide[a]

Acetone cyanohydrin
Acrylonitrile
Adiponitrile
Aminopolycarboxylic acids
Barium cyanide
β-amines
Cyanuric chloride
Diaminomaleonitrile
Lactic acid
Methionine
Sodium cyanide
Tertiary alkyl amines

[a] From *Chemical Origins Markets*, 5th ed.[14]

production of acetone cyanohydrin which is used as an intermediate for methyl methacrylate (marketed as Plexiglas). Hydrogen cyanide is also used in the production of cyanuric chloride (an intermediate in pesticide manufacturing), miscellaneous chelating agents, and sodium cyanide. Table 8.4.4 lists a number of products that are manufactured using hydrogen cyanide.[14]

Most reported accidents with hydrogen cyanide involve operators inhaling or being splashed with liquid hydrogen cyanide. Some of these accidents are due to equipment failure (blocked lines, frozen valves, etc.), and some are due to operator error.

8.5 HYDROGEN FLUORIDE

Anhydrous hydrogen fluoride is a clear, colorless, corrosive liquid with a pungent, irritating odor. Hydrogen fluoride is hygroscopic and fumes upon exposure to moist air. Hydrogen fluoride in aqueous solutions is hydrofluoric acid, a highly corrosive liquid. Concentrated aqueous solutions also boil at relatively low temperatures and fume upon contact with moist air. The physical properties of anhydrous and aqueous hydrofluoric acid are listed in Table 8.5.1. As a result of the relatively low atmospheric boiling point of hydrogen fluoride (67°C), spills and leaks of liquid can result in hazardous releases to the atmosphere nearly as severe as direct gas or vapor releases. In addition, since the vapor density of hydrogen fluoride is greater than that of air, releases will remain close to the ground and could create potentially dangerous situations for workers and the surrounding communities.

Liquid hydrogen fluoride has a large coefficient of thermal expansion, and temperature increases can result in containment failure if there is no room for thermal expansion of the liquid. Thus liquid-full equipment presents a special hazard. A liquid-full vessel is a vessel that is not vented and has little or

TABLE 8.5.1   Physical Properties of Hydrogen Fluoride: CAS Registry
               Number 07664-39-3

Property	Value	Ref.
Molecular Weight	20.01	
Normal Boiling Point	67.12°F at 14.7 psia	29
Melting Point	-118.4°F	29
Liquid specific gravity (H_2O = 1)	0.991 at 67.15°F	6
Vapor specific gravity (air = 1)	2.4 at 68°F	3
Vapor pressure	17.8 psia at 77°F	29
Vapor pressure equation		

$$\log P' = A - \frac{B}{T + C}$$

Where P' = vapor pressure (mmHg)
        T = Temperature (°C)
        A = 7.68098, a constant
        B = 1,475.60, a constant
        C = 287.88, a constant

Liquid Viscosity	0.256 centipoise at 32°F	5
Solubility in water	Complete	29
Specific heat of vapor at constant volume	0.55 Btu/(lb-°F) at 68°F	29
Specific heat of vapor at constant pressure	2.99 Btu/(lb-°F) at 69°F	29
Specific heat of liquid at constant pressure	0.62 Btu/(lb-°F) at 68°F	29
Latent heat of vaporization	1.62 Btu/lb at 67.15°F	5
Liquid surface tension	10.1 dynes/cm at 32°F	29
Average coefficient of thermal expansion, 0-60°F	0.00112°F	31

Additional properties useful in determining other properties from physical property correlations:

Critical temperature	370°F	29
Critical pressure	940 psia	29
Critical density	18.10 lb/ft^3	29
Energy of molecular interaction	355 kJ	7
Effective molecular diameter	3.240	7

no vapor space present above the liquid. A liquid-full line is a section of pipe that is sealed off at both ends and is full of liquid with little or no vapor space.

Hydrogen fluoride, whether anhydrous or in aqueous solutions, is a highly reactive chemical. The most significant chemical properties contributing to the potential for releases are as follows.[29,30]

1. Anhydrous hydrogen fluoride rapidly absorbs moisture to form hydrofluoric acid, which is corrosive to most metals and results in the formation of hydrogen gas in the presence of moisture. This corrosiveness can lead to equipment failure, and the potential buildup of hydrogen gas in confined areas makes for a fire and explosion hazard.
2. Hydrogen fluoride reacts with metal carbonates, oxides, and hydroxides. Accumulation of these fluoride compounds can render valves and other close-fitting moving parts inoperable in a process system, causing possible equipment or process failures. Hydrogen fluoride also attacks glass, silicate ceramics, leather, natural rubber, and wood, but does not promote their combustion.
3. Considerable heat evolves when anhydrous hydrogen fluoride or concentrated hydrofluoric acid is diluted with water. Violent reactions can result from the inappropriate addition of water or caustic solutions to these materials.
4. Anhydrous hydrogen fluoride and hydrofluoric acid react with substances containing silica and silicon oxide to form silicon tetrafluoride and fluorosilic acid. SiF, a colorless gas at ambient temperature, is highly toxic. An equilibrium mixture of SiF_4 in the presence of moisture also contains hydrogen fluoride and hydrofluoric acid.
5. Anhydrous hydrogen fluoride and hydrofluoric acid react exothermally with organic and inorganic reducing agents but do not promote their combustion.
6. Anhydrous hydrogen fluoride reacts with cyanides and sulfides to produce toxic hydrogen cyanide and hydrogen sulfide, respectively. Since both these compounds are flammable, their formation in confined areas can result in potentially explosive mixtures.

Hydrogen fluoride is highly toxic, highly corrosive, and a severe irritant to the skin, eyes, and respiratory system. The toxicology of hydrogen fluoride has been investigated through accidental human exposure and through animal studies.[18,23,32-34] The acute effects of very-short-term exposure to elevated concentrations of hydrogen fluoride, however, are not well documented. Table 8.5.2 shows the exposure limits for hydrogen fluoride. The concentrations at which various acute effects occur vary significantly with time of exposure and with individuals. For instance, inhalation of 50 ppm hydrogen fluoride for 30-60 minutes might be fatal, while a concentration of 110 ppm inhaled for 1 minute might be tolerated with only the initial onset of toxic effects. Less severe exposures cause irritation of the nose and eyes, irritation of the skin, and some degree of conjunctival and respiratory irritation. More severe exposures can lead to severe irritation of the eyes and eyelids, ulceration of the skin, inflammation and congestion of the lungs, and eventual cardiovascular collapse and death. Additional effects may include dyspnea, bronchopneumonia, cyanosis, shock, muscle spasms, convulsions, paresthesia, jaundice, oliguria, albuminuria, hematuria, nausea, vomiting, abdominal pain, diarrhea, and burns

TABLE 8.5.2 Exposure Limits for Hydrogen Fluoride

Limit	Concentration (ppm)	Description	Ref.
IDLH	20	This concentration poses an immediate danger to life and health (i.e., causes of irreversible toxic effects for a 30-minute exposure).	12
PEL	3	A time-weighted 8-hour exposure to this concentration, as set by OSHA, should result in no adverse effects for the average worker.	18
LC_{10}	50	This concentration is the lowest published lethal concentration for a 30-minute exposure.	18
TC_{10}	110	This concentration is the lowest published concentration causing toxic effects (irritation) for a 1-minute exposure.	18

TABLE 8.5.3 Predicted Effects on Human Health of Exposure to Various Concentrations of Hydrogen Fluoride[a]

Concentration (ppm)	Predicted Effect
0.5-3	Odor threshold
2	Repeated 6-hour exposures can result in stinging eyes, and facial skin and nasal irritation
>10	Possible lung injury
>50	Vapor is intolerably irritating and causes damage to the lungs; inhalation may result in serious injury

[a] From *Industrial Hygiene*, Vol. 1, 2nd ed.[2]

of the mouth, esophagus, and digestive tract. The designation of a concentration of 20 ppm as IDLH (immediately dangerous to life and health) is based on a 30-minute exposure. Predicted effects on human health of hydrogen fluoride are presented in Table 8.5.3.

Hydrogen fluoride is manufactured by the reaction with sulfuric acid of fluorspar, a fluorine-containing mineral.

$$CaF_2 + H_2SO_4 \rightarrow CaSO_4 + 2HF \qquad (18.5.1)$$

Figure 8.5.1 presents a block diagram of a typical hydrogen fluoride manufacturing process. The degree to which this reaction proceeds is dependent on the purity and fineness of the fluorspar, the concentration of the sulfuric acid used, the ratio of sulfuric acid to fluorspar, the temperature of the reaction, the time allowed for completion of the reaction, and the intimacy of mixing of the

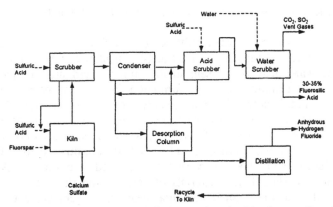

Figure 8.5.1. Conceptual diagram of typical hydrogen fluoride manufacturing process.

acid and spar.

Finely ground acid grade fluorspar (>97% CaF) is reacted in a heated rotating steel kiln with sulfuric acid to form calcium sulfate and hydrogen fluoride. The reaction is endothermic, and heat is supplied either externally by direct fire to the rotary kiln or by the addition of sulfur trioxide and steam to the reaction.[35,36] The heat absorbed is 603 Btu/lb. In a typical process, fluorspar and sulfuric acid are fed continuously and cocurrently to the kiln by a screw conveyor. The reaction is typically carried out at a temperature in the range of 392-482°F. Crude product gas, which exits the reactor at approximately 212-338°F, consists primarily of hydrogen fluoride saturated with sulfuric acid and a variety of impurities, which vary depending on the composition of the raw materials used in the reaction. The gas is fed to a gas scrubber, where it is scrubbed with sulfuric acid to remove small particles of fluorspar and/or calcium sulfate. After leaving the gas scrubber, the crude hydrogen fluoride gas is cooled and liquefied. In a typical process, the gas is cooled in shell and tube heat exchangers from 300 to 70°F. The gas is then contacted with cold liquid hydrogen fluoride at 10°F in contact condensers to produce a liquid hydrogen fluoride product.[35] The uncondensed gases are scrubbed with sulfuric acid to recover additional hydrogen fluoride. The final effluent gases are absorbed in water and recovered as fluorosilic acid. After liquefaction, the crude hydrogen fluoride is distilled to produce anhydrous hydrogen fluoride with a purity greater than 99.9%. Anhydrous hydrogen fluoride has a sizable vapor pressure at room temperature and should be regarded as a liquefied gas under pressure.

Of course when the process of interest is the manufacture of hydrogen fluoride, this chemical is present in high concentrations or in relatively pure form in all areas following the reactor or kiln. Thus, the possibility of a large release of the chemical is greater than it is in a process that consumes hydrogen fluoride as a reactant. Specific high hazard areas in the manufacturing process, excluding bulk storage and transfer, include the reactor (kiln), the hydrogen fluoride scrubber and condensers, the desorption column, and the hydrogen fluoride distillation apparatus.

Although the reaction between sulfuric acid and calcium fluoride (fluorspar) is endothermic, the reactor or kiln can be considered to be a high hazard area because water may be present in the sulfuric acid used in the manufacturing process, resulting in the formation of highly corrosive hydrofluoric acid. Additionally, the corrosiveness of hydrogen fluoride increases with temperature. Undetected corrosion could lead to equipment failure and a possible release of hydrogen fluoride. In addition, other portions of the process, including the scrubbing units, could be affected by similar corrosion problems if hydrofluoric acid vapor is carried to these other portions. A properly designed system should use materials of construction that take this corrosion problem into account. Shell and tube heat exchangers present a potential hazard from tube leakage where water is used as the cooling medium. Undetected small leaks over time could cause corrosion and eventually a failure. An additional concern associated with cooling equipment in this process is the buildup of sulfur deposits on cooling surfaces resulting from sulfur and sulfur-forming impurities present in the initial reaction products. These deposits can lead to clogging of piping and heat transfer equipment and loss of cooling. The desorption and stripping unit operations, which have a thermal energy input, are subject to potential overheating and overpressure. Loss of cooling in condensers could be a cause for overpressure. The reboilers and bottoms pumps are potential weak points in these systems because the operating conditions are severe.

The primary uses for hydrogen fluoride in the United States are in the manufacture of chlorofluorocarbons, aluminum fluoride, sodium aluminum fluoride (cryolite), and uranium tetrafluoride, and in petroleum alkylation.

8.6 SULFURIC ACID

Sulfur dioxide (SO_2) is a colorless gas with a characteristic pungent odor and taste. Although the gas is relatively inert and stable, it is toxic and highly irritating. Its more important physical properties are presented in Table 8.6.1. Because of the low boiling point and because the gas is considerably heavier than air, spills and leaks of liquid sulfur dioxide could result in a vapor cloud or plume that will remain close to the ground, posing a threat to workers and surrounding communities.

Sulfur dioxide is extremely stable to heat, even up to 3600°F.[37] It does not form flammable or explosive mixtures with air. It will, however, react with water or steam to produce toxic and corrosive fumes.[40] When the gas dissolves in water it forms a weak acid solution of sulfurous acid (H_2SO_3), which is corrosive and unstable when exposed to heat.[41]

TABLE 8.6.1 Physical Properties of Sulfur Dioxide (SO₂): CAS Registry Number 7446-09-5

Property	Value	Ref.
Molecular Weight	64.06	
Normal Boiling Point	14.0°F at 14.7 psia	37
Melting Point	-98.9°F	37
Liquid specific gravity (H_2O = 1)	1.436 at 32°F	38
Vapor specific gravity (air = 1)	2.263 at 32°F	37
Vapor pressure	49.1 psia at 70°F	38
Vapor pressure equation		

$$\log P' = A - \frac{B}{T+C}$$

where P' = vapor pressure (mmHg)
T = Temperature (°C)
A = 7.28228, a constant
B = 999.900, a constant
C = 237.190, a constant

Liquid Viscosity	0.49 centipoise at –4°F and 14.22 psai	39
Solubility in water at 1 atm, g/100g H_2O	32°F 22.971	37
	50°F 16.413	
	68°F 11.571	
	86°F 8.247	
	104°F 5.881	
Specific heat of vapor at constant pressure	0.149 Btu/(lb-°F) at 77°F	37
Specific heat of liquid at constant pressure	0.327 Btu/(lb-°F) at 68°F	39
Latent heat of vaporization	167.24 Btu/lb at 14.0°F	37
Liquid surface tenstion	28.59 dynes/cm at 14°F	39

Additional properties useful in determining other properties from physical property correlations:

Critical temperature	315.7°F	37
Critical pressure	1147 psia	37
Critical density	0.51 lb/ft³	37
Energy of molecular interaction	252 kJ	7
Effective molecular diameter	4.29	7

The deleterious effect of sulfur dioxide and sulfites in domestic water is increased corrosivity owing to the lowered pH. However, oxidation of sulfite to sulfate in aqueous solutions uses dissolved oxygen, and this may retard corrosion.[40] While the oxidation of sulfite and sulfurous acid to sulfate and sulfuric acid in the atmosphere is an environmental concern, this reaction is too slow to significantly reduce the concentration of sulfur dioxide in the short time that elapses during a large release. Most metals (e.g., iron, steel, copper, aluminum, and brass) are resistant to commercial dry sulfur dioxide, dry gaseous sulfur dioxide, and hot gaseous sulfur dioxide containing water vapors above the dew point.[37] However, these materials are readily corroded by wet sulfur dioxide gas below the dew point. Zinc is also readily oxidized by sulfur dioxide to form ZnS_2O_4.

Sulfur dioxide is a toxic, highly irritating gas that can have immediate effects on the eyes, throat, lungs, and skin. The toxicology of sulfur dioxide has been studied through cases of accidental human exposure and through animal studies.[42] A concentration of 100 ppm has been designated as the IDLH limit, which is based on a 30-minute exposure.[43] Table 8.6.2 summarizes some of the relevant exposure limits for sulfur dioxide. The primary health effects from exposure to sulfur dioxide occur in the upper respiratory tract and the bronchi. Chronic exposure may result in nasopharyngitis, fatigue, altered sense of smell, and chronic bronchitis symptoms such as dyspnea on exertion, cough, and increased mucous excretion.[43] Chronic exposure also may cause edema of the lungs or glottis and can produce respiratory paralysis.[40] In concentrations greater than 20 ppm, sulfur dioxide is an eye irritant and will cause pain, tearing, inflammation, swelling of tissue, and possible destruction of the eye.[41] Acclimation to the effects of sulfur dioxide has been reported to develop quickly as a result of the depression of the tracheobronchial nerve reflexes; this adjustment is not considered to be a beneficial effect because of the possibility that the absence of discomfort merely removes one measure of protection.[42] The physical effects of increasing levels of gas concentrations on humans are summarized in Table 8.6.3.

Sulfur dioxide is not listed in the National Toxicology Program, the Registry of Toxic Effects of Chemical Substances (1981-1982), nor by the International Agency for Research on Cancer as a carcinogen or potential carcinogen.[45] However, sulfur dioxide has been implicated as a cocarcinogen (promoter) with arsenic.[42] Contact with liquid sulfur dioxide may cause cryogenic burns to the skin in addition to conjunctivitis and corneal opacity of the eye.[41,43] It is also reported that high concentrations of sulfite ion in water may cause eczema.[40]

The reactions for production of sulfuric acid are as follows:

$$S + O_2 \rightarrow SO_2$$
$$SO_2 + 0.5\,O_2 \rightarrow SO_3$$
$$SO_3 + H_2O \rightarrow H_2SO_4 \qquad\qquad (18.6.1)$$

The heats of reaction at 25°C for these reactions are −298, -98, and -130 kJ/gmol, respectively.

A block diagram for the double-absorption sulfuric acid process is given in Fig. 8.6.1. Atomized molten sulfur is burned in a horizontal, brick-

TABLE 8.6.2 Exposure Limits for Sulfur Dioxide

Limit	Concentration (ppm)	Description	Ref.
IDLH	100	This concentration poses an immediate danger to life and health (i.e., causes of irreversible toxic effects for a 30-minute exposure).	43
PEL	5	A time-weighted 8-hour exposure to this concentration, as set by OSHA, should result in no adverse effects for the average worker.	43
LC_{10}	400	This concentration is the lowest published lethal concentration for a human over a 5-minute exposure.	40
TC_{10}	4	This concentration is the lowest published concentration causing toxic effects (irritation).	40

TABLE 8.6.3 Predicted Effects on Human Health of Exposure to Various Concentrations of Sulfur Dioxide[a]

Concentration (ppm)	Predicted Effect
0.3-1	Can be detected by taste and smell
3	Easily noticeable odor
6-12	Immediate irritation of nose and throat
20	Eye irritation-ill effects if exposure is prolonged
50-100	Maximum permissible concentration for exposure lasting 30-60 minutes
>400	Immediately dangerous to life

[a] From the National Joint Health and Safety Committee for Water Service.[44]

lined combustion chamber to produce sulfur dioxide. Since atomization of the molten sulfur is critical to the combustion process, nozzle pressures will approach 150 psi. Sulfur feed lines, storage, and nozzle must be steam jacketed and held at 135-155°C, where viscosity is at a minimum. Gas exit temperature depends on the sulfur dioxide concentration and ranges from 1050 to 1130°C. To drive the reaction equilibrium (sulfur dioxide plus oxygen provides sulfur trioxide) toward the SO_3, current double-absorption plants contain four catalyst passes. Oxidation occurs at 1.2-1.5 atm. Approximately 90-95% of the total sulfur trioxide produced is removed from the converter after three passes. The remaining sulfur trioxide produced in the fourth converter is absorbed in the final absorbing tower. Heat removed from the process gases is recovered in the steam system by a boiler feed water heater. Acid gases flow to a packed tower, where sulfur trioxide is absorbed. Acid temperature rise is due to the heat of hydration of sulfur trioxide and absorbed sensible heat from the gas. It has been

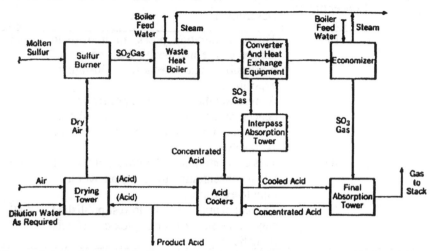

Figure 8.6.1. Conceptual diagram of typical double-absorption sulfuric acid process.

shown that conversion of sulfur dioxide decreases with an increase in temperature and, for that reason, it is desirable to carry out the reaction at as low a temperature as practicable. At 400°C, where the equilibrium condition seems to be very favorable (almost 100% conversion), the rate of approach toward equilibrium is slow. The reaction rate at 500°C can be several orders of magnitude faster than at 400°C.

The potential for a hazardous release of liquid or gaseous sulfur dioxide exists in any type of chemical plant that handles this material. The possible sources of such a release are numerous. Large-scale releases may result from leaks or ruptures of large storage vessels (including tank cars on site). Smaller releases may occur as a result of ruptured lines, broken gauge glasses, or leaking valves, fittings, flanges, valve packing, or gaskets. The properties of sulfur dioxide that can promote equipment failure are a high coefficient of expansion and the corrosiveness of sulfurous acid, which is formed when dry sulfur dioxide comes into contact with moisture.

Process causes are related to the fundamentals of process chemistry, control, and general operation. Examples of possible sulfur dioxide releases include

1. Excess sulfur dioxide feed to a chlorine dioxide reactor, leading to excessive exothermic reaction, combined with failure of the cooling system
2. Backflow of process reactants to a sulfur dioxide feed tank, resulting in the formation of corrosive sulfurous acid or explosive reactions with incompatible materials
3. Inadequate water removal from hydrocarbon feeds in a sulfur dioxide

extraction process over time, leading to progressive corrosion
4. Excess feeds in any part of a process, leading to overfilling or over-pressuring of equipment
5. Loss of condenser cooling to distillation units
6. Overpressure in sulfur dioxide storage vessels from overheating or over-filling

Equipment causes of accidental releases include hardware failures resulting, for example, from excessive stress on materials of construction owing to improper fabrication, construction, or installation. Failure of vessels at normal operating conditions due to weakening of equipment can result from excessive external loadings or thermal cycling. Thermal fatigue and/or shock can occur in reaction vessels, heat exchangers, distillation columns, and process pumps. Failure can also result from mechanical fatigue due to vibration, stress cycling, or hydrogen embrittlement.

About 60% of the sulfuric acid produced in the United States is used in the production of chemical fertilizers. Because of its high boiling point (337°C), sulfuric acid is used to prepare inorganic acids such as hydrochloric acid. Sulfuric acid is also used in the manufacture of detergents, drugs, dyes and pigments, explosives, and paper. It is used in petroleum refining (a process to separate different components in petroleum) and in metallurgical processes. Sulfuric acid is the electrolyte in the lead storage battery commonly used in automobiles. Table 8.6.4 lists some of the end uses of sulfuric acid.

8.7 OTHER CHEMICAL PROCESSES

Three short presentations on other chemical processes are provided below. Detailed information on these and other chemicals are available in the literature.[46,47]

Urea

Urea is made by a process that combines ammonia with carbon dioxide under pressure to form ammonium carbamate, which is then decomposed into urea and water. The unreacted carbon dioxide and ammonia are recovered and recycled to the synthesis operation.

$$2NH_3 + CO_2 \leftrightarrow NH_2COONH_4$$
$$NH_2COONH_4 \leftrightarrow CO(NH_2)_2 + H_2O \qquad (8.7.1)$$

The main difference between the various urea synthesis processes are in the methods used to handle the converter effluent, to decompose the carbamate and carbonate, to recover the urea, and to recover the unreacted ammonia and carbon dioxide for recycle with a maximum recovery of heat. The annual production rate is approximately 4.5 million metric tons. End use is

TABLE 8.6.4 Sulfur Dioxide Reaction Products[a]

Sulfur Dioxide	Sulfonamides
Ammonium sulfite	Sulfadiazine
Ammonium thiosulfite	Sulfanilimide
Hydroxylamine sulfate	Sulfapyridine
Potassium bisulfite	Sulfathiazole
Potassium sulfate	Sulfolane
Potassium thiosulfate	Sulfurous acid
Sodium bisulfite	Sulfuryl chloride
Sodium persulfate	Pyrosulfuryl chloride
Sodium sulfite	Zinc sulfide
Sodium bisulfate	Zinc hydrosulfite
Sodium thiosulfate	Zinc formaldehyde sulfoxylate
Sodium hydrosulfite	

[a] From *Reigals' Handbook of Industrial Chemistry*, 8th ed.[20]

primarily as a fertilizer, although urea is also used in resin and plastics manufacture, and in the synthesis of organic materials.

Plastics

The manufacture processes for polymers are nearly as diverse as the number of polymers, but it is possible to classify many of them as follows:

1. Addition polymerization processes
    a. Mass
    b. Emulsion
    c. Suspension
    d. Solvent
    e. Solvent-nonsolvent
2. Condensation polymerization processes
3. Ring-opening and other "addition" polymerization in which the polymer grows in a manner other than by a chain reaction

Polyolefins, particularly polyethylene and polypropylene, have by far the largest market. In the manufacture of polyethylene, ethylene enters the process through a compressor, which raises the pressure. The initiator is metered into the compressed ethylene stream just before it enters the reactor, which may be either a tubular type or an autoclave. The reaction is exothermic, and a large fraction of the generated heat is removed with heat transfer oil or water. The discharge from the reactor passes through separators to remove the polymer from other components. Molten polyethylene is drawn off from the final separator with gear pumps, screw pumps, or extruders.

Organic Chemicals

Synthetic organic chemical production is approximately 200 billion pounds per year.[20] There are more than 8 million known organic compounds, and of these, the compounds that are commercially produced are too numerous to mention. Only a brief overview of commercially produced synthetic organic chemicals can be given here.

Commercially produced organic chemical products are basically derived from five major raw material categories: methane, ethylene, propylene, C_4, and higher aliphatics and aromatics. The origins of almost all organic chemicals produced commercially fall into one of these five categories. For instance, methane is used directly or indirectly to produce acetylene, methanol, formaldehyde, and silicons. Acetaldehyde, vinyl chloride, aspirin, and polyethylene are produced from ethylene. Propylene gives isopropanol, phenol, acrylic acid and esters, and glycol ethers. Gasoline, butadiene, and LPG are derived from C_4 hydrocarbons.

8.8 ILLUSTRATIVE EXAMPLES

The reader is encouraged to select other chemical process condensation processes and obtain information on physical and chemical properties, health effects, and methods of manufacture, and causes for potential accidents. This is obviously an open-ended question, therefore no solution is provided.

8.9 SUMMARY

1. Releases can originate from many sources, including ruptures in process equipment, separated flanges, actuated relief valves or rupture disks, and failed pumps or compressors. In addition, losses may occur through leaks at joints and connections such as flanges, valves, and fittings, where failure of gaskets or packing might occur.
2. A material's chemical properties, such as reactivity and coefficient of thermal expansion, can promote failure. Reactivity manifested in corrosion is a likely general cause of equipment failure. Evaporators and metering and control equipment are especially sensitive if they are in intermittent use where moisture might enter the system.
3. Process chemistry problems leading to releases are, of course, unique to each commercial process. On the other hand, equipment problems are not unique and can occur in any process. For instance, excessive stress may be due to improper fabrication, construction, or installation, or to mechanical fatigue, vibration, or thermal shock. Other accidental releases may be related to operational causes such as overfilling vessels, errors in loading and unloading, inadequate maintenance, or incomplete knowledge of the process or chemical system.
4. The hazardous and/or toxic chemicals processes reviewed in this chapter

include chlorine, ammonia, hydrogen cyanide, hydrogen fluoride, and sulfuric acid. Physical and chemical properties, health effects, methods of manufacture, process diagrams, and equipment needs are detailed in conjunction with potential causes of accidents.

PROBLEMS

1. What types of cell are used for the manufacture of chlorine? Discuss a typical cell.
2. Discuss how ammonia is prepared (reaction of hydrogen and nitrogen).
3. Discuss possible causes of overheating in a cyanide reactor.
4. What are the major uses of hydrogen cyanide in the United States?
5. State examples of process causes of sulfur dioxide release.
6. State the two most significant chemical properties of hydrogen fluoride contributing to the potential for release.

REFERENCES

1. D. M. Kiefer, "Profits Up Sharply as Producers Reap Benefits of Restructuring," Chem. Eng. News, p. 26, December 14, 1987.
2. G. C. White, Ed., The Handbook of Chlorination, 2nd ed., Van Nostrand Reinhold, New York, 1986.
3. R. Perry and D. Green, Perry's Chemical Engineers' Handbook, 7th ed., McGraw-Hill, New York, 1997.
4. Chlorine Manual, The Chlorine Institute, New York, 1983.
5. J. Dean, Ed., Lange's Handbook of Chemistry, 12th ed., McGraw-Hill, New York, 1979.
6. R. C. Weast, Ed., CRC Handbook of Chemistry and Physics, 63rd ed., CRC Press. Boca Raton, FL, 1982.
7. R. B. Bird, W. E. Stewart, and E. N. Lightfoot, Transport Phenomena, Wiley, New York, 1960.
8. Kirk-Othmer Encyclopedia of Chemical Technology, Vol. I, 2rd ed., Wiley, New York, 1978.
9. W. Braker et al., Effects of Exposure to Toxic Gases—First Aid and Medical Treatment, 2nd ed., Matheson, Lindhurst, NJ, 1977.
10. R. J. S. Lewis and R. L. Tatken, Eds., Registry of Toxic Effects of Chemical Substances, U.S. Department of Health and Human Services. National Institute of Occupational Safety and Health Publication No. 79-100, January 1985 Update, NIOSH, Cincinnati, OH.
11. Chemical Products Synopsis, Mannsville Chemical Products, Cortland, NY, 1985.
12. NIOSH/OSHA Pocket Guide to Chemical Hazards, U.S. Department of Health and Human Services, National Institute of Occupational Safety and Health Publication No. 78-210, 1995, NIOSH, Cincinnati, OH.
13. H. A. Sommers, Chem. Eng Prog., p. 97, March 1965.

14. G. M. Lawler, Ed., *Chemical Origins Markets*, 5th ed., Chemical Information Services, Stanford Research Institute, Menlow Park, CA, 1977.

15. J. M. Blanken, "Behavior of Ammonia in the Event of a Spillage," in *CEP Technical Manual, Ammonia Plant Safety and Related Facilities*, Vol. 22, AIChE, New York, NY, 1980.

16. *Kirk-Othmer Encyclopedia of Chemical Technology*, Vol. 2, 3rd ed., Wiley, New York, 1983.

17. *Anhydrous Ammonia*, Pamphlet G-2, 7th ed., Compressed Gas Association, Arlington, VA, 1984.

18. R. L. Tatken and R. J. Lewis, Ed., *Registry of Toxic Effects of Chemical Substances* (RTECS), 1981-1982 ed., 3 vols., NIOSH Contract No. 218-81-8101, U.S. Department of Health and Human Services, National Institute of Occupational Safety and Health, Publication No. 83-107, NIOSH, Cincinnati, OH, June 1983.

19. Air Products and Chemicals, Inc., *Specialty Gas Material Safety Data Sheet*, Allentown, PA, rev. February 1984.

20. J. A. Kent, Ed., *Reigal's Handbook of Industrial Chemistry*, 8th ed., Van Nostrand Reinhold, New York, 1983.

21. Kirk-*Othmer Encyclopedia of Chemical Technology*, Vol. 7, 3rd ed., Wiley, New York, 1983.

22. *Hydrogen Cyanide Storage and Handling*, E. I. Du Pont de Nemours and Company, Wilmington, DE, 1983.

23. *Chemical Emergency Preparedness Program Interim Guidance, Chemical Profiles*, 2 vols., U.S. Environmental Protection Agency, Washington, DC, 1985.

24. Du Pont de Nemours, E. 1., and Co., U.S. Patent No. 3,360,335, December 24, 1967.

25. Du Pont de Nemours, E. 1., and Co., U.S. Patent No. 3,718,731, February 27, 1973

26. Du Pont de Nemours, E. 1., and Co., U.S. Patent No. 3,104, 945, September 24, 1963.

27. Du Pont de Nemours, E. 1., and Co., U.S. Patent No. 3,215,495, November 2, 1965.

28. *Chemical Profile*, Hydrogen Cyanide, Chemical Marketing Reporter, Schnell Publishing, New York, June 4, 1984.

29. Kirk-*Othmer Encyclopedia of Chemical Technology*, Vol. 10, 3rd ed., Wiley, New York, 1980.

30. *Hydrochloric Acid—Properties, Uses, Storage and Handling*, E. 1. Du Pont de Nemours and Company, Wilmington, DE, September 1984.

31. *Hydrogen Fluoride Product Data Brochure*, Pennwalt Chemical Corporation, Philadelphia, July 1979.

32 G. D. Clayton and F. E. Clayton, ed., *Patty's Industrial Hygiene and Technology*, 3rd ed., Vols. 1 and 2, Wiley-Interscience, New York, 1978

33. *Toxic and Hazardous Industrial Chemicals Safety Manual*, International

Technical Information Institute of Japan, Tokyo, 1976.
34. *Effects of Exposure to Toxic Gases—First Aid and Medical Treatment,* Matheson Gas Products, Secaucus, NJ, 1984.
35. H. O. Burris, U.S. Patent No. 4,031,191, June 21, 1977.
36. U.S. Environmental Protection Agency, *Industrial Process Profiles for Environmental Use,* Publication No. 600/2-77-020, EPA, Cincinnati, OH, February 1977.
37. Kirk-*Othmer Encyclopedia of Chemical Technology,* Vol. 22, 3rd ed., Wiley, New York, 1983.
38. *Sulfur Dioxide,* Pamphlet G-3, 3rd ed.. Compressed Gas Association, New York, 1964.
39. *Sulfur Dioxide Technical Handbook,* 5th ed. Tennessee Chemical Company, Atlanta, 1979.
40. N. I. Sax, ed., *Sulfur Dioxide, Dangerous Properties of Industrial Materials Report,* Vol. 1, No. 3, January-February, New York, 1981.
41. *Material Safety Data Sheet,* Liquid Air Corporation, Alphagaz Division, Walnut Creek, CA, October 1985.
42. U.S. Department of Health, Education, and Welfare, *Criteria for a Recommended Standard. . . Occupational Exposure to Sulfur Dioxide,* HEW (NIOSH) Publication No. 74-111, NTIS Order No. PB-228152, 1974.
43. M. Sittig, *Handbook of Toxic and Hazardous Chemicals and Carcinogens,* 2nd ed., Noyes, Park Ridge, NJ, 1985.
44. National Joint Health and Safety Committee for Water Service, *Safety Aspects of Storage, Handling and Use of Chlorine and Sulfur Dioxide,* London, April 1982.
45. *Material Safety Data Sheet,* Tennessee Chemical Company, Atlanta, revised June 1984.
46. J. Spero, B. Devito, and L. Theodore, "Regulatory Chemicals Handbook", Marcel Dekker, New York City, 2000.
47. H. Beim, J. Spero, L. Theodore, "Rapid Guide to Hazardous Air Pollutants", John Wiley, New York, 1998.

Part III

Health Risk Assessment

Since 1970 the field of health risk assessment has received widespread attention within both the scientific and regulatory communities. It has also attracted the attention of the public. Properly conducted risk assessments have received fairly broad acceptance, in part because they put into perspective the terms toxic, hazard, and risk. Toxicity is an inherent property of all substances. It states that all chemical and physical agents can produce adverse health effects at some dose or under specific exposure conditions. In contrast, exposure to a chemical that has the capacity to produce a particular type of adverse effect, represents a health hazard. Risk, however, is the probability or likelihood that an adverse outcome will occur in a person or a group that is exposed to a particular concentration or dose of the hazardous agent. Therefore, risk is generally a function of exposure and dose. Consequently, health risk assessment can be defined as the process or procedure used to estimate the likelihood that humans or ecological systems will be adversely affected by a chemical or physical agent under a specific set of conditions.

Most human or environmental health hazards can be evaluated by dissecting the analysis into four parts: hazard identification, dose-response assessment or hazard assessment, exposure assessment, and risk characterization. For some perceived health hazards, the risk assessment might stop with the first step, hazard identification, if no adverse effect is identified or if an agency elects to take regulatory action without further analysis. Regarding hazard identification, a hazard is defined as a toxic agent or a set of conditions that has the potential to cause adverse effects to human health or the environment. Health hazard identification involves an evaluation of various forms of information in order to identify the different hazards. Dose-response or toxicity assessment is required in an overall assessment; responses/effects can vary widely since all chemicals and contaminants vary in their capacity to cause adverse effects. This step frequently requires that assumptions be made to relate experimental data from animals to humans. Exposure assessment is the determination of the magnitude, frequency, duration, and routes of exposure of human populations and ecosystems. Finally, in risk characterization, toxicology and exposure data/information are combined to obtain a qualitative or quantitative expression of risk. Thus, risk assessment involves the integration of the information and analysis associated with the above four steps to provide a complete characterization of the nature and magnitude of risk and the degree of confidence associated with this characterization.

In terms of the contents of Part III, Chapter 9 serves to introduce the general subject of health risk assessment (HRA). An expanded presentation on each of the four health risk assessment steps follows, as detailed below:

Chapter 10: Health Hazard Identification
Chapter 11: Dose-response/Toxicity Assessment
Chapter 12: Exposure Assessment
Chapter 13: Risk Analysis and Characterization

The reader should note that two general types of potential health risk exist. These are classified as:

1. *Acute*. Exposures occur for relatively short periods of time, generally from minutes to one to two days. Concentrations of (toxic) air contaminants are usually high relative to their protection criteria. In addition to inhalation, airborne substances might directly contact the skin, or liquids and sludges may be splashed on the skin or into the eyes, leading to adverse health effects. This subject area falls, in a general sense, in the domain of hazard risk assessment (HZRA) and is addressed in the next two Parts (IV and V) of this book.
2. *Chronic*. Continuous exposure occurs over long periods of time, generally several months to years. Concentrations of inhaled (toxic) contaminants are usually relatively low. This subject area falls in the general domain of health risk assessment (HRA) and it is this subject that is addressed in the next five chapters. Thus, in contrast to the acute (short-term) exposures that predominate in hazard risk assessments, chronic (long-term) exposures are the major concern in health risk assessments.

9

Introduction to Health Risk Assessment

9.1 INTRODUCTION

There are many definitions of the word risk. It is a combination of uncertainty and damage; a ratio of hazards to safeguards; a triplet combination of event, probability, and consequences; or even a measure of economic loss or human injury in terms of both the incident likelihood and the magnitude of the loss or injury (AIChE, 1989). People face all kinds of risks everyday, some voluntarily and others involuntarily. Therefore, risk plays a very important role in today's world. Studies on cancer caused a turning point in the world of risk because it opened the eyes of risk scientists and health professionals to the world of risk assessments.

Since 1970 the field of health risk assessment has received widespread attention within both the scientific and regulatory committees. It has also attracted the attention of the public. Properly conducted risk assessments have received fairly broad acceptance, in part because they put into perspective the terms toxic, hazard, and risk. Toxicity is an inherent property of all substances. It states that all chemical and physical agents can produce adverse health effects at some dose or under specific exposure conditions. In contrast, exposure to a chemical that has the capacity to produce a particular type of adverse effect, represents a health hazard. Risk, however, is the probability or likelihood that an adverse outcome will occur in a person or a group that is exposed to a particular concentration or dose of the hazardous agent. Therefore, risk can be generally a function of exposure and dose. Consequently, health risk assessment is defined as the process or procedure used to estimate the likelihood that

The authors gratefully acknowledge the assistance of Christopher Ruocco in researching, reviewing, and editing this chapter.

humans or ecological systems will be adversely affected by a chemical or physical agent under a specific set of conditions.

The term risk assessment is not only used to describe the likelihood of an adverse response to a chemical or physical agent, but it has also been used to describe the likelihood of any unwanted event. This subject is treated in more detail in the next Part. These include risks such as: explosions or injuries in the workplace; natural catastrophes; injury or death due to various voluntary activities such as skiing, sky diving, flying, and bungee jumping; diseases; death due to natural causes; and many others.[1]

Risk assessment and risk management are two different processes, but they are intertwined. Risk assessment and risk management give a framework not only for setting regulatory priorities, but also for making decisions that cut across different environmental areas. Risk management, refers to a decision making process that involves such considerations as risk assessment, technology feasibility, economic information about costs and benefits, statutory requirements, public concerns, and other factors. Therefore, risk assessment supports risk management in that the choices on whether and how much to control future exposure to the suspected hazards may be determined.[2] Regarding both risk assessment and risk management, this chapter and the four chapters to follow will primarily address this subject from a health perspective.

Before leaving this introductory section, the reader is again reminded of the difference between health risk assessment (HRA) and hazard risk assessment (HZRA). Unfortunately, both terms have been used interchangeably by researchers and industrial personnel. As indicated above, this Part of the book will address chronic health problems (HRA) while Part IV and V will be primarily on acute health problems (HZRA).

9.2   THE HEALTH RISK EVALUATION PROCESS

Health risk assessments provide an orderly, explicit and consistent way to deal with issues in evaluating whether a hazard exists and what the magnitude of the hazard may be. This evaluation typically involves large uncertainties because the available scientific data are limited, and the mechanisms for adverse health impacts or environmental damage are only imperfectly understood. When one examines risk, how does one decide how safe is safe, or how clean is clean? To begin with, one has to look at both sides of the risk equation, i.e., both the toxicity of a pollutant and the extent of public exposure. Information is required at both the current and potential exposure, considering all possible exposure pathways. In addition to human health risks, one needs to look at potential ecological or other environmental effects. In conducting a comprehensive risk assessment, one should remember that there are always uncertainties, and these assumptions must be included in the analysis.[2]

In recent years, several guidelines and handbooks have been produced to help explain approaches for doing health risk assessments. As discussed by a special National Academy of Sciences committee convened in 1983, most human or environmental health hazards can be evaluated by dissecting the analysis into four parts: hazard identification, dose-response assessment or hazard assessment, exposure assessment, and risk characterization (see Figure 9.2.1). For some perceived health hazards, the risk assessment might stop with the first step, hazard identification, if no adverse effect is identified or if an agency elects to take regulatory action without further analysis.[1] Regarding hazard identification, a hazard is defined as a toxic agent or a set of conditions that has the potential to cause adverse effects to human health or the environment. Health hazard identification involves an evaluation of various forms of information in order to identify the different hazards. Dose-response or toxicity assessment is required in an overall assessment; responses/effects can vary widely since all chemicals and contaminants vary in their capacity to cause adverse effects. This step frequently requires that assumptions be made to relate experimental data for animals and humans. Exposure assessment is the determination of the magnitude, frequency, duration, and routes of exposure of human populations and ecosystems. Finally, in risk characterization, toxicology and exposure data/information are combined to obtain a qualitative or quantitative expression of risk.

Risk assessment involves the integration of the information and analysis associated with the above four steps to provide a complete characterization of the nature and magnitude of risk and the degree of confidence associated with this characterization. A critical component of the assessment is a full elucidation of the uncertainties associated with each of the major steps. Under this broad concept of risk assessment are encompassed all of the essential problems of toxicology. Risk assessment takes into account all of the available dose-response data. It should treat uncertainty not by the application of arbitrary safety factors, but by stating them in quantitatively and qualitatively explicit terms, so that they are not hidden from decision makers. Risk assessment defined in this broad way, forces an assessor to confront all the scientific uncertainties and to set forth in explicit terms the means used in specific cases to deal with these uncertainties. An expanded presentation on each of the four health risk assessment steps is provided below.

9.3   HEALTH HAZARD IDENTIFICATION

Hazard identification is the most easily recognized of the actions of regulatory agencies. It is defined as the process of determining whether human exposure to an agent could cause an increase in the incidence of a health condition (cancer, birth defect, etc.) or whether exposure by a nonhuman receptor, for example, fish, birds, or other wildlife, might adversely be affected. It involves characterizing the nature and strength of the evidence of causation. Although the question of whether a substance causes cancer or other adverse health effects in humans is theoretically a yes-no question, there are few chemicals or physical agents on which the human data are definitive. Therefore, the question is often restated in terms of effects in laboratory animals or other test systems: "Does the

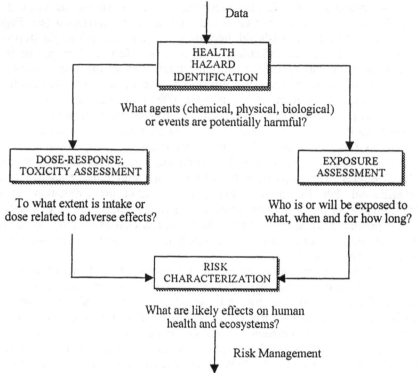

Figure 9.2.1. The Health Risk Evaluation Process

agent induce cancer in test animals?" Positive answers to such questions are typically taken as evidence that an agent may pose a cancer risk for any exposed human. Information for short-term in vitro tests and structural similarity to known chemical hazards may, in certain circumstances, also be considered as adequate information for identifying a hazard.[1]

A health hazards identification for a chemical plant or industrial application can include information about:

1. Chemical identities
2. The locations of facilities that use, produce, process, or store chemical materials that can be classified as health hazards
3. The design of the chemical plant
4. The quantity of material that is normally involved in a release
5. The nature of the problem most likely to accompany the health hazard

An important aspect of hazards identification is a description of the pervasiveness of the hazard. For example, most environmental assessments require knowledge of the concentration of material in the environment, weighted in some way to account for the geographical magnitude of the site affected; that is, a 1-acre or 300-acre site, a 1,000 gal/min or 1,000,000 gal/min stream. All too often environmental incidents regarding chemical emission have been described by statements like "concentrations as high as 150 ppm" of a chemical were measured at a 1,000-acre waste site. However, following closer examination, one may find that only 1 of 200 samples collected on a 20-acre portion of a 1,000-acre site showed this concentration and that 2 ppm was the geometric mean level of contamination in the 200 samples.

An appropriate sampling program is critical in the conduct of a health risk assessment. This topic could arguably be part of the exposure assessment, but it has been placed within hazard identification because, if the degree of contamination is small, no further work may be necessary. Not only is it important that samples be collected in a random or representative manner, but the number of samples must be sufficient to conduct a statistically valid analysis. The number needed to insure statistical validity will be dictated by the variability between the results. The larger the variance, the greater the number of samples needed to define the problem.[1]

The means of identifying health hazards is complex. Different methods are used to collect and evaluate toxic properties (those properties that indicate the potential to cause biological injury, disease or death under certain exposure conditions). One method is the use of epidemiological studies that deal with the incidence of disease among groups of people. Epidemiological studies attempts to correlate the incidence of cancer from an emission by an evaluation of people with a particular disease and people without the disease. Long-term animal bioassays are the most common method of hazard determination. (A bioassay as referred to here is an evaluation of disease in a laboratory animal). Increased tumor incidence in laboratory animals is the primary health effect considered in animal bioassays. Exposure testing for a major portion of an animal's lifetime (2 to 3 years for rats and mice) provides information on disease and susceptibility, primarily for carcinogenicity (the development of cancer).

The understanding of how a substance is handled in the body, transported, changed, and excreted, and of the response of both animals and humans, has advanced remarkably. There are many questions concerning these animal tests as to what information they provide, which kinds of studies are the best, and how the animal data compares with human data. In an attempt to answer these questions, epidemiological studies and animal bioassays are compared to each other to determine if a particular chemical is likely to pose a health hazard to humans. Many assumptions are made in hazard assessments. For example, it is assumed that the chemical administered in a bioassay is in a form similar to that present in the environment. Another assumption is that animal carcinogens are also human carcinogens. An example is that there is a similarity between animal and human metabolism, and so on. With these and other assumptions, and by analyzing hazard identification procedures, lists of hazardous chemicals have been developed.[2]

9.4   DOSE-RESPONSE

Dose-response assessment is the process of characterizing the relationship between the dose of an agent administered or received and the incidence of an adverse health effect in exposed populations, as well as estimating the incidence of the effect as a function of exposure to the agent. This process considers such important factors as intensity of exposure, age pattern of exposure, and possibly other variables that might affect response, such as sex, lifestyle, and other modifying factors. A dose-response assessment usually requires extrapolation from high to low doses and extrapolation from animals to humans, or from one laboratory animal species to a wildlife species. A dose-response assessment should describe and justify the methods of extrapolation used to predict incidence, and it should characterize the statistical and biological uncertainties in these methods. When possible, the uncertainties should be described numerically rather than qualitatively.

Toxicologists tend to focus their attention primarily on extrapolations from cancer bioassays. However, there is also a need to evaluate the risks of lower doses to see how they affect the various organs and systems in the body. Many scientific papers focused on the use of a safety factor or uncertainty factor approach, since all adverse effects other than cancer and mutation-based developmental effects are believed to have a threshold i.e., a dose below which no adverse effect should occur. Several researchers have discussed various approaches to setting acceptable daily intakes or exposure limits for developmental and reproductive toxicants. It is thought that an acceptable limit of exposure could be determined using cancer models, but today they are considered inappropriate because of thresholds.[1]

For a variety of reasons, it is difficult to precisely evaluate toxic responses caused by acute exposures to hazardous materials. First, humans experience a wide range of acute adverse health effects, including irritation, narcosis, asphyxiation, sensitization, blindness, organ system damage, and death. In addition, the severity of many of these effects varies with intensity and duration of exposure. Second, there is a high degree of variation in response among individuals in a typical population. Third, for the overwhelming majority of substances encountered in industry, there are not enough data on toxic responses of humans to permit an accurate or precise assessment of the substance's hazard potential. Fourth, many releases involve multi-components. There are presently no rules on how these types of releases should be evaluated. Fifth, there are no toxicology testing protocols that exist for studying episodic releases on animals. In general, this has been a neglected area of toxicology research. There are many useful measures available to use as benchmarks for predicting the likelihood that a release event will result in serious injury or death. Several references review various toxic effects and discuss the use of various established toxicological criteria.[4,5,6]

Dangers are not necessarily defined by the presence of a particular chemical, but rather by the amount of that substance one is exposed to, also known as the dose.   A dose is usually expressed in milligrams of chemical received per kilogram of body weight per day.   For toxic substances other than carcinogens, a threshold dose must be exceeded before a health effect will occur, and for many substances, there is a dosage below which there is no harm.   A health effect will occur or at least be detected at the threshold.   For carcinogens, it is assumed that there is no threshold, and, therefore, any substance that produces cancer is assumed to produce cancer at any concentration.   It is vital to establish the link to cancer and to determine if that risk is acceptable.   Analyses of cancer risks are much more complex than non-cancer risks.[2]

Not all contaminants or chemicals are created equal in their capacity to cause adverse effects.   Thus, cleanup standards or action levels are based in part on the compounds' toxicological properties.   Toxicity data are derived largely from animal experiments in which the animals (primarily mice and rats) are exposed to increasingly higher concentrations or doses.   Responses or effects can vary widely from no observable effect to temporary and reversible effects, to permanent injury to organs, to chronic functional impairment to ultimately, death.

9.5   EXPOSURE ASSESSMENT

Exposure assessment is the process of measuring or estimating the intensity, frequency, and duration of human or animal exposure to an agent currently present in the environment or of estimating hypothetical exposures that might arise from the release of new chemicals into the environment.   In its most complete form, an exposure assessment should describe the magnitude, duration, schedule, and route of exposure; the size, nature, and classes of the human, animal, aquatic, or wildlife populations exposed; and the uncertainties in all estimates.   The exposure assessment can often be used to identify feasible prospective control options and to predict the effects of available control technologies for controlling or limiting exposure.[1]

Much of the attention focused on exposure assessment has come recently.   This is because many of the risk assessments done in the past used too many conservative assumptions, which caused an overestimation of the actual exposure.   Without exposures there are no risks.   To experience adverse effects, one must first come into contact with the toxic agent(s).   Exposures to chemicals can be via inhalation of air (breathing), ingestion of water and food (eating and drinking), or absorption through the skin.   These are all pathways to the human body.

Generally, the main pathways of exposure considered in this step are atmospheric surface and groundwater transport, ingestion of toxic materials that have passed through the aquatic and terrestrial food chain, and dermal absorption.   Once an exposure assessment determines the quantity of a chemical with which human populations may come in contact, the information can be combined with toxicity data (from the hazard identification process) to estimate potential health risks.[2]   The primary purpose of an exposure assessment is to determine the concentration levels over time and space in each environmental media where human and other environmental receptors may come into contact with chemicals of concern. There are four components of an exposure assessment: potential sources, significant exposure pathways, populations potentially at risk, and exposure estimates.[1]

The two primary methods of determining the concentration of a pollutant to which target populations are exposed are direct measurement and computer analysis also known as computer dispersion modeling. Measurement of the pollutant concentration in the environment is used for determining the risk associated with an exiting discharge source. Receptors are placed at regular intervals from the source, and the concentration of the pollutant is measured over a certain period of time (usually several months or a year). The results are then related to the size of the local population. This kind of monitoring, however, is expensive and time-consuming. Many measurements must be taken because exposure levels can vary under different atmospheric conditions or at different times of the year. Computer dispersion modeling predicts environmental concentrations of pollutants (see Chapters 13 in this Part for more information on dispersion modeling). In the prediction of exposure, computer dispersion modeling focuses on discharge of a pollutant and the dispersion of that discharge by the time it reaches the receptor. This method is primarily used for assessing risk from a proposed facility or discharge. Sophisticated techniques are employed to relate reported or measured emissions to atmospheric, climatological, demographic, geographic, and other data in order to predict a population's potential exposure to a given chemical.[2]

9.6   HEALTH RISK CHARACTERIZATION

Risk characterization is the process of estimating the incidence of a health effect under the various conditions of human or animal exposure described in the exposure assessment. It is performed by combining the exposure and dose-response assessments. The summary effects of the uncertainties in the preceding steps should be described in this step. The quantitative estimate of the risk is the principal interest to the regulatory agency or risk manager making the decision. The risk manager must consider the results of the risk characterization when evaluating the economics, societal aspects, and various benefits of the risk assessment. Factors such as societal pressure, technical uncertainties, and severity of the potential hazard influence how the decision makers respond to the risk assessment. There is room for a lot of improvement in this step of the risk assessment.[1,6,7]

A risk estimate indicates the likelihood of occurrence of the different types of health or environmental effects in exposed populations. Risk assessment should include both human health and environmental evaluations (i.e., impacts on ecosystems). Ecological impacts include actual or potential effects on plants and animals (other than domesticated species). The number produced from the risk characterization, representing the probability of adverse health effects being caused, must be evaluated. This is done because certain agencies will only look at risks of specific numbers and act on them.

There are two major types of risk: maximum individual risk and population risk. Maximum risk is defined exactly as it implies, that is the maximum risk to an individual person. This person is considered to have a 70-year lifetime of exposure to a process or a chemical. Population risk is the risk to a population. It is expressed as a certain number of deaths per thousand or per million people. For example, a fatal annual risk of 2×10^{-6} refers to 2 deaths per year for every million individuals. These risks are based on very conservative assumptions, which may yield too high a risk.

9.7   ILLUSTRATIVE EXAMPLES

Example 9.1

What are the general duties of state, territorial, and local air pollution control agencies as they relate to health risk assessments?

Solution 9.1

References:

1.   "Air Pollution Information Activities at State and Local Agencies – United States, 1992".
2.   W. Matystik, L. Theodore, R. Diaz, "An Internet Guide to State Regulatory Agencies", Government Institutes, 1992

Example 9.2

Risk Assessment (RA) is routinely used in setting air standards. Give the sequence of steps in the risk assessment process.

Solution 9.2

The sequence of steps in the risk assessment process are:

1.   Evaluation of the source
2.   Quantification of the release
3.   Calculations of dispersion to the receptor
4.   Quantification of exposure at designated receptors
5.   Assessment of the toxicity of the chemical
6.   Quantification of health effects associated with the release and subsequent exposure

Example 9.3

What is the role of human health risk assessment in risk management?

Solution 9.3

Human health risk assessment estimates the likelihood of health problems occurring if no cleanup action were taken at the site. To estimate the baseline risk at a site, the following four-step process should be taken to determine the possible human risk which will then determine what sort of corrective action should be employed:

1. Health Hazard Identification
2. Dose-Response; Toxicity Assessment
3. Exposure Assessment
4. Risk Characterization

Example 9.4

What are some of the important complications in the "hazard identification" step of the health risk assessment process?

Solution 9.4

The following factors should be reviewed in the hazard identification step of a risk assessment effort:

1. The sufficiency of the epidemiological database for most chemicals
2. The nature of adverse health effects: cancer versus non-cancer
3. The preponderant role of cigarette smoking affecting the overall outcome in the test population
4. Multi-source, multi-pollutant synergistic impacts
5. The questionable validity if animal studies and short-term tests with respect to the prediction of long-term, adverse health impacts in a human population

9.8   SUMMARY

1. Health risk assessment is defined as the process or procedure used to estimate the likelihood that humans or ecological systems will be adversely affected by a chemical or physical agent under a specific set of conditions.
2. The health risk evaluation process consists of four steps: hazard identification, dose-response assessment or hazard assessment, exposure assessment, and risk characterization.
3. In hazard identification, a hazard is a toxic agent or a set of conditions that

has the potential to cause adverse effects to human health or the environment.

4. In dose-response assessment, effects are evaluated and these effects vary widely because their capacities to cause adverse effects differ.
5. Exposure assessment is the determination of the magnitude, frequency, duration, and routes of exposure to human populations and ecosystems.
6. In risk characterization, the toxicology and exposure data are combined to obtain a quantitative or qualitative expression of risk.

PROBLEMS

1. List the four main components of health risk assessment and explain them briefly.
2. Discuss how the dose response/toxicological assessment step can be improved.
3. Discuss how the exposure assessment step can be improved.
4. Discuss how the "risk assessment" procedure can be improved.

REFERENCES

1. D. Paustenbach, *The Risk Assessment of Environmental and Human Health Hazards: A Textbook of Case Studies,* New York City, John Wiley & Sons, 1989.
2. G. Burke, B. Singh, and L. Theodore, *Handbook of Environmental Management and Technology,* 2nd ed., New York City, John Wiley & Sons, 2000.
3. J. Rodricks, and R. Tardiff, *Assessment and Management of Chemical Risks.* Washington D.C., American Chemical Society, 1984.
4. D. B. Clayson, D. Krewski, and I. Munro, *Toxicological Risk Assessment*, Boca Raton, FL., CRC Press, Inc., 1985.
5. V. Foa, E.A. Emmett, M. Maron, and A. Colombi, *Occupational and Environmental Chemical Hazards,* Chichester, England, Ellis Horwood Limited, 1987.
6. R. Bethea, *"Incorporation of Occupational Safety and Health into Unit Operations Laboratory Courses",* NIOSH, Cincinnati, 1991.
7. USEPA, location unknown, 2000.

10

Health Hazard Identification

10.1 INTRODUCTION

Hazard identification is defined as the process of determining whether human exposure to an agent could cause an increase in the incidence of a health condition (cancer, birth defect, etc.) or whether exposure to nonhumans, such as fish, birds, and other forms of wildlife, could cause adverse effects. Hazard identification characterizes the hazard in terms of the agent and dose of the agent. Since there are few hazardous chemicals or hazardous agents for which definitive exposure data in humans exists, the identification of health hazards is often characterized by the effects of health hazards on laboratory test animals or other test systems [1].

There are numerous methods available to identify the potential for chemicals to cause both health conditions and adverse effects on the environment. These can include, but are not limited to, toxicology, epidemiology, molecular and atomic structural analysis, MSDS sheets, engineering approaches to problem solving, fate of chemicals, and carcinogenic versus non-carcinogenic health hazards.

The remaining sections of this chapter will address the above key issues in the following order:

Section 10.2: Toxicology Principles
Section 10.3: Epidemiology Principles
Section 10.4: Molecular/Atomic Structural Analysis
Section 10.5: Material Safety Data Sheet (MSDS)
Section 10.6: Engineering Problem Solving

The authors gratefully acknowledge the assistance of Mary Keane in researching, reviewing, and editing this chapter.

10.2   TOXICOLOGY PRINCIPLES

What makes a waste hazardous? From a regulatory stance, a waste is hazardous if it exhibits reactivity, corrosivity, ignitability, or toxicity. The potential effect of toxicity on living things, especially humans, has driven regulatory initiatives in hazardous waste management. A primary objective of hazardous waste management, and in particular, hazardous waste incineration, is to protect human health by reducing the risk associated with substances in hazardous chemicals. Thus, an understanding of toxicology is needed in order to determine if a chemical substance is a health hazard.

Chemical hazards yield toxic effects when these hazards enter the human body and other living organisms. Adverse health effects become evident in both the short and long term with symptoms ranging from mild allergic reactions to death.

The general subject of toxicology deals with the adverse effects of chemical substances on living things. The science of toxicology is not a basic science like mathematics or physics. Toxicology is a relatively new discipline that continues to develop. It evolved from other sciences such as physiology, pharmacology, biochemistry, molecular biology, and epidemiology.

Many engineers do not fully understand toxicology, and their education and experiences usually do not prepare them to make intelligent toxicological evaluations. However, engineers can often assist in an overall health risk study with the identification of a chemical hazard.

Since the fundamental mechanisms that cause toxic responses are not fully understood, toxicological findings are largely based on observations. Although some of the toxicological information relating to humans is based on human experience, the majority of this information is derived from animal experimentation.

Health effects of chemical hazards observed in laboratory experiments with animals are used to predict health effects of these same chemical hazards in humans. Extrapolation procedures account for the differences in exposure levels between laboratory animals and humans. Risk estimates may vary by an order of magnitude depending on the mathematical method. Thus, it is evident that the quantification of the toxicity of a chemical has a high degree of uncertainty associated with it. However, in most cases, the available toxicological data is sufficient to predict the risks associated with a chemical. In cases where the toxicological data for a chemical is insufficient, alternative approaches can determine if the chemical is a health hazard [2].

This subject will receive additional treatment in the next chapter; an extensive section is devoted to toxicology. Technical details and definitions are provided in that material.

10.3   EPIDEMIOLOGY PRINCIPLES [3]

Epidemiology is an important method employed in the identification of human toxicity and health hazards resulting from exposure to chemicals. It is the only method that provides direct human evidence to identify toxicity and health hazards in humans.

Epidemiology analyzes statistical data to determine the relationship between a chemical and the exposed population. However, positive statistical analysis does not always translate into a cause and effect relationship between the chemical and disease in humans. When a cause and effect relationship between the chemical and disease has been established, further statistical analysis helps define the upper limit of human toxic risk.

Those epidemiological studies that establish a cause and effect relationship between a chemical and disease in humans, such as cancer or reproductive toxicity, have been drawn from studies of relatively small population groups exposed to high dosages of the chemical. As noted by Tomatis, the identification of human carcinogens has occurred under conditions of exposure similar to those used in experimental carcinogenesis[4]. (Carcinogenesis is the process that occurs between exposure to a carcinogenic chemical and the development of a malignancy.) In experimental carcinogenesis, a limited number of experimental animals are exposed to high levels of a chemical or mixture of chemicals to increase the sensitivity of the experimental animal to elicit a response.

Although valuable in its own right and in combination with other scientific methods, epidemiology information is limited. Epidemiology methods are not sensitive to increases in chemically induced disease at low exposure, and most human exposures are low level exposures. As a result, a negative relationship determined by epidemiology does not necessarily demonstrate the absence of a health hazard, nor does it demonstrate the potential presence of this hazard in the long term.

Epidemiology is useful in the detection of an increase in rare human disease in an exposed population. However, the usefulness of epidemiology in the detection of an increase in common diseases is limited due to the long latency period between exposure and detection of disease. In order to detect an increase in more common diseases, such as cancers of the lung, breast and colon, special circumstances must exist where the exposure dose is unusually high or the incidence of the disease is unusually high. Control group studies involving unexposed humans provide data for epidemiological analysis to detect increases in common diseases, but these control groups may be difficult to identify if exposure is widespread.

As with toxicology, this subject area is treated in more extensive detail in the next chapter. An entire section is devoted to epidemiology, and technical details and information are provided.

10.4   MOLECULAR/ ATOMIC STRUCTURAL ANALYSIS

Chemistry is the science of the combination of atoms, and physics is the science of the forces between atoms. Simply stated, chemistry deals with matter and its transformations, and physics deals with energy and its transformations. These transformations may be temporary, such as a change in phase, or seemingly permanent, such as a change in the form of matter resulting from a chemical reaction.   The study of atomic and molecular structure deals with these transformations, and can be used to make a preliminary identification of a health hazard.

Molecular structural analysis is a developing method. The objective of a molecular structural analysis is to demonstrate a physical, structural, or chemical similarity between the chemical in question and a known toxic chemical that produces toxic and health effects in experimental animals and/or humans. Unfortunately, scientists do not fully understand the effects of slight changes in the chemical structure and their biological effect on humans. As a result, this type of analysis is useful in preliminary studies to identify potential health hazards for further examination with more established methods in short-term tests or tests in experimental animals. In its present stage of development, molecular structural analysis cannot be used to make absolute decisions about the appropriate levels of exposure of humans to chemicals [3].

There are two logical approaches for applying atomic and molecular structural analysis for the purpose of identifying health hazards, the synthetic approach and the analytical approach. A synthetic approach begins with an analysis of the structure and behavior of matter in its simplest state. This approach then progresses to more complex states of matter, from electrons to atoms to molecules, then to combinations of atoms and molecules, then to combinations resulting from chemical reactions. The analytical approach begins with known matter or chemicals found in the laboratory, and works towards simpler states of subdivision in order to explain experimental results. Obviously, both approaches have merit, but subdivision presumably addresses the former topic.

10.5   MATERIAL SAFETY DATA SHEET (MSDS) [5]

The Material Safety Data Sheet (MSDS) is a detailed information bulletin prepared by the manufacturer or importer of a chemical that describes the physical and health hazards, routes of exposure, precautions for safe handling and use, emergency and first-aid procedures, and control measures. Information on an MSDS aids in the selection of safe products and helps prepare employers and employees to respond effectively to daily exposure situations as well as to emergency situations. It is also a source of information for identifying chemical hazards.

In line with The Occupational Safety and Health Administration (OSHA) requirements, employers must maintain a complete and accurate MSDS for each hazardous chemical that is used in their facility. They are entitled to obtain this information automatically upon purchase of the material. When new and significant information becomes available concerning a product's hazards or ways to protect against the hazards, chemical manufacturers, importers, or distributors must add it to their MSDS within three months and provide it to their customers with the next shipment of the chemical. Thus, employers must have an MSDS for each hazardous chemical used in the workplace. If there are multiple suppliers of the same chemical, there is no need to retain multiple MSDS's for that chemical.

While MSDS's are not required to be physically attached to a shipment, they must accompany or precede the shipment. When the manufacturer/supplier fails to send an MSDS with a shipment labeled as a hazardous chemical, the employer must obtain one from the chemical manufacturer, importer, or distributor as soon as possible. Similarly, if the MSDS is incomplete or unclear, the employer should contact the manufacturer or importer to get clarification or obtain missing information.

When an employer is unable to obtain an MSDS from a supplier or manufacturer, he/she should submit a written complaint, with complete background information, to the nearest OSHA area office. OSHA will then, at the same time, call and send a certified letter to he supplier or manufacturer to obtain the needed information. If the supplier or manufacturer still fails to respond within a reasonable time, OSHA will inspect the supplier or manufacturer and take appropriate enforcement action.

It is important to note that OSHA specifies the information to be included on an MSDS, but does not prescribe the precise format for an MSDS. A non-mandatory MSDS form that meets the Hazard Communication Standard requirements has been issued and can be used as is or expanded as needed. The MSDS must be in English and must include at least the following information presented in Table 10.5.1. In reviewing this material, the reader should understand the effect and importance of each subsection in helping to identify a chemical hazard, particularly the section on health hazards.

10.6   ENGINEERING PROBLEM SOLVING

Perhaps the most important job that the engineer faces on a day-to-day basis is the need to solve problems. Several problem solving approaches are presented in Part IV in an attempt to identify hazards (particularly of an accidental nature). These problem solving methods include, but are not limited to:

- "What if" Approaches
- Hazard and Operability (HAZOP) Studies
- Preliminary Hazard Analysis (PLHA)
- Process Hazard Analysis (PHA)
- Safety Reviews
- Safety Audits

Table 10.5.1   MSDS Information

Chemical Identity	• The chemical and common name(s) must be provided for single chemical substances. • An identity on the MSDS must be cross-referenced to the identity found on the label.
Hazardous Ingredients	• For a hazardous chemical mixture that has been tested as a whole to determine its hazards, the chemical and common names of the ingredients that are associated with the hazards, and the common name of the mixture must be listed. • If the chemical is a mixture that has not been tested as a whole, the chemical and common names of all ingredients determined to be health hazards and compromising 1 percent or greater of the composition must be listed. • Chemical and common names of carcinogens must be listed if they are present in the mixture at levels of 0.1 percent or greater. • All components of a mixture that have been determined to present a physical hazard must be listed. • Chemical and common names of all ingredients determined to be health hazards and comprising less than 1 percent (0.1 percent for carcinogens) of the mixture must also be listed if they can still exceed an established OSHA Permissible Exposure Limit (PEL) or the ACGIH Threshold Limit Value (TLV) or present a health risk to exposed employees in these concentrations. The next chapter will address PELs and TLVs in further detail.
Physical and Chemical Characteristics	• The physical and chemical characteristics of the hazardous substance must be listed. These include items such as boiling and freezing points, density, vapor pressure, specific gravity, solubility, volatility, and product's general appearance and odor. These characteristics provide important information for designing safe and healthful work practices.
Fire and Explosion Hazard Data	• The compound's potential for fire and explosion must be described. Also, the fire hazards of the chemical and the conditions under which it could ignite or explode must be identified. Recommended extinguishing agents and fire-fighting methods must be described.
Reactivity Data	• This section presents information about other chemicals and substances with which the chemical is incompatible, or with which it reacts. Information on any hazardous

decomposition products, such as carbon monoxide, must be included.

Health Hazards
- The acute and chronic health hazards of the chemical, together with signs and symptoms of exposure, must be listed. In addition, any medical conditions that are aggravated by exposure to the compound must be included. The specific types of chemical health hazards defined in the standard include carcinogens, corrosives, toxins, irritants, sensitizers, mutagens, teratogens, and effects on target organs (i.e., liver, kidney, nervous system, blood, lungs, mucous membranes, reproductive system, skin, eyes, etc.).
- The route of entry section describes the primary pathway by which the chemical enters the body. There are three principal routes of entry: inhalation, skin, and ingestion.
- This section of the MSDS supplies the PEL, the TLV, and other exposure levels used or recommended by the chemical manufacturer.
- If OSHA, the National Toxicology Program (NTP), or the International Agency for Research on Cancer (IARC) listed the compound as a carcinogen (cancer causing agent), it must be indicated as such on the MSDS.

Precautions for Safe Handling and Use
- The standard requires the preparer to describe the precautions for safe handling and use. These include recommended industrial hygiene practices, precautions to be taken during repair and maintenance of equipment, and procedures for cleaning up spills and leaks. Some manufacturers also use this section to include useful information not specifically required by the standard, such as EPA waste disposal methods and state and local requirements.

Control Measures
- The standard requires the preparer of the MSDS to list any generally applicable control measures. These include engineering controls, safe handling procedures, and personal protective equipment. Information is often included on the use of goggles, gloves, body suits, respirators, and face shields.

Employer   • Employers must ensure that each employee has basic
Responsibilities   knowledge of how to find information on an MSDS and
                        how to properly make use of that information.
                        Employers also must ensure the following:
                      • Complete and accurate MSDS's are made available
                        during each work shift to employees when they are in
                        their work areas.
                      • Information is provided for each hazardous chemical.

These problem solving methods often involve a need to determine the following:

• What is known
• What is unknown
• What is desired or required

Additionally, questions often arise regarding the identity, location, timing, and magnitude of the problem.

Although the above approaches cannot be directly applied to identifying health hazards, the general methodology is applicable to identifying these health hazards since these are basically problem solving approaches [6]. The following procedure is a sample problem solving approach for engineers [7]. For the purposes of this Section, this procedure is presented as a problem solving approach for chemical health hazard identification; however, this procedure can also be used to solve a near infinite number of engineering problems.

1. Identify the problem. Understand the information available and the requirements for the answer.
2. Determine what additional information is required and obtain it.
3. If applicable, draw a simplified picture or diagram of what is taking place and include the available data.
4. Determine a basis on which to start the problem.
5. If chemical equations are involved, write those equations and verify that they are balanced.
6. Decide which formulas or relations govern the particular health hazard.
7. If applicable, perform the necessary calculations.
8. Determine whether the conclusion regarding the status of the health hazard seems reasonable.

10.7   FATE OF CHEMICAL HEALTH HAZARDS

In the process of identifying chemical health hazards, the near term and long term fate of the hazard should be incorporated into the analysis. Near-term concerns relate primarily to the release of the chemical into the environment. This leads to the general subject area of exposure assessment, including routes of exposure - a topic that is treated in extensive detail in Chapter 12. However, the fate of the chemical (hazard) following the point of human entry is another consideration when attempting to identify health hazards. An overview of this topic is presented here [8].

As one might expect, exposure to a chemical compels a response by the human body. The body responds to a chemical with physiological (metabolic) processes in order to absorb, distribute, store, transform, or eliminate that chemical. To become a chemical health hazard, the chemical or the transformation of that chemical by the body must reach a target organ for a sufficient length of time and at a sufficient concentration to produce toxic effects. A target organ is the "preferential anatomical site" for the expression of toxic effects by a chemical substance in the human body.

Consider the routes of exposure of the chemical benzene and the physiological response by the human body to that chemical. The route of exposure of a dose of benzene vapors into the body is inhalation. The body will absorb 30% of the inhaled dose into the lungs and blood. The body will eliminate the remaining 70% of the inhaled amount by exhalation. Another route of exposure exhibited in laboratory animals is ingestion. The body will absorb more than 90% of an ingested dose of benzene. The least effective route of exposure is dermal exposure, with approximately 0.2% of the dose absorbed transdermally into the body. In the body, more than half of the benzene that has been absorbed is distributed to organs with a rich blood supply, such as the liver and kidneys, and to tissues with a rich fat supply, such as adipose tissue, the brain, and bone marrow. In pregnant women, benzene is distributed to the placenta and fetus. Fatty tissues slowly release benzene, and the body transforms the non-polar, fat soluble benzene into polar metabolites that are eliminated in the urine by excretion. Although a substantial database exists for several chemical compounds such as benzene, the knowledge for most substances is incomplete.

In the blood stream, most chemicals distribute throughout the body. In fact, few agents attack locally at entry; rather, most agents use the flow of blood to reach other organs and tissues. Distribution is affected by absorption, perfusion, exposure route, and tissue affinity. As a result, the chemical distributes throughout the body in different locations in different amounts rather than in one location in the total amount. Absorption determines the passage of the chemical into the blood and from the blood into tissues and cells. Perfusion is the movement of blood through organ tissue. Both the liver and brain are well perfused. The brain is protected by the barrier between the blood and the brain.

On the other hand, the liver is not protected, and its total potential absorption of a chemical is much greater.

The exposure route partly determines the distribution of the chemical in the body. Like the chemical benzene, a single chemical may follow multiple routes of exposure. The liver, like the skin, acts as a filter. The liver is the primary detoxification site. Toxicants that are absorbed into the lungs, skin, mouth, and esophagus may temporarily bypass the liver; however, toxicants absorbed through the stomach and intestines follow the blood's direct path to the liver.

The term storage is used to describe a site or an organ that is not a target organ that exhibits a concentration of a chemical substance. Tissue affinity allows chemicals to store in tissues at these sites. The concentration at a storage site is sometimes high or higher than the concentration of the chemical in the target organ. The storage site concentration and location depends on the chemical.

Storage may be a defense mechanism of the body. Chemicals are slowly released from the storage site. Through storage, the body eliminates the chemical substance, thus preventing or reducing a distribution of the chemical that may cause toxic effects in the body. Storage acts in equilibrium with other processes in the body. This equilibrium can be reversed, thus allowing elimination of the stored chemical over time, even after the exposure has ended.

The site of accumulation may define the point of toxic action. Inorganic mercury accumulation in the kidneys causes sever functional impairment [9]. Kidney damage has been shown to occur when the accumulated total of cadmium in the kidney cortex reaches 100-200 ppm [10].

Biotransformation, elimination, and formation of a chemical-receptor complex are alternatives to storage. Biotransformation is the metabolism of the toxicant by enzyme-rich organs. The toxicant transforms into other molecular species known as metabolites, which may or may not be less toxic than the parent toxicant. Toxicants and metabolites that are not stored in the body are eliminated by the body. These substances, particularly polar compounds, are eliminated through urine, bile, feces, or secretions. Non-polar and non-volatile compounds are more difficult for the body to eliminate. These substances can often be eliminated by urine after the toxic agent has been metabolically transformed into a more polar, and thus more water soluble substance. Toxic agents attack one or more organs, known as target organs, by forming a chemical-receptor complex.

These mechanisms can affect the near-term and ultimate fate of a chemical hazard. Recognition of these mechanisms can significantly assist in the identification of a chemical agent as a health hazard. In recent years, the understanding of chemical transport, chemical manipulation in the body, and response by animals and humans to chemicals has advanced to a point where it is possible to determine whether a chemical is indeed a health hazard.

Paustenbach has provided an excellent review of the physical and chemical properties of substances and how this information is used to predict the fate of the substance in the environment [1]. Although the details are beyond the scope of this subsection, the key factors are listed below:

1. Water solubility
2. Photodegradation (direct and indirect)
3. Biodegradation in soil
4. Vapor pressure and density
5. Dissociation content
6. Ultraviolet – Visible (UV – Vis) absorption spectrum
7. Sorption and desorption
8. Partition coefficient
9. Bioconcentration factor (BCF)
10. Hydrolysis
11. Algae assay
12. Cellulose decomposition
13. Nitrogen transformation
14. Seed germination
15. Sulfur transformation
16. Microbial growth inhibition

10.8 CARCINOGENS VERSUS NONCARCINOGENS

The classification as to whether a chemical agent is a carcinogen or a non-carcinogen can help identify whether it is a health hazard. Both topics are briefly reviewed in this section [8]. More extensive information is provided in Chapter 15.

Non-Carcinogens

Toxicity, i.e., the degree to which a chemical is considered a health hazard, is characterized by its threshold. A threshold or lower limit below which effects cannot be observed characterizes the dose-effect or dose-response relationship. These effects can occur at the cellular, subcellular, and molecular level. The body protects itself against toxic chemicals with repair mechanisms and by attributing critical functions to large numbers of the same units. This way, in order for a toxic effect to occur, a number of these units greater than the threshold for the target dose must be affected. For example, carbon tetrachloride is a solvent that causes disease in liver tissue. The body's repair mechanism allows it to replace lost cells with new cells, thus allowing the liver to continue to function. Beyond a certain threshold however, the liver cannot function and the damage may not be reversed.

Noncarcinogenic effects include all toxicological responses except tumors. Toxicological responses and mechanisms vary widely, and examples of these include interference with normal cell processes by displacing elements out of the cell and binding with a cell to reduce membrane permeability. However, the majority of noncarcinogenic effects involve enzymes. In the body, different enzymes perform specific functions. When an enzyme binds with a toxic substance, the enzyme may be prohibited from performing its function properly, thereby exhibiting a toxic response.

Carcinogens

Carcinogens cause cancer. Cancerous cells are normal cells that become abnormally altered and divide uncontrollably. The disease of cancer is characterized by tumors or neoplasms (meaning "new growth"); however, not all tumors are cancerous. Benign tumors are not cancerous and do not spread. Malignant tumors are cancerous and spread, or metastasize, to surrounding structures. This invasion of surrounding structures by malignant tumors occurs because the abnormal alteration of the cells prevent them from responding to the body's regulatory signals that control cell growth.

As previously noted, carcinogenesis is the process that occurs between exposure to a carcinogenic chemical and the development of malignancy. The three stages of carcinogenesis are initiation, promotion, and progression. Initiation is the alteration or mutation of a normal cell into an abnormal cancerous cell. Promotion is the increase in the replication rate and number of initiated cells. Promotion is caused by promoting carcinogens. The promoter is not usually the same carcinogen that initiated the first stage of carcinogenesis. All cells that have been initiated and promoted do not develop into malignant cells. The body's defense mechanism against foreign substances - the immune system - recognizes and rejects some of these cells. In progression, the third stage of carcinogenesis, the abnormal cells invade surrounding tissues and spread to distant organ sites. Progression involves more genetic mutations than those required in initiation and promotion.

10.9   ILLUSTRATIVE EXAMPLES

Example 10.1

What is the difference between hazardous chemicals and toxic chemicals?

Solution 10.1

Hazardous chemicals is a broad category that includes chemicals that may be toxic, flammable, corrosive, explosive, or harmful to the environment. A toxic chemical is one type of a hazardous chemical. Toxic chemicals cause adverse health effects, such as severe illness or death, when ingested, inhaled, or absorbed by a living organism.

Example 10.2

Qualitatively describe in "layman's language" the information that is provided on an MSDS sheet.

Solution 10.2

An MSDS sheet serves as a reference source for information on a hazardous substance. The MSDS sheet identifies the substance, identifies the producer or seller of the substance, the location of the producer or seller, explains why the substance is hazardous, explains how a person can be exposed to the substance, identifies conditions that increase the hazard, explains safe handling procedures, identifies proper protective clothing or devices to be used when working with the substance, explains the steps that should be taken if a person is exposed to the substance, and explains the steps that should be taken if there is a spill or emergency situation.

Example 10.3

Are the chances of getting cancer for someone who is exposed to a small amount of a toxic chemical once the same as for someone who is exposed to a small amount of that same toxic chemical everyday?

Solution 10.3

The chances are generally not the same. Exposure depends on both the amount of chemical exposure and the frequency of chemical exposure. Repeated exposure to low levels of a mix of chemicals may be linked to health problems. However, a single incident at a higher level, if below a toxic threshold, may not be linked to health problems.

Example 10.4

There are four methods that can be used to determine a chemical health hazard. These methods are emission factors, mass balance considerations, engineering calculations, and direct emission measurements. Describe each of these approaches.

Solution 10.4

Emission factors are emission rates determined by regulatory agencies based on data generated from a given source that are normalized to some unit of production or rate of chemical use. These factors are compiled for industries, processes or sources. This information is used to estimate emission rates without a detailed analysis. The emission factor must be closely related to the industry, process or source in question. In some cases, it may be difficult to predict emission rates for a new process or product accurately with conventional emission factors.

Mass balance considerations apply the law of the conservation of mass to account for each constituent entering and leaving a system. Constituents that do not comprise the product are either retained by the system or released from the system as waste. This method requires a quantitative analysis of the influent and effluent streams and an understanding of chemical reactions occurring within the system.

Engineering calculations predict emission rates without the use of emission factors. These calculations use basic science and engineering principles, chemical property data, and operating conditions to provide a detailed analysis of the emissions for a specific process. This is a more sophisticated approach than emission factors, and is useful for evaluating various operational and control alternatives.

Direct emission measurements involve the direct measurement of emission rates from specific sources. Direct emission measurements provide the data for emission factor and engineering calculations. This is the only method that provides emission rates for a given source for a given set of conditions.

Example 10.5

The metabolism of ethyl alcohol may be considered to occur via a zero-order reaction (i.e., its elimination occurs linearly with time.) If a person is able to metabolize approximately 10 mL of the alcohol per hour, how long a time period is required to eliminate 8 pints of beer containing 3.2% alcohol? Assume that the volume of a pint of beer is 530 mL.

Solution 10.5

t = total dose/ rate of elimination
t = (8 pints of beer) (530 mL/ pint of beer) (3.2% alcohol) / 10 mL per hour
t = 13.6 hours

10.10   SUMMARY

1. Hazard identification is defined as the process of determining whether human exposure to an agent could cause an increase in the incidence of a health condition, or whether exposure to nonhumans could cause adverse effects.
2. The general subject of toxicology deals with the adverse effects of chemical substances on living things.
3. Epidemiology is an important method employed in the identification of human toxicity and health hazards resulting from exposure to chemicals. It

is the only method that provides direct human evidence to identify toxicity and health hazards in humans.

4. Molecular structural analysis is a developing method that demonstrates physical, structural, or chemical similarities between a known toxic chemical and a chemical in question in order to determine if that chemical may also be toxic.

5. The Material Safety Data Sheet (MSDS) is a detailed information bulletin prepared by the manufacturer or importer of a chemical that describes the physical and health hazards, routes of exposure, precautions for safe handling and use, emergency and first-aid procedures, and control measures.

6. Engineers use problem solving methods to identify hazards.

7. The process of identifying chemical health hazards should also incorporate the near term (release into the environment) and long term fate of the chemical health hazard following entry into the human body.

8. Non-carcinogenic effects include all toxicological responses except tumors. Not all tumors are cancerous. Malignant tumors are cancerous and spread, or metastasize, to surrounding structures.

PROBLEMS

1. Briefly describe toxicology and epidemiology.
2. Describe the physical and chemical characteristics included on an MSDS. Explain why this information should be included in an MSDS.
3. Explain why first aid measures should be included on an MSDS.
4. Describe storage of a chemical in the human body and its role in the fate of a chemical.
5. Describe carcinogenesis and name the three stages.

REFERENCES

1. D. Paustenbach, *The Risk Assessment of Environmental and Human Health Hazards: A Textbook of Case Studies*, John Wiley & Sons, New York City, 1989.

2. M.D. LaGrega, P.L. Buckingham, J.C. Evans, and the Environmental Resources Management Group, *Hazardous Waste Management*, McGraw Hill, Inc., USA, 1994.

3. "Risk Analysis in the Chemical Industry," Chemical Manufacturers Association, Government Institution, Rockville, Maryland, 1985.

4. L. Tomatis, "Long-term and Short-term Assays for Carcinogens," IARC Monographs Supplement 2, Lyon, France, 1980.

5. U.S. Department of Labor: "Hazard Communication-A Compliance Kit," Publ. No. OSHA 3104, Washington, DC, 1988.

6. Personal notes: L. Theodore, 2001.

7. Adapted from D. Himmelblau, *Basic Principles and Calculations in Chemical Engineering*, Fifth ed., Prentice Hall, Inc., Englewood Cliffs, NJ, 1993.

8. Adapted from M.D. LaGrega, P.L. Buckingham, J.C. Evans, and the Environmental Resources Management Group, *Hazardous Waste Management*, McGraw Hill, Inc., USA, 1994.

9. T.W. Clarkson, "Effects – General Principles Underlying the Toxic Action of Metals," Handbook on the Toxicology of Metals, L. Friberg, G.F. Nordberg, and V.B. Vouk, eds. 2nd ed., Elsevier, Amsterdam, Holland, 1986.

10. World Health Organization: "Recommended Health-Based Limits in Occupational Exposure to Heavy Metals," Report of a WHO Study Group, WHO, Geneva, Switzerland, 1980.

11

Dose-Response

11.1 INTRODUCTION

Dose-response assessment is the process of characterizing the relationship between the dose of an agent administered or received and the incidence of an adverse health effect in exposed populations, and estimating the incidence of the effect as a function of exposure to the agent. This process considers such important factors as intensity of exposure, age pattern of exposure, and possibly other variables that might affect response, such as sex, lifestyle, and other modifying factors. A dose-response assessment usually requires extrapolation from high to low doses and extrapolation from animal to humans, or one laboratory animal species to a wildlife species. A dose-response assessment should describe and justify the methods of extrapolation used to predict incidence, and it should characterize the statistical and biological uncertainties in these methods. When possible, the uncertainties should be described numerically rather than qualitatively.

Dangers are not necessarily defined by the presence of a particular chemical, but rather by the amount of that substance one is exposed to, also known as the dose. A dose is usually expressed in milligrams of chemical received per kilogram of body weight per day. For toxic substances other than noncarcinogens, a threshold dose must be exceeded before a health effect will occur. For many substances, there is a dosage below which there is no harm, i.e., a health effect will occur or at least be detected at the threshold. For carcinogens, it is assumed that there is no threshold, and therefore any substance that produces cancer is assumed to produce cancer at any concentration. It is vital to establish the link to cancer for carcinogens and to

The authors gratefully acknowledge the assistance of Joseph Flesche in researching, reviewing, and editing this chapter.

determine if that risk is acceptable; analyses of cancer risks are much more complex than those for noncancer risks.

This chapter focuses on the general subject of toxicology and its companion topic, dose-response. The following section headings and subject areas are addressed following this introductory section.

11.2 Definitions
11.3 Toxicology
11.4 Epidemiology
11.5 Noncarcinogens
11.6 Carcinogens
11.7 Uncertainties/Limitations

11.2 DEFINITIONS

Before proceeding to some of the more technical aspects of toxicology and the general subject of dose-response, several important definitions used by the profession and appearing in the literature are provided below (in alphabetical order).

Acceptable Daily Intake (ADI) An estimate similar in concept to the RfD, but derived using a less strictly defined methodology. RfDs have replaced ADIs as the USEPA's (Agency) preferred values for use in evaluating potential noncarcinogenic health effects resulting from exposure to a chemical.

Acceptable Intake for Chronic Exposure (AIC) An estimate similar in concept to the RfD but derived using a less strictly defined methodology. Chronic RfDs have replaced AICs as the Agency's preferred values for use in evaluating potential noncarcinogenic health effects resulting from chronic exposure to a chemical.

Acceptable Intake for Subchronic Exposure (AIS) An estimate similar in concept to the subchronic RfD, but derived using a less strictly defined methodology. Subchronic RfDs have replaced AISs as the Agency's preferred values for use in evaluating potential noncarcinogenic health effects resulting from subchronic exposure to a chemical.

Acute Toxicity The adverse effect occurring within a short time of (oral) administration of a single dose of a substance or multiple doses given within 24 hours.

Carcinogen Potency Factor (CPF) A CPF is the slope of the dose-response curve at very low exposures. The dimensions of a CPF, are expressed as the inverse of daily dose $(mg/kg\text{-}day)^{-1}$.

Ceiling Recommended Exposure Limit (CREL) Exposure limit which should not be exceeded at any time.

Ceiling Value (CV) The airborne concentration of a potentially toxic substance which should never be exceeded in the breathing zone.

Chronic Reference Dose An estimate (with an uncertainty spanning perhaps an order of magnitude or greater) of a daily exposure level for the human population, including sensitive subpopulation, that is likely to be without an appreciable risk of deleterious effects during a lifetime. Chronic RfDs are specifically developed to be protective for long-term exposure to a compound (the Superfund program guideline is seven years to lifetime).

Developmental Reference Dose (RfD_{dt}) An estimate (with an uncertainty spanning perhaps an order of magnitude or greater) of an exposure level for the human population, including sensitive subpopulations, that is likely to be without an appreciable risk of developmental effects. Developmental RfDs are used to evaluate the effects of a single exposure event.

Dose The amount of a substance available for interaction with metabolic processes or biologically significant receptors after crossing the outer boundary of an organism. The potential dose is the amount ingested, inhaled, or applied to the skin. The applied dose is the amount of a substance presented to an absorption barrier and available for absorption (although not necessarily having yet crossed the outer boundary of the organism), The absorbed dose is the amount crossing a specific absorption barrier (e.g., the exchange boundaries of skin, lung, and digestive tract) through uptake processes. Internal dose is a more general term denoting the amount absorbed without respect to specific absorption barriers or exchange boundaries. The amount of the chemical available for interaction by any particular organ or cell is termed the deliverable dose for that organ or cell.

Dose Rate This represents the dose per unit time, for example in mg/day, sometimes also called dosage. Dose rates are often expressed on a per-unit-bodyweight-basis, yielding units such as mg/kg-d. They are often expressed as averages over some time period, e.g., a lifetime.

Dose-Response Curve A graphical representation of the quantitative relationship between the administered, applied, or internal dose of a chemical or agent, and a specific biological response to that chemical or agent.

Dose-Response Evaluation The process of quantitatively evaluating toxicity information and characterizing the relationship between the dose a contaminant administered or received, and the incidence of adverse health effects in the exposed population. From a quantitative dose-response relationship, toxicity values can be derived that are used in the risk characterization step to estimate the likelihood of adverse effects occurring in humans at different exposure levels.

Inhalation Reference Concentration (RfC) An estimate (with an uncertainty spanning perhaps an order of magnitude) of the daily exposure of the human population to a chemical, through inhalation, that is likely to be without risk of deleterious effects during a lifetime.

Integrated Risk Information System (IRIS) A USEPA data base containing verified RfDs and slope factors and up-to-date health risk and EPA regulatory information for numerous chemicals. IRIS is the USEPA's preferred source for toxicity information for Superfund studies/projects.

Immediately Dangerous to Life and Health (IDLH) The concentration representing the maximum level of a pollutant from which an individual could escape within 30 min without escape-impairing symptoms or irreversible health effects.

LC_n The concentration of a toxicant lethal to n% of a test population.

LD_n The dose of a toxicant lethal to n% of a test population.

Lethal Concentration 50 (LC_{50}) A calculated concentration of a chemical in air to which exposure for a specific length of time is expected to cause death in 50% of a defined experimental animal population.

Lethal Dose 50 (LD_{50}) A calculated dose of a chemical in water to which exposure for a specific length of time is expected to cause death in 50% of a defined experimental animal population.

Limit of Detection (LOD) The minimum concentration of a substance being measured that, in a given matrix and with a specific method, has a 99% probability of being identified, qualitatively or quantitatively measured, and reported to be greater than zero.

Lowest-Observed-Adverse-Effect-Level   (LOAEL) In   dose-response experiments, the lowest exposure level at which there are statistically or biologically significant increases in frequency or severity of adverse effects between the exposed population and its appropriate control group.

Lowest-Observed-Effect-Level (LOEL) In dose-response experiments, the lowest exposure level at which there are statistically or biologically significant increases in the frequency or severity of any effect between the exposed population and its appropriate control group.

Maximally Exposed Individual (MEI) The single individual with the highest exposure in a given population (also, maximum exposed individual). This term has historically been defined in various ways, including as defined here and also synonymously with worst case or bounding estimate.

Median Effective Concentration (EC) The concentration of toxicant or intensity of other stimulus which produces some selected response in one half of a test population.

Median Effective Dose (ED) The statistically derived single dose of a substance that can be expected to cause a defined nonlethal effect in 50% of a given population of organisms under a defined set of experimental conditions.

Median Lethal Concentration (LC) The concentration of a toxicant lethal to one half of a test population.

Median Lethal Dose (LD) The statistically derived single dose of a chemical that can be expected to cause death in 50% of a given population of organisms under a defined set of experimental conditions. This figure has often been used to classify and compare toxicity among chemicals but its value for this purpose is doubtful. One commonly used classification of this kind is as follows:

Category	LD_{50} Orally to Rat, mg/kg body weight
Very Toxic	<25
Toxic	>25 to 200
Harmful	>200 to 2000

Method Detection Limit (MDL) See Limit of Detection (LOD)

NIOSH Ceiling Limit (NIOSH CL) NIOSH-recommended 15-min exposure limit, which should not be exceeded.

No-Observed-Adverse-Effect-Level (NOAEL) In dose-response experiments, an exposure level at which there are no statistically or biologically significant increases in the frequency or severity of adverse effects between the exposed population and its appropriate control; some effects may be produced at this level, but they are not considered to be adverse, nor precursors to specific adverse effects. In an experiment with more than one NOAEL, the regulatory focus is primarily on the highest one, leading to the common usage often term NOAEL to mean the highest exposure level without adverse effect.

No-Observed-Effect-Level (NOEL) In dose-response experiments, an exposure level at which there are no statistically or biologically significant increases in the frequency or severity of any effect between the exposed population and its appropriate control group.

Oral Reference Dose (RfD) An estimate (with an uncertainty spanning perhaps an order of magnitude) of the daily exposure of the human population to a chemical, through ingestion, that is likely to be without risk of deleterious effects during a lifetime.

Permissible Exposure Limit (PEL) expressed as a time-weighted average, is the concentration of a substance to which most workers can be exposed without adverse effects, averaged over a normal 8-h workday or a 40-h workweek.

Recommended Exposure Limit (REL) NIOSH-recommended exposure limit for an 8- or 10-h time-weighted-average exposure and/or ceiling.

Reference Dose (RfD) The USEPA's preferred toxicity value for evaluating noncarcinogenic effects resulting from exposures at Superfund sites.

Representativeness The degree to which a sample is, or samples are, characteristic of the whole medium, exposure, or dose for which the samples are being used to make inferences.

Short Term Exposure Limit (STEL) The time weighted average (TWA) airborne concentration to which workers may be exposed for periods up to 15 minutes, with no more than 4 such excursions per day and at least 60 minutes between them.

Slope Factor The slope factor is used to estimate an upper-bound lifetime probability of an individual developing cancer as a result of exposure to a particular level of a potential carcinogen. Also see Carcinogen Potency Factor (CPF)

Suggested No Adverse Response Level (SNARL) The maximum dose or concentration which, on the basis of current knowledge, is likely to be tolerated by an organism without producing any adverse effect.

Threshold Limit Value (TLV) The airborne concentration of a potentially toxic substance to which it is believed healthy working adults may be exposed safely through a 40 hour working week and a full working life. This concentration is measured as a time weighted average concentration.

Threshold Limit Value Ceiling (TLV-C) The concentration that RDA should not be exceeded during any part of the working exposure. If conventional industrial hygiene instantaneous monitoring is not feasible, then the TLV-C can be assessed by sampling over a 15-minute period except for those substances that may cause immediate irritation when exposures are short.

Threshold Limit Value-Short Term Exposure Limit (TLV-STEL) The concentration to which workers can be exposed continuously for a short period of time without suffering from 1) irritation, 2) chronic or irreversible tissue damage, or 3) narcosis to a sufficient degree which may increase the likelihood of accidental injury, impair self-rescue or materially reduce work efficiency, and provided that the daily TLV-TWA is not exceeded. It is not a separate independent exposure limit; rather, it supplements the time-weighted average (TWA) limit where there are recognized acute effects from a substance whose toxic effects are primarily of a chronic nature. STELs are recommended only where toxic effects have been reported from high short-term exposures in either humans or animals. Exposures above the TLV-TWA up to the STEL should not be longer than 15 minutes and should not occur more than four times per day. There should be at least 60 minutes between successive exposures in this range. An averaging period other than 15 minutes may be recommended when this is warranted by observed biological effects.

Threshold Limit Value-Time Weighted Average (TLV-TWA) The time-weighted average concentration for a conventional 8-hour workday and a 40-hour workweek, to which nearly all workers may be repeatedly exposed, day after day, without adverse effect.

Time-Weighted Average (TWA) An allowable exposure concentration averaged over a normal 8-h workday or a 40-h workweek.

Time Weighted Average Concentration (TWA) The concentration of a substance to which a person is exposed in the ambient air, averaged over a period, usually 8 hours.

Regarding units for some of the above terms, doses generally are expressed in terms of the quantity administered per unit body weight, or quantity per skin surface area, or quantity per unit volume of the respired air. In addition, doses are also expressed over the duration of time which the dose was administered. Dose amounts are generally expressed as milligrams (one thousandth of one gram) per kilogram (mg/kg). In some cases, grams per kilogram (gm/kg), micrograms (one millionth of a gram) per kilogram (µg/kg), or nanograms (one billionth of a gram) per kilogram (ng/kg) are used. Volume measurements of dose can be converted to weight units by appropriate calculations. Densities can be obtained from standard reference texts. Where densities are not available, liquids are assumed to have a density of one gram per milliliter. All body weights are converted to kilograms (kg) for uniformity. Concentrations of a gaseous substance in air are generally listed as parts of vapor or gas per million by volume (ppmv). Concentrations of liquid or solid substance are usually expressed as parts per million by weight or mass (ppmw or ppmm). Other units include any mass per unit volume combination of units.

11.3 TOXICOLOGY

Toxicology is the science of poisons, i.e., the study of chemical or physical agents that produce adverse responses in biological systems. Together with other scientific disciplines (such as epidemiology, the study of the cause and distribution of disease in human populations, and risk assessment), toxicology can be used to determine the relationship between an agent of interest and a group of people or a community. Of the many different types of toxicology (see Table 11.3.1), all types, or different applications of the science, start from a common nomenclature and set of cardinal principles.

Of interest to the engineer and scientist are the regulatory and environmental applications of the discipline. The former is of use in interpreting the setting of standards for allowable exposure levels of a given contaminant or agent in an ambient or occupational environment; the latter is of use in estimating the persistence and movement of an agent in a given environment. Both applications can be of direct use to risk assessment activities, and both regulatory toxicology and environmental toxicology closely involve other branches of the discipline. The relationship is particularly close for the regulatory toxicologist who depends largely on the products of descriptive toxicology when making decisions on the risk posed by a specific agent.

There is a wide spectrum of doses among chemical agents needed to produce some adverse health effect. Although dose and exposure are sometimes used interchangeably, this is technically incorrect. As described earlier, the dose is the concentration or amount of an agent that becomes biologically available to the body at an anatomic site or target organ, and that is capable of inducing an adverse health effect. Exposure, on the other hand, represents in a very broad sense the amount of the agent in the environment of concern. Exposure levels only translate to dose if the agent becomes available to the body through one of three principal routes of exposure: respiration, ingestion, or absorption through the skin.

TABLE 11.3.1 Types of Toxicology [1]

Type	Purpose
Clinical toxicology	To determine the effects of chemical poisoning and the treatment of poisoned people
Descriptive toxicology	To test the toxicity of chemicals
Environmental toxicology	To determine the environmental fate of chemicals and their ecological and health effects
Forensic toxicology	To answer medicolegal questions about health effects
Industrial toxicology	To determine health effects of occupational exposures
Mechanistic toxicology	To describe the biochemical mechanisms that cause health effects
Regulatory toxicology	To assess the risk involved in marketing chemicals and products, and establish their subsequent regulation by government agencies

Returning to dose, toxicologists employ quantitative measures of toxicity or the ability of an agent to cause some health effect. Health effects can range from the minor, skin irritation, to the major, death. A standard measure of toxicology employs death as the outcome. As described in the previous section, one measure is the dosage of an agent needed to produce death in 50 percent of the treated animals (LD_{50}), or lethal dose. The primary source of data for such measures are tests administered to laboratory animals, commonly the mouse and/or rat. Some chemicals considered extremely poisonous or toxic will achieve the LD_{50} with only a few micrograms of dose. Other agents will only cause harm if the host is challenged with large concentrations. The range of dose for some common agents is provided expressed in Table 11.3.2. As noted earlier, most characterizations of dose are expressed as an amount relative to body weight, e.g., in milligrams/kilograms of body weight of the test animal. Thus, the LD_{50} attempts to answer the question, "How toxic is the compound or agent?"

Toxicity may also be viewed as a relative concept depending on the type of agent and the amount of the agent (dose). Toxicologists classify agents as to their toxicity by arranging the universe of all potential agents into categories based on the results of laboratory tests similar in nature to the LD_{50} results. The LD_{50} has an analog in the field of pharmacology where the effective dose for 50 percent of the test population, or ED_{50}, is routinely calculated for medicines. Categories for toxic agents range from practically nontoxic to extremely and super toxic, each with relevant specific dosages.

TABLE 11.3.2 Approximate LD$_{50}$'s of Some Chemical Agents(2)

Agent	LD$_{50}$ (mg/kg)
Ethyl alcohol	10,000
Sodium chloride	4,000
Morphine sulfate	900
Strychnine sulfate	2
Nicotine	1
Dioxin (TCDD)	0.001
Botulinum toxin	0.00001

In conclusion, the purpose of the toxicity assessment is to weigh available evidence regarding the potential for particular contaminants to cause adverse effects in exposed individuals and to provide, where possible, an estimate of the relationship between the extent of exposure to a contaminant and the increased likelihood and/or severity of adverse effects.

11.4 EPIDEMIOLOGY [3]

Epidemiology is a discipline within the health sciences that deals with the study of the occurrence of disease in human populations. The term is derived from the Greek words "Epi" (upon) and "Demos" (people) or diseases upon people. Whereas physicians are generally concerned with the single patient, epidemiologists are generally concerned with groups of people who share certain characteristics. A good example would be the interest epidemiologists show in characteristics associated with adverse health effects, e.g. smoking and lung cancer, asbestos exposure and asbestosis, or noise and hearing loss.

Epidemiology operates within the context of public health with a strong emphasis on the prevention of disease through the reduction of factors that may increase the likelihood that an individual or group will suffer a given disease. Implicit in the practice of epidemiology is the need for different disciplines in studying the influence of occupation on human health.

Epidemiologic data come from many different sources. Acquiring reliable, accurate, and complete data describing occupational health problems is a key concern of the epidemiologist. A primary and continuing problem is the ascertainment of occupational disease. Ascertainment is the identification of diseases that are, in this case, of occupational origin.

Occupational disease is not a new phenomenon. Ample historical evidence exists recounting the effects of lead poisoning, chronic respiratory problems associated with mining, and hazards of manufacturing (including traumatic injury).

Although it has been known for a long time that occupational exposures can induce human disease, as in the above examples, the fact remains that diseases of occupational origin are underreported. This can be attributed to three major factors. The first is that health professionals generally do not gather enough information concerning the patient's occupational history or the various jobs and duties carried out by the patient to possibly link employment with his/her symptoms.[4] The second is that many of the diseases associated with occupational causes could have been caused by other risk factors. Therefore, the occupationally caused case of lung cancer does not appear with some distinct marker to differentiate it from a lung tumor caused by personal risk factors such as smoking. Exceptions do, of course, exist: mesothelioma - a relatively rare cancer of the lining of the lung - generally only occurs with exposure to asbestos. A third factor, particularly for chronic diseases, is the long time interval that can exist between initial exposure to an occupational agent and the development of disease. This long time interval can make the recognition of the occupational origin of a disease quite difficult. This is in stark contrast to the relative ease of associating injuries with job-related causes.

Latency refers to the period of time that elapses between the first contact of a harmful agent and a host, and the development of identifiable symptoms or disease. Latency may be as short as a few hours, the time required for photochemical smog to induce watery eyes. Or it may stretch to 20 - 30 years for a chronic condition such as asbestosis or malignant neoplasm of the lung. The association between a given exposure and a disease is all that more difficult because of the passage of time.

The types of epidemiologic studies that attempt to note the number of cases of specific disease in a specific time period are generally known as descriptive studies. Descriptive studies attempt to provide investigators with information concerning the distribution of the disease in time and space as well as to identify attributes that may increase the chances of an individual contracting the disease. These attributes, called risk factors, include factors subject to change such as physical inactivity as well as those that are immutable, such as gender or age. For example, well-established risk factors for occupationally induced lung cancer include asbestos and coke oven emissions. Descriptive studies are also helpful in the formation of hypotheses regarding exposure and disease. Studies seeking to prove or disprove specific hypotheses are called analytical studies. These are briefly discussed below.

The two basic types of analytical studies are the cohort and the case-control study. Each has strengths and weaknesses as well as different resource and time requirements. The cohort study involves the study of individuals classified by exposure characteristics, e.g., a group of welders. The study then follows the development of disease in the welders' group as well as in an unexposed comparison population. The measure that assesses the magnitude of association between the exposure and disease and that indicates the likelihood of developing the disease in the exposed group relative to the unexposed is the relative risk. A relative risk of 1.0 indicates no difference between the disease experience in the two groups. A relative risk of greater than 1.0 indicates a positive association between the exposure and the disease, and an increased risk in those who are subject to the exposure.

In the case-control design, a group with a disease (cases) is compared with a selected group of nondiseased (control) individuals with respect to exposure. The relative risk in control studies can only be estimated as the incidence rate among exposed individuals and cannot be calculated. The estimator used is the odds ratio, which is the ratio of the odds of exposure among the cases to that among the controls.

The main difference between the case-control and the cohort type of study is that in the case-control format, the investigator begins by classifying study subjects as to disease status. With the cohort study, the investigator begins by separating study subjects by exposure status. There are major resource consumption differences between the types of study. Cohort studies generally consume more resources and take longer to complete than do case-control studies.

A pressing challenge for epidemiologists interested in occupational health is to derive an accurate picture of disease frequency. This challenge is met by two broad types of measurement: prevalence and incidence. These are briefly described below.

Disease refers to the number of cases existing in a population. Point-prevalence identifies the prevalence estimated at a given time, e.g., the number of workers with abnormal chest films from a survey. Prevalence is computed as the number of cases divided by the number of study subjects at a given point in time.

$$\text{Prevalence} = \frac{\text{number of persons with a disease}}{\text{total number in the study}} \qquad (11.4.1)$$

Prevalence is thus not a true rate but simply a proportion, although the term prevalence rate is used.

Incidence, a true rate, refers to the number of new cases of a disease in a defined population in a given period of time. Thus, the incidence rate can be expressed as:

$$\text{Incidence Rate} = \frac{\text{number of new cases of disease during time period}}{\text{total number at risk}} \qquad (11.4.2)$$

Central to epidemiology is the use of rates to express the health experience of populations. Rates are important because epidemiology is inherently a comparative discipline. An epidemiologist is constantly attempting to compare the disease experience of a study population with that of a comparison population. A rate is nothing more than a specialized proportion in which the counts of people with a particular disease are placed over a denominator that is composed of people who are at risk, i.e., who have a chance of developing the disease. Men, for example, would not be included in the denominator used to calculate the prevalence or incidence of uterine cancer.

11.5 NONCARCINOGENS [5]

This Section summarizes how toxicity information is considered in the toxicity assessment for noncarcinogenic effects. The reference dose, or RfD, is the toxicity value used most often in evaluating noncarcinogenic effects. The methods the USEPA uses for developing RfDs were presented earlier but are described again because of the number of different RfDs. Various types of RfDs are available depending on the exposure route (oral or inhalation), the critical effect (developmental or other), and the length of exposure being evaluated (chronic, subchronic, or single effect).

A chronic RfD is defined as an estimate (with an uncertainty spanning perhaps an order of magnitude or greater) of a daily exposure level for the human population, including sensitive subpopulations, that is likely to be without an appreciable risk of deleterious effects during a lifetime. Chronic RfDs are specifically developed to be protective for long-term exposure to a compound. As a guideline for Superfund program risk assessments, chronic RfDs generally are often used to evaluate the potential noncarcinogenic effects associated with exposure periods between 7 years (approximately 10 percent of a human lifetime) and a lifetime. Many chronic RfDs have been reviewed and verified by an intra-Agency RfD Workgroup and entered into the USEPA's Integrated Risk Information System (IRIS).

More recently, the USEPA has begun developing subchronic RfDs (RfD_{ss}), which are useful for characterizing potential noncarcinogenic effects associated with shorter-term exposures, and developmental RfD_{dt}s, which are useful specifically for assessing potential developmental effects resulting from exposure to a compound. For example, as a guideline for Superfund program risk assessments, subchronic RfDs should be used to evaluate the potential noncarcinogenic effects of exposure periods between two weeks and seven years. Such short-term exposures can result when a particular activity is performed for a limited number of years or when a chemical with a short half-life degrades to negligible concentrations within several months. Developmental RfDs are used to evaluate the potential effects on a developing organism following a single exposure event.

Concept of Threshold

For many noncarcinogenic effects, protective mechanisms are believed to exist that must be overcome before the adverse effect is manifested. For example, where a large number of cells perform the same or similar function, the cell population may have to be significantly depleted before the effect is seen. As a result, a range of exposures exists from zero to some finite value that can be tolerated by the organism with essentially no chance of an expression of adverse effects. In developing a toxicity value for evaluating noncarcinogenic effects (i.e., an RfD), the approach is to identify the upper bound of this tolerance range (i.e., the maximum subthreshold level). Because variability exists in the human population, attempts are made to identify a subthreshold level protective of sensitive individuals in the population. For most chemicals, this level can only be estimated; the RfD incorporates uncertainty factors indicating the degree or extrapolation used to derive the estimated value. RfD summaries in IRIS also contain a statement expressing the overall confidence that the evaluators have in the RfD (high, medium, or low). The RfD is generally considered to have an uncertainty spanning an order of magnitude or more, and therefore the RfD should not be viewed as a strict scientific demarcation between what level is toxic and nontoxic.

Derivation of an Oral RfD

In the development of oral RfDs, all available studies examining the toxicity of a chemical following exposure by the oral route are gathered and judged for scientific merit. Occasionally, studies based on other exposure routes (e.g., inhalation) are considered, and the data are adjusted for application to the oral route. Any differences between studies are reconciled and an overall evaluation is reached. If adequate human data are available, this information is used as the basis for the RfD. Otherwise, animal study data are used; in these cases, a series of professional judgments are made that involve, among other considerations, an assessment of the relevance and scientific quality of the experimental studies. If data from several animal studies are being evaluated, the USEPA first seeks to identify the animal model that is most relevant to humans based on a defensible biological rationale, e.g., using comparative metabolic data. In the absence of a species that is clearly the most relevant, the USEPA assumes that humans are at least as sensitive to the substance as the most sensitive animal species tested. Therefore, as a matter of science policy, the study on the most sensitive species (the species showing a toxic effect at the lowest administered dose) is selected as the critical study for the basis of the RfD. The effect characterized by the "lowest-observed-adverse-effect-level" (LOAEL) after accurate conversions, to adjust for species differences is referred to as the critical toxic effect.

After the critical study and toxic effect have been selected, the USEPA identifies the experimental exposure level representing the highest level tested at which no adverse effects (including the critical toxic effect) were demonstrated. This highest "no-observed-adverse-effect-level" (NOAEL) is the key datum obtained from the study of the dose-response relationship. A NOAEL observed in an animal study in which the exposure was intermittent (such as five days per week) is adjusted to reflect continuous exposure.

The NOAEL is selected based in part on the assumption that, if the critical toxic effect is prevented, then all toxic effects are prevented. The NOAEL for the critical toxic effect should not be confused with the "no-observed-effect-level" (NOEL). In some studies, only LOAEL rather than a NOAEL is available. The use of a LOAEL, however, requires the use of an additional uncertainty factor (as seen below).

The RfD is derived from the NOAEL (or LOAEL) for the critical toxic effect by consistent application of uncertainty factors (UFs) and a modifying factor (MF). The uncertainty factors generally consist of multiples of 10 (although values less than 10 are sometimes used), with each factor representing a specific area of uncertainty inherent in the extrapolation from the available data. The bases for application of different uncertainty factors are explained below.

1. A UF of 10 is used to account for variation in the general population and is intended to protect sensitive subpopulations (e.g., the elderly and children).

2. A UF of 10 is used when extrapolating from animals to humans. This factor is intended to account for the interspecies variability between humans and other mammals.

3. A UF of 10 is used when a NOAEL derived from a subchronic instead of a chronic study is used as the basis for a chronic RfD.

4. A UF of 10 is used when a LOAEL is used instead of a NOAEL. This factor is intended to account for the uncertainty associated with extrapolating from LOAELs to NOAELs.

5. In addition to the UFs listed above, a modifying factor (MF) is also applied. An MF ranging from >0 to 10 is included to reflect a qualitative professional assessment of additional uncertainties in the critical study and in the entire data base for the chemical not explicitly addressed by the preceding uncertainty factors. The default value for the MF is 1.0.

To calculate the RfD, the appropriate NOAEL (or the LOAEL if a suitable NOAEL is not available) is divided by the product of all of the applicable uncertainty factors and the modifying factor.

$$RfD = \frac{NOAEL \text{ or } LOAEL}{\left(UF_1 \cdot UF_2 \ldots \cdot MF\right)}$$

(11.5.1)

Oral RfDs typically are expressed as one significant figure in units of mg/kg-day.

Derivation of an Inhalation RfD (RfD$_i$)

The methods the USEPA uses in the derivation of inhalation RfDs are similar in concept to those used for oral RfDs; however, the actual analysis of inhalation exposures is more complex than oral exposures due to (1) the dynamics of the respiratory system and its diversity across species, and (2) differences in the physiochemical (both physical and chemical) properties of contaminants. Although the identification of the critical study and the determination of the NOAEL in theory are similar for oral and inhalation exposures, several important differences should be noted. In selecting the most appropriate study, the USEPA considers differences in respiratory anatomy and physiology, as well as differences in the physicochemical characteristics of the contaminant. Differences in respiratory anatomy and physiology may affect the pattern of contaminant deposition in the respiratory tract, and the clearance and redistribution of the agent. Consequently, the different species may not receive the same dose of the contaminant at the same locations within the respiratory tract even though both species were exposed to the same particle or gas concentration. Differences in the physicochemical characteristics of the contaminants, such as the size and shape of a particle or whether the contaminant is an aerosol or a gas, also influence deposition, clearance, and redistribution.

In inhalation exposures, the target tissue may be a portion of the respiratory tract or, if the contaminant can be absorbed and distributed through the body, some extrarespiratory organ. Because the pattern of deposition may influence concentrations at different tissues of the lung, the toxic health effect observed may be more directly related to the pattern of deposition than to the exposure concentration. Consequently, the USEPA considers the deposition, clearance mechanisms, and the physicochemical properties of the inhaled agent in determining the effective dose delivered to the target organ.

Doses calculated in animals are converted to equivalent doses in humans on the basis of comparative physiological considerations (e.g., ventilatory parameters and regional lung surface areas). Additionally, if the exposure period was discontinuous, it is adjusted to reflect continuous exposure.

The inhalation RfD is derived from the NOAEL by applying uncertainty factors similar to those listed above for oral RfDs. A UF of 10 is used when extrapolating from animals to humans in addition to the calculation of the human equivalent dose, to account for interspecific variability in sensitivity to the toxicant. The resulting RfD value for inhalation exposure is generally reported as a concentration in air in mg/m^3 for continuous, 24 hour/day exposure, although it may be reported as a corresponding inhaled intake (in mg/kg-day). A human body weight of 70 kg and an inhalation rate of 20 m^3/day are used to convert between an inhaled intake expressed in units of mg/kg-day and a concentration in air expressed in mg/m^3.

Derivation of a Subchronic RfD

The chronic RfDs described above pertain to lifetime or other long-term exposures and may be overly protective if used to evaluate the potential for adverse health effects resulting from substantially less-than-lifetime exposures. For such situations, the USEPA has begun calculating toxicity values specifically for subchronic exposure durations, using a method similar to that outlined above for chronic RfDs. The USEPA's Environmental Criteria and Assessment Office develops subchronic RfDs and, although they have been peer-reviewed by Agency and outside reviewers, subchronic RfDs values have not undergone verification by an intra-Agency workgroup. As a result, subchronic RfDs are considered interim rather than verified toxicity values and have yet to be placed in IRIS.

Development of subchronic RfDs parallels the development of chronic reference doses in concept; the distinction is one of exposure duration. Appropriate studies are evaluated and a subchronic NOAEL is identified. The RfD is derived from the NOAEL by the application of the UFs and MF, as outlined above. When experimental data are available only for shorter exposure durations than desired, an additional uncertainty factor is applied. This is similar to the application of the uncertainty factor for duration differences when a chronic RfD is estimated from subchronic animal data. On the other hand, if subchronic data are missing and a chronic oral RfD derived from chronic data exists, the chronic oral RfD is adopted as the subchronic oral RfD. In this instance, there is no application of an uncertainty factor to account for differences in exposure duration.

Derivation of Developmental Toxicant RfD (RfD$_{dt}$)

In developing an RfD$_{dt}$, evidence is gathered regarding the potential of a substance to cause adverse effects in a developing organism as a result of exposure prior to conception (either parent), during prenatal development, or postnatally to the time of sexual maturation. Adverse effects can include death, structural abnormality, altered growth, and functional deficiencies. Maternal toxicity is also considered. The evidence is assessed, and the substance is assigned a weight-of-evidence designation according to scheme. In this scheme, three levels are used to indicate the assessor's degree of confidence in the data: definitive evidence, adequate evidence, and inadequate evidence. The definitive and adequate evidence categories are subdivided as to whether the evidence demonstrates the occurrence or the absence of adverse effects.

After the weight-of-evidence designation is assigned, a study is selected for the identification of a NOAEL. The NOAEL is converted to an equivalent human dose, if necessary, and divided by uncertainty factors similar to those used in the development of an oral RfD. It should be remembered that the RfD$_{dt}$, is based on a short duration of exposure because even a single exposure at a critical time (e.g., during gestation) may be sufficient to produce adverse developmental effects, and that chronic exposure is not a prerequisite for developmental toxicity to be manifested. Therefore, RfD$_{dt}$ values are appropriate for evaluating single event exposures, which usually are not adjusted based on the duration of exposure.

Calculation Scheme for Noncarcinogens

The preliminary assessment of noncarcinogenic risk associated with a hazardous waste site as recommended by the USEPA is typically calculated in four major steps (6).

1. Identify discrete exposure conditions:
    - Exposure route
    - Frequency
    - Duration
    - Administered dose

2. Derive appropriate reference doses for each discrete set of conditions.

3. Evaluate the hazard for noncarcinogenic effects as a ratio of exposure dose to the recommended RfD.

4. Aggregate the hazard for multiple chemical agents and exposure pathways as a hazard index, where appropriate.

The ratio referred to in the third step is utilized to quantify risk from noncarcinogens. As the fourth step indicates, the hazard index for individual chemicals may be summed for chemicals affecting a particular target organ or acting by a common mechanism in order to provide a final measure of noncarcinogenic toxic risk. If the sum of hazard indices is less than one, then the risk of adverse health effects is considered acceptable.

Dose-Response Relationship

A dose-response relationship is presented between an agent and an effect (response) when, as the concentration of the agent at the reactive site increases, the probability that the effect or response in the host also increases. A characteristic dose-response curve is presented in Figure 11.5.2. A threshold would exist if there was a level of dose for which no apparent effect would be discerned as presented in Figure 11.5.3. This is a strongly debated topic for carcinogens. The regulatory community, in the interests of protecting the health of the public, usually assumes that no thresholds exist (for carcinogens) and performs its functions accordingly. This is discussed further in the next section.

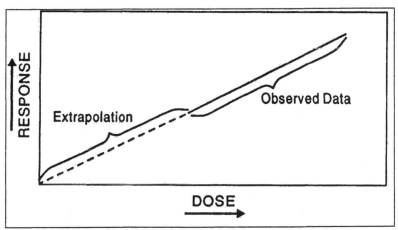

Figure 11.5.2. Characteristic dose-response curve.

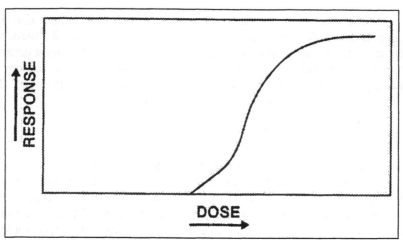

Figure 11.5.3. Dose-response curve containing a threshold.

11.6 CARCINOGENS

This section describes how the types of toxicity information are considered in the toxicity assessment for carcinogenic effects. A slope factor and the accompanying weight of evidence determination are the toxicity data most commonly used to evaluate potential human carcinogenic risks. The methods the USEPA uses to derive these values are outlined below.

Concept of Nonthreshold Effects

Carcinogenesis, unlike many noncarcinogenic health effects, is generally thought to be a phenomenon for which risk evaluation based on the presumption of a threshold is inappropriate. For carcinogens, the USEPA assumes that a small number of molecular events can evoke changes in a single cell that can lead to uncontrolled cellular proliferation and eventually to a clinical state of disease. This hypothesized mechanism for carcinogenisis is referred to as "non-threshold" because there is believed to be essentially no level of exposure to such a chemical that does not pose a finite probability, however small, of generating a carcinogenic effects. The USEPA uses a two-part evaluation in which the substance first is assigned a weight-of-evidence classification, and then a slope factor is calculated.

Assigning a Weight of Evidence

In the first step of the evaluation, the available data are evaluated to determine the likelihood that the agent is a human carcinogen. The evidence is characterized separately for human studies and animal studies as sufficient, limited, inadequate, no data, or evidence of no effect. The characterizations of these two types of data are combined, and based on the extent to which the agent has been shown to be a carcinogen in experimental animals or humans, or both, the agent is given a provisional weight-of-evidence classification. USEPA scientists then adjust the provisional classification upward or downward, based on other supporting evidence of carcinogenicity

The USEPA classification system for weight of evidence is shown in Table 11.6.1. This system is adapted from the approach taken by the International Agency for Research on Cancer (IARC).

Generating a Slope Factor [7]

A slope factor is generated in the second part of the evaluation. Based on the evaluation that the chemical is a known or probable human carcinogen, a toxicity value that defines quantitatively the relationship between dose and response (i.e., the slope factor) is calculated. Slope factors are typically calculated for potential carcinogens in classes A, B1, and B2. Quantitative estimation of slope factors for the chemicals in class C proceeds on a case-by-case basis.

Generally, the slope factor is a plausible upper bound estimate of the probability of a response per unit intake of a chemical over a lifetime. The slope factor is used in risk assessments to estimate an upper-bound lifetime probability of an individual developing cancer as a result of exposure to a particular level of a potential carcinogen. Slope factors should always be accompanied by the weight-of-evidence classification to indicate the strength of the evidence that the agent is a human carcinogen. Calculational details are presented below.

Identifying the Appropriate Data Set

In deriving slope factors, the available information about a chemical is evaluated and an appropriate data set is selected. In choosing appropriate data sets, human data of high quality are preferable to animal data. If animal data are used, the species that responds most similarly to humans (with respect to factors such as metabolism, physiology, and pharmacokinetics) is preferred. When no clear choice is possible, the most sensitive species is given the greatest emphasis. Occasionally, in situations where no single study is judged most appropriate, yet several studies collectively support the estimate, the geometric mean of estimates from all studies may be adopted as the slope. This practice ensures the inclusion of all relevant data.

Because risk at low exposure levels is difficult to measure directly either by animal experiments or by epidemiologic studies, the development of a slope factor generally entails applying a model to the available data set and

TABLE 11.6.1   EPA Weight-of-Evidence Classification System for Carcinogenicity

Group	Description
A	Human carcinogen
B1 or B2	Probable human carcinogen
	B1 indicates that limited human data are available
	B2 indicates sufficient evidence in animals and inadequate or no evidence in human
C	Possible human carcinogen
D	Not classifiable as to human carcinogenicity
E	Evidence of noncarcinogenicity for humans

using the model to extrapolate from the relatively high doses administered to experimental animals (or the exposures noted in epidemiologic studies) to the lower exposure levels expected for human contact in the environment.

A number of mathematical models and procedures have been developed to extrapolate from carcinogenic responses observed at high doses to responses expected at low doses. Different extrapolation methods may provide a reasonable fit to the observed data but may lead to large differences in the projected risk at low doses. The choice of a low-dose extrapolation model is governed by consistency with current understanding of the mechanism of carcinogenesis, and not solely on goodness-of-fit to the observed tumor data. When data are limited and when uncertainty exists regarding the mechanisms of carcinogenic action, the USEPA guidelines suggest that models or procedures that incorporate low-dose linearity are preferred when compatible with the limited information available. The USEPA's guidelines recommend that the linearized multistage model be employed in the absence of adequate information to the contrary.

In general, after the data are fit to the appropriate model, the upper 95th percent confidence limit of the slope of the resulting dose-response curve is calculated. This value is known as the slope factor and represents an upper 95th percent confidence limit on the probability of a response per unit intake of a chemical over a lifetime (i.e., there is only a 5 percent chance that the probability of a response could be greater than the estimated value on the basis of the experimental data and model used). In some cases, slope factors based on human dose-response data are based on the "best" estimate instead of the upper 95th percent confidence limits. Because the dose-response curve generally is linear only in the low-dose region, the slope factor estimate only holds true for low doses. Information concerning the limitations on use of slope factors can be found in IRIS.

When animal data are used as a basis for extrapolation, the human dose that is equivalent to the dose in the animal study is calculated using the assumption that different species are equally sensitive to the effects of a toxicant if they absorb the same amount of the agent (in milligrams) per unit of body surface area. This assumption is made only in the absence of specific information about the equivalent doses for the chemical in question. Because surface area is approximately proportional to the 2/3 power of body weight, the equivalent human dose (in mg/day, or other units of mass per unit time) is calculated by multiplying the animal dose (in identical units) by the ratio of human to animal body weights raised to the 2/3 power. (For animal doses expressed as mg/kg-day, the equivalent human dose, in the same units, is calculated by multiplying the animal dose by the ratio of animal to human body weights raised to the 1/3 power.)

When using animal inhalation experiments to estimate lifetime human risks for partially soluble vapors or gases, the air concentration (ppm) is generally considered to be the equivalent dose between species based on equivalent exposure times (measured as fractions of a lifetime). For inhalation of particulates or completely absorbed gases, the amount absorbed per unit of body surface area is considered to be the equivalent dose between species.

Toxicity values for carcinogenic effects can be expressed in several ways. The slope factor is usually, but not always, the upper 95th percent confidence limit of the slope of the dose-response curve and is expressed as $(\text{mg/kg-day})^{-1}$. If the extrapolation model selected is the linearized multistage model, this value is also known as the q1. That is:

$$\text{Slope factor} \quad = \text{risk per unit dose} \qquad (11.6.1)$$
$$= \text{risk per mg/kg-day}$$

Where data permit, slope factors listed in IRIS are based on absorbed doses, although to date many of them have been based on administered doses.

Toxicity values for carcinogenic effects also can be expressed in terms of risk per unit concentration of the substance in the medium where human contact occurs. These measures, called unit risks, are calculated by dividing the slope factor by 70 kg and multiplying by the inhalation rate (20 m^3/day) or the water consumption rate (2 L/day), respectively, for risk associated with unit concentration in air or water. Where an absorption fraction less than 1.0 has been applied in deriving the slope factor, an additional conversion factor is necessary in the calculation of unit risk so that the unit risk will be on an administered dose basis. The standardized duration assumption for unit risks is understood to be continuous lifetime exposure. Hence, when there is no absorption conversion required:

$$\text{air unit risk} \quad \begin{aligned} &= \text{risk per ug/m}^3 \\ &= \text{slope factor x } 1/70 \text{ kg x } 20\text{m}^3/\text{day x } 10^{-3} \end{aligned} \qquad (11.6.2)$$

$$\begin{aligned} \text{water unit} &= \text{risk per } \mu\text{g/L} \\ \text{risk} &= \text{slope factor x } 1/70 \text{ kg x } 2\text{L/day x } 10^{-3} \end{aligned} \qquad (11.6.3)$$

The multiplication by 10^{-3} is necessary to convert from mg (the slope factor, or q1, is given in (mg/kg-day)$^{-1}$) to μg (the unit risk is given in $(\mu g/m3)^{-1}$ or $(\mu g/L)^{-1}$).

The number of clear human epidemiologic studies is small. A total of approximately 50 compounds (e.g., benzene, vinyl chloride) and complex exposures (e.g., aluminum production, tobacco smoke) have sufficient data available to permit their classification as human carcinogens.[8] The most potent human carcinogens known, the aflatoxins, are of natural origin. Their presence in food products through infestation by toxin-producing fungi constitute a serious problem in several tropical and subtropical countries.

The USEPA utilizes the following description for Group A - Human Carcinogens: [Compounds for which] "there is sufficient evidence from epidemiologic studies to support a causal association between exposure to the agent and cancer." This designation is based on the weight of evidence. The designation as a human carcinogen represents a consensus opinion with an expert group rather than a scientific fact. The weight of evidence system is shown in Table 11.6.1.

Because of limited epidemiological data, most of our understanding about the potential for a toxic substance to cause cancer is based on animal testing. Substances not having an apparent epidemiological association with human cancer but which have induced cancer in test animals are referred to as potential human carcinogens. Several hundred substances for which there is inadequate human epidemiological data have been assigned as potential human carcinogens based upon testing data in laboratory animals, usually rats and mice.

Dose-Response Relationships

To predict the likely outcome from human exposure to carcinogenic chemicals requires the development of dose-response relationships. To develop these relationships, toxicologists ordinarily depend upon animal data. A typical dose-response curve in the high dose region is given in Figure 11.6.1 for a hypothetical carcinogen. The approach to developing dose-response relationships for carcinogens differs somewhat from the approach for noncarcinogens. For ease of comparisons, doses (the abscissa) are almost always plotted as the daily administered dose per unit of body weight (mg/kg-day). Dose is plotted against response (the ordinate) which is the incidence of

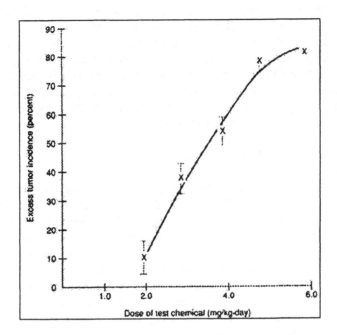

Figure 11.6.1 Hypothetical dose-response curve for a typical carcinogen

the total number of tested animals with tumors (dimensionless). The above diagram represents the best fit of mean values. It should be noted that the ordinate is reported as excess incidence of cancer, or the incidence in the exposed animals minus the normal spontaneous (background) incidence in the control population.

When assessing lifetime cancer risk to humans, it is widely accepted that carcinogenesis works in a manner such that it is possible, however remote, that exposure to a single molecule of a genotoxic carcinogen could result in one of the two mutations necessary to initiate cancer (i.e., genotoxins do not exhibit thresholds). In theory, therefore, the dose-response curve is asymptotic to "zero" incidence. Because there is no threshold, there is no safe level, only "acceptable levels." The public has expressed concern with any level, and regulatory agencies tend to establish a regulatory goal of one-in-one million (1 x 10^{-6} or 0.0001%) excess lifetime cancer risk.

As stated before, it is not practical to test in the range of such a low incidence. In fact, laboratories conduct animal tests at high doses to increase the likelihood of inducing cancer in a portion of a test population. Thus, to apply the results of animal tests to human exposures, the data from the high doses used in the tests must be extrapolated to the low doses of public concern.

The calculation of carcinogenic risk involves the use of a carcinogen potency factor (CPF). A CPF is the slope of the dose-response curve at very low exposures, and is now referred to as slope factor. The dimensions of a CPF, or a slope factor, are expressed as the inverse of daily dose $(mg/kg\text{-}day)^{-1}$. Having derived a CPF, the calculation of carcinogenic risk is straight forward. Quantification of carcinogenic risk of an exposure simply requires converting the dose to appropriate terms (mg/kg-day) and multiplying it by the slope factor is demonstrated in Illustrative Example 6.

11.7 UNCERTAINTIES/LIMITATIONS

It is important to keep in mind that statistically based studies by themselves can never prove the existence of a cause and effect relationship. However, such observations may be used to generate or to test a hypothesis. Many possibilities exist for introducing bias in this type of investigation, and statistical correlations may be fortuitous. It is, therefore, of utmost importance that any such correlation or conclusion be supported by a biological plausibility based on existing toxicology data and other information. Here, common sense is a logical first step. Even for the case of a well established cause and effect relationship, confounding factors often impede the establishment of a reliable dose-response (effect) relationship.[6]

For a variety of reasons it is difficult to precisely evaluate toxic responses caused by exposures (particularly acute ones) to hazardous chemicals. Five of these toxic responses are briefly discussed below.

1. Humans experience a wide range of acute adverse health effects, including irritation, narcosis, asphyxiation, sensitization, blindness, organ system damage, and death. In addition, the severity of many of these effects varies with intensity and duration of exposure. For example, exposure to a substance at an intensity that is sufficient to cause only mild throat irritation is of less concern than one that causes severe eye irritation, lacrimation, or dizziness, since the latter effects are likely to impede escape from the area of contamination.

2. There is a high degree of variation in response among individuals in a typical population. Generally, sensitive populations include the elderly, children, and individuals with diseases that compromise the respiratory or cardiovascular system.

3. For the overwhelming majority of substances encountered in industry, there are not enough data on toxic responses of humans to permit an accurate or precise assessment of the substance's hazard potential. Frequently, the only data available are from controlled experiments conducted with laboratory animals. In such cases, it is necessary to extrapolate from effects observed in animals to effects likely to occur in

humans. Thus, extrapolation requires the professional judgment of a toxicologist.

4. Many releases involve multicomponents. There are presently no "rules" or "guidelines" on how these types of releases should be evaluated. Are they additive synergistic, or antagonistic in their effect on population? As more information is developed on the characterization of multicomponent releases from source and dispersion experimentation/modeling, corresponding information is needed in the toxicology arena. Unfortunately, even toxic response data of humans to single component exposures are woefully inadequate for a large number of chemical species.

5. There are no toxicology testing protocols that exist for studying episodic releases on animals. This has been in general a neglected aspect of toxicology research. There are experimental problems associated with the testing of toxic chemicals at high concentrations for very short durations in establishing the concentration/time profile. In testing involving fatal concentration/time exposures, a further question exists of how to incorporate early and delayed fatalities into the study results.

Uncertainties Related to Toxicity Information

Toxicity information for many of the chemicals found at Superfund sites is often limited. Consequently, there are varying degrees of uncertainty associated with the toxicity values calculated. Sources of uncertainty associated with toxicity values may include:

1. Using dose-response information from effects observed at high doses to predict the adverse health effects that may occur following exposure to the low levels expected from human contact with the agent in the environment

2. Using dose-response information from short-term exposure studies to predict the effects of long-term exposures, and vice versa

3. Using dose-response information from animal studies to predict effects in humans

4. Using dose-response information from homogeneous animal populations or healthy human populations to predict the effects likely to be observed in the general population consisting of individuals with a wide range of sensitivities

An understanding of the degree of uncertainty associated with toxicity values is an important part of interpreting and using these values. Therefore, as part of the toxicity assessment for Superfund sites, a discussion of the strength of the evidence of the entire range of principal and supporting studies should be included. The degree of confidence ascribed to a toxicity value is a function of both the quality of the individual study from which it was derived and the completeness of the supporting data base. The USEPA-verified RfDs found in IRIS are accompanied by a statement of the confidence that the evaluators have in the RfD itself, the critical study, and the overall data base. All USEPA-verified slope factors are accompanied by a weight-of evidence classification, which indicates the likelihood that the agent is a human carcinogen. The weight-of-evidence classification is based on the completeness of the evidence that the agent causes cancer in experimental animals and humans. These designations should be used as one basis for the discussion of uncertainty.

11.8 ILLUSTRATIVE EXAMPLES

Example 11.1

Define and compare the following pairs of parameters used in health risk analysis:

a. NOEL and NOAEL
b. LOEL and LOAEL
c. ADI and RfD

Solution 11.1

a. NOEL is an acronym for "No Observed Effect Level," the heighest dose of a toxic substance that will not cause an effect. NOAEL is an acronym for "No Observed Adverse Effect Level," the highest dose of the toxic substance that will not cause an adverse effect.
b. LOEL is an acronym for "Lowest Observed Effect Level," the lowest dose tested for which effects were shown. LOAEL is an acronym for "Lowest Observed Adverse Effect Level," the lowest dose tested for which adverse effects were shown.

   Both NOEL and LOEL are based on a wide range of toxic substance effects, while NOAEL and LOAEL are based on only adverse effects. Neither set of parameters take into account the individual variation in susceptibility.

c. ADI, the Acceptable Daily Intake, is the level of daily intake of a particular substance that will not produce an adverse effect.

   RfD, the Reference Dose, is an estimate of a daily exposure level for the human population that will not produce an adverse effect.

The RfD development follows a stricter procedure than that used for the ADI. This sometimes results in a lower value for ADI.

Comment: When using data for LOELs, LOAELs, NOELs, or NOAELs, it is important to be aware of their limitations. As discussed in the chapter, statistical uncertainty exists in the determination of these parameters due to the limited number of animals used in the studies to determine the values. However, any toxic effect might be used for the NOAEL and LOAEL so long as it is the most sensitive toxic effect and considered like it to occur in humans.

Example 11.2

Certain assumptions are usually made about an "average" person's attributes for risk assessments applied to large groups of individuals. List the standard values usually assigned to represent the "average" values for the following.

1. Body weight
2. Daily drinking water intake
3. Amount of air breathed per day
4. Expected life span
5. Dermal contact area

Solution 11.2

The standard values for an "average" person are:

1. A body weight of 70kg for an adult or 10kg for a child.
2. An adult drinks 2 L per day of water or 1L per day for a child
3. An adult breathes 20 m^3 of air per day or 10 m^3 per day for a child
4. A life span of 70 years for an adult
5. An adult dermal contact area of 1000 cm^2 or 300 cm^2 for a child

Example 11.3

The drinking water maximum contaminant level (MCL) set by the USEPA for altrazine is 0.003 mg/L and its Reference Dose (RfD) is 3.5 mg/kg-d. How many liters of water containing atrazine at its MCL would a person have to drink each day to exceed the RfD for this triazine herbicide?

Solution 11.3

Assuming that those exposed can be represented by a 70 kg individual, the volume of drinking water at the MCL to reach the RfD for atrazine is:

$$\left(3.5\frac{mg}{kg \cdot day}\right)(70kg)\left(\frac{1}{0.003\frac{mg}{L}}\right) = 81,667\frac{L}{day}$$

This large volume indicates that there is either no health problem or there is considerable uncertainty (i.e., the product of the uncertainty factors is large) in estimating a reference dose for atrazine.

Example 11.4

Due to contamination from a metal plating facility, the water from a nearby community water supply well was shown to contain cyanide at a concentration of 20 µg/L, nickel at 95 µg/L, and chromium(III) at 10,200 µg/L. If the daily water intake is assumed to be 0.2L, and the body weight of an adult is 70kg, do these noncarcinogenic chemicals pose a health hazard?

Solution 11.4

The dose, D, for cyanide is:

$$D = \frac{\left(0.02\frac{mg}{L}\right)\left(0.2\frac{L}{day}\right)}{70kg}$$

$$= 5.71 \cdot 10^{-5}\frac{mg}{kg \cdot day}$$

The remaining calculations are shown in the following table, with RfD values from the IRIS database:

Substance	Concentration C (mg/L)	Dose (mg/kg-day)	RfD (mg/kg-day)	Hazard Ratio Dose/RfD
Cyanide	0.020	5.71E-5	0.02	0.003
Nickel	0.095	2.71E-4	0.02	0.014
Chromium (III)	10.2	2.91E-2	1.0	0.029
Total				0.046

The individual ratios are well below 1, and this preliminary evaluation does not indicate an unacceptable hazard.

Example 11.5

The odor perception threshold for benzene in water is 2 mg/L. The benzene drinking water unit risk is 8.3 x 10^{-7} L/µg. Calculate the potential benzene intake rate (mg benzene/kg-d) and the cumulative cancer risk from drinking water with benzene concentrations at half of its odor threshold for a 30 year exposure duration.

Solution 11.5

The following benzene intake rate is determined using the ingestion rate equation given in the problem statement

$$\text{Ingestion Rate} = \frac{\left(1\frac{mg}{L}\right)\left(2\frac{L}{day}\right)(30\,years)}{(70kg)(70\,years)}$$

$$= 0.012\frac{mg}{kg\cdot day}$$

The cancer risk from this ingestion at half of the odor threshold of 1 mg/L is calculated based on the benzene unit risk of 8.3 x 10^{-7} L/µg or:

$$\text{Cancer Risk} = \left(1000\frac{\mu g}{L}\right)\left(8.3\cdot10^{-7}\frac{L}{\mu g}\right)$$

$$= 8.3\cdot10^{-4}$$

This risk is high relative to the widely accepted standard range of environmental risk of 1×10^{-6} to 1×10^{-4}.

Example 11.6

What are the maximum number of excess lifetime cancer cases expected for a population of 5,000 adults with a daily intake of 0.20mg of benzene? The slope factor for benzene may be assumed to be 0.029 $(mg/kg\text{-}day)^{-1}$.

Solution 11.6

Assume an adult weight of 70kg. The lifetime cancer risk is calculated as:

$$\text{Individual Cancer Risk} = \left(0.20\frac{mg}{day}\right)\left(\frac{1}{70kg}\right)\left(0.029\frac{kg \cdot day}{mg}\right)$$

$$= 8.29 \cdot 10^{-5} \text{ (fractional basis)}$$

$$= 8.29 \cdot 10^{-3}\%$$

$$\text{Maximum Cases} = (\text{risk})(\text{exposed population})$$

$$= \left(8.29 \cdot 10^{-5}\right)(5000)$$

$$= 0.41 \text{ lifetime cancer cases}$$

Fractional numbers can pose a problem, but should be rounded to the next integer.

Example 11.7

The air in a factory contains 500 ppm of butane (TLV = 800 ppm), 100 ppm of cyclohexane (TLV = 300 ppm), 100 ppm of ethyl ether (TLV = 400 ppm) and 500 ppm of liquid petroleum gas (TLV = 1000 ppm). Is this a safe work place?

Solution 11.7

When two or more hazardous substances which act upon the same organ system are present, their combined effect, rather than the isolated individual effects, must be combined. If the sum of the following fractions exceeds unity, the threshold limit of the mixture has been exceeded. Thus,

$$\frac{C_1}{T_1}+\frac{C_2}{T_2}+\frac{C_3}{T_3}+\frac{C_4}{T_4}+...\frac{C_n}{T_n}>1;\quad \text{Not a safe working place}$$

$$=1;\quad \text{Caution}$$
$$<1;\quad \text{A safe working place}$$

where C = concentration measured in work area, ppm;
    T = corresponding threshold limit values, TLVS, ppm.

Based on the equation presented in the problem statement, the mixture of butane, cyclohexane, ethyl ether and liquid petroleum gas is evaluated in terms of the combined TLV as follows:

butane + cyclohexane + ethyl ether + liquid petroleum gas

$$\frac{500\text{ppm}}{800\text{ppm}}+\frac{100\text{ppm}}{300\text{ppm}}+\frac{100\text{ppm}}{400\text{ppm}}+\frac{500\text{ppm}}{1000\text{ppm}}=1.708>1$$

Therefore, this room is not a safe working place and should be controlled to prevent life and health-threatening situations at this facility.

For Threshold Limit Values (TLV) of other chemicals, refer to the American Conference of Industrial Hygienist Handbook.

Example 11.8

Two large bottles of flammable solvent were ignited by an undetermined ignition source after being knocked over and broken by a janitor while cleaning a 10 ft x 10 ft x 10 ft research laboratory. The laboratory ventilator was shut off and the fire was fought with a 10 lb CO_2 fire extinguisher. As the burning solvent had covered much of the floor area, the fire extinguisher was completely emptied in putting the fire out.

The Immediately Danger to Life or Health (IDLH) level for CO_2, set by the National Institute for Occupational Safety and Health (NIOSH) is 50,000 ppm. At that level, vomiting, dizziness, disorientation, and breathing difficulties occur after a 30-minute exposure; at a 100,000 ppm, death can occur after a few minutes, even if the oxygen in the atmosphere would otherwise support life.

Calculate the concentration of CO_2 in the room after the fire extinguisher is emptied. Does it exceed the IDLH value? Assume that the gas mixture in the room is uniformly mixed, that the temperature in the room is 300°C (warmed by the fire above normal room temperature of 200°C) and that the ambient pressure is 1 atm.

Solution 11.8

First calculate the number of moles of CO_2 discharged by the fire extinguisher.

$$\text{moles of } CO_2 = \left(10lb\ CO_2\right)\left(\frac{454g}{lb}\right)\left(\frac{gmol\ CO_2}{44\ g\ CO_2}\right)$$

$$= 103 \text{ gmol } CO_2$$

Calculate the volume of the room.

$$\text{Room volume} = \left(10ft\right)\left(10ft\right)\left(10ft\right)\left(0.0283\frac{m^3}{ft^3}\right)$$

$$= 28.3 \text{ m}^3 = 28300L$$

Next, calculate the total number of moles of gas in the room.

$$\text{moles of gas} = \frac{PV}{RT}$$

$$= \frac{\left(1atm\right)\left(28300L\right)}{\left(0.08206\frac{atm-L}{gmol-K}\right)\left(303K\right)}$$

$$= 1138 \text{ gmol of gas}$$

Calculate the concentration, or mole fraction, of CO_2 in the room.

$$\text{mole fraction} = \frac{gmol\ CO_2}{gmol\ gas}$$

$$= \frac{103\ gmol\ CO_2}{1138\ gmol\ gas}$$

$$= 0.0905$$

Convert this fraction to a percent and compare to the IDLH and lethal levels.

$$\%CO_2 \quad = (\text{mole fraction})(100\%)$$
$$= (0.0905)(100\%)$$
$$= 9.05\%$$

The IDLH level is 5.0% and the lethal level is 10.0%. Therefore, the level in the room of 9.05% does exceed the IDLH level for CO_2. It is also dangerously close to the lethal level. The person extinguishing the fire is in great danger and should take appropriate safety measures.

11.9 SUMMARY

1. Dose-response assessment is the process of characterizing the relationship between the dose of an agent administered or received and the incidence of an adverse health effect in exposed populations, and estimating the incidence of the effect as a function of exposure to the agent. This process considers such important factors as intensity of exposure, age pattern of exposure, and possibly other variables that might affect response, such as sex, lifestyle, and other modifying factors.
2. Before proceeding to some of the more technical aspects of toxicology and the general subject of dose-response, several important definitions used by the profession and appearing in the literature are provided.
3. Toxicology is the science of poisons, i.e., the study of chemical or physical agents that produce adverse responses in biological systems. Together with other scientific disciplines (such as epidemiology, the study of the cause and distribution of disease in human populations, and risk assessment), toxicology can be used to determine the relationship between an agent of interest and a group of people or a community.
4. Epidemiology is a discipline within the health sciences that deals with the study of the occurrence of disease in human populations. Epidemiologists are generally concerned with groups of people who share certain characteristics.
5. The reference dose, or RfD, is the toxicity value used most often in evaluating noncarcinogenic effects. The methods the USEPA uses for developing RfDs were presented earlier but are described again because of the number of different RfD. Various types of RfDs are available depending on the exposure route (oral or inhalation), the critical effect (developmental or other), and the length of exposure being evaluated (chronic, subchronic, or single effect).
6. A slope factor and the accompanying weight of evidence determination are the toxicity data most commonly used to evaluate potential human

carcinogenic risks. The methods the USEPA uses to derive these values are outlined below.

7. It is important to keep in mind that statistically based studies by themselves can never prove the existence of a cause and effect relationship. However, such observations may be used to generate or to test a hypothesis. Many possibilities exist for introducing bias in this type of investigation, and statistical correlations may be fortuitous.

PROBLEMS

1. What is the definition of carcinogenic "unit risk"?
2. Explain how the time weighted average, TWA, is defined.
3. Toxicological tests assess the effects of toxicants, including air toxicants, on animals, humans, microorganisms, or living cells. These tests have been used to determine the relative toxicity of various toxic compounds (including air toxins), i.e., categorizing chemicals in order from "extremely toxic" to "relatively harmless." The dose and concentration of a toxicant which cause the deaths of exposed animals are represented by the LD_{50} and LC_{50} values, respectively. Explain in the terms LD_{50} and LC_{50}, and what routes of exposure are taken into consideration while conducting these toxicity tests.
4. Describe and illustrate the process of setting a reference dose (RfD) using a schematic dose response curve. Correctly label the axis and all other important information.
5. A dose-response relationship provides a mathematical formula or graph for estimating a person's risk of illness at each exposure level for air toxins. To estimate a dose-response relationship, measurements of health risks are needed for at least one dose level of the air toxic compared to an unexposed group. However, there is one important difference between the dose-response curve commonly used for estimating the risk of cancer and the ones used for estimating the risk of all other illnesses: the existence of a threshold dose; that is, the highest dose at which there is no risk of illness. Because a single cancerous cell may be sufficient to cause a clinical case of cancer, EPA's and many others' dose-response models for cancer assume that the threshold dose level for cancer is zero. In other words, people's risk of cancer is increased even at very low doses. However, the increased cancer risk at very low doses is likely to be very low.
   a. Draw a straight line model showing the level of cancer risk increasing at a constant rate as the dose level increases. The model should illustrate increasing risk of cancer for the air toxic.
   b. Also develop a straight line model to show the USEPA's methodology in which the EPA adjusts the observed threshold downward by dividing by uncertainty factors that range from I to 10,000, known as

the human threshold. Information is available from the EPA Journal (September, 1990), Hazardous Substances in our Environment, Chapter 3, page 223.

REFERENCES

1. D. Gute and N. B. Hanes, "An Applied Approved to Epidemiology and Toxicology for Engineers", NIOSH, Cincinnati, Ohio, 1993.
2. C. A. Wengt, "Hazardous Waste Management", McGraw Hill, New York City, 1989.
3. L.B. and Hadden, S. G. Hauling Hazardous Materials: The Regultory Picture. Health and Environ. Dig. 2 (12): 1 – 3, 1989.
4. C.D. Klassen, M. O. Amdur and J. Doull, "ASA Casarett and Doull's Toxicology", Macmillan, New York City, 1986.
5. Adopted from USEPS Webpage, 2001.
6. D. LaGreda, P. L. Buckingham and J. C. Evans, "Hazardous Waste Management", McGraw Hill, NY, 1994.
7. Personal Notes: A. M. Flynn and L. Theodore, 2001.
8. IARC Monographs on the Evaluation of the Carcinogenic Risks to Humans - Overall Evaluations of Carcinogenicity: An Updating of IARC Monographs, Lyon, vol. 1 to 42 supl. 7, 1987.
9. "Guidelines for Chemical Process Quantitative Risk Analysis", 2nd Edition CCPS, AIChE, New York City, 2000.

12

Exposure Assessment

12.1 INTRODUCTION

Exposure assessment is the process of measuring or estimating the intensity, frequency, and duration of human or animal exposure to an agent currently present in the environment or of estimating hypothetical exposures that might arise from the release of new chemicals into the environment. In its most complete form, an exposure assessment should describe the magnitude, duration, schedule, and route of exposure, size, nature, and classes of the human, animal, aquatic, or wildlife populations exposed, and the uncertainties in all estimates. The exposure assessment can often be used to identify feasible prospective control options and to predict the effects of available treatment technologies for controlling or limiting exposure [1].

More attention has been recently focused on exposure assessment. This is because many of the risk assessments performed in the past used too many and overly conservative assumptions. This in turn, caused an overestimation of the actual exposure.

Obviously, without exposure(s) there are no risks. To experience adverse effects, one must first come into contact with the toxic agent(s). Exposures to chemicals can occur via inhalation of air(breathing), ingestion of water and food (eating and drinking), or adsorption through the skin. These are all pathways to the human body.

Generally, the main pathways of exposure considered in this step are atmospheric transport, surface and groundwater transport, ingestion of toxic materials that have passed through the aquatic and terrestrial food chain, and dermal absorption. Once an exposure assessment determines the quantity of a chemical with which human populations may come in contact, the information

The authors gratefully acknowledge the assistance of Anne Scarpinito in researching, reviewing, and editing this chapter.

can be combined with toxicity data (from the hazard identification process) to estimate potential health risks [2]. Thus, the primary purpose of an exposure assessment is to determine the concentration levels over time and space in each environmental media where human and other environmental receptors may come into contact with the chemicals of concern.

This chapter focuses on some of the practical considerations of exposure assessment, with particular emphasis on atmospheric dispersion. Following this Introduction section, the reader is introduced to OSHA's "Components of an Exposure Assessment Program". The next two sections deal with the development of the describing equations for dispersion in water systems and soils. Since the bulk of the work in this area is concerned with exposure to air health hazards/pollutants, the remainder of the chapter focuses on atmospheric dispersion applications. The effective height of an emission is next considered in view of the various equations and correlations currently in use. Atmospheric dispersion equations for continuous sources are reviewed; the effects of multiple sources as well as discharges (including particulates) from line and area sources are also briefly discussed. Both continuous and instantaneous discharges are of concern in accident and emergency management. Although, the bulk of the material here has been presented for continuous emissions for a point source (e.g., a stack), the chapter includes what has been referred to in the literature as a "puff" model-namely, an equation that can be used for estimating the effect of discharges from instantaneous (as opposed to continuous) sources. This chapter concludes with a short section on available computer models.

12.2 COMPONENTS OF AN EXPOSURE ASSESSMENT

Exposure is defined as the contact of an organism (humans in the case of health risk assessment) with a chemical or physical agent [3]. The magnitude of exposure is determined by measuring or estimating the amount of an agent available at the exchange boundaries (i.e., the lungs, gut, skin) during a specified time period. Exposure assessment is the determination or estimation (qualitative or quantitative) of the magnitude, frequency, duration, and route of exposure. Exposure assessments may consider past, present, and future exposures, using varying assessment techniques for each phase. Estimates of current exposures can be based on measurements or models of existing conditions; those of future exposures can be based on models of future conditions; and, those of past exposures can be based on measured or modeled past concentrations, or measured chemical concentrations in tissues.

The exposure assessment can be described by the following four steps.

Step 1 – Characterization of exposure setting

In this step, the assessor characterizes the exposure setting with respect to the general physical characteristics of the site and the characterizations of the populations on and near the site. Basic site characteristics such as climate, vegetation, ground-water hydrology, and the presence and location of surface waters are identified in this step. Populations also are identified and are described with respect to those characteristics that influence exposure, such as location relative to the site, activity patterns, and the presence of sensitive subpopulations. This step considers the characteristics of the current population, as well as those of any potential future populations that may differ under an alternate land use.

Sources of this information include site descriptions and data from the preliminary assessment (PA), site inspection (SI), and remedial investigation (RI) reports. Other sources include local soil surveys, wetland maps, aerial photographs, and reports by the National Oceanographic and Atmospheric Association (NOAA) and the U.S. Geological Survey (USGS). One can also consult with appropriate technical experts (e.g., hydrogeologists, air modelers) as needed to characterize the site.

Step 2 – Identification of exposure pathways

In this step, the exposure assessor identifies those pathways by which the previously identified populations may be exposed. Each exposure pathway describes a unique mechanism by which a population may be exposed to the chemical at or originating from the site. Exposure pathways are identified based on consideration of the sources, releases, types, and location of chemicals at the site, the likely environmental fate (including persistence, partitioning, transport, and intermedia transfer) of these chemicals, and the location and activities of the potentially exposed populations. Exposure points (points of potential contact with the chemical) and routes of exposure (e.g., ingestion, inhalation) are identified for each exposure pathway.

To determine possible release sources for a site in the absence of remedial action, all available site description and data from the PA, SI, and RI reports should be engaged. One should attempt to identify potential release mechanisms and obtain receiving media for past, current, and future releases. Monitoring data in conjunction with information on source locations to support the analysis of past, continuing, or threatened releases should be employed. For example, soil contamination near an old tank would suggest the tank (source) ruptured or leaked (release mechanism) to the ground (receiving media). Any source that could be an exposure point in addition to a release source (e.g., open barrels or tanks, surface waste piles or lagoons, contaminated soil) should also be noted.

After a chemical is released to the environment it may be:

1. transported (e.g., convected downstream in water or onto suspended sediment or through the atmosphere)
2. physically transformed (e.g., volatilization, precipitation);

3. chemically transformed (e.g., photolysis, hydrolysis, oxidation, reduction, etc.);
4. biologically transformed (e.g., biodegradation); and
5. accumulated in one or more media (including the receiving medium).

Step 3 – Quantification of exposure

In this step, the assessor quantifies the magnitude, frequency and duration of exposure for each pathway identified in Step 2. This step is most often conducted in two stages: estimation of exposure concentrations and calculation of intakes. The later estimation is considered in Step 4. In this part of step 3, the exposure assessor determines the concentration of chemicals that will be contacted over the exposure period. Exposure concentrations are estimated using monitoring data and/or chemical transport and environmental fate models. Modeling may be used to estimate future chemical concentrations in media that are currently contaminated or that may become contaminated, and current concentrations in media and/or at locations for which there are no monitoring data. The bulk of the material in this chapter is concerned with this step.

Step 4 – Qualification of intakes

In this step, the exposure assessor calculates chemical-specific exposures for each exposure pathway identified in step 2. As described in the last chapter, exposure estimates are expressed in terms of the mass of substance in contact with the body per unit body weight per unit time (e.g., mg chemical per kg body weight per day, also expressed as mg/kg-day). These exposure estimates are termed "intakes" (for the purposes of this text) and represent the normalized exposure rate. Several terms common in other EPA documents and the literature are equivalent or related to intake are provided below.

- Normalized Exposure Rate Equivalent to intake
- Administered Dose Equivalent to intake
- Applied Dose Equivalent to intake
- Absorbed Dose Equivalent to intake multiplied by an absorption factor

Chemical intakes are calculated using equations that include variables for exposure concentration, contact rate, exposure frequency, exposure duration, body weight, and exposure averaging time. The values of some of these variables depend on site conditions and the characteristics of the potentially exposed population.

After intakes have been estimated, they are organized by population, as appropriate. Then, the sources of uncertainty (e.g., variability in analytical data, modeling results, parameter assumptions) and their effect on the exposure estimates are evaluated and summarized. This information on uncertainty is important to site decision-makers who must evaluate the results of the exposure and risk assessment, and make decisions regarding the degree of remediation required at a site. The exposure assessment concludes with a summary of the estimated intakes for each pathway evaluated.

In summary, a preliminary description of the exposure scenario should be obtained which answers the following questions:

1. Where, when and how will the release of the toxicant occur?
2. What is in the immediate vicinity of the release?
3. What is the quantity, physical state, and chemical identity of the released material?
4. What are the concentrations and durations of exposure in the area of the toxicant's release?
5. Will the toxicant be distributed to a larger area, and if so what will be its form (physical and chemical), concentration, and duration of residence throughout the area of distribution? This description should include the concentrations at various locations and times throughout its residence, and it should include air and waterborne materials as well as those taken up by biological materials such as plants and animals.

Finally, knowing the quantity and type of emissions released by power plant stacks does not provide enough information to estimate the exposure of individuals to these substances.

To estimate exposure, the Environmental Protection Agency (EPA) has relied on a standard assumption called the maximally exposed individual (MEI). However, the EPA has acknowledged that the MEI standard considerably overestimates individual exposure. As a more realistic alternative, the Electric Power Research Institute (EPRI) developed the reasonably exposed individual (REI) measure of exposure. In both cases, the exposed individual lives in an area with the highest concentration of power plant emissions. The MEI measure assumes that the individual is sedentary, breathes at a steady rate, and lives outside any structure for his or her entire lifetime. The REI measure accounts for time spent indoors (where exposure to some pollutants is reduced), time spent working in distant areas, residential relocations, physical activity (and hence varying breathing rates), and even the replacement of fossil-fired generation units after 45 years of operation.

12.3 DISPERSION IN WATER SYSTEMS

Enormous amounts of waste dumped into water systems are degrading water quality and causing increased human health problems. In assessing this pollution, there are two distinct problem areas. The first, and worst, is in marine estuaries and associated coastal waters. As fewer and fewer alternatives remain for land disposal, wastes are finding their way more often into water. The second area consists of the oceans themselves, although it is believed that currently not much of a problem exists, because relatively little waste is dumped directly into the oceans. Municipal sewage and agricultural runoff are major sources of water pollution; yet industry gets the brunt of the blame for toxic pollution in marine waters. Nevertheless, in the aggregate, industrial discharges represent the largest source of toxic pollutants entering the marine environment. The major reason for this is that marine disposal, if available, is much cheaper for industry than other ways of disposing of wastes.

In addition to the normal, everyday pollutant emissions into water systems is the ever-present threat of a discharge resulting from an accident, an emergency, or a combination of these. The dispersion and ultimate fate of such pollutants is a major concern to the environmental engineer. It is for this reason that the present section on dispersion applications in water systems has been included. Much of this material has been "excerpted" and edited from one of the classic works in this field by Thomann and Mueller [4].

In general, the role of the water quality engineer and scientist is to analyze water quality problems by dividing each case into its principal components. These are:

1. Inputs: the discharge into the environment of residue from human and natural activities.
2. The Reactions and Physical Transport: the chemical and biological transformation, and water movement, that result in different levels of water quality at different locations in time in an aquatic ecosystem.
3. Output: the resulting concentration of a substance such as dissolved oxygen or nutrients at a particular location in a river or stream and during a particular time of the year or day.

The inputs are discharged into such ecological systems as rivers, lakes, estuaries, groundwater, or oceanic regions. After the occurrence of chemical, biological, and physical phenomena (e.g., bacterial biodegradation, chemical hydrolysis, physical sedimentation), these inputs result in a specific concentration of the substance in the given water body. Concurrently, through various mechanisms of public hearings, legislation, and evaluation, a desirable water use is being considered or has been established for the particular region of the water body under study. Such a desirable water use is translated into public health and/or ecological criteria, which are then compared to the concentration of the substance resulting from the discharge of the residue. This comparison may demonstrate the need for an environmental engineering control program, if the actual or forecasted concentration is not equal to that desired. Environmental engineering controls are then instituted on the inputs to ensure the reduction necessary to reach the desired concentration. The same basic philosophy is applied in treating accidental and/or emergency and/or catastrophic discharges into water systems.

The principal inputs can be divided into two broad categories: point sources and nonpoint sources. The point inputs are considered to have a well-defined point of discharge, which under most circumstances is continuous. A discharge pipe or group of pipes can be located and identified with a particular discharger. The two principal point source groupings are municipal point sources and industrial discharges. The principal nonpoint sources are agricultural, silvicultural, atmospheric, and urban-suburban runoff. In each case, the distinguishing feature of the nonpoint source is that the origin of the discharge is diffuse; that is, it is not possible to relate the discharge to a specific, well-defined location.

Furthermore, the source may enter the given river or lake via overland runoff, as in the case of agriculture, or through the surface of the land and water as an atmospheric input. The urban and suburban runoff may enter the water body through a large number of smaller drainage pipes designed to carry not wastes but storm runoff. In some instances in urban runoff, the discharge may be a large pipe draining a similarly large area. Other nonpoint sources include pollution due to drainage from abandoned mines and from construction activities, as well as leaching from land disposal of solid (hazardous) wastes.

The best physical model that can be used to describe a stream or river is a tubular flow reactor [5]. The describing equations then become one dimensional in the Cartesian (rectangular) coordinate system in the direction of the moving water. Most streams and rivers are subjected to sources or sinks of a pollutant, which are distributed along the length of the stream. An example of an external source is runoff from agricultural areas, whereas oxygen-demanding material distributed over the bottom of the stream exemplifies an in-stream, or internal source. The concentration in a stream or river due to multiple point and/or distributed sources is the linear summation of the responses due to the individual sources plus the response due to any upstream boundary condition. For streams with multiple sources in which flows and velocities vary-but are constant for a given length (reach) solutions are available although this situation is sometimes treated analytically as a one-dimensional transport equation in spherical coordinates [4,5].

The region between the free-flowing river and the ocean is a fascinating diverse, and complex water system: the coastal regime of estuaries, bay, and harbors. The ebb and flow of the tides, the incursion of salinity from the ocean, the influx of nutrients from the upstream drainage all contribute to the generation of a unique aquatic ecosystem. (The estuarine and wetland regions are considered to be crucial to the maintenance of major fish stocks such as the striped bass and the bluefish, which to varying degrees use the estuarine areas as spawning and nursery grounds.) The movement of the tides into and out of estuaries and the associated density effects created by the incursion of salinity are of particular importance in describing the quality of such bodies of water. (Many major cities are located along estuaries, primarily as a result of the historical need for ready access to national and international commerce routes. Such cities discharged large quantities of waste for many years, but because of the large volumes of estuaries, effects were not immediately felt. Later, however, especially in the 1950s, the load on estuaries became very heavy, quality deteriorated rapidly, and great interest centered on the analysis of water quality in estuaries.)

Tides are the movement of water above and below a datum plane, usual mean sea level. Tidal currents are the associated horizontal movement of the water into and out of an estuary. Tides and tidal currents are due to the attractive force of the moon and sun on the waters of the earth. There is a "pulling and tugging," which raises the water at certain locations and lowers it at others. These motions occur on a more or less cyclical basis, reflecting the regularity of the lunar and solar cycles. Tides are also present in lakes and seas, produced principally by winds blowing across the lake surface and "piling up" the water, which in turn sets the lake into an oscillatory motion or *seiche*. The approximately regular motion of the lake results in a motion in lake tributaries similar to estuarine tides.

The interesting behavior of tidal currents in open offshore waters is due to the lack of physical boundaries. A tidal current here tends to move about a point in a rotary-type current. Therefore, this type of current will tend to move any wastes discharged offshore in an elliptical pattern on which may be superimposed a net current drift.

An important characteristic of estuarine hydrology is the net flow through the estuary over a tidal cycle or a given number of cycles. This is the flow that over a period of several days or weeks flushes material out of the estuary; it is a significant parameter in the estimation of the distribution of estuarine water quality. If the estuary is well-mixed from top to bottom and side to side (i.e., no significant gradients in velocity), the net flow at any location in the estuary is approximately equal to the sum of the upstream external flow inputs to the estuary. This is true because it is known that the flow inputs are not causing the estuary to overflow. Therefore, this flow must, on balance, be leaving the estuary at any cross section (4).

If the mixing is "perfect," the estuary behavior may be approximated by what chemical engineers define as a *continuous stirred tank reactor* (CSTR) (5). However, accurately estimating the time and spatial behavior of water quality in estuaries is complicated by the effects of tidal motion as just described. The upstream and downstream currents produce substantial variations of water quality at certain points in the estuary, and the calculation of such variation is indeed a complicated problem. However, the following simplifications provide some remarkably useful results in estimating the distribution of estuarine water quality.

1. The estuary is one-dimensional
2. Water quality is described as a type of average condition over a number of tidal cycles
3. The cross-sectional area and flow do not vary with downstream distance
4. Steady-state conditions prevail
5. Chemical reaction effects can be treated as first order

A water body is considered to be a one-dimensional estuary when it is subjected to tidal reversals (i.e., reversals in direction of the water quality parameter are dominant). Since the describing (differential) equations for the distribution of either reactive or conservative (nonreactive) pollutants are linear, second-order equations, the principle of superposition discussed previously also applies to estuaries. The principal additional parameter introduced in the describing equation is a tidal dispersion coefficient E. Methods for estimating this tidal coefficient are provided by Thomann and Mueller [4].

A major portion of our water-based recreational activities occurs in the thousands of lakes, reservoirs, and other small, relatively quiescent bodies of water. The ecosystems of lakes throughout the world are of primary concern in water quality management. The lakes and reservoirs vary from small ponds and dams to the magnificent and monumental large lakes of the world such as Lake Superior (one of the Great Lakes) and Lake Baikal in the Soviet Union, the deepest lake in the world (1620 m; 5310 ft).

Limnology is the study of the physical, chemical, and biological behavior of lakes and reservoirs. Recreation, sport fishing (and for the larger lakes, commercial fishing), and water supply for municipal and industrial uses are all intimately related to the quality of these water bodies. The distinguishing physical features of lakes include relatively low to zero flow and development of significant vertical gradients in temperature and other water quality variables. Lakes therefore often become sinks for nutrients, toxicants, and other pollutants in incoming rivers.

This section provides a general overview of the properties of lake systems and presents the basic tools needed for modeling of lake water quality. The principal physical features of a lake are length, depth (i.e., water level), area (both of the water surface and of the drainage area), and volume. The relationship between the flow of a lake or reservoir and the volume is also an important characteristic. The ratio of the volume to the (volumetric) flow represents the hydraulic retention time (i.e., the time it would take to empty out the lake or reservoir if all inputs of water to the lake ceased). This retention time is given by the ratio of the water body volume and the volumetric flow rate.

As with rivers and estuaries, an understanding of the water balance and circulation of lakes is of considerable importance in water quality analysis and engineering. Inflows may include surface inflow, subsurface inflow, and water imported into the lake. Outflows may include surface and subsurface outflow from the reservoir and exported water. The change in storage in the lake or reservoir may also include subsurface storage or "bank" storage of water. In determining the hydrological balance of a lake, the change in volume and surface inflow and outflow usually can be measured easily. Precipitation also can be measured without difficulty except for large lakes, where it must be estimated for the open water. The remaining unknowns include subsurface water movements and evaporation.

The most important single factor in the evaporative loss of water is the incoming solar radiation. The vapor pressure (i.e., the maximum gaseous pressure of water vapor at a given temperature and 100% humidity) is also a primary variable and depends on the water temperature. Wind, air temperature, and water quality also contribute to the evaporative flux. In addition, lakes in general are not well mixed. Gradients develop along shore as well as with depth. Many lakes, during summer heating, develop a warmer layer of water at the surface overlying a colder deeper layer of water. The vertical temperature profile at the end of winter is often homogeneous from top to bottom. As spring warming begins, the surface layer begins to heat and because of its lower density, begins to stratify and become a distinct layer from the deeper layer beneath it. By midsummer, a strong stratification may have formed and often three distinct vertical regions can be identified. During the fall, surface temperatures begin to cool and subsequently the thermocline penetrates deeper into the lake; fall mixing begins. Isothermal conditions then prevail again during late fall. As surface temperatures continue to decline and reach levels below the temperature of maximum density, a winter inversion may develop.

The major reasons for the behavior of vertical temperature in water bodies are the low thermal conductivity and the absorption of heat in the first few meters. As the surface waters begin to heat, transfer to lower layers is reduced and a stability condition develops. The prediction of thermal behavior in lakes and reservoirs is an important power plant siting consideration and also is a major factor in preventing excessive thermal effects on sensitive ecosystems. Furthermore, the extent of thermal stratification influences the vertical dissolved oxygen (DO) profiles where reduced DO often results from minimal exchange with aerated water [4].

It is often useful to describe lakes and reservoirs under the assumption that the body of water is completely mixed horizontally and vertically. This "completely mixed" assumption, which is similar to the one sometimes made in the cases of rivers and estuaries, is justified on the basis of wind mixing and internal density variations. An assumption of this type should be recognized as a gross approximation to the actual lake, since variations in concentration of many substances will exist throughout the lake. However, the assumption permits many useful estimates to be made of the behavior of such systems. For this condition, the lake may be physically described as a perfectly mixed, stirred reactor. If overall flows are small and/or can be neglected, the system reduces to a batch reactor [5]. If the overall flows need to be considered, the describing equations may be made to take the form of a continuously stirred tank flow reactor under either steady or transient conditions [5].

Treating a body of water as a completely mixed system can be a valuable approach for estimating the effects of human activities. It can be applied to a number of pollutants, including suspended and dissolved substances, as well as to heat balance computations. As noted earlier in this section, the vertical behavior of lakes is of particular importance because surface and bottom waters exhibit quite different quality during periods of stratification. The estimation of vertical mixing is therefore of importance and, for some situations, a simple method can be used based on the completely mixed analysis.

For the effect and quantification of the vertical mixing phenomena, however, see Thomann and Mueller [4].

The modeling of a groundwater chemical pollution problem may be one-, two-, or three-dimensional. The proper approach is dependent on the problem context. For example, the vertical migration of a chemical from a surface source to the water table is generally treated as a one-dimensional problem. Within an aquifer, this type of analysis may be valid if the chemical rapidly penetrates the aquifer so that concentrations are uniform vertically and laterally. This is likely to be the case when the vertical and lateral dimensions of the aquifer are small relative to the longitudinal scale of the problem or when the source fully penetrates the aquifer and forms a strip source.

Groundwater pollution problems, however, are modeled using a two dimensional analysis. A typical aquifer has area dimensions that are much larger than the vertical dimension. Therefore, chemicals dissolved in the groundwater achieve vertical uniformity a short distance from the source and the chemical plume will move in the lateral and longitudinal directions.

A three-dimensional analysis is generally used when a significant vertical concentration gradient exists. Depending on their density, chemicals that are highly immiscible in water and are released to the ground in concentrated form will tend either to float on the surface of the aquifer or to fall to the bottom. These pollutants may then slowly dissolve into the groundwater. Leaky underground tanks for the storage of gasoline, oils, or solvents are the most important sources of such contamination [6].

Dispersion modeling equations for water systems take the same form as those presented later in this chapter for the atmosphere. Analytical solutions are not nearly as complicated or difficult, since the bulk motion of the fluid (in this case, water) is a weak variable with respect to magnitude, direction, time, and position as it is when the fluid is air.

12.4 DISPERSION IN SOILS

Few states had regulatory programs for land disposal of hazardous wastes before 1976. However, national awareness of hazardous waste problems increased dramatically in the mid- to late 1970s as it became evident that mismanagement and indiscriminant dumping of hazardous wastes at many sites had led to the release of toxic materials throughout the environment. Uncontrolled hazardous waste sites are distributed throughout the United States. They occur in various geological settings and in urban as well as rural areas. Uncontrolled sites may be operational, inactive, or abandoned. A wide range of chemical wastes has been deposited at uncontrolled land sites and the extent and severity of the resulting environmental contamination varies greatly across sites.

Many of the factors that influence the extent of contamination are site specific, either climatic or hydrogeological. Other factors that influence the extent of contamination relate to land surface features such as topography or development, which determine exposure routes. Additional important considerations are the types of chemical present on site, the potential for migration, the degree of contamination, and the extent of the area affected. Other significant issues include the location of the site with respect to drinking water sources, population centers' potential social and economic impacts of contamination, and the potential for land redevelopment and reuse.

In the past the presence of hazardous substances in soils was not a major public concern. In spite of the large number of documented hazardous waste sites in the United States, relatively few sites have been cleaned up with specific redevelopment in mind. Remedial actions usually are undertaken to contain or remove chemical contaminants; little or no consideration is given to the ultimate use of the site. If land reuse is decided before the cleanup there may be an opportunity to tailor the cleanup activities to best suit the site redevelopment.

The extent of cleanup that is necessary to protect human health and welfare varies with different use categories. Residential development is probably the most sensitive type of land use because of the long-term and multiple exposure routes and because of potential exposure to the most sensitive population segments (e.g., children and elderly persons). Excavation and removal appears to be the remedial action alternative selected at most sites where there is redevelopment. This is because no one can guarantee that a site is safe (i.e., offers zero risk) unless all contaminants are removed. Neither a developer nor a municipality can accept responsibility for site safety as long as hazardous materials remain there. In situ treatment approaches are seldom viewed as the best option because they are unproven and because 100% detoxification or stabilization cannot be achieved.

As part of each hazardous waste remedial action, the contamination at the site must be assessed. The options for remedial action to remove or otherwise deal with hazardous materials will depend on the nature of the contamination that will be allowed to remain on site after cleanup. Thus there must be plans for site remediation to cover any accidental or emergency discharges to land or soils that might develop.

To determine acceptable contaminant levels in soils, two primary exposure routes are usually considered: (1) inhalation of gases, vapors, or airborne particulate emanating from the site, and (2) ingestion of contaminated drinking water. Other routes that can contribute to exposure include absorption of pollutants through direct skin contact and uptake of water or soil contaminants by plants that are part of the food chain.

The potential for hazardous contaminants in soils to migrate to groundwater or to surface water is often of major concern. Detailed evaluation of cleanup levels frequently involves modeling the movement of contaminants to groundwater or surface water and estimating the maximum levels in soil that will not interfere with acceptable water quality characteristics. Drinking water standards and water quality criteria developed by the USEPA are widely used as guidance for acceptable levels in water. Water quality standards or criteria developed by individual state environmental agencies also may be applied.

One measure of the significance of contaminants in soil or solid waste samples may be determined by comparing the levels with reported naturally occurring concentrations. Provided levels are within the range that may occur naturally, one might conclude that the sample contaminant levels are of little consequence.

A rigorous treatment of dispersion in soils is beyond the scope of this book. However, some qualitative discussion is warranted because of the potential and existing problems already described. Two main problems arise because dispersion in soil (or land) is anisotropic (i.e., it varies with direction) and the permeability is not only a variable but also an unknown.

The variation in permeability with direction reflects the differences in path length through which a fluid element moves and the forces it experiences in moving through the porous media in a given direction. The directional nature of the permeability can be expressed as a symmetric second-order tensor with six independent components. This term can be reduced to three principal components (as with the diffusion coefficient) by choosing a coordinate system that corresponds to the principal axes of the soil. The anisotropy of a porous medium (like a soil) is undoubtedly related to the internal structure of the medium, and the structure is related to the circumstances under which the bed was formed. A porous medium may be expected to be anisotropic if the elementary particles are asymmetric and have, on the average, a particular orientation. Even beds of symmetric particles can be anisotropic if they are present in certain regular patterns. In naturally occurring soils, however, the packing pattern is sufficiently random to preclude the occurrence of anisotropy in beds made up of symmetric particles. The size distribution of the particles making up the bed also can affect the magnitude of the permeability, but ordinarily this does not contribute to the anisotropy.

For beds made up of randomly oriented symmetric particles, the permeability may be estimated by

$$k = \left[\frac{\varepsilon^3}{(1-\varepsilon)^2} \right] \left(\frac{d_e^2}{36k} \right) \qquad (12.4.1)$$

with k equal to 5. The equation predicts quite well the permeability if the particles are approximately the same size. The hydraulic radius of the bed d_e, can be estimated from measurements of pore size distribution. The porosity that must be used in this equation is the "effective porosity" (i.e., the porosity that reflects only the interconnecting flow channels). Typical porosities are in the 0.4-0.5 range. Vertical and horizontal permeabilities in soils can range from 1 to 50×10^{-7} cm^2, while average or equivalent particle diameters, d_e, can be as small as 10 μm.

For one-dimensional dispersion in soils, the describing equation for a conservative species and/or pollutant, c, is a cartesian (rectangular) coordinate system moving with velocity v_x is

$$\frac{\partial c}{\partial t} + v_x \frac{\partial c}{\partial x} = D_L \frac{\partial^2 c}{\partial x^2}$$

(12.4.2)

where D is the "effective" dispersion coefficient. For two-dimensional dispersion, the equation is

$$\frac{\partial c}{\partial t} + v_x \frac{\partial c}{\partial x} = D_L \frac{\partial^2 c}{\partial x^2} + D_T \frac{\partial^2 c}{\partial y^2}$$

(12.4.3)

where D_L and D_T are the longitudinal and transverse dispersion coefficients. Solutions to these equations have been presented earlier. However, real-world applications are rarely possible at this time because accurate values for the dispersion coefficients are not available. This is mainly because there are many "dispersion" mechanisms occurring in the soil and it is difficult even to quantify each effect. A short description of some of these mechanisms is given below.

1. *Molecular Diffusion:* If time scales are sufficiently long, dispersion results from molecular diffusion.
2. *Eddies:* If the flow within the individual flow channels of the porous medium (soil) becomes turbulent, dispersion results from eddy migration.
3. *Mixing Due to Obstructions:* The tortuosity of the flow channels in a porous medium means that fluid elements starting a given distance from each other and proceeding at the same velocity will not remain the same distance apart.
4. *Presence of Autocorrelation in Flow Paths:* Dispersion can result because all pores in the porous medium are not accessible to a fluid element after it has entered a particular flow path. In order words, the connectivity of the medium is not complete.
5. *Recirculation Due to Local Regions of Reduced Pressure:* Dispersion can be caused by a recirculation arising from flow restrictions. The converiosn of pressure energy into kinetic energy gives a local region of low pressure, and if this region is accessible to fluid that has passed through the region previously, a recirculation is set up.
6. *Dead-end Pores:* Dead-end volumes cause dispersion in unsteady flow (concentration profiles varying) because, as a solute-rich front passes the pore, transport occurs by molecular diffusion into the pore. After the front has passed, this solute will diffuse back out, thus dispersing.
7. *Adsorption:* Dipsersion by adsorption is another unsteady-state phenomenon. As with dead-end pores, a concentration front deposits or

removes material and therefore tends to flatten concentration profiles in the interstitial fluid.

8. *Hydrodynamic Dispersion:* Macroscopic dispersion is produced in a capillary even in the absence of molecular diffusion because of the velocity profile produced by the adherence of the fluid to the wall. This causes fluid particles at different radial positions to move relative to one another, with the result that a series of mixing-cup samples at the end of the capillary exhibits dispersion.

12.5 EFFECTIVE HEIGHT OF EMISSION

For an emission height, the calculational sequence begins by estimating the "effective" height of the emission, employing an applicable plume rise equation. The maximum GLC may then be determined using an appropriate atmospheric diffusion equation (considered in the next Section).

The effective height of an emission rarely corresponds to the physical height of the source or the stack. If the plume is caught in the turbulent wake of the stack or of buildings in the vicinity of the source or stack, the effluent will be mixed rapidly downward toward the ground. If the plume is emitted free of these turbulent zones, a number of emission factors and meteorological factors influence the rise of the plume.

For the remainder of this section, we focus on the effective height of *stack* emissions that develops because of plume rise considerations. This effective height depends on a number of factors. The emission factors include the gas flow rate and temperature of the effluent at the top of the stack and the diameter of the stack opening. The meteorological factors influencing plume rise are wind speed, air temperature, shear of the wind speed with height, and atmospheric stability. No theory on plume rise presently takes into account all these variables, and it appears that the number of formulas for calculating plume rise varies inversely with our understanding of the process. Most of the equations that have been formulated for computing the effective height of an emission are semiempirical. When considering these "plume rise" equations, it is important to evaluate each one in terms of the assumptions made and the circumstances existing when the particular correlation was formulated. Depending on the circumstances, some equations may be more applicable than others.

The effective stack height (equivalent to the effective height of the emission) is usually considered as the sum of the actual stack height, the plume rise due to the velocity (momentum) of the issuing gases, and the buoyancy rise, which is a function of the temperature of the gases being emitted and the atmospheric conditions. Three key equations are provided below.

The Davidson-Bryant method is empirical because it is based on Bryant's wind tunnel experiments [7]. It is restricted to gases below 125°F and to stacks of moderate height and larger. It does not give maximum plume rise because it is a function of momentum only, not buoyancy. Although its values for plume rise are low, those for root-mean-square differences are also low. The method applies when the atmosphere neither resists nor assists the vertical motion of the plume and when wind velocities are 20 mph or higher-above this velocity the vertical vector is insignificant compared with the horizontal. The describing equation is

$$\Delta h = d_s \left(v_s / u \right)^{1.4} \left[1.0 + \left(\frac{T_G - T_a}{T_G} \right) \right] \qquad (12.5.1)$$

where: Δh = rise of the plume above stack (m)

d_s = inside diameter of stack (m)

v_s = stack exit velocity (m/s)

u = wind speed (m/s)

T_a = ambient air temperature (K)

T_G = stack gas temperature (K)

The Holland equation is valid for effluent gases hotter than 125°F and for neutral conditions [8]:

$$\Delta h = \frac{1}{u} \left(1.5 v_s d_s + 0.04 Q_H \right) \qquad (12.5.2)$$

$$Q_H = \left(\pi d_s v_s / 4 \right) \left[\frac{P}{R / MW} \right] c_p \left(\frac{T_G - T_a}{T_G} \right) \qquad (12.5.3)$$

where: Δh = rise of plume above stack (m)

Q_H = heat emission rate (kcal/s)

P = atmospheric pressure (dynes/m^2)

R = universal gas constant (1833.35 dyne - m/K - gmol)

MW = molecular weight of the effluent (g/gmol)

c_p = heat capacity of the effluent gas

at constant pressure (kcal/g - K)

A more popular form of Eq. (12.5.2) is

$$\Delta h = (v_s d_s / u) \left[1.5 + 2.68 \times 10^{-3} p \left(\frac{T_G - T_a}{T_G} \right) d_s \right] \qquad (12.5.4)$$

where p is the atmospheric pressure in millibars. Holland's equation frequently underestimates the effective height of emission; therefore, its use often provides a slight safety factor. Holland also suggests that a value between 1.1 and 1.2 times Δh from the equation should be used for unstable conditions; a value between 0.8 and 0.9 times the Δh from the equation should be used for stable conditions. Since the plume rise from a stack occurs over some distance downwind, Eq (12.5.4) should not be applied within the first few hundred meters of the stack.

TABLE 12.5.1 Stability Class Definitions

Stability	Type	Potential Temperature Lapse Rate, (K/100 m)[a] $\Delta\theta/\Delta z$
0	Unstable	$\Delta\theta/\Delta z < -0.22$
1	Neutral	$-0.22 \le \Delta\theta/\Delta z < 0.15$
2	Neutral	$0.15 \le \Delta\theta/\Delta z .85$
3	Stable	$0.85 \le \Delta\theta/\Delta z$

[a] $\theta = T + 0.0098z$, where T is in Kelvin and z in meters.

In view of the stability class definitions given in Table 12.5.1, Briggs used the following equations to calculate the plume rise [9].

$$\Delta h = 1.6(F)^{1/3}(u)^{-1}(x)^{2/3} ; \text{if } x < x_f \qquad (12.5.5)$$

$$\Delta h = 1.6(F)^{1/3}(u)^{-1}(x_f)^{2/3} ; \text{if } x > x_f \qquad (12.5.6)$$

$$x^* = 14(F)^{5/8} ; \text{when } F < 55 \text{ m}^4 / s^3 \qquad (12.5.7)$$

$$x^* = 34(F)^{2/5} ; \text{when } F > 55 \text{ m}^4 / s^3 \qquad (12.5.8)$$

$$x_f = 3.5 x^* \qquad (12.5.9)$$

where: Δh = plume rise (m)

Q_H = heat emission (cal/s)

F = buoyancy flux = $3.7 \cdot 10^{-5}(Q_H)(m^4/s^3)$

u = wind speed (m/s)

x = downwind distance (m)

x^* = distance of transition from first stage of
rise to the second stage of rise (m)

x_f = distance to final rise (m)

The term F may be estimated (if Q_H is not available) by

$$F = \frac{(g/\pi)(V)(T_S - T)}{T_S}$$ (12.5.10)

where: g = 9.8 m
V = stack gas volume flow (m^3/s)
T_s, T = stack gas and ambient air temperature (K)

Equations (12.5.5) – (12.5.9) apply under stable or neutral conditions. For stable conditions, $\Delta\theta/\Delta z$ is needed. If $\Delta\theta/\Delta z$ is not given, one may use 0.02 K/m for stability E and 0.035 K/m for stability F. (Stability categories and conditions are described in the next section.) The term

$$s = \frac{g(\Delta\theta/\Delta z)}{T}$$ (12.5.11)

can be used to calculate both Eq. (12.5.12) and (12.5.13)

$$\Delta h = 2.4(F/us)^{1/3}$$ (12.5.12)

$$\Delta h = 5F^{1/4}/s^{3/8}$$ (12.5.13)

The smaller of these two Δh's should be used; it represents the final plume rise. The distance to final rise is given by

$$x_f = \frac{3.14u}{s^{1/2}}$$ (12.5.14)

To calculate the rise for a downwind distance x less than x_f, one may employ

$$\Delta h = \frac{1.6F^{1/3}x^{2/3}}{u} \qquad (12.5.15)$$

which is the same equation used for unstable and neutral conditions.

$$\Delta h = \frac{3.75\bar{x}^{0.49}F^{1/3}}{\bar{u}} \qquad (12.5.16)$$

For very stable atmospheric conditions, $(0.70 < \Delta\theta / \Delta z < 1.87$, average $= 1.06°C/100$ m), for distance, $\bar{x}$, up to 1960 meters use

$$\Delta h = \frac{13.8\bar{x}^{0.26}F^{1/3}}{\bar{u}} \qquad (12.5.17)$$

Many more plume rise equations may be found in the literature. Table 12.5.2 briefly summarizes some of the equations. For additional information, consult the references.

12.6 ATMOSPHERIC DISPERSION EQUATIONS FOR CONTINUOUS SOURCES

There are many dispersion equations available, most of them semiempirical. It is not the intent of this book to develop each in detail but rather to look at the one that has found the greatest applicability today. (In the authors' opinion, the best atmospheric dispersion workbook published to date is that by Turner (8))

The coordinate system used in making the atmospheric dispersion estimates as suggested by Pasquill and modified by Gifford, is shown in Figure 12.6.1 (Note that this is the coordinate system used by most engineers.) The origin is at ground level at or beneath the point of emission, with the x axis extending horizontally in the direction of the mean wind. The y axis is in the horizontal plane perpendicular to the x axis, and the z axis extends vertically. The plume travels along or parallel to the x axis (in the mean wind direction).

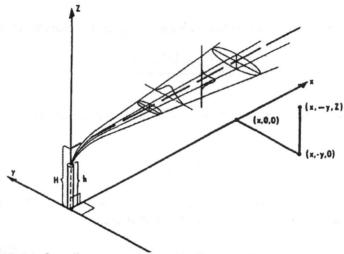

Figure 12.6.1. Coordinate system showing Gaussian distributions in the horizontal and vertical.

TABLE 12.5.2 Plume Rise Equations

Name	Equation	Reference
Concawe No. 1	$\Delta h = 2.58[Q_H^{0.58}/u^{0.70}]$	9
Concawe No. 2	$\Delta h = 5.53[Q_H^{0.50}/u^{0.75}]$	9
Lucas, Moore, and Spurr	$\Delta h = 135\, Q_H^{1/4}/u$	10
Rauch	$\Delta h = 47.2\, Q_H^{1/4}/u$	11
Stone and Clark ASME	$\Delta h = [104.2 + 0.171 H_s] Q_H^{1/4}/u$	12
Momentum sources	$\Delta h = d_s (v_s/u)^{1.4}$	13
Neutral and unstable	$\Delta h = 5.7\, Q_H/u^3]$	13
Stable	$\Delta h = 0.672\left[Q_H / \left(u\dfrac{g}{T_a}\dfrac{\Delta\theta}{\Delta t}\right)\right]^{1/3}$	13
Moses and Carson All data	$\Delta h = -0.029 v_s (d_s/u) + (5.35 Q_H^{1/2}/u)$	14
Unstable	$\Delta h = 3.47 v_s (d_s/u) + (10.53 Q_H^{1/2}/u)$	14

Name	Equation	Reference
Neutral	$\Delta h = 0.35 v_s (d_s / u) + (5.41 Q_H^{1/2} / u)$	14
Stable	$\Delta h = -1.04 v_s (d_s / u) + (4.58 Q_H^{1/2} / u)$	14
Csanady	$\Delta h = 9.5 Q_H / u^3$	15

The concentration c of gas or aerosols at (x, y, z) from a continuous source with an effective emission height, H^* is given by:

$$c\left(x, y, z; H^*\right) = \frac{Q'}{2\pi\sigma_y\sigma_z\bar{u}} \left\{ \exp\left[-\frac{1}{2}\left(\frac{y}{\sigma_y}\right)^2 \right] \right\}$$

$$\left\{ \left[-\frac{1}{2}\left(\frac{z-H^*}{\sigma_z}\right)^2 \right] + \exp\left[-\frac{1}{2}\left(\frac{z+H^*}{\sigma_z}\right)^2 \right] \right\} \quad (12.6.1)$$

The assumptions made in the development of Eq. 12.6.1 are: (1) the plume spread has a Gaussian distribution in both the horizontal and vertical planes with standard deviations of plume concentration distribution in the horizontal and vertical of σ_y and σ_x respectively; (2) the emission rate of pollutants Q' is uniform; (3) total reflection of the plume takes place at the earth's surface; and (4) the plume moves downwind with mean wind speed $\bar{u}$. Although any consistent set of units may be used, the cgs system is preferred.

Equation is valid where diffusion in the direction of the plume travel can be neglected (i.e., no diffusion in the x direction). This is a valid assumption if the release is continuous or if the duration of release is equal to or greater than the travel time $x / \bar{u}$ from the source to the location of interest.

For concentrations calculated at ground level (z=0), Eq. (12.6.1) simplifies to

$$c\left(x, y, 0; H^*\right) = \frac{Q'}{\pi\sigma_y\sigma_z\bar{u}} \left\{ \exp\left[-\frac{1}{2}\left(\frac{y}{\sigma_y}\right)^2 \right] \right\}$$

$$\left\{ \exp\left[-\frac{1}{2}\left(\frac{H^*}{\sigma_z}\right)^2 \right] \right\} \quad (12.6.2)$$

where the concentration is to be calculated along the centerline of the plume (y = 0). Further simplification gives

$$c\left(x, 0, 0; H^*\right) = \frac{Q'}{\pi \sigma_y \sigma_z \bar{u}} \left\{ \exp\left[-\frac{1}{2}\left(\frac{H^*}{\sigma_z} \right)^2 \right] \right\} \qquad (12.6.3)$$

In the case of a ground level source with no effective plume rise ($H^* = 0$), then

$$c(x, 0, 0; 0) = \frac{Q'}{\pi \sigma_y \sigma_z \bar{u}} \qquad (12.6.4)$$

The values of σ_y and σ_z vary with the turbulent structure of the atmosphere, the height above the surface, the surface roughness, the sampling time over which the concentration is to be estimated, the wind speed, and the distance from the source. For the parameter values that follow, the sampling time was originally assumed to be about 10 minutes, the height to be the lowest several hundred meters of the atmosphere, and the surface to be relatively open country. The parameters are estimated from the stability of the atmosphere,

TABLE 12.6.1 Stability Categories[a]

Surface Wind Speed at 10 m (m/s)	Day			Night	
	Incoming Solar Radiation			Thinly Overcast Or ≥ 4/8 Low Cloud	≤ 3/8 Cloud
	Strong	Moderate	Slight		
2	A	A-B	B		
2-3	A-B	B	C	E	F
3-5	B	B-C	C	D	E
5-6	C	C-D	D	D	D
6	C	D	D	D	D

[a] Class A is the most unstable, and class F is the most stable. The neutral class, D, should be assumed for overcast conditions during the day or the night, regardless of wind speed. Night refers to the period from 1 hour before sunset to 1 hour after sunrise.

which is in turn estimated from the wind speed at the height of about 10 meters and, during the day, the incoming solar radiation or, during the night, the cloud cover. Stability categories are given in Table 12.6.1 Note that A, B, and C refer to daytime with unstable conditions; D refers to overcast or neutral conditions at night or during the day; E and F refer to nighttime stable conditions and are based on the amount of cloud cover. "Strong" incoming solar radiation

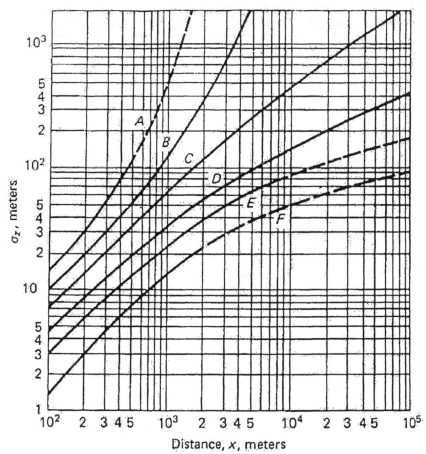

Figure 12.6.2. Vertical dispersion coefficient as a function if downwind distance from the source; A-F from Table 12.6.1.

corresponds to a solar altitude greater than 60' with clear skies (e.g., sunny midday in midsummer); "slight" insulation (rate of radiation from the sun received per unit of earth's surface) corresponds to a solar altitude from 15 to 35' with clear skies (e.g., sunny midday in midwinter). For the A-B, B-C, and C-D stability categories, use the average of the A and B values, B and C values, and C and D values, respectively.

Having determined the stability class, Figs. 12.6.2. and 12.6.3. may be used to evaluate σ_y and σ_z as a function of downwind distance from the source. Figures 12.6.2. and 12.6.3. apply strictly to open, level country and probably underestimate the plume dispersion potential from low level sources in built-up areas. Although the vertical spread may be less than the values for class F with

TABLE 12.6.2 Multiple-Stack Source Factors

Number of Stacks of a Given Height	Multiplication Factor
2	1.7
3	2.41
4	3.0
5	3.6
6	4.2
7	4.71
8	5.3
9	5.8
10	6.3

very light winds on a clear night, quantitative estimates of concentrations are nearly impossible for this condition. With very light winds on a clear night forground level sources free of topographic influences, frequent shifts in wind direction usually occur, which serve to spread the plume horizontally. For elevated sources under these extremely stable situations, significant concentrations usually do not reach the ground level until the stability changes. It is important to note that Eq. 12.6.1, as well as the emergency dispersion models and equations presented later in the Chapter, can be used to calculate the concentration profile of a pollutant, including hazardous and/or toxic substances. If the pollutant is explosive, a "vapor cloud profile" can be calculated where the vapor cloud would contain concentrations in the explosive range.

The effect of multiple-stack sources has been handled in the past by simply treating each as a distinct source and adding the resulting concentrations to obtain the total concentration. Probably the best method now available for estimating the effect of multiple sources is that suggested by the TVA [10]. Their procedure is as follows. For the case of a series of stacks, all having the same height, estimate the concentration for a single stack and then multiply this result by the factor indicated in Table 12.6.2 which corresponds to the total number of stacks of this size.

Stack emissions can include particulates as well as dense gases (heavier than air, e.g., chlorine). These are subjected to a downwash settling through the atmosphere due to the action of gravity. For the particles, especially large ones, an additional external force term must be included in the analysis.

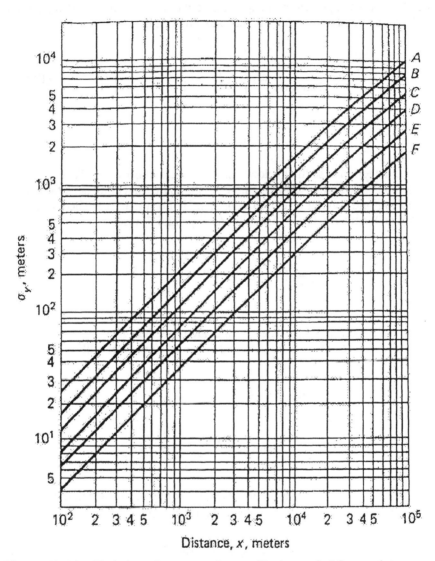

Figure 12.6.3. Variation of concentration profile downwind from point source; A-F designate stability categories listed in Table 12.6.1

A rather simple approach is to superimpose the settling velocity of the particle on the initial point of discharge in the following manner:

$$c = \frac{Q}{2\pi u \sigma_y \sigma_z} \exp\left\{-\frac{1}{2}\left[\left(\frac{y}{\sigma_y}\right)^2 + \left(\frac{z-H^*}{\sigma_z}\right)^2\right]\right\}$$

(12.6.5)

$$H^* = H - vx/u; \quad H^* > 0$$

(12.6.6)

and v is the terminal settling velocity of the particle in question. This effectively "repositions" the particle downstream from the source and eliminates the need of developing and solving a revised equation. If Eq. (12.6.5) applied to ground level (z = 0) centerline (y = 0) conditions, Eq. (12.6.7) results.

$$c = \frac{Q}{2\pi u \sigma_y \sigma_z} \exp\left[-\frac{1}{2}\left(\frac{H^*}{\sigma_z}\right)^2\right]$$

(12.6.7)

The rate of deposition of particles per unit area R is then given by the product of the local ground level concentration and the terminal vertical settling velocity of the particle.

$$R = cv$$

$$= \frac{Qv}{2\pi u \sigma_y \sigma_z} \exp\left[-\frac{1}{2}\left(\frac{H^*}{\sigma_z}\right)^2\right]$$

(12.6.8)

A consistent set of SI units is g/s-m² for R, g/s for Q, m/s for v and u, and m for σ and H^*. The terminal settling velocity is usually given by

$$v = \left(g d_p^2 \rho_p / 18\mu\right) C$$

(12.6.9)

where: g = acceleration due to gravity
         d_p = particle diameter
         ρ_p = particle density
         μ = viscosity of air
         C = Cunningham correction factor

It is important to note that the deposition rate is a strong function of particle diameter through the term v, which appears twice in the deposition flux equation. Equation (9.7.10) must be modified to treat process gas streams discharging particles of a given size distribution. The suggested procedure is somewhat similar to that for calculating overall collection efficiencies for particulate control equipment (12). For this condition, the overall rate is given by

$$R = \sum_{i=1}^{n} \left(\frac{Q_i v_i}{2\pi u \sigma_y \sigma_z} \right) \exp\left[-\frac{1}{2}\left(\frac{H^*}{\sigma_z}\right)^2 \right] \qquad (12.6.10)$$

where: i = size range in question

Q_i = discharge rate of particles in size range i

v_i = settling velocity of particles evaluated at the average particle size in i

H_i^* = corrected effective height evaluated at the average particle size in range i

Note that the particle size distribution has been divided into n size ranges

The entire development on atmospheric dispersion has been limited to emissions from a "point" (e.g., stack) source. Although most dispersion applications involve point sources, in some instances the location of the emission can be more accurately described physically and mathematically by either a line source or an area source.

Line sources are generally confined to roadways and streets along which there are well-defined movements of motor vehicles. For these sources, data are required on the width of the roadway and its center strip, the types and amounts (grams per second per meter) of pollutant emissions, the number of lanes, the emissions from each lane, and the height of emissions. In some situations (e.g., a traffic jam at a tollbooth, a series of industries located along a river, heavy traffic along a straight stretch of highway), the pollution problem may be modeled as a continuous emitting infinite line source. Concentrations downwind of a continuously emitting infinite line source, when the wind direction is normal to the line, can be calculated from

$$c(x,y,0;H) = \frac{2q}{\sqrt{2\pi}\sigma_z u} \exp\left[-\frac{1}{2}\left(\frac{H}{\sigma_z}\right)^2\right] \qquad (12.6.11)$$

Here q is the source strength per unit distance (e.g., g/s-m). Note that the horizontal dispersion parameter σ_y does not appear in this equation, since it is assumed that lateral dispersion from one segment of the line is compensated by dispersion in the opposite direction from adjacent segments. Also, y does not appear, since the concentration at a given x is the same for any value of y. Concentrations from infinite line sources, when the wind is not perpendicular to the line, can also be approximated. If the angle between the wind direction and line source is ϕ, we may write

$$c(x,y,0;H) = \frac{2q}{\sin\phi\sqrt{2\pi}\sigma_z u} \exp\left[-\frac{1}{2}\left(\frac{H}{\sigma_z}\right)^2\right] \qquad (12.6.12)$$

This equation should not be used where ϕ is less than 45°C. When the continuously emitting line source is reasonably short or "finite," one can account for the edge effects caused by the two ends of the source. If the line source is perpendicular to the wind direction, it is convenient to define the x axis in the direction of the wind and also passing through the sampling point downwind. The ends of the line source then are at two positions in the crosswind direction, y_1 and y_2, where y_1 is less than y_2. The concentration along the x axis at ground level is then given by

$$c(x,0,0;H) = \frac{2q}{\sqrt{2\pi}\sigma_z u}\left\{\exp\left[-\frac{1}{2}\left(\frac{H}{\sigma_z}\right)^2\right]\right\}$$

$$\times \int_{p_1}^{p_2} \frac{1}{\sqrt{2\pi}\exp}\left(-\frac{1}{2}p^2\right)dp \qquad (12.6.13)$$

where $p_1 = y_1/\sigma_y$ and $p_2 = y_2/\sigma_y$. Once the limits of integration have been established, the value of the integral may be determined from standard tables of integrals.

Area sources include the multitude of minor sources with individually small emissions that are impractical to consider as separate point or line sources. Area sources are typically treated as a grid network of square areas, with pollutant emissions distributed uniformly within each grid square. Area source information required includes types and amounts of pollutant emissions, the physical size of the area over which emissions are prorated, and representative height for the area. In dealing with dispersion of pollutants in areas having large numbers of sources [e.g., as in fugitive dust from (coal) piles, a large number of automobiles in a parking lot, or a multistack situation], there may be too many sources to consider each source individually [13]. Often an approximation can be made by combining all the emissions in a given area and treating this area as a source having an initial horizontal standard deviation, σ_{yo}. A virtual distance x_y can then be found that will give this standard deviation. This is just the distance that will yield the appropriate value for σ_y from Fig 12.6.3. Values of x_y will vary with stability. The equations for point sources may be used, determining σ_y as a function of $x + x_y$. This procedure effectively treats the area source as a crosswind line source with a normal distribution; this is a fairly good approximation for the distribution across an area source. The initial standard deviation for a square area source can be approximated by $\sigma_{yo} = s/4.3$, where s is the length of a side of the area. If the emissions within an area are from varying effective stack heights, the variation may be approximated by using a σ_{zo}. Thus, H would be the mean effective height of release and σ_{zo}, the standard deviation of the initial vertical distribution of sources. A virtual distance x_z and point source equations used for estimating concentrations can be found, determining σ_z as a function of $x+x_z$.

12.7 ATMOSPHERIC DISPERSION EQUATION FOR INSTANTANEOUS SOURCES

The previous section considered only sources that emit continuously. A rather significant amount of data and information is presented there. Unfortunately, little is available on instantaneous or "puff" sources. Turner's Workbook (8) provides an equation that may be used for estimation purposes. This approach is presented below.

Only sources emitting continuously or for periods equal to or greater than the travel times from the source to the point of interest were treated earlier. Cases of instantaneous release, as from an explosion, or short-term releases on the order of seconds, are also and often of practical concern. To determine concentrations at any position downwind, one must consider the time interval after the time of release and diffusion in the downwind direction as well as lateral and vertical diffusion. Of considerable importance, but very difficult, is the determination of the path or trajectory of the "puff." This is most important if concentrations are to be determined at specific points. Determining the trajectory is of less importance if knowledge of the magnitude of the concentrations for particular downwind distances or travel times is required but the exact points at which these concentrations occur need not be known. An equation that may be used for estimates of concentration downwind from an instantaneous release from height H is

$$c(x, y, 0; H) = \frac{2Q_T}{(2\pi)^{1.5}\sigma_x\sigma_y\sigma_z} \exp\left[-\frac{1}{2}\left(\frac{x-ut}{\sigma x}\right)^2\right]$$

$$\times \exp\left[-\frac{1}{2}\left(\frac{H}{\sigma_z}\right)^2\right] \times \exp\left[-\frac{1}{2}\left(\frac{y}{\sigma_y}\right)^2\right] \qquad (12.7.1)$$

the numerical value of $(2\pi)^{1.5}$ is 15.75. The notations have the usual meaning, with the important exceptions that Q_T represents the total mass of the release and the σ's are not necessarily those evaluated with respect to the dispersion of a continuous source at a fixed point in space. This equation can be simplified for centerline concentrations and ground level emissions by setting $y = 0$ and $H = 0$, respectively.

The σ's in Eq. (12.7.1) refer to dispersion statistics following the motion of the expanding puff. The σ_x is the standard deviation of the concentration distribution in the puff in the downwind direction, and t is the time after release. Note that there is no dilution in the downwind direction by wind speed. The speed of the wind mainly serves to give the downwind position of the center of the puff, as shown by examination of the exponential involving σ_x. Wind speed may influence the dispersion indirectly because the dispersion parameters σ_x, σ_x and σ_z may be functions of wind speed. The σ_y's and σ_z's for an instantaneous source are less than those given in Figures 12.6.2 and 12.6.3. Others have suggested values for a σ_y and σ_z for quasi-instantaneous sources. These are given in Table 12.7.1. The problem remains to make best estimates of σ_x. Much less is known of diffusion in the downwind direction than is known of lateral and vertical dispersion. In general, one should expect the σ_x value to be about the same as σ_y.

Initial dimensions of the puff (e.g., from an explosion) may be approximated by finding a virtual distance similar to that for area sources to give the appropriate initial standard deviation for each direction. Then σ_y will be determined as a function of $x + x_y$, σ_z as a function of $x + x_z$ and σ_x as a function of $x + x_z$.

TABLE 12.7.1 Estimation of Dispersion Parameters for Quasi Instantaneous Sources

Source	x = 100 m		x = 4 km	
	σ_y	σ_z	σ_y	σ_z
Unstable	10	15	300	220
Neutral	4	3.8	120	50
Very Stable	1.3	0.75	35	7

Unless another model is available for treating instantaneous sources, it is recommended that Eq. 12.7.1 be employed. The use of appropriate values of σ for this equation is not as clear-cut. As a first approximation, the reader may consider employing the values of σ provided in Figures 12.6.2. and 12.6.3. Quantifying the magnitude of the source term in Eq. 12.7.1 is a major problem that has yet to be resolved by industry and researchers.

12.8 STACK DESIGN CONSIDERATIONS

As experience in designing stacks has accumulated over the years, several rules of thumb have evolved [14].

1. Stack heights should be at least 2.5 times the height of any surrounding buildings or obstacles so that significant turbulence is not introduced by these factors.
2. The stack gas exit velocity should be greater than 60 ft/s so that stack gases will escape the turbulent wake of the stack. In many cases, it is good practice to have the gas exit velocity on the order of 90 or 100 ft/s. 3. A stack located on a building should be located in a position that will assure that the exhaust escapes the wakes of nearby structures.
3. Gases from stacks with diameters less than 5 feet and heights less than 200 feet will hit the ground part of the time, and ground concentrations may be excessive. In such cases, the plume becomes unpredictable.
4. The maximum ground concentration of stack gases subjected to atmospheric diffusion occurs about 5-10 effective stack heights downwind from the point of emission.
5. When stack gases are subjected to atmospheric diffusion and building turbulence is not a factor, ground level concentrations on the order of 0.001-1% of the stack concentration are possible for a properly designed stack.
6. Ground concentrations can be reduced by the use of higher stacks. The ground concentration varies inversely as the square of the effective stack height.
7. Average concentrations of a contaminant downwind from a stack are directly proportional to the discharge rate. An increase in discharge rate by a given factor increases ground level concentrations at all points by the same factor.
8. In general, increasing the dilution of stack gases by the addition of excess air in the stack does not affect ground level concentrations appreciably. Practical stack dilutions are usually insignificant in comparison to later atmospheric dilution by plume diffusion. Addition of diluting air will increase the effective stack height, however, by increasing the stack exit velocity. This effect may be important at low wind speeds. On the other hand, if the stack temperature is decreased appreciably by the dilution, the effective stack height may be reduced. Stack dilution will have an appreciable effect on the concentration in the plume close to the stack.

These nine rules of thumb represent the basic design elements of a pollution control system. The engineering approach suggests that each element be evaluated independently and as part of the whole control system. However, the engineering design and evaluation must be an integrated part of the complete pollution control program.

12.9 USEPA AIR QUALITY MODELS

Models are available, individually, from the National Technical Information Service (NTIS) on microcomputer disks. Regulatory and Air Quality models will also be available as a package. Most of the disks must be transported (uploaded) to IBM 3090 machines; however, some models do have PC executable code.

USEPA Regulatory Models/Programs

Several of the numerous USEPA regulatory models and programs available are listed below in alphabetical order.

1. BLP (Buoyant Line and Point Source Dispersion Model) is a Gaussian plume dispersion model associated with aluminum reduction plants.
2. CALINE3 (California Line Source Model) is a line source dispersion model that can be used to predict carbon monoxide concentrations near highways and arterial streets given traffic emissions, site geometry, and meteorology.
3. CDM2 (Climatological Dispersion Model) is a climatological dispersion model that determines long term quasi-stable pollutant concentrations.
4. COMPLEX I is a multiple point-source code with terrain adjustment representing a sequential modeling bridge between VALLEY and COMPLEX II.
5. CRSTER estimates ground-level concentrations resulting from up to 19 co-located elevated stack emissions.
6. EKMA was developed for relating concentrations of photochemically formed ozone to levels of organic compounds and oxides of nitrogen.
7. LONGZ-SHORTZ is designed to calculate the long and short-term pollutant concentrations produced at a large number of receptors by emissions from multiple stacks, buildings, and area sources.
8. MPRM 1.2 (Meteorological Processor for Regulatory Models) provides a general purpose computer processor for organizing available meteorological data into a format suitable for use by air quality dispersion models. Specifically, the processor is designed to accommodate those dispersion models that have gained USEPA approval for use in regulatory decision making.
9. MPTER is a multiple point-source Gaussian model with optional terrain adjustments.

10. PTPLU is a point-source dispersion Gaussian screening model for estimating maximum surface concentrations for one-hour concentrations.
11. RAM is a short-term Gaussian steady-state algorithm that estimates concentrations of stable pollutants.
12. RTDM (Rough Terrain Diffusion Model) is a sequential Gaussian plume model designed to estimate ground-level concentrations in rough (or flat) terrain in the vicinity of one or more co-located point sources.
13. Urban Airshed Model (UAM) is a three-dimensional grid based photochemical simualtion model for urban scale domains.
14. VALLEY is a steady-state, univariate Gaussian plume dispersion algorithm designed for estimating either 24-hour or annual concentrations resulting from emissions from up to 50 (total) point and area sources.
15. WRPLOT is an interactive program that generates wind rose statistics and plots for selected meteorological stations for user-specified date and time ranges. A wind rose depicts the frequency of occurrence of winds in each of 16 direction sectors (north, north-northeast, northeast, etc.) and six wind speed classes for a given location and time period.

Office of Research and Development (ORD) Air Quality Models

Other air quality models from ORD are also listed below.

1. APRAC-3 contains the emission factor computation methodology and treats traffic links in the primary network with low vehicle miles traveled as area sources.
2. CTDMPLUS is a refined air quality model for use in all stability conditions for complex terrain applications.
3. CTSCREEN is the screening mode of CTDMPLUS. Refer to CTDMPLUS above for all document information for this model.
4. HIWAY-ROADWAY are two models which compute the hourly concentrations of non-reactive pollutants downwind of roadways and predict pollutant concentrations within two hundred meters of a highway, respectively.
5. INPUFF is a Gaussian integrated puff model which is capable of addressing the accidental release of a substance over several minutes or of modeling the more typical continuous plume from a stack.
6. MESOPUFF is a Lagrangian model suitable for modeling the transport, diffusion and removal of air pollutants from multiple point and area sources at transport distances beyond 10-50 KM.
7. MPTDS is a modification of MPTER that explicitly accounts for gravitational settling and/or deposition loss of a pollutant.
8. PAL (Point, Area and Line Source Algorithm Model) is a short-term Gaussian steady state algorithm that estimates concentrations of stable pollutants from point, area and line sources.

9. PBM (Photochemical Box Model) is a simple stationary single-cell model with a variable height lid designed to provide volume-integrated hour averages of ozone and other photochemical smog pollutants for an urban area for a single day of simulation.

10. PEM (Pollution Episodic Model) is an urban scale air pollution model capable of predicting short-term average surface concentrations and deposition fluxes of two gaseous or particulate pollutants.

11. PLUVUE is a model that predicts the transport, atmospheric diffusion, chemical conversion, optical effects and surface deposition of point-source emissions.

12. The Shoreline Dispersion Model (SDM) is a multipoint Gaussian dispersion model that can be used to determine ground-level concentrations from tall stationary point source emissions near a shoreline environment. SDM is used in conjunction with MPTER algorithms to calculate concentrations when fumigation conditions do not exist.

13. TUPOS is a Gaussian model that estimates dispersion directly from fluctuation statistics at plume level.

14. UTIL-1 is a disk containing utility programs including CALMPRO, RUNAVG, UTMCON and CHAVG.

12.10 ILLUSTRATIVE EXAMPLES

Example 12.1

An oil tanker has collided with a freighter at sea. A rupture on the side of the tanker has released 450,000 gallons of crude oil (specific gravity = 0.88) into a "rough" sea. Estimate the spill area of oil resulting from the accident.

Solution 12.1

The oil will float, since it is lighter than water. First convert the volume V of oil from gallons to cubic feet.

$$V = 450,000 / 7.48 = 60,160 \, \text{ft}^3$$

From Table 12.2.1, assume the oil layer height h in a rough sea to be 0.05 inch. If the oil is distributed radially from the point of discharge in the form of a cylinder 0.05 inch high, the radius of spread r_s, is given by

$$V = \pi (r_s)^2 h$$

Solving for r_s yields

$$r_s = \left(V/\pi h\right)^{0.5} = \left[\frac{60,160}{\pi(0.05/12)}\right]^{0.5}$$

$$= 2144 \text{ ft, or nearly half a mile}$$

TABLE 12.10.1   Summary of Stack Height h (m) for Nine Wind Speeds u (m/s) and Class D and B Stabilities

	Class D		Class B	
u	Δh	$h + \Delta h$	$1.15\Delta h$	$h + 1.15\Delta h$
0.5	97.6	127.6	112.2	142.2
1.0	48.8	78.8	56.1	86.1
1.5	32.6	62.6	37.5	67.5
2	24.4	54.4	28.1	58.1
3	16.3	46.3	18.7	48.7
5	9.8	39.8	11.3	41.3
7	7.0	37.0	8.0	38.0
10	4.9	34.9		
20	2.4	32.4		

Example 12.2

A proposed source is to emit 72 g/s of a toxic pollutant from a stack 30 meters high with a diameter of 1.5 meters. The effluent gases are emitted at a temperature of 250°F (394K) with an exit velocity of 13 m/s. Using Holland's plume rise equation, obtain the plume rise as a function of wind speed for stability classes B and D. Assume that the design atmospheric pressure is 970 mbar and that the design ambient air temperature is 20°C (293K).

Solution 12.2

Holland's equation gives

$$\Delta H = \frac{v_s d}{u}\left[1.5 + 2.68 \times 10^{-3} \, p\left(\frac{T_s - T_a}{T_s}\right)d\right]$$

$$= \frac{(13)(1.5)}{u}\left[1.5 + 2.68 \times 10^{-3}\,(970)\left(\frac{394 - 293}{394}\right)(1.5)\right] = \frac{48.8}{u}$$

The effective stack heights for various wind speeds and stabilities are summarized in Table 12.10.1.

Example 12.3

An estimate is required of the total hydrocarbon concentration 300 meters downwind of an expressway on an overcast day with wind speed 4 m/s. The expressway runs north-south and the wind is from the west.   The measured traffic flow is 8000 vehicles per hour during this rush hour, and the average speed of the vehicles is 40 mph. At this speed the average vehicle is expected to emit 2×10^{-2} g/s of total hydrocarbons.

Solution 12.3

The expressway may be considered to be a continuous infinite line source. To obtain a source strength q (in g/s-m), the number of vehicles per meter of highway must be calculated and multiplied by the emission per vehicle.

$$vehicles/\,meter = 8000\,/(40)(1600) = 1.25 \times 10^{-1}$$

$$q = (1.25 \times 10^{-1})(2 \times 10^{-2})$$

where 2×10^{-2} is in (g/s-vehicle); thus

$$q = 2.5 \times 10^{-3}\, g/s - m$$

Stability class D applies under overcast conditions with wind speed 4 m/s. Under D, at x = 300 meters, $\sigma_z = 12$ meters . From Eq. (12.7.14),

$$c(300, 0, 0; 0) = \frac{2q}{(\sqrt{2\pi})(\sigma_z)(u)}$$

$$= \frac{(2)(2.5 \times 10^{-3})}{(2.507)(12)(4)}$$

$$= 4.2 \times 10^{-5}\ g/m^3 \text{ of total hydrocarbons}$$

Example 12.4

An inventory of SO_2 emissions has been conducted in an urban area by square areas, 5000 feet (1524 meters) on a side. The emissions from one such area are estimated to be 6 g/s for the entire area. This square is composed of residences and a few small commercial establishments. What is the concentration resulting from this area at the center of the adjacent square to the north when the wind at 2.5 m/s? The average effective stack height of these sources is assumed to be 20 meters.

Solution 12.4

A thinly overcast night with a wind speed of 2.5 m/s suggests stability class E. (It may actually be more unstable, since this is a built-up area.) To allow for the area source, let $\sigma_{yo} = 1524/4.3 = 354$ meters. For class E the virtual distance $x_y = 85$ km. For $x = 1524$ m, $\sigma_y = 28.5$ m. For $x + x_y = 10{,}024$ m, $\sigma_y = 410$ m. Therefore,

$$c = \frac{Q}{\pi \sigma_y \sigma_z u} \exp\left[-(0.5)\left(\frac{H}{\sigma_z}\right)^2\right]$$

$$= \left(\frac{6}{410 \times 28.5 \times 2.5}\right) \exp\left[-(0.5)\left(\frac{20}{28.5}\right)^2\right]$$

$$= 6.54 \times 10^{-5} (0.783)$$

$$= 5.1 \times 10^{-5} \text{ g/m}^3$$

Example 12.5

A tank 30 meters high in a plant containing a toxic gas suddenly explodes, resulting in an emission of 200 g/s for 2 minutes. A school is located 500 meters east and 100 meters north of the plant. If the wind velocity is 4.0 m/s from the west, how many seconds after the explosion will the concentration reach a maximum in the school? Humans will be adversely affected if the concentration of the gas is greater that 1.0 µg/L. Is there any impact on the students in the school?

Solution 12.5

For this problem,

$$x = 500 \text{ m} \qquad y = 100 \text{ m}$$
$$H = 30 \text{ m} \qquad u = 4 \text{ m/s}$$
$$Q_T = (200)(2)(60) = 24,000 \text{ g}$$

Stability class D is applicable. From Figures 12.6.2 and 12.6.3,

$$\sigma_y = 37 \text{ m} \qquad \sigma_x = 37 \text{ m} \qquad \sigma_z = 19.5 \text{ m}$$

There is a maximum concentration at the school when

$$(u)(5) = x = 500 \text{ m}$$

Therefore,

$$t = 500/4 = 125 \text{ s}$$

Equation 12.7.1 applies.

$$c(x, y, 0; H) = \frac{2Q_T}{(2\pi)^{1.5}\sigma_x\sigma_y\sigma_z} \exp\left[-\frac{1}{2}\left(\frac{x-ut}{\sigma_x}\right)^2\right]$$

$$\times \exp\left[-\frac{1}{2}\left(\frac{H}{\sigma_z}\right)^2\right]\exp\left[-\frac{1}{2}\left(\frac{y}{\sigma_y}\right)^2\right]$$

$$= \frac{(2)(24,000)}{(15.75)(37)(37)(19.5)}\exp\left[-\frac{1}{2}\left(\frac{30}{19.5}\right)^2\right]\exp\left[-\frac{1}{2}\left(\frac{100}{37}\right)^2\right] \quad (1)$$

$$= 9.065\times10^{-4} \text{ g/m}^3$$
$$= 0.9065 \text{ }\mu\text{g/L}$$

Since,

$$0.908 \text{ }\mu\text{g/L} < 1.0 \text{ }\mu\text{g/L}$$

There is no impact on the students in the school.

Example 12.6

A truck carrying two tanks containing a very unstable and hazardous gas is involved in an accident that results in the consecutive explosion of the tanks – one immediately, the second approximately a minute later. The total mass of the emission resulting from the explosion of each tank is 30,000 g. The wind velocity is 1 m/s from the north, and the effective height of emission is 30 meters at the time of the accident. Calculate the concentration of this gas at 500 meters south but 100 meters east from the site 10 minutes after the explosion of the first tank. Assume that stability category D applies.

Solution 12.6

For this problem,

$$x = 500 \text{ m} \qquad y = 100 \text{ m}$$
$$H = 30 \text{ m} \qquad u = 1 \text{ m/s}$$
$$Q_T = 30,000 \text{ g (for each tank)}$$

Since stability class D is applicable, Figures 12.6.2 and 12.6.3 yield

$$\sigma_y = 37 \text{ m} \qquad \sigma_x = 37 \text{ m} \qquad \sigma_z = 19.5 \text{ m}$$

Equation (12.7.1) again applies.

$$c(x, y, 0; H) = \frac{2Q_T}{(2\pi)^{1.5}\sigma_x\sigma_y\sigma_z} \exp\left[-\frac{1}{2}\left(\frac{x-ut}{\sigma_x}\right)^2\right]$$

$$\times \exp\left[-\frac{1}{2}\left(\frac{H}{\sigma_z}\right)^2\right]\exp\left[-\frac{1}{2}\left(\frac{y}{\sigma_y}\right)^2\right]$$

$$t = (10)(60) = 600 \text{ s}$$

$$c_1(x, y, 0; H) = (0.143)(0.026)(0.306)(0.026) = 2.96 \times 10^{-5} \text{ g/m}^3$$

For the second tank,

$$t = (9)(60) = 540 \text{ s}$$

$$c_2(x, y, 0; H) = (0.143)(0.557)(0.306)(0.026) = 6.33 \times 10^{-4} \text{ g/m}^3$$

Assuming that the concentrations are additive,

$$c_T = c_1 + c_2 = 2.96 \times 10^{-5} \text{ g/m}^3 + 6.33 \times 10^{-4} \text{ g/m}^3$$

$$= 664 \ \mu\text{g/m}^3$$

12.11   SUMMARY

1. Exposure assessment is the process of measuring or estimating the intensity, frequency, and duration of human or animal exposure to an agent present in the environment
2. Estimating hypothetical exposures that might arise from the release of new chemicals into the environment.
3. Exposure is defined as the contact of an organism (humans in the case of health risk assessment) with a chemical or physical agent (3).
4. The magnitude of exposure is determined by measuring or estimating the amount of an agent available at the exchange boundaries (i.e., the lungs, gut, skin) during a specified time period.
5. Exposure assessments may consider past, present, and future exposures, using varying assessment techniques for each phase
6. Enormous amounts of waste dumped into water systems are degrading water quality and causing increased human health problems.
7. Everyday pollutant emissions into water systems is an ever-present threat of a discharge resulting from an accident, an emergency, or a combination of these.
8. Treating a body of water as a completely mixed system can be a valuable approach for estimating the effects of human activities.
9. This method can be applied to a number of pollutants, including suspended and dissolved substances, as well as to heat balance computations.
10. A wide range of chemical wastes have been deposited at uncontrolled land sites and the extent and severity of the resulting environmental contamination varies greatly across sites.
11. Many of the factors that influence the extent of contamination are site specific; are climatic or hydrogeological; and others relate to land surface features such as topography or development, which determine exposure routes.
12. There are many dispersion equations available, most of them semiempirical.
13. Cases of instantaneous release, as from an explosion, or short-term releases on the order of seconds, are also and often of practical concern.

To determine concentrations at any position downwind, one must consider the time interval after the time of release and diffusion in the downwind direction as well as lateral and vertical diffusion.

14. Of considerable importance, but very difficult, is the determination of the path or trajectory of the "puff."

15. There are nine rules of thumb that represent the basic design elements of a pollution control system. The engineering approach suggests that each element be evaluated independently and as part of the whole control system.

PROBLEMS

1. Sulfur dioxide is being emitted from a stack at a rate of 50 scfs, and the prevailing wind velocity is 15 ft/s. The stack exit gas velocity is 60 ft/s. The stack diameter is 3.5 feet. Estimate the physical stack height necessary to meet an air pollution standard of 5 ppmv maximum SO_2 (30 min average concentration) GLC.

2. An incinerator stack is emitting fly ash at a rate of 2 tons per hour. Natural processes are capable of removing these particles from the affected ground

TABLE 12.12.1 Emergency Tolerance Limits for UDMH VaporVersus Exposure Time

Time (min)	Emergency Tolerance Limits (g/m^3)
5	1.2×10^{-1}
15	8.6×10^{-2}
30	4.9×10^{-2}
60	2.5×10^{-2}

surface at a steady rate, provided no more than 0.02% of the ground is covered by them per hour. The particles are, on the average, spheres of radius 10^{-4} foot and have an average density of 120 lb/ft^3. The wind speed is 5 mph. If L is the distance through which an average particle is carried by the wind, the particles will settle out uniformly over a wedge-shaped area whose central angle is 20o, at distances ranging from 0.5L to 20L. Determine the minimum stack height H required to prevent ground level accumulation. Assume monolayer deposition, with the ground area covered by a particle equal to its cross-sectional area [i.e., $\pi(dp)^2/4$].

3. A line of burning agricultural waste can be considered to be a finite line source 150 meters long. It is estimated that the total emission of organics is a t a rate of 90 g/s. What is the 3- to 15-minute concentration of organics 400 meters directly downwind from the center of the line when the wind is blowing at 3 m/s perpendicular to the line? Assume a sunny fall afternoon. What is the concentration directly downwind from one end of the source?

4. A spill estimated at 2.9 * 10^6 grams of unsymmetrical dimethyl hydrazine (UDMH) occurs at 1:15 a.m. on a clear night while a rocket is being fueled. A circular area 60 meters in diameter built around the launch pad is revetted into squares 20 feet on a side to confine to as small an area as possible any spilled any toxic liquids. In this spill only one such 20 by 20 foot area is involved. At the current wind speed of 2 m/s, it is predicted to vary ± 15° from centerline conditions for the next hour. Given the data in Table 12.12.1, what position downwind from the sources should be evacuated?

5. A 40-meter elevated vapor transfer line in a petroleum refinery suddenly ruptures. It is estimated that 375 g/s of a hazardous gas is being emitted. If the emission from the rupture is brought under control 2 minutes later, estimate the concentration 1000 meters downwind and 100 meters displaced from centerline conditions 5 minutes later. The wind velocity is 4.0 m/s and the stability category is D.

REFERENCES

1. D.Paustenbach, "The Risk Assessment of Environmental and Human Health Hazards", John Wiley and Sons, New York City, 1989.
2. L.Theodore, J.Reynolds and K.Morris, "Health, Safety and Accident Prevention: Industrial Applications", A Theodore Tutorial, East Wilkston, NY, 1996.
3. USEPA Web Page, 2001.
4. R. Thomman and J. Mueller, "Principles of Surface Water Quality and Control", Harper and Row, New York, 1987.
5. L. Theodore, "Transport Phenomena for Engineers", Intext, Scranton, PA, 1971.
6. John P. Connelly, Manhattan College, personal notes, 1989.
7. W. F. Davidson, Transactions of the Conference on Industrial Wastes, 14th Annual Meeting of the Industrial Hygiene Foundation of America 1949, p. 38.
8. D. B. Turner, *Workbook of Atmospheric Dispersion Estimates,* rev., Environmental Protection Agency, Publication No. AP-26, Research Triangle Park, NC, 1970.
9. G. A. Briggs, *Plume Rise,* Atomic Energy Commission Critical Review Series, U.S. AEC, Division of Technical Information.
10. W. Baasel, *J. Air Pollut. Control Assoc.,* **38**(8), 866 (1981).
11. L. Theodore and A. Buonicore, *Air Pollution Control Equipment,* CRC Press, Boca Raton, FL, 1988.
12. L.Theodore and R. Allen, "Air Pollution Control Equipment", A Theodore Tutorial, East Wilkston, NY, 1997.
13. C. H. Bosanquet, W.F. Carey and E. M. Halton, *Proc. Inst. Mech. Eng. (London),* **162**, 355 (1950).
14. L. Theodore, Manhattan College, personal notes, 2001.

13

Health Risk Analysis and Characterization

13.1   INTRODUCTION

Risk characterization is the process of estimating the incidence of a health effect under the various conditions of human or animal exposure described in the exposure assessment. It is performed by combining the exposure (see Chapter 12) and dose response (see Chapter 11) assessments. The summary effects of the uncertainties in the preceding steps should also be described in this step.

The quantitative estimate of the risk is the principal interest to the regulatory agency or risk manager making a decision. The risk manager must consider the results of the risk characterization when evaluating the economics, societal aspects, and various benefits of the assessment. Factors such as societal pressure, technical uncertainties, and severity of the potential hazard influence how the decision makers respond to the risk assessment. As one might suppose, there is room for improvement in this step of the risk assessment.[1]

A risk estimate indicates the likelihood of occurrence of the different types of health or environmental effects in exposed populations. Risk assessment should include both human health and environmental evaluations (i.e., impacts on ecosystems). Ecological impacts include actual and potential effects on plants and animal (other than domesticated species). The numbers produced from the risk characterization, representing the probability of adverse health effects being caused, must be evaluated.

This brief introduction section is followed by a host of topics related to health risk analysis and characterization. The subject matter includes:

The authors gratefully acknowledge the assistance of Lou Carrea in researching, reviewing, and editing this chapter.

The reader should note that the general risk subject areas, uncertainties and the public perception are treated again in Part IV, Chapter 18.

13.2   QUALITATIVE RISK SCENARIOS

Although the technical community has come a long way in understanding how to do a better job in hazard identification, dose-response assessment, and exposure assessment portions of risk assessment, it has only begun to understand how to best characterize health risks and how to present these risks most appropriately to both the public and decision makers. The next three sections specifically address these issues. This section deals with qualitative risk assessment while the next two sections deal with quantitative risk assessment.

Regarding numerical values assigned to health risk, Paustenback provides the following comment for Rodricks et al :[1, 2]

Examination of the risks of common human activities demonstrates…a lifetime risk of 1 in 100,000 or more is within the realm of, or orders of magnitude below, everyday risks that generally do not cause undue concern. These are risks that people, while they are aware of them and may have some concern or fear over them, do not in general alter their behavior to avoid… the risks from many activities greatly exceed the level of 1 in 100,000. In comparison to these background risks of "everyday activities", a lifetime risk of 1 in 100,000 is relatively small. Accordingly, regulatory action will not generally be justifiable unless risks are substantially higher than this 1 in 100,000 "benchmark."

Numerous qualitative approaches to risk assessment have been employed. Some sample categorizations follow. Health risks may be divided into risk levels as provided below:

Risk Level 1: Does not cause a health hazard
Risk Level 2: Unlikely to cause a health hazard
Risk Level 3: May cause a health hazard
Risk Level 4: May cause a severe health hazard
Risk Level 5: Will cause a severe health hazard

Health risks may also be set in terms of a logarithmic scale of risk levels as seen in Table 13.2.1:

TABLE 13.2.1 Risk Level and Risk Range

Risk Level	Risk Range	
1	1 in 1 - 1 in 9	10^0 - 10^{-1}
2	1 in 10 - 1 in 99	10^{-1} - 10^{-2}
3	1 in 100 - 1 in 999	10^{-2} - 10^{-3}
4	1 in 1000 - 1 in 9,999	10^{-3} - 10^{-4}
5	1 in 10 000 - 1 in 99,999	10^{-4} - 10^{-5}
6	1 in 100 000 - 1 in 999,999	10^{-5} - 10^{-6}
7	1 in 1 000 000 - 1 in 9,999,999	10^{-6} - 10^{-7}

The above risks can apply to either an annual or lifetime basis.

Qualitative health risk policies for companies, e.g., Flynn and Theodore Enterprises (FATE), could take the following form:

1. FATE will not knowingly pose a greater health risk to the public than it does to its own employees.

2. FATE will not expose its employees or neighbors to health risks that are considered unacceptable, based on industry practice and available technology.

3. FATE will comply with all applicable regulations and industry guidelines related to health risks, and will adopt its own standards where regulations do not exist or are inadequate.

4. FATE will neither undertake nor continue any operations whose associated health risks it does not understand or cannot control at a safe level.

Other possible health risk company policies could take the following form:

- The average individual health risk level for the public should be less than_____.
- The maximum individual health risk for FATE employees should be less than _____.
- The probability of one or more public deaths should be less than_____.
- The probability of 100 or more public deaths should be less than_____.
- The probability of one or more public illnesses should be less than_____.
- The probability of 100 or more public illnesses should be less than_____.

Once again, the above can apply to either an annual or lifetime basis.

13.3 QUANTITATIVE RISK: NONCARCINOGENS

The measure used to describe the potential for noncarcinogenic toxicity to occur in an individual is not expressed as the probability of an individual suffering an adverse effect. The EPA does not at the present time use a probabilistic approach to estimate the potential for noncarcinogenic health effects. Instead, the potential for non carcinogenic effects is evaluated by comparing an exposure level over a specified time period (e.g., lifetime) with a reference dose derived for a similar exposure period. This ratio of exposure to toxicity is called a hazard quotient and is described below. (The reader is referred to Chapter 11 for additional details on the material that follows). The noncancer hazard quotient assumes that there is a level of exposure (i.e., RfD) below which it is unlikely for even sensitive populations to experience adverse health effects.

$$\text{Noncancer Hazard Quotient} = E/RfD \qquad (13.3.1)$$

where: E = exposure level (or intake)
RfD = reference dose
E and Rfd are expressed in the same units

If the exposure level (E) exceeds this threshold (i.e., E/RfD exceeds unity), there may be concern for potential noncancer effects. As a rule, the greater the value of E/RfD above unity, the greater the level of concern. However, one should not interpret ratios of E/RfD as statistical probabilities; a ratio of 0.001 does not mean that there is a one in one thousand chance of the effect occurring. Further, it is important to emphasize that the level of concern does not increase linearly as the RfD is approached or exceeded because RfDs do not have equal accuracy or precision and are not based on the same severity of toxic effects. Thus, the slopes of the dose–response curve in excess of the RfD can range widely depending on the substance.

Three exposure durations that will need separate consideration for the possibility of adverse noncarcinogenic health effects are chronic, subchronic, and shorter-term exposures. *Chronic* exposures for humans range in duration from seven years to a lifetime; such long term exposures are almost always of concern for Superfund sites (e.g., inhabitants of nearby residences, year-round users of specified drinking water sources). *Subchronic* human exposures typically range in duration from two weeks to seven years and are often of concern at Superfund sites. For example, children might attend a junior high school near the site for no more than two or three years. Exposures less than two weeks in duration are occasionally of concern at Superfund sites. Also, if chemicals known to be developmental toxicants are present at a site, *short-term* exposures of only a day or two can be of concern.

Risks for Multiple Substances

To assess the overall potential for noncarcinogenic effects posed by more than one chemical, a hazard index (HI) approach has been developed based on EPA's Guidelines for Health Risk Assessment of Chemical Mixtures. This approach assumes that simultaneous subthreshold exposures to several chemicals could result in an adverse health effect. It also assumes that the magnitude of the adverse effect will be proportional to the sum of the ratios of the subthreshold exposures to acceptable exposures. The non cancer hazard index is equal to the sum of the hazard quotients, as described below, where E and the RfD represent the same exposure period (e.g., subchronic, chronic, or shorter-term).

$$\text{Hazard Index} = E_1/Rfd_1 + E_2/RfD_2 + \ldots + E_i/RfD_i) \qquad (13.3.2)$$

where: E_I = exposure level (or intake) for the ith toxican
    RfD_i = reference dose for the ith toxicant

It should be noted that E and RfD are expressed in the same units and represent the same exposure period (i.e., chronic, subchronic, or shorter term)

When the hazard index exceeds unity, there may be concern for potential health effects. While any single chemical with an exposure level greater than the toxicity value will cause the hazard index to exceed unity, the reader should note that for multiple chemical exposures, the hazard index can also exceed unity even if no single chemical exposure exceeds its RfD.

It is important to calculate the hazard index separately for chronic, subchronic, and short-term exposure periods as described below. It is also important to remember to include RfDs for the noncancer effects of carcinogenic substances.

Noncarcinogenic Effects—Chronic Exposures

For each chronic exposure pathway (i.e., seven years to lifetime exposure), calculate a separate chronic hazard index from the ratios of the chronic daily intake (CDI) to the chronic reference dose (RfD) for individual chemicals as described below:

$$\text{Chronic Hazard Index} = CDI_1/RfD_1 + CDI_2/RfD_2 + \ldots CDI_i/RfD_i \qquad (13.3.3)$$

where: CDI = chronic daily intake for the ith toxicant in mg/kg-day
    RfDi = chronic reference dose for the ith toxicant in mg/kg-day

Noncarcinogenic Effects—Subchronic Exposures

For each subchronic exposure pathway (i.e., two weeks to seven year exposure), calculate a separate subchronic hazard index from the ratios of subchronic daily intake (SDI) to the subchronic reference dose (RfDs) for individual chemicals as described below.

$$\text{Subchronic Hazard Index} = SDI_1/RfDs_1 + SDI_2/RfDs_2 + \quad\quad (13.3.4)$$
$$\quad\quad\quad ......SDL_i/RfDs_i$$

where: SDI   = subchronic daily intake for the ith toxicant in mg/kg-day
       RfDsi = subchronic reference dose for the ith toxicant in mg/kg-day

Noncarcinogenic Effects—Less Than Two Week Exposures

The same procedure above may be applied for simultaneous shorter-term exposures to several chemicals. For drinking water exposures, 1- and 10-day Health Advisories can be used as reference toxicity values. Depending on available data, a separate hazard index might also be calculated for developmental toxicants (using $RfD_{dt}s$), which might cause adverse effects following exposures of only a few days.

There are several limitations to this approach that must be acknowledged. As mentioned earlier, the level of concern does not increase linearly as the reference dose is approached or exceeded because the RfDs do not have equal accuracy or precision and are not based on the same severity of effects. Moreover, hazard quotients are combined for substances with RfDs based on critical effects of varying toxicological significance. Also, it will often be the case that RfDs of varying levels of confidence that include different uncertainty adjustments and modifying factors will be combined (e.g., extrapolation from animals to humans, from LOAELs to NOAELs, or from one exposure duration to another).

Another limitation with the HI approach is that the assumption of dose additivity is most properly applied to compounds that induce the same effect by the same mechanism of action. Consequently, application of the hazard index equation to a number of compounds that are not expected to induce the same type of effects or that do not act by the same mechanism could overestimate the potential for effects, although such an approach is appropriate at a screening level. This possibility is generally not of concern if only one or two substances are responsible for driving the HI above unity.

If the HI is greater than unity as a consequence of summing several hazard quotients of similar value, it would be appropriate to segregate the compounds by effect and mechanism of action and to derive separate hazard indices for each group.

Segregation of Hazard Indices

Segregation of hazard indices by effect and mechanism of action can be complex and time consuming because it is necessary to identify all of the major effects and target organs for each chemical and then to classify the chemicals according to target organ(s) or mechanism of action. This analysis is not simple and should be performed by a toxicologist. If the segregation is not carefully done, an underestimate of a true hazard could result. Agency review of particularly complex or controversial cases can be requested through the regional risk assessment support staff.

The procedure for recalculating the HI by effect and by mechanism of action is briefly described later. If one of the effect-specific hazard indices exceeds unity, consideration of the mechanism of action might be warranted. A strong case is required, however, to indicate that two compounds which produce adverse effects on the same organ system (e.g., liver), although by different mechanisms, should not be treated as dose additive. Any such determination should be reviewed.

If there are specific data germane to the assumption of dose-additivity (e.g., if two compounds are present at the same site and it is known that the combination is five times more toxic than the sum of the toxicities for the two compounds), then the development of the hazard index should be modified accordingly. The reader can refer to the EPA (1986b) mixture guidelines for discussion of a hazard index equation that incorporates quantitative interaction data. If data on chemical interactions are available, but are not adequate to support a quantitative assessment, note the information in the "assumptions" being documented for the risk assessment.

Combining Risks Across Exposure Pathways

In some situations, an individual might be exposed to a substance or combination of substances through several pathways. For example, a single individual might be exposed to substance(s) from a hazardous waste site by consuming contaminated drinking water from a well, eating contaminated fish caught near the site, and through inhalation of dust originating from the site. The total exposure to various chemicals will equal the sum of the exposures by all pathways. One should not automatically sum risks from all exposure pathways evaluated for a site, however.

There are two steps required to determine whether risks or hazard indices for two or more pathways should be combined for a single exposed individual or group of individuals. The first is to identify reasonable exposure pathway combinations. The second is to examine whether it is likely that the same individuals would consistently face the "reasonable maximum exposure" (RME) by more than one pathway.

One should identify exposure pathways that have the potential to expose the same individual or sub-population at the key exposure areas evaluated in the exposure assessment, making sure to consider areas of highest exposure for each pathway for both current and future land-uses (e.g., nearest down-gradient well, nearest downwind receptor). For each pathway, the risk estimates and hazard indices have been developed for a particular exposure area and time period; they do not necessarily apply to other locations or time periods. Hence, if two pathways do not affect the same individual or subpopulation, neither pathway's individual risk estimate or hazard index affects the other, and risks should not be combined.

Once reasonable exposure pathways combinations have been identified, it is necessary to examine whether it is likely that the same individuals would consistently face the RME. Note that the RME estimate for each exposure pathway includes many conservative and upper-bound parameter values and assumptions (e.g., upper 95th confidence limit on amount of water ingested, upper-bound duration of occupancy of a single residence). Also, some of the exposure parameters are not predictable in either space or time (e.g., maximum downwind concentration may shift compass direction, maximum ground-water plume concentration may move past a well). For real word situations in which contaminant concentrations vary over time and space, the same individual may or may not experience the RME for more than one pathway over the same period of time. One individual might face the RME through a different pathway. The RME risks for more than one pathway can be combined only if one can explain why the key RME assumptions for more than one pathway should apply to the same individual or subpopulation.

In some situations, it may be appropriate to combine one pathway's RME risks with other pathway's risk estimates that have been derived from more typical exposure parameter values. In this way, resulting estimates of combined pathway risks may better relate to RME conditions.

If it is deemed appropriate to sum risks and hazard indices across pathways, the risk assessor should clearly identify those exposure pathway combinations for which a total risk estimate or hazard index is being developed. The rationale supporting such combinations should also be clearly stated.

To assess the overall potential for non carcinogenic effects posed by several exposure pathways, the total hazard index for each exposure duration (i.e., chronic, subchronic, and shorter-term) should be calculated separately. This equation is described below:

Total Exposure
Hazard Index =   Hazard Index (exposure pathway$_1$) +
                   Hazard Index (exposure pathway$_2$) +   (13.3.5)
                   ....Hazard Index (exposure pathway$_i$)

Note that the total exposure hazard index is calculated separately for chronic, subchronic, and shorter-term exposure periods.

When the total hazard index for an exposed individual or group of individuals exceeds unity, there may be concern for potential non cancer health effects. As indicated before, for multiple exposure pathways, the hazard index can exceed unity even if no single exposure pathway hazard index exceeds unity. If the total hazard index exceeds unity and if combining exposure pathways has resulted in combining hazard indices based on different chemicals, one may need to consider segregating the contributions of the different chemicals according to their major effect.

13.4 QUANTITATIVE RISK: CARCINOGENS

For carcinogens, risks are estimated as the incremental probability of an individual developing cancer over a lifetime as a result of exposure to the potential carcinogen (i.e., incremental or excess individual lifetime cancer risk). The guidelines provided in this section are consistent with EPA's *Guidelines for Carcinogen Risk Assessment.* For some carcinogens, there may be sufficient information on mechanism of action that a modification of the approach outlined below is warranted. (Once again, the reader is referenced to Chapter 11 for additional details on the material presented below.)

The slope factor (SF) converts estimated daily intakes averaged over a lifetime of exposure directly to incremental risk of an individual developing cancer. Because relatively low intakes (compared to those experienced by test animals) are most likely from environmental exposures at most sites, it generally can be assumed that the dose-response relationship will be linear in the low-dose portion of the multistage model dose-response curve. Under this assumption, the slope factor is a constant, and risk will be directly related to intake. Thus, the linear form of the carcinogenic risk equation is usually applicable for estimating risks. This linear low-dose risk equation is described below.

$$\text{Risk} = (\text{CDI})\,(\text{SF}) \qquad\qquad (13.4.1)$$

Where: Risk = a unitless probability (e.g., 2×10^{-5}) of an individual developing cancer
CDI = chronic daily intake averaged over 70 years (mg/kg-day)
SF = slope factor, expressed in $(\text{mg/kg-day})^{-1}$

However, this linear equation is valid only at low risk levels (i.e., below estimated risks of 0.01). For situations where chemical intakes might be high (i.e., risk above 0.01), an alternate equation should be used. The one-hit equation, which is consistent with the linear low-dose model given above and described below, should be used instead.

$$\text{Risk} = 1 - \exp(-\text{CDI} \times \text{SF}) \qquad\qquad (13.4.2)$$

where: Risk = a unitless probability (e.g., 2×10^{-5}) of an individual developing cancer
Exp = the exponential term
CDI = chronic daily intake averaged over 70 years (mg/kg-day)

Because the slope factor is often an upper 95th percentile confidence limit of the probability of response based on experimental animal data used in the multistage model, the carcinogenic risk estimate will generally be an upper-bound estimate. This means that the EPA is reasonably confident that the "true risk" will not exceed the risk estimate derived through use of this model and is likely to be less than that predicted.

Risks for Multiple Substances

The cancer risk equation described below estimates the incremental individual lifetime cancer risk for simultaneous exposure to several carcinogens and is based on EPA's risk assessment guidelines. This equation represents an approximation of the precise equation for combining risks which accounts for the joint probabilities of the same individual developing cancer as a consequence of exposure to two or more carcinogens. The difference between the precise equation and the approximation described is negligible for total cancer risks less than 0.1. Thus, the simple additive equation is appropriate for most risk assessments. The cancer risk equation for multiple substances is given by:

$$Risk_T = \sum Risk_i \qquad\qquad (13.4.3)$$

where: $Risk_T$ = the total cancer risk, expressed as a unitless probability
$Risk_i$ = the risk estimate for the ith substance.

The risk summation techniques assume that intakes of individual substances are small. They also assume independence of action by the compounds involved (i.e., that there are no synergistic or antagonistic chemical interactions and that all chemicals produce the same effect, i.e., cancer). If these assumptions are incorrect, over- or under-estimation of the actual multiple-substance risk could result.

There are several limitations to this approach that must be acknowledged. First, because each slope factor is an upper 95th percentile estimate of potency, and because upper 95th percentiles of probability distributions are not strictly additive, the total cancer risk estimate might become artificially more conservative as risks from a number of different carcinogens are summed. If one or two carcinogens drive the risk, however, this problem is not of concern. Second, it often will be the case that substances with different weights of evidence for human carcinogenity are included. The cancer risk equation for multiple substances sums all carcinogens equally. In addition, slope factors derived from animal data will be given the same weight as slope factors derived from human data. Finally, the action of two different carcinogens might not be independent.

Combining Risk Across Exposure Pathways

The reader should note that the introductory comments in the similarly titled subsections of the previous section applies to carcinogens as well. The calculation proceeds as follows. First, sum the cancer risks for each exposure pathway contributing to exposure of the same individual or subpopulation. For Superfund risk assessments, cancer risks from various exposure pathways are assumed to be additive, as long as the risks are for the same individuals and time period (i.e., less-than-lifetime exposures have all been converted to equivalent lifetime exposures). This summation procedure is described below:

Total Exposure Cancer Risk = Risk (exposure pathway$_1$) +
                               Risk (exposure pathway$_2$) +   (13.3.5)
                               ....Risk (exposure pathway$_i$)

Although the exact equation for combining risk probabilities includes terms for joint risks, the difference between the exact equation and the approximation described above is negligible for total cancer risks of less than 0.1.

13.5 RISK UNCERTAINTIES/ LIMITATIONS

Although great controversy can surround results of risk assessments, especially quantitative risk assessments, they are useful in particular applications. They can help establish priorities for regulatory action or interventions of any type. A uniform risk assessment performed across a range of substances can create a spectrum of the health risk to humans. The limits of risk assessment can be tested when government agencies (faced with the absence of other types of data and the need for action) must rely on risk assessment methods to establish health-based standards or guidelines to prevent human exposure to hazardous substances. Because of risk assessment shortcomings and the desire for greater specificity in measuring exposure, increasing interest is shown in understanding pathologic changes at the molecular level with the hope that these investigators will permit toxicologic and epidemiologic analyses of greater accuracy and sensitivity (see Chapter 11).[4, 5] In a general sense, problems in this area arise because:

1. Uncertainty associated with available data
2. Concerns associated with assumed information
3. Uncertainty associated with describing equations
4. Concerns associated with limited and/constrained describing equations
5. Concerns associated with overall quality analysis

Uncertainty and Variability

In the risk characterization, conclusions about hazard and dose response are integrated with those from the exposure assessment. In addition, confidence about these conclusions, including information about the uncertainties associated with each aspect of the assessment in the final risk summary, should be highlighted. In the previous assessment steps and in the risk characterization, the risk assessor should also distinguish between variability and uncertainty.

Variability arises from true heterogeneity in characteristics such as dose-response differences within a population, or differences in contaminant levels in the environment The values of some variables used in an assessment change with time and space, or across the population whose exposure is being estimated. Assessments should address the resulting variability in doses received by members of the target population. Individual exposure, dose, and risk can vary widely in a large population. The central tendency and high end individual risk descriptors are intended to capture the variability in exposure, lifestyles, and other factors that lead to a distribution of risk across a population.

Uncertainty on the other hand, represents lack of knowledge about factors such as adverse effects or contaminant levels which may be reduced with additional study. Generally, risk assessments carry several categories of uncertainty, and each merits consideration. Measurement uncertainty refers to the usual error that accompanies scientific measurements—standard statistical techniques can often be used to express measurement uncertainty. A substantial amount of uncertainty is often inherent in environmental sampling, and assessments should address these uncertainties. There are likewise uncertainties associated with the use of scientific models, e.g., dose-response models, and models of environmental fate and transport. Evaluation of model uncertainty would consider the scientific basis for the model and available empirical validation.

Assessment and Presentation of Uncertainty

This subsection discusses practical approaches to assessing uncertainty in Superfund site risk assessments and describes ways to present key information bearing on the level of confidence in quantitative risk estimates for a site. The risk measures used in Superfund site risk assessments usually are not fully probabilistic estimates of risk, but conditional estimates given a considerable number of assumptions about exposure and toxicity (e.g., risk given a particular future land use). Thus, it is important to fully specify the assumptions and uncertainties inherent in the risk assessment to place the risk estimates in proper perspective. Another use of uncertainty characterization can be to identify areas where a moderate amount of additional data collection might significantly improve the basis for selection of a remedial alternative.

Highly quantitative statistical uncertainty analysis is usually not practical or necessary for Superfund site risk assessments for a number of reasons, not the least of which are the resource requirements to collect and analyze site data in such a way that the results can be presented as valid probability distributions. As in all environmental risk assessments, it is already known that uncertainty about the numerical results is generally large (i.e., on the range of at least an order of magnitude or greater). Consequently, it is more important to identify the key site-related variables and assumptions that contribute most to the uncertainty than to precisely quantify the degree of uncertainty in the risk assessment.

Thus, the focus of this subsection is on qualitative/semiquantitative approaches that can yield useful information to decision-makers for a limited resource investment. There are several categories of uncertainties associated with site risk assessments. One is the initial selection of substances used to characterize exposures and risk on the basis of the sampling data and available toxicity information. Other sources of uncertainty are inherent in the toxicity values for each substance used to characterize risk. Additional uncertainties are inherent in the exposure assessment for individual substances and individual exposures. These uncertainties are usually driven by uncertainty in the chemical monitoring data and the models used to estimate exposure concentrations in the absence of monitoring data, but can also be driven by population intake parameters. As described earlier, additional uncertainties are incorporated in the risk assessment when exposures to several substances across multiple pathways are summed.

13.6 RISK-BASED DECISION MAKING

The use of the risk-based decision making process allows for efficient allocation of limited resources, such as time, money, regulatory oversight, and qualified professionals. Advantages of using this process include:

1. Decisions are based on reducing the risk of adverse human or environmental impacts
2. Site assessment activities are focused on collecting only that information that is necessary to make risk-based corrective action decisions
3. Limited resources are focused on those sites that pose the greatest risk to human health and the environment at any time
4. Compliance can be evaluated relative to site-specific standards applied at site-specific point(s) of compliance
5. Higher quality, and in some cases, faster clean-ups may be achieved than are currently possible
6. Documentation is developed that can demonstrate that the remedial action is protective of human health, safety, and the environment

By using risk-based decision making, decisions are made in a consistent manner. Protection of both human health and the environment can be accounted for. [6]

A variety of EPA programs involved in the protection of groundwater and cleanup of environmental contamination utilize the risk-based decision making approach. Under the EPA's regulations dealing with cleanup of underground storage tank (UST) sites, regulators are expected to establish goals for clean-up of UST releases based on consideration of factors that could influence human and environmental exposure to contamination. Where UST releases affect groundwater being used as public or private drinking water sources, EPA generally recommends that clean-up goals be based on health-based drinking water standards; even in such cases, however, risk-based decision making can be employed to focus corrective action.[6] Finally, the reader should note that the bulk of this material also applies to hazardous risk analysis. That topic is treated in Chapter 18, Part IV.

Risk-based decision making and risk-based corrective action are decision making processes for assessing and responding to a health hazard. The processes take into account effects on human health and the environment, inasmuch as chemical releases vary greatly in terms of complexity, physical and chemical characteristics, and in the risk that they may pose. Risk-based corrective action (RBCA) was initially designed by the American Society for Testing and Materials (ASTM) to assess petroleum releases, but the process may be tailored for use with any hazard.

The United States Environmental Protection Agency (the EPA) and several state environmental agencies have developed similar decision-making tools. The EPA refers to the process as "risk-based decision making". While the ASTM RBCA standard deals exclusively with human health risk, the EPA advises that, in some cases, ecological goals must also be considered in establishing clean-up goals.[6]

13.7 PUBLIC PERCEPTION OF RISK

Public concern about risk ranges from earthquakes, fires, and hurricanes to asbestos, radon emissions, ozone depletion, toxins in our food and water, and so on. Many of the public's worries are out of proportion, because the fear is either overestimated or at times underestimated. The risks given the most publicity and attention receive the greatest concern, while the ones that are more familiar and accepted are given less thought.

A large part of what the public knows about risk comes from the media. Whether it is newspapers, magazines, radio or television, the media provides information about the nature and extent of specific risks. It also helps shape the perception of the danger involved within a given risk.

Lay people and experts disagree on risk estimates for many environmental problems. This creates a problem, since the public generally does not trust the experts. This chapter concentrates on how the public views risk and what the future of public risk perception will be. The reader should note that much of this material, as with the previous Section, applies to hazard risk assessment- a topic that is treated in Chapter 19, Part IV.

Everyday Risks

The public often worries about the largely publicized risks and thinks little about those that they face regularly. A study was recently performed that compared the responses of two groups, 15 national risk assessment experts and 40 members of the League of Women Voters on the risks of 30 activities and technologies.[7] This search produced striking discrepancies, as presented in Table 13.7.1. The League members rated nuclear power as the number 1 risk, while experts numbered it at 20 and the League ranked x-rays at 22 while the experts gave it a rank of 7.

There are various reasons for the differences in risk perception. Government regulators and industry officials look at different aspects in assessing a given risk than would members of the community.

The other working groups had greater difficulty when ranking the 31 environmental problem issues because there are no accepted guidelines for quantitatively assessing relative risks. As noted in the EPA's study, the following general results were produced for each of the four types of risks described earlier in this section.

1. No problems rank high in all four types of risk, or relatively low in all four
2. Problems that rank relatively high in three of the four types, or at least medium in all four, include criteria air pollutants, stratospheric ozone depletion, pesticide residues on food, and other pesticide risks (runoff and air deposition of pesticides)
3. Problems that rank relatively high in cancer and non-cancer health risks but low in ecological and welfare risks include hazardous air pollutants, indoor radon, indoor air pollution other than radon, pesticide application, exposure to consumer products, and worker exposures to chemicals
4. Problems that rank relatively high in ecological and welfare risk but low in both health risks include global warming, point and non-point sources of surface water pollution, physical alteration of aquatic habitats (including estuaries and wetlands), and mining waste
5. Areas related to groundwater consistently rank medium or low

Although there were great uncertainties involved in making these assessments, the divergence between the EPA effort and relative risks is noteworthy. From this study, areas of relative high risk but low EPA effort/concern include indoor radon, indoor air pollution, stratospheric ozone depletion, global warming, non-point sources, discharges to estuaries, coastal waters and oceans, other pesticide risks, accidental releases of toxics, consumer products, and worker exposures. The EPA gives high effort but relatively medium or low risks to RCRA sites, Superfund sites, underground storage tanks, and municipal non-hazardous waste sites.

TABLE 13.7.1 Consensus Ranking of Environmental Problem Areas on the Basis of Population Cancer Risk

RANK	PROBLEM AREA	SELECTED COMMENTS
1 (tied)	Worker exposed to chemicals	About 250 cancer cases per year estimated based on exposure to 4 chemicals; but workers face potential exposures to over 20,000 substances. Very high individual risk possible.
1 (tied)	Indoor Radon	Estimated 5000-20,000 lung cancers annually from exposure in homes.
3	Pesticide residues on foods	Estimated 6000 cancers annually, based on exposure to 200 potential oncogens.
4 (tied)	Indoor air pollutants (non radon)	Estimated 3500-6500 cancers annually, mostly due to tobacco smoke.
4(tied)	Consumer exposure to chemicals	Risk from 4 chemicals investigated is about 100-135 cancers annually, an estimated 10,000 chemicals in consumer products. Cleaning fluids, pestisticides, particleboard, and asbestos-containing products especially noted.
6	Hazardous/toxic air pollutants	Estimated 2000 cancers annually based on an assessment of 20 substances.
7	Depletion of stratospheric ozone	Ozone depletion projected to result in 10,000 additional annual deaths not ranked higher because of the uncertainties in future risk.
8	Hazardous waste sites, inactive	Cancer incidence of 1000 annually from 6 chemicals assessed. Considerable uncertainty since risk based on extrapolation from 35 sites to about 25,000 sites.
9	Drinking water	Estimated 400-1000 annual cancers, mostly from radon and trihalomethanes.

TABLE 13.7.1 (cont'd)

RANK	PROBLEM AREA	SELECTED COMMENTS
10	Application of pesticides	Approximately 100 cancers annually small population exposed but high individual risks.
11	Radiation other than radon	Estimated 360 cancers per year. Mostly from building materials. Medical exposure and natural background levels not included.
12	Other pesticide risks	Consumer and professional Exterminator uses estimated cancers of 150 annually. Poor data.
13	Hazardous waste sites, active	Probably fewer than 100 cancers annually: estimates sensitive to assumptions regarding proximity of future wells to waste sites.
14	Non hazardous waste sites Industrials	No real analysis done banking based on consensus of professional opinion
15	New toxic chemicals	Difficult to assess;done by consensus.
16	Non hazardous waste sites	Estimated 40 cancers annually not including municipal surface impoundments
17	Contaminated sludge	Preliminary results estimate 40 cancers annually, mostly from incineration and landfilling.
18	Mining Waste	Estimated 10-20 cancers annually, largely due to arsenic. Remote locations and small population exposure reduce overall risk though individual risk may be high.
19	Releases from storage tanks	Preliminary analysis based on benzene, indicates low cancer incidence (<1).
20	Non point source discharges to surface water	No quantitative analysis available; judgment.
21	Other ground water	Lack of information; individual risks considered less than 10^{-6}, with rough estimate of total population risk at <1

TABLE 13.7.1 (cont'd)

RANK	PROBLEM AREA	SELECTED COMMENTS
22	Criteria air pollutants	Excluding carcinogenic particles and VOC's (included under hazardous/ toxic air pollutants); ranked low because remaining criteria pollutants have not been shown to be carcinogens.
23	Direct point-source discharges To surface water	No quantitative assessment available. Only ingestion of contaminated seafood was considered.
24	Indirect point source discharges to surface water.	Same as above
25	Accidental release, toxics	Short duration exposure yields low cancer risk; non-cancer health effects of much greater concern.
26	Accidental releases, oil spills	See above. Greater concerns for welfare and ecological effects.

Not ranked: Biotechnology, global warming, other air pollutants, discharges to estuaries, coastal waters and oceans; discharge to wetlands.

Outrage Factors

The perception of a given risk is amplified by what are known as "outrage" factors. These factors can make people feel that even small risks are unacceptable. More than twenty outrage factors have been identified; a few of the main ones are defined below.[6]

1. *Voluntariness*. A voluntary risk is much more acceptable to people than an imposed risk. People will accept the risk from skiing, but not from food preservatives, although the potential for injury from skiing is 1,000 times greater than from preservatives
2. *Control*. Risks that people can take steps to control are more acceptable than those they feel are beyond their control. When prevention is in the hands of the individual the risk is perceived much lower than when it is in the hands of the government. You can choose what you eat, but you cannot control what is in your drinking water
3. *Fairness*. Risks that seem to be unfairly shared are believed to be more hazardous. People who endure greater risk than their neighbors and do not attain anything from it are generally out-raged by this. If I am not getting anything from it, why should others benefit?
4. *Process*. The public views the agency: Is it trustworthy or dishonest, concerned or arrogant? If the agency tells the community what's going on

before decisions are made, the public feels more at ease. They also favor a company that listens and responds to community concerns

5. *Morality.* Society has decided that pollution is not only harmful, it is evil. Talking about cost-risk tradeoffs sounds cold-hearted when the risk is morally relevant

6. *Familiarity.* Risks from exotic technologies create more dread than do those involving familiar ones. "A train wreck that takes many lives has less impact on people's trust of trains than would a smaller, hypothetical accident involving recombinant DNA, which is only perceived to have catastrophic mishaps".[7]

7. *Memorability.* An incident that remains in the public's memory makes the risk easier to imagine and is, therefore, more risky.

8. *Dread.* There are some illnesses that are feared more than others. Today there is greater fear given to AIDS and cancer than there is to asthma.

These outrage factors are not distortions in the public's perception of risk. They are inborn parts of what is interpreted as risk. They are explanations of why the public fears pollutants in the air and water more than they do geological radon. The problem is that many risk experts resist the use of the public's "irrational fear" in their risk management.

A problem exists in the perception of risk because the experts' and lay people's views differ. The experts usually base their assessment on mortality rates, while the lay people's fears are based on "outrage" factors. In order to help solve this problem, in the future, risk managers must work to make truly serious hazards more outrageous. One example is the ongoing concern for the risk involved in cigarette smoke. Another effort must be made to decrease the public's concern with low to modest hazards, i.e., risk managers must diminish "outrage" in these areas. In addition, people must be treated fairly and honestly.

13.8 ILLUSTRATIVE EXAMPLES

Example 13.1

At the maximum allowable EPA level of 5 parts per billion (ppb) of PCE (tetrachloroethylene), a person would have to drink 2 L of water per day for 70 years to increase the risk of getting cancer by 1 in 10,000. If one drinks 2 L of water per day for 70 years, what is the incresed risk of getting cancer when the concentration of PCE in water is 0.01 parts per million (ppm)?

Solution 13.1

The chance of getting cancer by drinking 2 L of water per day for 70 years that contains 5 ppb of PCE is 1/10,000. (This probability is the same as 0.0001). To convert 0.01 parts per million (ppm) to parts per billion (ppb):

0.01ppm (1,000 ppb/1 ppm) = 10 ppb

10 ppb (0.0001/5 ppb) = 2 (0.0001) = 0.0002 = 2/10,000 or 1/5,000

Therefore,the risk is twice as great, i.e., 1 in 5,000.

Example 13.2

List the four principal agencies that regulate risk issues.

Solution 13.2

1. The USEPA administers air , water, and toxic substance legislation.

2. The OSHA, a part of the Department of Labor, sets exposure standards and
    safety rules for work places.

3. The Food and Drug Administration (FDA) regulates foods, drugs, and
    cosmetics; it is housed in the Department of Health and Human Services;
    and reports to the Assistant Secretary for Health.

4. The Consumer Product Safety Commission (CPSC), an independent
    agency, controls the packaging, labeling, and distribution of a broad range
    of toys, clothes, electronics, and other products.

Example 13.3

Xylene is one of the organic compounds found in gasoline, and high
concentrations have been measured at gasoline stations. If the yearly average
concentration of xylene at a neighborhood gas station is 220 $\mu g/m^3$, what is the
hazard index associated with long term exposure? Likewise, if the peak 1-hour
concentration near the pumps is 1,000 $\mu g/m^3$, what is the hazard index for short
term exposure? Note: the chronic and acute response exposure levels for xylene
are 300 $\mu g/m^3$ and 4,400 $\mu g/m^3$, respectively.

Solution 13.3

In health risk assessments, non carcinogenic risks are estimated via "Hazard
Indices". A general equation for a hazard index (HI) is as follows:

$$HI = C/REL$$

where: C = the pollutant concentration, $\mu g/m^3$
           REL = the reference exposure level, in $\mu g/m^3$

If the pollutant causes an acute non carcinogenic risk, the maximum one hour concentration is used for C, and the acute reference exposure limit is used for the REL. Likewise, if the pollutant causes a chronic non carcinogenic risk, the one year average concentration is used, as is the chronic reference exposure limit. In this procedure, a hazard index is calculated for each pollutant separately, and then the indices are summed for each toxicological endpoint (i.e., the respiratory system, the central nervous system, etc.). Finally, the total hazard index is then compared to a value which is considered significant.

The hazard index for chronic exposure to xylene concentrations at the gas station is :

$$HI = (220 \ \mu g/m^3)/(300 \ \mu g/m^3)$$
$$= 0.73$$

The hazard index for acute exposure to the emissions near the pumps is:

$$HI = (1,000 \ \mu g/m^3)/(4,400 \ \mu g/m^3)$$
$$= 0.23$$

In California, a hazard index greater than 0.5 is considered to be significant. Therefore, if xylene was the only contaminant at the gas station, the long term exposure could be considered hazardous, but the short term exposure to the emissions near the pump might not.

Example 13.4

Several industrial facilities near a residential area emit the inhalable pollutants ethylene oxide, polychlorobiphenyls (PCBs) and polycyclic aromatic hydrocarbons (PAHs). The annual average concentration of ethylene oxide, PCBs, and PAHs are 10 $\mu g/m^3$, 2 $\mu g/m^3$, and 5 $\mu g/m^3$, respectively.

Calculate the cancer risk caused by each pollutant and the total cancer risk. Express the results in additional cancer cases per million people. Use the following data to solve this problem:

Cancer risk of pollutant = (Annual Average Concentration)(Unit Risk)

Pollutant	Unit Risk ($m^3/\mu g$)
Ethylene Oxide	8.8×10^{-5}
PCBs	1.4×10^{-3}
PAHs	1.7×10^{-3}

Solution 13.4

The cancer risk caused by ethylene oxide is calculated using the equation and data presented in the problem statement:

Cancer risk of ethylene oxide $= (10 \ \mu g/m^3)(8.8 \times 10^{-5} \ m^3/\mu g) = 88 \times 10^{-5}$
$= 880 \times 10^{-6}$
$= 880$ excess cancer cases per million people

The cancer risk caused by PCBs is estimated to be:

Cancer risk of PCBs $= 2 \ \mu g/m^3 \ (1.4 \times 10^{-3} \ m^3/\mu g)$
$= 2.8 \times 10^{-3}$
$= 2,800 \times 10^{-6}$
$= 2,800$ excess cancer cases per million people

The cancer risk caused by PAHs is estimated to be :

Cancer risk of PAHs $= 5 \ \mu g/m^3 \ (1.7 \times 10^{-3} \ m^3/\mu g) = 8.5 \times 10^{-3}$
$= 8500 \times 10^{-6}$
$= 8,500$ excess cancer cases per million people

The total cancer risk by inhalation is calculated from the arithmetic sum of individual cancer risks, assuming no interaction of pollutants in terms of carcinogenic effects:

Total Cancer Risk $= 880 + 2,800 + 8,500$
$= 12,180$ excess cancer cases per million people

The rate is extremely high, since it is 12,180 times higher than the 1 in a million cancer risk normally used as a basis for management of air toxics. This situation should be rectified as soon as possible by a reduction in air toxics. Also note that the PAH's contribute the majority of risk at this facility.

Example 13.5

Environmental tobacco smoke and gasoline vapors both contain mixtures of trace amounts of many of the individual compounds regulated as Air Toxics under Title III, section 112 of the 1990 Clean Air Act Amendmnts. Much of the general public is more likely to be exposed to these mixtures during the course of their lives than to specific compounds on the air toxics list. Hence, estimation of the cancer risk resulting from exposure to these mixtures is a useful and relevant exercise.

Which of the following mixture of air toxics imposes the greatest cancer risk to an individual breathing it for 70 years at an average concentration of 5 $\mu g/m^3$?

1. Environmental tobacco smoke [unit cancer risk 2.80×10^{-5} $(\mu g/m^3)^{-1}$]

2. Gasoline vapors [unit cancer risk 1.60×10^{-6} $(\mu g/m^3)^{-1}$]

Note: The source for the above risk values is Table III-7, Preliminary Cancer Potency Values for the Air Toxics "Hot Spots" act, found in California Air Pollution Control Officers Association, "*Air Toxics 'Hot Spots' Program, Revised 1192 Risk Assessment Guidelines,*" page III-28, published October 1993.

Solution 13.5

To solve this problem, compute the cancer risk for each air toxic mixture by multiplying the average concentration by the unit cancer risk.

1. For environmental tobacco smoke:

$$Risk = (5 \ \mu g/m^3) \ (2.80 \times 10^{-5}) = 1.40 \times 10^{-4}$$
$$= 140 \text{ in } 10^6 \ (140 \text{ in a million})$$

2. For gasoline vapors:

$$Risk = (5 \ \mu g/m^3) \ (1.60 \times 10^{-5}) = 8.00 \times 10^{-5}$$
$$= 80 \text{ in } 10^6 \ (80 \text{ in a million})$$

Since 140 in 10^6 exceeds 80 in 10^6, the conclusion is, that for identical durations of exposure to identical concentrations in air, environmental tobacco smoke poses a greater cancer risk than gasoline vapors. However, both mixtures exhibit high cancer risks at trace levels.

Example 13.6

A ground water plume has developed from a leaking underground storage tank (UST) and fumes have migrated to the basement of nearby homes. A sample of indoor air indicates toluene and benzene concentrations of 50 and 70 $\mu g/m^3$, respectively. Determine the resulting health risk for a 70 kg adult living in the residence who is exposed to these vapors for 15 years, assuming a breathing rate of 15 m^3/d and 75% absorption of both toluene and benzene. The carcinogenic slope factor-risk specific doses (RSDs) via the inhalation route are 0.021 $(mg/kg-d)^{-1}$ and 0.028 $(mg/kg-d)^{-1}$ for toluene and benzene respectively. Assume the environmental health risks for toluene and benzene are additive.

To calculate the environmental risk use the equation:

$$RISK = \frac{(C)(CSF)(BR)(DUR)}{(ABW)(LIFE)}$$

Where: RISK = probability of cancer, dimensionless
C = Air concentration, mg/m3
CSF = Carcinogenic slope factor, RSDs(mg/kg-d)$^{-1}$
BR = Breathing rate, m^3/d
DUR = Exposure duration period, yr
ABW = Average body weight, kg
LIFE = Lifetime exposure of 70 yr

Note: The RSD is an estimate of the average daily dose of a carcinogen that corresponds to a specific excess cancer risk for a 70-yr lifetime exposure.

Solution 13.6

Based on the problem statement,

$$Total\ Risk = RISK_{toluene} + RISK_{benzene}$$

Using the formula presented in the problem statement, the environmental risks of toluene and benzene are determined as follows

$$RISK_{toluene} = \frac{\left(50\,mg/m^3\right)\left(0.021(mg/kg-d)^{-1}\right)\left(15m^3/d\right)(15yr)}{(70kg)(70yr)} \times \frac{(1mg)}{(1000mg)}$$

$$RISK_{toluene} = 4.82 \times 10^{-5}$$

$$RISK_{benzene} = \frac{\left(70\,mg/m^3\right)\left(0.028(mg/kg-d)^{-1}\right)\left(15m^3/d\right)(15yr)}{(70kg)(70yr)} \times \frac{(1mg)}{(1000mg)}$$

$$RISK_{benzene} = 9.0 \times 10^{-5}$$

$$\begin{aligned}
TOTAL\ RISK &= RISK_{toluene} + RISK_{benzene} \\
&= 4.82x10^{-5} + 9.0x10^{-5} \\
&= 1.38x10^{-4}
\end{aligned}$$

Under these conditions, one could expect 1.38 additional cases of cancer per 10,000, or 138 additional cases of cancer per 1,000,000 people exposed to toluene and benzene from this source.

13.9 SUMMARY

1. Risk characterization is the process of estimating the incidence of a health effect under the various conditions of human or animal exposure as described in the exposure assessment. It evolves from both dose exposure assessment and toxicity response assessment. The data are then combined to obtain qualitative and quantitative expression of risk.

2. Qualitative approaches to risk assessment have been employed in order to best characterize health risks.

3. The potential for noncarcinogenic health effects is evaluated by comparing an exposure level over a specified time period (e.g., lifetime) with a reference dose derived for a similar exposure period. The ratio of exposure to toxicity in called a hazard quotient; and, when it is greater then unity there is a higher level of concern for potential noncancer effects.

4. For carcinogens, risks are estimated as the incremental probability of an individual developing cancer over a lifetime as a result of exposure to the potential carcinogen. The slope factor (SF) converts estimated daily intakes averaged over a lifetime of exposure directly to incremental risk of an individual developing cancer.

5. The risk assessment steps and the risk characterization are influenced by uncertainty and variability. Variability arise from heterogeneity such as dose-response differences within a population, or differences in contaminant levels in the environment. Uncertainty on the other hand, represents lack of knowledge about factors such as adverse effects or contaminant levels.

6. The use of a risk-based decision making process allows for efficient allocation of limited resources, such as time, money, regulatory oversight, and qualified professionals..

7. The media provides information regarding the nature and extend of risks. Every day risks, are underestimated by the public, because much attention is directed to largely publicized risks. The perception of a given risk is amplified by "outrage factors"which can make people feel even small risks are unacceptable. Experts view the perception of risk differently than lay people as a result a serious hazard may be underestimated.

PROBLEMS

1. Where can one go to get more information about chemical risks?
2. Do chemical companies and other industries keep track of employees' cancer and respiratory illnesses and deaths?

3. If xylene and toluene both affect the human reproduction system in response to chronic exposure, what is the hazard index after adding the effect of an average toluene concentration of 1,400 $\mu g/m^3$? Note: the chronic response exposure level for toluene is 200 $\mu g/m^3$.

4. The following equation can be used in risk assessment studies for carcinogens:

$$C_m = \frac{(R)(W)(L)}{(P)(I)(A)(ED)}$$

where: C_m = action level, i.e., the concentration of carcinogen above which remedial action should be taken
R = risk or the probability of contracting cancer
W = body weight
L = assumed lifetime
P = potency factor
I = intake rate
A = absorption factor, i.e.,the fraction of carcinogen absorbed by the human body
ED = exposure duration

Determine the action level in $\mu g/m^3$ for an 80 kg person with a life expectancy of 70 years exposed to benzene over a 15-year period. The "acceptable" risk is one incident of cancer per 1 million persons or 10^{-6}. Assume a breathing (intake) rate of 15 m^3/d and an absorption factor of 75%. The potency factor for benzene is 1.80 $(mg/kg\text{-}d)^{-1}$

5. In health risk assessment, the carcinogenic risk calculation by inhalation (IR) can be calculated by :

$$IR_i = C_i \, UR_i \, L/70$$

Where: IR_i = individual inhalation excess lifetime cancer risk from pollutant i
C_i = annual average air concentration of pollutant i
UR_i = unit risk value by inhalation for pollutant i
L = operational lifetime of a facility, yr; usually assumed to be 70 year

Compare the inhalation risk for the emission of 100 $\mu g/m^3$ of arsenic with unit risk of 3.30×10^{-3} $(\mu g/m^3)^{-1}$) and dioxin/dibenzofuran with unit risk of $3.80 \times 10^{+1}$ $(\mu g/m^3)^{-1}$). Also how much arsenic is equivalent to 100 lb of dioxin/dibenzofuran from an inhalation risk perspective?

REFERENCES

1. D. Paustenbach, "The Risk Assessment, Environmental and Human Health," John Wiley & Sons, New York City, 1985.
2. "Introduction to Environmental Engineering and Science," G. Master, Upper Saddle River, NJ, Prentice Hall, 1991.
3. USEPA WebSite, 2001.
4. D. Gute and N.B. Hanes "An Applied Approach & Epidemiology and Toxicology for Engineers," NIOSH, Cincinnati, Ohio, 1993.
5. P.G. Shields and N.B. Hanes "Molecular Epidemiology and the Genetics of Environmental Cancer," JAMA, 681-87, 1991.
6. G. Holmes, B. Singh, and L. Theodore, " Handbook of Environmental Management and Technology." John Wiley and Sons, New York, 2000.
7. D. Goleman, "Assessing Risk: Why Fear May Outweigh Harm," New York Times, Feb. 1, 1994.

REFERENCES

1. D. Paustenbach, *The Risk Assessment of Environmental and Human Health Risks*, John Wiley & Sons, New York City, 1989.

2. "Environmental Compliance Monitoring and Research," G. Skumorz Springbook Information Associates, 1991.

Part IV

Hazard Risk Assessment

Risk analysis of accidents serves a dual purpose. It estimates the probability that an accident will occur and also assesses the severity of the consequences of an accident. Consequences may include damage to the surrounding environment, financial loss, injury to life and/or death. This Part of the book (Part IV) is primarily concerned with the methods used to identify hazards and causes and consequences of accidents. Issues dealing with health risks have been explored in the previous Part (III). Risk assessment of accidents provides an effective way to help ensure either that a mishap will not occur or reduces the likelihood of an accident. The result of the risk assessment also allows concerned parties to take precautions to prevent an accident before it happens.

As with Health Risk Assessment (HRA), there are four key steps involved in a hazard risk assessment (HZRA). These are detailed below (see figure) if the system in question is a chemical plant.

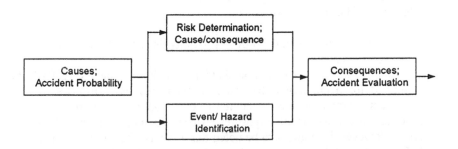

1.   The event or series of events that will initiate an accident has to be identified. An event could be a failure to follow correct safety procedures, improperly repaired equipment, or failure of a safety mechanism.

2. The probability that an accident will occur has to be determined. For example, if a chemical plant has a given life, what is the probability that the temperature in a reactor will exceed the specified temperature range? The probability can be ranked qualitatively from low to high. A low probability means that it is unlikely for the event to occur in the life of the plant. A medium probability suggests that there is a probability that the event will occur. A high probability means that the event will probably occur during the life of the plant.

3. The severity of the consequences of the accident must be determined. This will be described later in detail.

4. If the probability of the accident and the severity of its consequences are low, then the risk is usually deemed acceptable and the plant should be allowed to operate. If the probability of occurrence is too high or the damage to the surroundings is too great, then the risk is usually unacceptable and the system needs to be modified to minimize these defects.

This algorithm allows for reevaluation of the process if the risk is deemed unacceptable (the process is repeated starting with either step one or two).

In line with the above figure, the five chapters associated with this Part of the book are:

Chapter 14: Introduction to Hazard Risk Assessment
Chapter 15: Event/Hazard Identification
Chapter 16: Accident Causes and Probability
Chapter 17: Accident Consequences and Evaluation
Chapter 18: Hazard Risk Analysis

Note that the last chapter provides details in risk determination and cause/consequence analysis.

Finally, the reader should note that hazard risk assessments (HZRA) are examined for acute rather than chronic exposures. For purposes of this book, acute exposures are considered to occur for a short period of time. Material on chronic health exposures (HRA) is available in Part III of the book. However, it should also be noted that HRA is often an integral part of HZRA, particularly with any (and in particular) accidental chemical release.

14

Introduction to Hazard Risk Assessment

14.1 INTRODUCTION

Risk evaluation of accidents serves a dual purpose. It estimates the probability that an accident will occur and also assesses the severity of the consequences of an accident. Consequences may include damage to the surrounding environment, financial loss, or injury to life. This chapter is primarily concerned with the methods used to identify hazards and the causes and consequences of accidents. Issues dealing with health risks have been explored in the previous chapter. Risk assessment of accidents provides an effective way to help ensure either that a mishap does not occur or reduces the likelihood of an accident. The result of the risk assessment allows concerned parties to take precautions to prevent an accident before it happens.

Regarding definitions, the first thing an individual needs to know is what exactly is an accident. An accident is an unexpected event that has undesirable consequences.[1] The causes of accidents have to be identified in order to help prevent accidents from occurring. Any situation or characteristic of a system, plant, or process that has the potential to cause damage to life, property, or the environment is considered a hazard. A hazard can also be defined as any characteristic that has the potential to cause an accident. The severity of a hazard plays a large part in the potential amount of damage a hazard can cause if it occurs. Risk is the probability that human injury, damage to property, damage to the environment, or financial loss will occur. An acceptable risk is a risk whose probability is unlikely to occur during the lifetime of the plant or process. An acceptable risk can also be defined as an accident that has a high probability of occurring, with negligible consequences. Risks can be ranked qualitatively in categories of high, medium, and low. Risk can also be ranked quantitatively as an annual number of fatalities per million affected individuals. This is normally denoted as a number times one millionth, e.g., 3×10^{-6} indicates that on average 3 workers will die every year out of one million individuals. Another quantitative approach that has become popular in industry is the Fatal Accident Rate (FAR) concept (this is discussed in more detail in Chapter 18). This determines or estimates the number of fatalities over the lifetime of 1000 workers. The lifetime of a worker is defined as 10^5 hours, which is based on a 40-hour work week for 50 years. A reasonable FAR for a chemical plant is 3.0 with 4.0 usually taken as a maximum. The FAR for an individual at home is approximately 3.0. A FAR of 3.0 means that there are 3 deaths for every 1000 workers over a 50-year period.

14.2 RISK EVALUATION PROCESS FOR ACCIDENTS

There are several steps in evaluating the risk of an accident (see Figure 14.2.1).

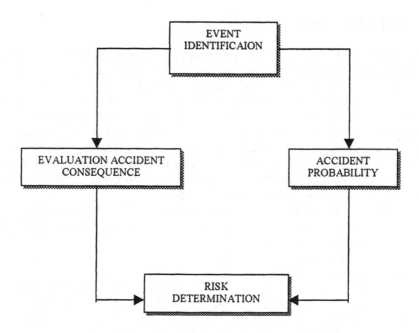

Figure 14.2.1. Hazard risk assessment flowchart.

A more detailed figure is presented below if the system in question is a chemical plant:

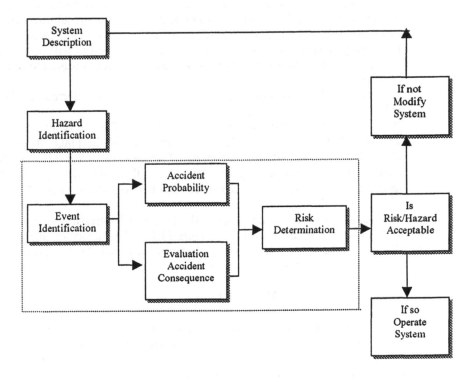

Figure 14.2.2   Hazard risk assessment flowchart.

1. A brief description of the equipment and chemicals used in the plant is needed
2. Any hazard in the system has to be identified. Hazards that may occur in a chemical plant include:

    a. Fire
    b. Toxic vapor release
    c. Slippage
    d. Corrosion
    e. Explosions
    f. Rupture of pressurized vessel
    g. Runaway reactions

3. The event or series of events that will initiate an accident has to be identified. An event could be a failure to follow correct safety procedures, improperly repaired equipment, or a safety mechanism
4. The probability that the accident will occur has to be determined. For example, if a chemical plant has a 10-year life, what is the probability that

the temperature in a reactor will exceed the specified temperature range? The probability can be ranked from low to high. A low probability means that it is unlikely for the event to occur in the life of the plant. A medium probability suggests that there is a possibility that the event will occur. A high probability means that the event will probably occur during the life of the plant

5.   The severity of the consequences of the accident must be determined. This will be described later in detail

6.   If the probability of the accident and the severity of its consequences are low, then the risk is usually deemed acceptable and the plant should be allowed to operate. If the probability of occurrence is too high or the damage to the surroundings is too great, then the risk is usually unacceptable and the system needs to be modified to minimize these effects

As indicated above, the heart of the hazard risk assessment algorithm provided is enclosed in the dashed box (Figure 14.2.2). The algorithm allows for reevaluation of the process if the risk is deemed unacceptable (the process is repeated starting with either step one or two).

14.3 HAZARD IDENTIFICATION

Hazard or event identification provides information on situations or chemicals and their releases that can potentially harm the environment, life, or property. Information that is required to identify hazards includes chemical identities, quantities and location of chemicals in question, chemical properties such as boiling points, ignition temperatures, and toxicity to humans. There are several methods used to identify hazards. The methods that will be discussed later in this Part will include the process checklist and the hazard and operability study (HAZOP).

A process checklist evaluates equipment, materials and safety procedures.[1] A checklist is composed of a series of questions prepared by an engineer who knows the procedure being evaluated. It compares what is in the actual plant to a set of safety and company standards. Some questions that may be on a typical checklist are:

1.   Was the equipment designed with a safety factor?
2.   Does the spacing of the equipment allow for ease of maintenance?
3.   Are the pressure relief valves on the equipment in question?
4.   How toxic are the materials that are being used in the process and is there adequate ventilation?
5.   Will any of the materials cause corrosion to the pipe(s)/reactor(s)/system?
6.   What precautions are necessary for flammable material?
7.   Is there an alternate exit in case of fire?
8.   If there is a power failure what fail-safe procedure(s) does the process contain?

9. What hazard is created if any piece of equipment malfunctions?

These questions and others are answered and analyzed. Changes are then made to reduce the risk of an accident. Process checklists are updated and audited at regular intervals.

A hazard and operability study is a systematic approach to recognizing and identifying possible hazards that may cause failure of a piece of equipment.[2] This method utilizes a team of diverse professional backgrounds to detect and minimize hazards in a plant. The process in question is divided into smaller processes (subprocesses). Guidewords are used to relay the degree of deviation from the subprocesses' intended operation. The guidewords can be found in Table 14.3.1. The causes and consequences of the deviation from the process are determined. If there are any recommendations for revision they are recorded and a report is made. A summary of the basic steps of a HAZOP study is[2]:

1. Define objectives
2. Define plant limits
3. Appoint and train a team
4. Obtain complete preparative work (i.e., flow diagrams, sequence of events)
5. Conduct examination meetings that select subprocesses, agree on intention of subprocesses, state and record intentions, use guide words to find deviations from the intended purpose, determine the causes and consequences of deviation, and recommend revisions
6. Issue meeting reports
7. Follow up on revisions

There are other methods of hazard identification. A "what-if" analysis presents certain questions about a particular hazard and then tries to find the possible consequences of that hazard. The human-error analysis identifies potential human errors that will lead to an accident. They can be used in conjunction with the two previously described methods. .

Several of these hazard identification methods are discussed in more detail in the next chapter.

14.4 CAUSES OF ACCIDENTS

The primary causes of accidents are mechanical failure, operational failure (human error), unknown or miscellaneous, process upset, and design error. Figure 14.4.1 is the relative number of accidents that have occurred in the petrochemical field (on a percentage basis).[3] There are three steps that normally lead to an accident:

1. Initiation
2. Propagation
3. Termination

TABLE 14.3.1 Guide Words Used to Relay the Degree of Deviation from Intended Subprocess Operation

Guide Word	Meaning
No	No part of intended function is accomplished
Less	Quantitative decrease in intended activity
More	Quantitative increase in intended activity
Part of	The intention is achieved to a certain percent
As well as	The intention is accomplished along with side effects
Reverse	The opposite of the intention is achieved
Other than	A different activity replaces the intended activity

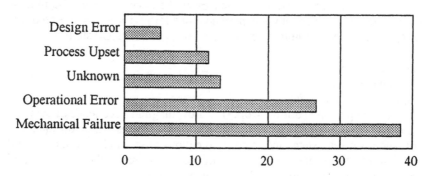

Figure 14.4.1. Causes of accidents in the petrochemical field.

The path that an accident takes through the three steps can be determined by means of a fault tree analysis.[1] A fault tree is a diagram that shows the path that a specific accident takes. The first item needed to construct a fault tree is the definition of the initial event. The initial event is a hazard or action that will cause the process to deviate from normal operation. The next step is to define the existing conditions needed to be present in order for the accident to occur. The propagation (e.g., the mechanical failure of equipment related to the accident) is discussed. Any other equipment or components that need to be studied have to be defined. This includes safety equipment that will bring about the termination of the accident. Finally, the normal state of the system in question is determined. The termination of an accident is the event that brings the system back to its normal operation. An example of an accident would be the failure of a thermometer in a reactor. The temperature in the reactor could rise and a runaway reaction might take place. Stopping the flow to the reactor and/or cooling the contents of the reactor could terminate the accident.

14.5 CONSEQUENCES OF ACCIDENTS

Consequences of accidents can be classified qualitatively by the degree of severity. A quantitative assessment is beyond the scope of the text; however information is available in the literature (Belandi, 1988). Factors that help to determine the degree of severity are the concentration of the hazard that is released, the length of time that a person or the environment is exposed to a hazard, and the toxicity of the hazard. The worst-case consequence or scenario is defined as a conservatively high estimate of the most severe accident identified.[1] On this basis one can rank the consequences of accidents into low, medium, and high degrees of severity.[3] A low degree of severity means that the hazard is nearly negligible, and the injury to person, property, or the environment is observed only after an extended period of time. The degree of severity is considered to be medium when the accident is serious, but not catastrophic, the toxicity of the chemical released is great, or the concentration of a less toxic chemical is large enough to cause injury or death to persons and damage to the environment unless immediate action is taken. There is a high degree of severity when the accident is catastrophic or the concentrations and toxicity of a hazard is large enough to cause injury or death to many persons, and there is long-term damage to the surround environment. Figure 14.5.1 provides a graphical, qualitative representation of the severity of the consequences.[3]

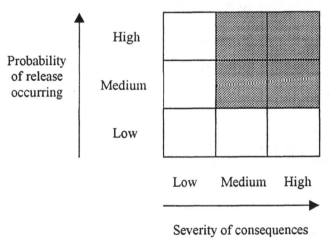

Figure 14.5.1. Qualitative probability-consequence analysis.

14.6 CAUSE-CONSEQUENCE ANALYSIS

Cause-consequence risk evaluation combines event tree and fault tree analysis to relate specific accident consequences to causes.[1] The process of cause-consequence evaluation usually proceeds as follows:

1. Select an event to be evaluated
2. Describe the safety system(s)/procedure(s)/factor(s) that interfere with the path of the accident
3. Perform an event tree analysis to find the path(s) an accident may follow
4. Perform a fault tree analysis to determine the safety function that failed
5. Rank the results on a basis of severity of consequences

As its name implies, cause-consequence analysis allows one to see how the possible causes of an accident and the possible consequences that result from that event interact with each other.

14.7 FUTURE TRENDS

For the most part, future trends will be found in hazard accident prevention, not hazard analysis. To help promote hazard accident prevention, companies should start employee-training programs. These programs should be designed to alert staff and employees about the hazards they are exposed to on the job. Training should also cover company safety policies and the proper procedures to follow in case an accident does occur. A major avenue to reducing risk will involve source reduction of hazardous materials. Risk education and communication are two other areas that will need improvement.

14.8 ILLUSTRATIVE EXAMPLES

Example 14.1

In attempting to solve problems, engineers usually apply what has come to be defined as the "engineering approach". List some of the key considerations that are employed in this methodology.

Solution 14.1

Identify: What is the problem?
Location: Where is the problem?
Timing: When does the problem occur?
              When was it first observed?
Magnitude: How far does the problem extend?
              How many units are affected?
              How much of any one unit is affected?

Example 14.2

In case of an accident at a chemical plant, who will warn local residents about toxic emissions and provide for appropriate protection?

Solution 14.2

The LEPC (see Chapter 3) has developed warning systems, evacuation plans, and shelter-in-place instructions. Individual citizens can also ask the local plant to explain how their emergency response plans mesh with the LEPC. The plant must immediately report all incidents of chemical releases to the NRC, the SERC, and the LEPC.

Example 14.3

A consulting engineer has been hired by a state agency to perform an exposure assessment at a private dwelling. The dwelling is located over a shallow aquifer that has been contaminated by a gasoline leak from an underground storage tank at the corner gasoline station. List at least eight important environmental factors that the consulting engineer should consider when visiting the dwelling.

Solution 14.3

The following are some of the factors that a consulting engineer should consider in conducting the exposure and risk assessment.

1. Presence or absence of a basement, and if present, possible ingress routes for liquid and vapor into the basement
2. Ventilation of the dwelling
3. Concentration of the dissolved hydrocarbon fraction in groundwater near the dwelling
4. Presence or absence of separate-phase gasoline near the home
5. Water supply system used by the dwelling's occupants; i.e., are they using a well that taps into the contaminated aquifer, or are they on a municipal water distribution system?
6. Ages, genders, and lifestyles of the dwelling's occupants
7. Concentration of vapors in the zone near or under the house, and possible risk of explosion or intoxication
8. If not yet near the dwelling, the rate of transport of dissolved and separate phases towards the dwelling
9. Toxicity and cancer dose-response data for the constituents of the gasoline
10. Estimated additional cancer risk for dwelling's occupants when exposure data are combined with cancer dose-response data

Example 14.4

Outline a risk assessment for the accidental spill of battery acid in an auto repair shop.

Solution 14.4

1. Obtaining data on the chemical and toxicological properties of sulfuric acid
2. Training personnel in the proper procedures and precautions for handling automobile batteries
3. Training personnel in the proper methods for clean up of an acid spill
4. Having protective gear available for personnel who are designated to clean up any spill
5. Having spill clean-up equipment available in a prominent and unobstructed location in the shop

Example 14.5

Very often the actions of the first person on the scene of an accident can have a large impact on the final outcome of the incident. However, it is crucial that this person not subject him or herself to personal injury. The actions of an untrained person and trained person may differ greatly.

a. What sequence of actions would you recommend for the untrained individual who observes a large spill from a tank truck accident?
b. How would you, the trained leader of an emergency response team, behave in a similar circumstance?

Solution 14.5

a. In case of the observation of a potential chemical emergency, the untrained individual should immediately notify the authorities. Call 911 and report it, or call the police or fire department, or both. Give full information as to nature of the accident and spill, location, and any other observed details such as injury to persons or property.

Do not approach the spill. It could be hazardous, or may produce an explosion, fire, or toxic fumes. Contact nearby persons advising them to also be cautious or to evacuate their residence if necessary.

b. The proper procedures for emergency response at a chemical emergency should include the following:

1. Make a safe approach to the scene; use binoculars so you can survey the situation at a distance. See what you are getting into. Assume it is a spill of a hazardous substance, unless you have evidence otherwise
2. Protect yourself. Use proper protective equipment and clothing, depending upon the nature of the hazard. See literature for information on proper attire, etc., for a given type of hazard
3. Identify the hazard. The vehicle should have Department of Transportation

information readily seen on the vehicle or container indicating the general nature of the hazard. More specific information may be obtained from the manufacturer or Chem Trec of the Chemical Manufacturers Association as to the substance and safety measures in dealing with and containing the material. Chem Trec is available by phone everyday, 24 hours a day. The free phone number is 1-800-424-9300. When you call give the product identification number, the product name if known, and information from the shipping papers in the vehicle. Describe the situation and arrange for phone contact with the manufacturer

4. Secure the scene; get medical attention to the injured. Move bystanders away to a safe distance. Inform print and electronic media. Protect people and property as necessary. Be aware that explosive materials and toxic vapors can do damage at a great distance

5. With proper equipment and suiting attempt to control and contain the hazardous material, fire, or other event. Consider the nature and quantity of the material that could be involved in your response to the situation. See literature and Chem Trec or manufacturer on how to attack the problem with safety

6. Clean up in such a way that the hazardous material is removed and does not produce a future danger. It may be necessary to pump out a hazardous liquid or to remove contaminated soil, etc. Decontamination procedures may be necessary. See EPA's Standard Operating Safety Guides for instructions for decontamination required following exposure to different danger levels of hazards

7. After the emergency is over analyze the situation, discussing it with all concerned and knowledgeable persons. Write a report giving all lessons to be learned from the event in order to reduce the chance of such an event happening again

14.9 SUMMARY

1. Risk assessment of accidents estimates the probability that hazardous materials will be released and also assesses the severity of the consequences of an accident.

2. The risk evaluation process defines the equipment, hazards, and events leading to an accident. It determines the probability that an accident will occur. The severity and acceptability of the risk are also evaluated.

3. Hazard identification provides information on situations or chemicals that can potentially harm the environment, life, or property. The processes described are process checklist, event tree, hazard and operability study.

4. Accidents occur in three steps: initiation, propagation, and termination. The primary causes of accidents are mechanical failure, operational failure (human error), unknown or miscellaneous, process upset, and design error.

5. Consequences of accidents are classified by degree of severity into low, medium, and high.

6. Cause-consequence analysis allows one to see the possible causes of an accident and the possible accident that results from a certain event.
7. For the most part future trends can be found in hazard prevention.

PROBLEMS

1. The four basic dimensions of a problem are: what, when, where, and to what extent. Using these dimensions, generate questions that should be asked during a HAZOP.
2. How does one know what chemicals are used or made in an industrial plant near one's home and what amounts are being stored there?
3. Can a risk assessment provide information on exactly what to do about a specific hazard?
4. Compare individual hazard risk with population risk.
5. Locate newspaper and/or news magazine articles in the library (*New York Times, Newsweek, Time, etc.*) about a recent accident that involved evacuation of populations due to risk to their health and/or safety. From these sources, write a brief essay describing what happened, where the incident took place, the number of people killed and/or injured, the immediate impact in the community, etc.

REFERENCES

1. *Guidelines for Hazard Evaluation Procedures,* CCPS, AICHE, New York City, 1992.
2. L. Theodore and K. Morris, *"Health, Safety and Accident Prevention: Industrial Applications",* A Theodore Tutorial, East Willington, NY, 1998.
3. J. Crowl, and J. Louvar, *Chemical Safety Fundamentals With Applications,* Englewood, NJ, Prentice-Hall, 1990.

15

Event/Hazard Identification

15.1 INTRODUCTION

Hazard or event identification provides information on situations or chemicals and their releases that can potentially harm the environment, life, or property. Information that is required to identify hazards includes chemical identities, quantities and location of chemicals in question, chemical properties such as boiling points, ignition temperatures, and toxicity to humans.[1] There are several methods used to identify some of the hazards. Some of these methods are discussed in this chapter.

Chapter contents include the following:

Section 15.2. Event/Hazard Evaluation Techniques
Section 15.3. System Checklists
Section 15.4. Safety Review/Safety Audit
Section 15.5. "What If" Analysis
Section 15.6. Preliminary Hazard Analysis (PHA)
Section 15.7. Hazard and Operability Study (HAZOP)

An extensive treatment is provided in Section 15.7 since Hazard and Operability Studies (HAZOP) are perhaps the most often used and successful technique for identifying hazards and hazardous events.

Obviously, the key word in this chapter is "identify." Material on identifying health hazards can be found in Chapters 9 and 10 of Part III. The identifying concerns for this chapter primarily deal with accidents, and then accidents can take the form of either a chemical release or a disaster arising from a blast/fire/fragment problem.

The authors gratefully acknowledge the assistance of Elizabeth Thomas in researching, reviewing, and editing this chapter.

15.2   EVENT/HAZARD EVALUATION TECHNIQUES

Several hazard evaluation techniques have been developed; when applied properly to a given system, hidden system failure modes can be identified and techniques for their rectification can be recommended. Many occupational and environmental safety problems are recognized as the result of an emergency, and in many of these situations, once the emergency is over, the problem is considered resolved. Although solving safety problems once they have occurred is the domain of the design engineer, the true role of the design team must be to prevent accidents from occurring in the first place. The hazard evaluation techniques, when integrated with engineering design, provide the design engineer with the necessary tools to identify and modify those components of the system that have the potential to cause an accident. [2,3]

To properly apply the system safety techniques to the design and operation of potentially hazardous technologies, the design engineer must have a clear understanding of the system and be able to prepare a written response to questions such as:

1.   What is the intended function of the system?
2.   What are the raw materials, intermediates, final products and byproducts?
3.   What steps are taken to convert the raw materials to final products? (e.g., chemical reactions, physical operations, etc.)
4.   How does the system interact with the environment? (e.g., hazardous waste streams, toxic releases, etc.)
5.   How does the system interact with personnel? (e.g., the need for personal protective equipment.)
6.   What sources of energy does the system use and how is this energy supplied to the system?
7.   What are the maintenance requirements of the system?
8.   How does the system interact with other systems within the plant?

The above list is illustrative only and must be tailored or adjusted to the particular system design at hand.

Proper application of an event hazard evaluation technique also requires a sound knowledge of the types of hazards involved within the system. The design engineer should develop a list summarizing the types of hazards that warrant further evaluation within the system. This list could take the following hazards into account:

1.   Toxic chemicals
2.   Fire
3.   Explosion
4.   Runaway chemical reaction
5.   Temperature extremes or excursions
6.   Radiation

7. Equipment/instrumentation malfunction that can be a factor in the appearance of a hazard
8. System moving part
9. Electrical
10. Hazardous noise and vibration
11. Mechanical
12. Environmental pollution
13. Pressure

After the system has been defined, a hazard evaluation technique can be used to identify different types of hazards within the system components and to propose possible solutions to eliminate the hazards. This topic is treated in more detail in the next two chapters. These procedures are extremely useful in identifying system modes and failures that can contribute to the occurrence of accidents; they should be an integral part of different phases of process development from conceptual design to installation, operation, and maintenance.[4,5] The hazard evaluation techniques that are useful in the preliminary and detailed stages of the design process include:

1. Checklist approach
2. Safety review/safety audit
3. "What if" analysis
4. Preliminary hazard analysis
5. Hazard and operability study

Other procedures appearing in the literature include:

1. Relative ranking techniques
2. Failure modes effects criticality analysis (FMECA)
3. Fault tree analysis (FTA)
4. Event tree analysis (ETA)
5. Cause-consequence analysis (C-CA)

However, these latter procedures are more risk related topics, and are therefore treated later in the chapter.

15.3   SYSTEM CHECKLISTS

A system checklist is useful to identify compliance problems and also those areas of the system that require further hazard evaluation. The method is easy to use and can be applied to any component of a given system such as equipment, instrumentation, materials, and procedures. This method, which produces qualitative results, must be prepared by an engineer thoroughly experienced with the system; once the checklist is prepared, however, it can be used by engineers or managers who may have less technical experience with the system.[2]

The method of checklists can be applied to any phase of a project's life cycle from preliminary design to shipment of products and disposal of wastes. Since the safety requirements of a system are a strong function of the nature of the process, preparing a standard "checklist" format applicable to all systems may be difficult; therefore, the checklist must be streamlined to the specific problem at hand.   For example, in a preliminary plant design, the design engineer might prepare a checklist to cover the following areas:

- Raw materials
- Products
- Intermediate products
- Equipment
- Instrumentation
- Plant layout
- Start-up
- Shut down
- Emergency shutdown
- Personal protective equipment (PPE)
- Contingency planning (both personnel and community)
- Waste disposal

Each specific area mentioned above can be further expanded to provide more details for hazard evaluation. For example, the design engineer might prepare the following checklist to gain more insight into the hazards posed by the raw materials:

- Flammability:
    - What is the flash point?
    - What are the upper and lower flammability limits?
    - What is the autoignition temperature?
    - What is the fire point temperature?
    - What is the vapor pressure?
    - What are the products of combustion?
    - What is the evaporation rate?
    - What is the proper fire extinguishing agent?
    - Does the material undergo hazardous polymerization?
    - Is the material pyrophoric? (i.e., can it catch fire upon contact with air?)

- Toxicity:

- What are the exposure limits for the material? e.g., threshold limit values (TLV), permissible exposure limit (PEL), recommended exposure limit (REL).
- Is the material classified as "highly toxic" or "toxic" based upon the results of tests on laboratory animals? e.g., LD_{50} or LC_{50} data. NOTE: as described in Chapter 11, LD_{50} and LC_{50} are referred to as the dose of a chemical that is lethal to 50 percent of laboratory test animals.

- Storage:
    - What materials are incompatible with the raw material?
    - What monitoring devices are needed for the storage area? e.g., combustible gas meter, organic vapor analyzer, etc.
    - How should a spill of this material be cleaned up?
    - Based on the flammability data, does the storage area require ignition proof equipment?
    - Can the material undergo hazardous polymerization or decomposition under storage conditions?
    - Do containers of this material need grounding and/or bonding to protect against electrostatic hazards?

- Reactivity:
    - Is the material stable under storage conditions?
    - Is the material water reactive? (A water reactive material can violently react with water to produce a toxic or flammable gas.)

A similar checklist can be prepared for the other areas of interest mentioned above. Although the results of a checklist study are qualitative, these results can be used to identify design areas that require further hazard evaluation and to communicate the safety needs of the plant to the management.

These checklists may be used to indicate compliance with standard procedures. As indicated above, a checklist is easy to use and can be applied to each stage of a project of plant development. A checklist is a convenient means of communicating the minimal acceptable level of hazard evaluation that is required for any job, regardless of scope. As such, it is particularly useful for an inexperienced engineer to work through the various requirements in the checklist to reach a satisfactory conclusion. However, a system checklist should be audited and updated regularly.

Interestingly, many companies use standard checklists for controlling the development of a project from initial design through plant shutdown. It serves a form for approval by various staff and management functions before a new or retrofit project can move ahead.

15.4 SAFETY REVIEW/SAFETY AUDIT

A Safety Review/Safety Audit essentially consists of a walk-through on-site inspection that can include interviews with plant staff to identify plant conditions or operating procedures that could lead to an accident. The walk-through on-site inspection can vary from an informal routine function that is principally visual, with emphasis on housekeeping, to a formal week-long examination by a team with appropriate backgrounds and responsibilities. The emphasis in this section is on the latter and it is sometimes referred to as a "Safety Review." It has also been referred to as a Process Safety Review, a Loss Prevention Review, or a Process Review. As described above, such a program is intended to identify plant conditions or operating procedures that could lead to an accident and significant losses in life or property.

While this qualitative technique is most commonly applied to operating process plants, it is also applicable to pilot plants, laboratories, storage facilities, or support functions. This comprehensive review is intended to complement other safety efforts and routine visual inspections. It should be treated as a cooperative effort to improve the overall safety and performance of the plant rather than as a dreaded interference with normal operations. Cooperation is essential. People are likely to become defensive unless considerable effort is made to present the review as a benefit to each participant.

The review includes interviews with many people in the plant: operators, maintenance staff, engineers, management, safety staff, and others, depending upon the plant organization. Having the support and involvement of all these groups provides a thorough examination from many perspectives.

The review looks for major risk situations. General housekeeping and personnel attitude are not the objectives, although they can be significant indicators of where to look for real problems or places where meaningful improvements are needed. Various hazard evaluation techniques, such as checklists (see previous Section), "what-if" questions (see Section 15.5), and raw material evaluations, can be also used during the review.

At the end of the Safety Review/Safety Audit, recommendations are made for specific actions that are needed with justifications, recommended responsibilities, and completion dates. A follow-up evaluation or re-inspection should be planned to verify the acceptability of the corrective action.

For a complete review, data requirements that the team will need include:

1. Access to applicable codes and standards
2. Detailed plant descriptions such as piping and instrumentation drawings and flowcharts
3. Plant procedures for start-up, shutdown, normal operation and emergencies
4. Personnel injury reports
5. Hazardous incidents reports

6. Maintenance records such as critical instrument checks, pressure relief valve tests, pressure vessel inspections
7. Process material characteristics (e.g., Toxicity and reactivity information)

15.5 "WHAT IF" ANALYSIS[1]

The main purpose of the "What If" qualitative method is to identify the hazards associated with a process by asking questions that start with "What if...." This method can be extremely useful if the design team conducting the examination is experienced and knowledgeable about the operation; if not, the results are usually incomplete. The examination usually starts at the point of input and continues in a "railroad" manner according to the flow of the process.[2,7]

The first step of a "What If" analysis is to define the study boundaries. There are two types of study boundaries to be considered: the consequence category boundary, which includes public risk, employee risk, and economic risk, and the physical boundary which addresses the section of the plant that should be considered for analysis.

The second stop is to obtain all the information about the process that will be needed for a thorough evaluation including but not limited to: the process materials used and their physical properties, the chemistry and thermodynamics of the process, a plant layout, and a description of all the equipment used including controls and instrumentation. The last part of the information gathering step may be viewed as the preliminary formation of the "What If" questions.

The third step is to select a review team. The team is usually composed of two or three members that have combined experience in the process to be studied, knowledge in the consequence field, and experience in general hazard evaluation. If the team is inexperienced, results may be incomplete or incorrect.

Once the team has been established, the review is conducted. Typically, the review begins with the process inputs and proceeds through the system to the outputs. Each of the "What If" questions is addressed by identifying the hazard and it's consequence, and then recommending solutions or alternatives to alleviate the risk.[2]

The final step in the "What If" analysis is reporting the results in a systematic and easily understood format. An example of a common format is provided in Table 15.5.1, which includes the question, their consequences, and recommendations. An ethylene polymerization process is used to demonstrate the format for a "What If" analysis.

Although the "What If" procedure is not as structured as some other event/hazard and identification studies, it is a powerful procedure if the staff members are experienced. Otherwise (and as indicated above), the results from the procedure may neither complete nor accurate.

TABLE 15.5.1 "What If" Analysis on the Ethylene Polymerization Reactor

What if...	Consequence/Hazard	Recommendation
1. Cooling water pump breaks down	Runaway reaction/ explosion/fire	Stand-by pump/alarm system
2. Too much oxygen fed into reactor	Runaway reaction/ explosion/fire/flying debris	Alarm system/feed flow control/initiator flow control
3. Wrong initiator	None likely	-------------------------
4. Valve after reactor gets clogged	Pressure buildup/explosion/ fire/flying debris	Feed flow control/initiator flow control/alarm system
5. Compressor breaks down	None likely	------------------------- --------
6. Trauma to cooling jacket	Runaway reaction/ explosion/fire/flying debris	Temperature alarm/feed flow control

15.6 PRELIMINARY HAZARD ANALYSIS (PHA)

A preliminary hazard analysis is an overall qualitative study that yields a rough assessment of the potential hazards; it may also include the means for their rectification within a system. It is called "preliminary" because it is usually refined through additional studies. PHA, which is part of the U.S. Military Standard System Safety Program, contains a brief description of potential hazards in system development, operation, or disposal. This method focuses special attention on sources of energy for the system and on hazardous materials that might adversely affect the system or environment. Resources necessary to conduct a PHA include plant design criteria, equipment, and material specifications.

The results of a PHA study can be summarized in the form of a table (see Table 15.6.1) or a logic diagram (such as could be illustrated by completing Figure 15.6.1). In either format, potential hazards that pose a high risk, along with their cause and major effects, are identified. In addition, for each hazard identified, preliminary means of control are also often prescribed in the analysis. Thus, a PHA is not performed only to develop a list of possible hazards; it also is used to identify those hazardous features of a system that can result in unacceptable risks and to assist in developing preventive measures in the from of engineering or administrative controls, or use of personal protective equipment.[2,3,8,9]

TABLE 15.6.1 Summary Table to Be Completed for Preliminary Hazard Analysis

Hazard	Cause	Major Effects	Corrective/Preventive Measures
----------------	------------------	--------------------	---------------------------
----------------	------------------	--------------------	---------------------------
----------------	------------------	--------------------	---------------------------
----------------	------------------	--------------------	---------------------------

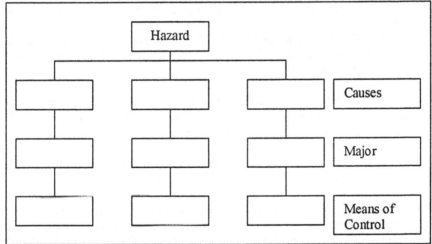

Figure 15.6.1.   Logic diagram to complete for Preliminary Hazard Analysis (PHA).

As indicated above, the PHA may serve as a precursor to further hazard analyses. It is included in this chapter because it can provide a cost effective, early-on plant method for hazard identification. As its title indicates, the PHA is really intended for use only in the preliminary phase of plant development for cases where past experience provides little or no insight into any potential safety problems, e.g., a new plant with a new process.

15.7   HAZARD AND OPERABILITY STUDY (HAZOP)

A hazard and operability study (HAZOP) is a systematic approach to recognizing and identifying possible hazards that may cause failure of a piece of equipment in new or existing facilities. This qualitative enterprise is conducted by a team of technical experts in plant design and operation, plus other experts, as required.

A HAZOP study may be applied to operating process plants, or it may be performed at various stages throughout the design. An early start will lead to a safer, more efficient design and, ultimately, higher profits.

Before any action is taken, the goals of the study should be defined. There are six objectives to a HAZOP study:

1. To identify areas of the design that may possess a significant hazard potential
2. To identify and study features of the design that influence the probability of a hazardous incident occurring
3. To familiarize the study team with the design information available
4. To ensure that a systematic study is made of the areas of significant hazard potential
5. To identify pertinent design information not currently available to the team
6. To provide a mechanism for feedback to the client of the study team's detailed comments

A relatively simple example of a HAZOP study using guidewords is shown for a boiler drum in Fig. 15.7.1 and Table 15.7.2. The intent of the operation is to maintain the water level in the horizontal drum between 30 and 40% of the volume.

Next, the method requires a determination of the plant limits (i.e., the areas of the plant that will be evaluated). Some hazards identified may be considered for on-site impacts. An experienced team of principal engineers, a HAZOP chairperson, and an external HAZOP expert is the recommended make-up of the foundation for the team. Other experts from other disciplines, such as instrumentation and process control, may be periodically called on to identify and evaluate deviations from normal operations.

Before any assessment can be performed, the team must be supplied with required documentation and process details. As with other hazard identification steps, the following materials are usually needed:

1. Process description
2. Process flow sheets
3. Data on the chemical, physical, and toxicological properties of all raw materials, intermediates, and products
4. Piping and instrumentation diagrams (P&IDs)
5. Equipment, piping, and instrument specifications
6. Process control logic diagrams
7. Layout drawings
8. Operating procedures
9. Maintenance procedures
10. Emergency response procedures
11. Safety and training manuals

TABLE 15.7.1   Guide Words Used in HAZOP Studies

Guide Words	Meaning	Examples
No or Not	No part of the intention is achieved, but nothing else happens.	No flow, no agitation, no reaction
More	Quantitative increases or	More flow, higher
Less	decreases to the intended activity.	pressure, lower temperature, less time
As well as	All of the intention is achieved, but some additional activity occurs.	Additional component, contaminant, extra phase
Part of	Only part of the intention is achieved; part is not.	Component omitted, part of multiple destinations omitted
Reverse	The opposite of the intention occurs.	Reverse flow, reverse order of addition
Other than	No part of the intention is achieved. Something different happens.	Wrong component, start-up, shutdown, utility failure

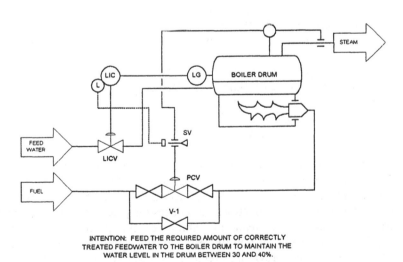

INTENTION: FEED THE REQUIRED AMOUNT OF CORRECTLY
TREATED FEEDWATER TO THE BOILER DRUM TO MAINTAIN THE
WATER LEVEL IN THE DRUM BETWEEN 30 AND 40%.

Figure 15.7.1.   Example of a HAZOP study.

Depending on the stage of the design being evaluated, some of this information may not be available.

Determining the method of assessment is the next step. The section of the process to be studied is first identified; generally the focus is on a major piece of equipment, although a pump or a valve may be chosen depending on the hazardous nature of the materials being handled and the operating conditions. Once the intended operation has been defined, a list of possible deviations from the intended operations is developed. The degree of deviation from normal operation are conveyed by the use of guide words, some of which are listed in Table 15.7.1. The purpose of these guide words is to develop the thought process and encourage discussion that is related to any potential deviations in the system. When a possible deviation is recognized, the possible causes and consequences are usually determined. Alterations and appropriate action to be taken are then recommended. Final steps in the methodology include issuing formal reports and following up recommendations.

The overall HAZOP method is summarized in the following steps:

1. Define objectives
2. Define plant limits
3. Appoint and train a team
4. Obtain complete preparative work
5. Conduct examination meetings to
   - select a manageable portion of the process
   - review the flow sheet and operating process
   - agree on how the process is intended to operate
   - state and record the intention
   - search for possible ways to deviate from the intention, using the guide words
   - determine possible causes for the deviation
   - determine possible consequences of the deviation
   - recommend action to be taken
6. Issue meeting reports
7. Follow up on recommendations

The HAZOP is a very useful technique, which may lead to a more reliable and safer process. Whether it is applied to preliminary design stages or to the detailed layout of an existing plant, it can lead to a better understanding of the process and possible malfunctions. It reduces the possibility of accidents for the process, improves on-stream availability of the process, and provides training for the evaluation of any process. Finally, HAZOP is also a way of optimizing a process and providing a reliable and cost-effective system.

TABLE 15.7.2 Guide Words for Fig. 15.7.1

Guide Word	Deviation	Cause	Consequence
NO	NO flow of feed water	LICV (Level Indicator Control Valve) closed	Loss of level in drum and explosion of drum by flame impingement on dry shell if flame continues
	NO level in drum	Loss of feed water pressure	Same as above
		Feed water flow stops	Same as above
		Massive leak from drum	Extinguish flame?
MORE	MORE level in drum than 40%	Level control fault	Excessive entrainment in stream
LESS	LESS level in drum than 30%	Similar to NO level	Loss of level in drum and explosion of drum by flame impingement on dry shell if flame continues
AS WELL AS	Contaminants AS WELL AS feed water	Water treatment fault	Fouling of boiler Corrosive steam
PART OF	PART OF feed water (treatment chemicals) omitted	Water treatment fault	Same as above
REVERSE	REVERSE flow in feed water line	Loss of feed water pressure	Steam in feed water system
OTHER THAN	Unplanned shutdown (OTHER THAN normal operation)	Utility failure	Does LICV fail closed or open?
	OTHER THAN feed water	Not possible	

15.8   ILLUSTRATIVE EXAMPLES

Example 15.1

Briefly describe what the term "system description" refers to relative to hazard risk assessment.

Solution 15.1

"System description" is the compilation of the process/plant information needed for the risk analysis. For example, site locations, environs, weather data, process flow diagrams (PFDs), piping an instrumentation diagrams (P&IDs), layout drawings, operating and maintenance procedures, technology documentation, process chemistry, and thermophysical property data may be required.

Example 15.2

The word "what" appears in numerous hazard identification procedures. List some questions/comments that are concerned with this term.

Solution 15.2

	Is	Is Not
What:	What do we know?	What don't we know?
	What was observed?	What was not observed?
	What is a related problem?	What is unrelated?
	What are the constraints?	What is not a constraint?
	What is expected?	What is unexpected?
	What is the same?	What is different?
	What is the importance?	What is not important?
	What resources are needed?	
	What are the criteria?	
	What is the purpose?	

Example 15.3

It is not unusual to find reaction flasks containing volatile solvents stored in research laboratory refrigerators. Consider the following: a 500 mL flask of diethyl ether (MW=74.14 g/gmol, specific gravity=0.713) is stored in a 15 ft^3, unventilated refrigerator at 41°F and 1 atm. (Note: the vapor pressure of diethyl ether at 41°F is 200 mm Hg).

a. How many grams of the diethyl ether must evaporate to achieve the minimum percent of ether vapors that can produce a flammable mixture? As defined earlier, this minimum percentage of vapors is the lower flammable limit (LFL) which for diethyl ether is 1.9% v/v.
b. How many grams of the ether must evaporate to reach the maximum percent of ether vapors that would still be flammable? This maximum percentage of vapor is was earlier defined as the upper flammable limit (UFL) which for diethyl ether is 35% v/v.

c. Comment on the hazards of such a storage practice and suggest safer alternatives.

Solution 15.3

a. The quantity of diethyl ether needed to reach a concentration of 1.9 vol% in the refrigerator may be found by first calculating the number of gmol of air in the refrigerator.

To find the gmol air in the refrigerator, use the Ideal Gas Law, PV=nRT, with P = 1.00 atm, T = 41 °F = 5 °C = 278 K

$$\text{Volume of the refrigerator} = \left(15.0\text{ft}^3\right)\left(0.0283\frac{m^3}{ft^3}\right) = 0.425m^3 = 425L$$

$$n = \frac{\left(0.0821\frac{atm-L}{gmol-K}\right)(278K)}{(1atm)(425L)} = 18.6 \text{ gmol air in the refrigerator}$$

To find the moles of diethyl ether needed to reach the 1.9% LFL, the gmol of air in the refrigerator is multiplied by 0.019:

(18.6 gmol air)(0.019) = 0.353 gmol diethyl ether

To convert to grams of diethyl ether, the following calculation is made:

(0.353 gmol diethyl ether)(74.1 g/gmol) = 26.2 g diethyl ether

The LFL is reached in the refrigerator when 26.2 g of diethyl ether in the flask have evaporated.

b. The quantity of diethyl ether needed to reach a concentration of 36 vol% in the refrigerator (its UFL) may be found by multiplying the gmol of air in the refrigerator by 0.36, and finally, converting gmol to g of diethyl ether.

The gmol air in the refrigerator = 18.6 gmol air

The number of gmol of diethyl ether needed to reach the 36% UFL is:

(18.6 gmol air)(0.36) = 6.70 gmol diethyl ether

To convert to g diethyl ether, the following calculation is made:

(6.70 gmol diethyl ether)(74.1 g/gmol) = 496 g diethyl ether

The mass of diethyl ether in the flask is determined as:

(500 mL diethyl ether)(0.713 g/mL) = 357 g diethyl ether in the flask

Since 496 g of diethyl ether represents more diethyl ether than is in the flask, the diethyl ether UFL cannot be reached in the refrigerator.

c. To better evaluate the chance of reaching the diethyl ether LFL in the refrigerator, it would be useful to find what percent of the diethyl ether in the flask would have to evaporate in order to produce this situation:

$$\frac{26.1\,\text{g diethyl ether evaporating}}{357\,\text{g diethyl ether in flask}} \times 100 = 7.3\%$$

Since a vapor pressure of 200 mm Hg is considerable, the chance of as little as 7.3% of the diethyl ether evaporating to reach the LFL is quite high. Those periods of time, such as weekends, when the refrigerator might not be opened, would result in situations of greatest risk.

As there is no possibility of exceeding the UFL, the hazard is not reduced by further evaporation. A storage practice such as this, which results in a significant chance of producing a flammable mixture of vapors within an enclosed space, is not advisable. An explosion proof ventilated refrigerator should be used, or the solvent should be changed/eliminated.

Example 15.4

Two large bottles of flammable solvent were ignited by an undetermined ignition source after being knocked over and broken by a janitor while cleaning a 10 ft x 10 ft x 10 ft research laboratory. The laboratory ventilator was shut off and the fire was extinguished with a 10 lb CO_2 fire extinguisher. As the burning solvent had covered much of the floor area, the fire extinguisher was completely emptied in putting the fire out. Will the concentration of the CO_2 in the room exceed lethal limits?

Solution 15.4

The mole fraction, or percent, of CO_2 in the room can be obtained through the following calculations:

$$\%CO_2 = \frac{gmol\ CO_2\ in\ 10\ lb\ extinguisher}{gmol\ air\ in\ room} \times 100$$

The number of gmol of CO_2 in the 10 lb fire extinguisher is:

$$\frac{(10\ lb\ CO_2)(454\ g/lb)}{44\ g/gmol\ CO_2} = 103\ gmol\ CO_2$$

The ideal gas law is utilized to determine the total number of moles of air in the room at 1 atm and 30°C (303 K):

Room volume = (10 ft) (10 ft) (10 ft) (0.0283 m3/ft3) = 28.3 m = 28,300 L

Thus,

$$n = \frac{(1\ atm)(28,300\ L)}{(0.0821\ atm\text{-}L/gmol\text{-}K)(303\ K)} = 1,137\ gmol\ air$$

Substituting into the above equation for $\%CO_2$ yields:

$$\%CO_2 = \frac{103\ gmol\ CO_2}{1,137\ gmol\ air} \times 100 - 9.06\%$$

Comparing this $\%CO_2$ composition with the IDLH and lethal levels shows that the IDLH is 50,000 ppm (5.0%), the percent of CO_2 in the room is 9.06%, while the lethal level is 10%.

The 9% CO_2 concentration in the room is very close to the lethal concentration (which is an IDLH of 10%) and a great danger exists to the person putting out the fire. Consideration must be given to providing and using a self-contained breathing apparatus in fighting chemical fires.

Example 15.5

The dynamic seal for a control valve suddenly starts leaking toluene-2, 4-diisocynate (TDI) vapor at a rate of 40 cm^3/h into a 12 ft x 12 ft x 8 ft room. The air in the room is uniformly mixed by a ceiling fan. The background TDI vapor concentration is 1.0 ppb. Air temperature and pressure are 77°F and 1 atm, respectively. Calculate the ppm value of leaking TDI vapor if its vapor pressure is 35 mm Hg at 25°C. Also, calculate the number of minutes after the leak starts that a person sleeping on the job would be at risk of being exposed to TDI vapor with respect to the STEL, ¼ LEL (lower exposure limit), and TLV.

TDI exposure limit values are as follows:

1. Short Term Exposure Limits (STEL) = 0.02 ppm
2. 25% of Lower Exposure Limit (¼ LEL) = 0.325 v/v
3. Threshold Limit Value (TLV) = 0.005 ppm

Solution 15.5

1. Calculate the maximum TDI vapor concentration in ppm based on its vapor pressure at the leak.

    Maximum vapor concentration = (35/760)) (1,000,000) = 46,100 ppm

2. Calculate the TDI vapor concentration (ppm) in the room as a function of time, t (min).

$$C\,(ppm) = 0.001 + \frac{(40)\left(\dfrac{1}{60}\right)\left(\dfrac{1}{1000}\right)(t)(46100)}{(12)(12)(8)(0.02823)(1000)}$$

    C= 0.001 + 0.95×10^{-3}t

    where t is the time in minutes measured from the time the leak starts.

3. Calculate the time, t, to reach the STEL of 0.02 ppm.

    Use the mean value theorem:

$$\int_{t1}^{t2} C\,dt = C_{avg}\,(t2 - t1)$$

    Note also that C_{avg} = 0.02 ppm, and (t2-t1) = 15 minutes (the range upon which the time-weighted average value is based).

Integrating the above reaction and solving yields:

$$\int_{t1}^{t2}\left(0.001+0.95\times10^{-5}\,t\right)dt = \left(0.001+0.95\times10^{-5}\right)\left(\frac{t^2}{2}\right)\Bigg|_{t1}^{t2} = C_{avg}\left(t_2 - t_1\right)$$

$(0.001)\,(t_2 - t_2) + (0.47 \times 10^{-5})\,(t_2{}^2 - t_1{}^2) = 0.05\,(15) = 0.30$
$(0.001)\,(15) + (0.47 \times 10^{-5})\,(t_2 - t_2)\,(t_2 + t_1) = 0.30$
$(0.001)\,(15) + (0.47 \times 10^{-5})\,(15)\,(t_2 + t_1) = 0.30$

Substitute $t_2 = 15 + t1$

$0.015 + (7.05 \times 10^{-5})\,(15 + 2t_1) = 0.30$
$(7.05 \times 10^{-5})\,(15 + 2t_1) = 0.285$
$(15 + 2t_1) = 4042.6$
$t_1 = 12.71$ minutes

The total time to reach the STEL of 0.02 ppm on a TWA basis is $t_1 + 15$, or $t_2 = 27.71$ minutes.

4. Calculate the time in minutes, t, to reach ¼ the Lel.

    $325,000$ ppm $= 0.001 + 0.94 \times 10^{-3} t$
    $t = 345,000,000$ minutes

5. Express the average concentration in terms of time.

    $C_{avg} = 0.001 + 0.001 + (0.94 \times 10^{-3} t/2)$

6. Calculate the average concentration after the TWA.

    For an eight hour averaging period, $t = 8h = 480$ minutes, and the mean TDI concentration from the equation is $C_{avg} = 0.227$ ppm. Therefore, the TLV is exceeded by the mean TDI vapor concentration in the room over the eight-hour averaging period by a factor of 45.4.

15.9 SUMMARY

1. Hazard or event identification provides information on possibly dangerous situations that can occur when dealing with chemicals. There are several different methods that are used in hazard or event identification.
2. In order to properly apply the system safety techniques to the design and operation of potentially hazardous technologies, a series of questions regarding safety must be asked and answered.

3. In order to determine what safety questions must be asked, a system checklist that can be used to identify possible safety issues should be created. There are a number of basic issues that can be included on the list, including flammability, toxicity, storage and reactivity.
4. A Safety Review/Safety Audit is a qualitative technique most commonly applied to operating process plants and includes a walk-through on-site inspection of the plant. This review looks for major risk situations and makes recommendations.
5. The "What If" is another qualitative analysis procedure and is conducted by an experienced review team that poses questions that start with "What if…"
6. A preliminary hazard analysis is an overall qualitative study that yields a rough assessment of the potential hazards of a system, and can provide possible solutions.
7. Another qualitative approach to recognizing and identifying possible hazards that may cause problems is a hazard and operability study (HAZOP). By identifying the goals of the study and by using guide words, a comprehensive report on the safety of a system can be created.

PROBLEMS

1. Briefly discuss why event/hazard identification is a critical step in any hazard risk assessment process.
2. The words "when" and "where" appear in numerous hazard identification procedures. List some questions/comments that are concerned with these two terms.
3. The words "who" and "why" and "how" appear in numerous hazard identification procedures. List some questions/comments that are concerned with these two terms.
4. A reagent bottle containing 1 kg of benzene accidentally falls to the floor of a small enclosed storage area measuring 10 m^3 in volume. In time, the spilled liquid benzene completely evaporates. With this information answer the following questions:
    a. If 5% of the volume of benzene vapor remains trapped in the storage room, what is the benzene concentration in mg/m^3?
    b. Using the answer from part a, what is the exposure duration, ED?
5. List and discuss the seven key guide words that are employed in a HAZOP study.

REFERENCES

1. Adapted from: H. Karraman, J.K. Rao and F.V. Bowr, "Application of Hazard Evaluation Techniques to the Design of Potential Hazardous Identification in Chemical Processes," NIOSH, Cincinnati, Ohio, 1992.
2. "Guidelines for Hazard Evaluation Procedures," CCPS, AIChE, New York, 1992.
3. E.J. Henley, and H. Kamamoto, "Reliability Engineering and Risk Assessment", 1st ed., Prentice Hall, Englewood Cliffs, NJ, 1981.
4. N.B. Hanes, and A.M. Rossignol, "Comprehensive Occupational Safety and Health Engineering Academic Program Development Strategy," U.S. Department of Health and Human Services, Springfield, VA: Nat. Tech. Info. PB 86-226453, 1984.
5. H.E. Roland, and B. Moriarty, "System Safety Engineering and Management", 1st ed., John Wiley & Sons, New York City, NY, 1983.
6. Adapted from: "Guidelines for Hazard Evaluation Procedures," CCPS, AIChE, New York City, NY, 1992.
7. M. Dollah-Kanan, Z.H. Mustaffa, and Z. Abidin, "Safety System Management for Design of Hazardous Technologies," School of Engineering, California State University, Long Beach, CA, 1988.
8. G.A. Page, "Hazard Evaluation Procedures," American Society of Safety Engineers, Professional Development Conference and Exposition, Las Vegas, NV. American Society of Safety Engineers, Des Plaines, IL, 1988.
9. H.R. Kavianian, R. Orr, R. Arbuckle, and A. Edwards, "Hazard Analysis and Safety Management of a Radioactive Gas Handlin Process," School of Engineering, California State University, Long Beach, CA, 1988.

16

Accident Causes and Probability

16.1 INTRODUCTION

There are a host of reasons why accidents occur in industry. The primary causes are mechanical failure; operational error (human error), process upset, and design error. Figure 16.1.1 provides the relative number of accidents that have occurred in the petrochemical field.[1]

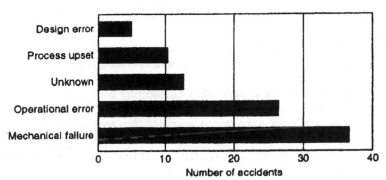

Figure 16.1.1. Causes of accidents in the petrochemical field.

There are three steps that normally lead to an accident.

1. Initiation
2. Propagation
3. Termination

The authors gratefully acknowledge the assistance of Felipe Martin in researching, reviewing, and editing this chapter.

The path that an accident takes through the above three steps can be determined by means of a fault tree analysis.[2]

System safety approaches have been employed in an attempt to understand the root cause of an accident. This is an extension of the safety review and safety audit material presented in the previous chapter. The system safety approach begins by defining a system and focusing on how accidents can occur within that system as a result of equipment failure, human error, environmental conditions, or a combination of these. The preventive measures to mitigate the hazards include design equipment, or development of procedural safeguards. Early identification, analysis, control, or elimination of the hazards in a process can eliminate the need for major design changes later in the project. System safety provides a thorough and systematic approach with which to address workplace hazards.

The methods used in system safety engineering are among the most effective and advanced methods to prevent system failures that result in accidents. Although system safety engineering is a relatively new field, it has been used extensively by the military and the aerospace and nuclear industries to improve the safety of highly complex systems. This approach is based on the concept that:

1. Accidents within a system are the result of a number of interacting causes
2. Each cause for an accident can be identified and analyzed in a logical manner
3. Control measures for the cause of each accident can be developed in terms of equipment, instrumentation, and/or standard operating procedures

Through logical application of system safety engineering concepts, an optimum degree of safety can be achieved throughout a system's life cycle. These techniques provide not only a unique opportunity for analysis of all safety aspects of a problem but a valuable tool for communicating safety information to management for designing and operating existing and new facilities, including retrofits.[3]

These concepts are based on the idea that an optimum degree of safety can be achieved within the constraints of system effectiveness. This optimum is attained through a logical reasoning process. Accidents, or potential accidents, are considered to be the result of a number of interacting causes within the system. Each accident cause and interaction is logically identified and evaluated, from a careful analysis of the system.

This chapter addresses many of the above issues. Included in the material to follow are:

Section 16.5   Other Causes
Section 16.6   Fault Tree Analysis (FTA)

16.2   CAUSES OF ACCIDENTS

Today the chemical industry is involved in a broad spectrum of manufacturing processes that range from biological preparations to plastics and explosives. Although the basic plans and designs for these processes may be similar, each individual plant will have its own unique set of potential hazards.

Basic chemical processes that are common in industry include acylation, alkaline fusion, alkylation, amination, aromatization, calcination, carboxylation, causticization, combustion, condensation, coupling, cracking, diazotation, electrolysis, esterification, fermentation, halogenation, hydroforming, hydrolysis, isomerization, neutralization, nitration, nitrosation, oxidation, polymerization, pyrolysis, reduction, and thermal decomposition (4). Some of these processes are considered to be more hazardous than others. Alkylation, amination, aromatization, combustion, condensation, diazotization, halogenation, nitration, oxidation, and polymerization are examples of the more hazardous processes. However, any process that exhibits one or more of the following characteristics should be considered to be extra hazardous:

1. The process is subject to explosive reaction or detonation under normal conditions
2. The process is subject to explosive reaction or detonation when exposed to shock or abnormally high temperatures or pressures
3. The process reacts violently with water
4. The process is subject to spontaneous polymerization
5. The process is subject to spontaneous heating
6. The process is subject to exothermic reactions with the development of excessive temperatures and pressures
7. The process normally operates at very high pressures or temperatures and may result in the massive release of flammable gases or vapors
8. The process operates in or near the explosive range of the reactants or products
9. The process is subject to a dust or mist explosion hazard
10. The process emits environmental contaminants or pollutants into the atmosphere
11. The process uses or produces toxic substances
12. The process uses or produces very corrosive materials
13. The process emits dangerous radiation
14. The process presents waste disposal problems

Deviations from normal process conditions, as manifested by the following circumstances, must be well understood if accidents are to be prevented:

1. Abnormal temperatures
2. Abnormal pressures
3. Material flow stoppage
4. Equipment leaks or spills
5. Failure of equipment

Chemical processing under "extreme conditions" of high temperatures and pressures requires more thorough analysis and extra safeguards. As discussed in Chapter 7, explosions at higher initial temperatures and pressures are much more severe. Therefore, chemical processes under extreme conditions require specialized equipment design and fabrication. Other factors that should be considered when evaluating a chemical process are rate and order of the reaction, stability of the reaction, and the health hazards of the raw materials used.

Under such circumstances, there is a need for high standards in equipment design, operation, and maintenance. Regardless of adequate safeguards and controls of highly technological processes, accidents do and will continue to occur. It is therefore important to examine the causes of such accidents associated with the specific pieces of equipment, supporting systems, and materials being handled. Generally, the sequence of events resulting in an accident can be traced back to one or a combination of the following causes:

1. Equipment failure
2. Control system failure
3. Utilities and ancillary equipment outage
4. Human error
5. Fire exposure/explosions
6. Natural causes
7. Plant layout

These causes are treated in the sections that follow.

16.3 EQUIPMENT FAILURE

The equipment of a processing system is designed according to process conditions with a view to containing the chemical(s) and maintaining the control parameters required to produce the desired product. Equipment failure can generally be attributed to one or more of the following hazards:

1. Process design limits exceeded
2. Defective fabrication

3. Corrosion and erosion failures
4. Material stress or fatigue

5. Poor maintenance/repair
6. Inability to contain toxic and/or hazardous materials

Most equipment failures occur under abnormal conditions, especially elevated pressures and temperatures. The design of equipment presents internal and external constraints. External limits may arise from physical laws, while internal limits may depend on the process and materials. In any case, if these limits are exceeded, the chance of an accident is greatly increased.

It is impossible to quantify the possible hazards associated with each piece of equipment. To supplement the brief introduction to process equipment presented in Chapter 5, possible failures and hazards associated with typical equipment are presented below.

Reactors

The reactor is often the most vital piece of equipment in a process. It is in this vessel that the raw material is converted into products. Usually the temperature controls the reaction. Depending on the kinetics, the reaction may require catalysts and very high temperatures. Exceeding temperature limits may result in undesirable side reactions, which produce hazardous substances and metallic creep failure of the vessels. The vessels are designed to accommodate endothermic or exothermic reactions and are usually equipped with heating or cooling jackets. A hazard associated with an exothermic reactor is a *runaway reaction*. In this type of accident, as illustrated in Fig. 16.3.1, heat is not adequately removed from the system and excessive temperature and/or pressure buildup result. An example of a hazard associated with catalytic reactions is the deactivation or composition change of the catalyst, either with age or because of contaminants from nonselective reactions. This could produce hazardous materials and cause a dramatic change in temperature and/or pressure.

Heat Exchangers

The function of the *heat exchanger* is to transfer heat from one process (of utility) stream to another. The desired heat exchange operation determines the design of the equipment. Heat transfer equipment includes heaters, coolers, condensers, evaporators, reboilers, and steam generators. The term "fired" exchanger refers to units in which process fluids are heated by combustion gases. There are many types of potential hazard, and these depend on the nature of the heat transfer (conduction, convection, radiation, boiling and condensation), the physical properties of the process fluids, the fluid flow rates, and the surface heat transfer. Regardless of the type of exchanger used, most process and service fluids will cause fouling on the heat transfer surface. The

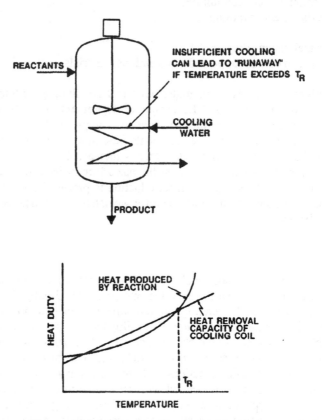

Figure 16.3.1 Exothermic chemical reaction.

accumulation of deposited materials effectively insulates the surface, thereby reducing heat transfer. If heat transfer is substantially reduced, however, tube rupture may result. Tube rupture may also occur from tube vibrations, corrosion, or erosion. Flow-induced tube vibrations attributable to the shell-side fluid flow can cause failure. For example, high fluid velocities on the shell side and inadequate supports can cause changes in the pressure distribution resulting in oscillations of the tube, which lead to rupture. Corrosion from moisture at pipe connections, leakage around baffles, and poor fabrication techniques leading to mechanical and thermal fatigue are common causes of failure. Corrosion of and leakage through internal joints can also lead to intermixing of process streams and utility fluids.

Vessels

Pressure vessels and tanks may be used for reactions, storage of raw materials, containment of chemical intermediates or products, and phase separations. The design of a vessel depends on the conditions necessary for handling the materials inside the vessel (pressure, temperature, phase, reactivity, corrosiveness, etc.). Pressure vessels may have internal components such as trays in a fractionation column or separation device. Reports of pressure vessel failures in the literature have been very rare (5). The possible causes of failure are corrosion, brittle fracture, creep, and mechanical and thermal fatigue. Leaks may occur from overfilling a tank, which can lead to a major failure if adequate drainage is not provided. Low-pressure vessels are easily damaged because they are more fragile. Also, a vessel may collapse if it is not designed to withstand vacuum when drained. Exceeding pressure limits is an obvious cause of failure that can result in rupture of the vessel. Overpressure protective equipment, such as relief valves and rupture disks, usually is installed as a safeguard. These devices were discussed earlier in chapter 5.

Storage tanks are simpler to operate than pressure vessels and present fewer potential hazards. This is because tanks usually are not operated under a wide range of pressure, temperature, and fluid characteristics. Storage tanks are subject to construction specifications on metal types, wall thickness, insulation, welding, supports, and other variables, depending on the material being stored. Cryogenic substances, hazardous chemical substances, and gases under extreme conditions determine the fabrication and support systems needed to prevent corrosion and mechanical-thermal stresses that could lead to vessel rupture. Low cycle fatigue, which arises from the effect of repeated temperature and pressure cycles, could cause disintegration of a vessel.

Phase separators basically follow the same design conditions. If corrosion due to incompatible materials should damage internal baffles, poor phase separation and thus composition change could result; this could cause an explosion. Figure 16.3.2 shows a failure of a type that is possible when baffle leakage results in a water-contaminated organic stream. Further processing (e.g., heating) of these streams could result in serious consequences.

Mass Transfer Unit Operations

Adsorbers, distillation columns, and packed towers are more complicated vessels and as a result, the potential exists for more serious hazards. These vessels are subject to the same potential hazards discussed previously in relation to leaks, corrosion, and stress. However, these separation columns contain a wide variety of internals or separation devices. Adsorbers or strippers usually contain packing, packing supports, liquid distributors, hold-down plates, and weirs. Depending on the physical and chemical properties of the fluids being passed through the tower, potential hazards may result if incompatible materials are used for the internals. Reactivity with the metals used may cause undesirable reactions, which may lead to elevated temperatures and pressures and, ultimately, to vessel rupture. Distillation columns may contain internals such as sieve trays, bubble caps, and valve plates, which are also in contact with the

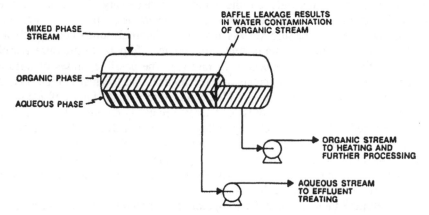

Figure 16.3.2. Possible failure mode for a phase separation vessel.

fluids. Their operation is based on thermal separation, which dictates specific operating requirements. Evaporation and extraction are also common unit operations involving gas and liquid separation. The basic principle of safety for all these operations is the containment of the process streams in the event of failure.

Materials of Construction

As described above, poor fabrication, wrong specifications, or incompatible materials cause many hazards and accidents related to equipment breakdown. In most cases, an equipment failure occurs when it is not expected. Before equipment is designed, experts should quantify the mechanical and chemical limits of the designated materials of construction. Obviously, a transient fluid and a broad operating range will lower the life of the equipment material. Usually a leak or pitting is an indication of corrosion. Depending on the properties of the material (metal, plastic, rubber, wood, etc.), different conditions may arise that ultimately will result in equipment failure.

　　　Determining the materials of construction is a critical design step that often is not given the attention it warrants. The materials for specific items such as mechanical seal parts, valve packing, seal rings, O-rings, and fasteners, as well as manufacturing contaminants such as oils and release agents, are very important. These items, which usually are not covered by standards and codes, vary with different equipment suppliers. Where special materials of construction are required, it is also necessary to pay close attention to fabrication details and procedures.

　　　To make a proper process evaluation or decision, it is important to have a detailed analysis of all anticipated process streams. Reliable data are available on recommended materials of construction for standard strength solutions of the more common liquid chemicals. Frequently, however, it is a component present in trace quantities that causes the greatest problem. Examples are the corrosion of steel in closed water systems containing oxygen, or the stress corrosion of austenitic stainless steel that is exposed to a steam phase containing chlorides.

Consideration must be given to possible equipment corrosion from such external sources as a corrosive atmosphere, spills, insulation, or gland leakage. For example, insulation containing trace quantities of chlorides can cause stress corrosion failure of 18-8 stainless steel vessels and piping.[6]

Control Systems

Automation and instrumentation are critical to the safe control of processes. Suitable measurement devices and control of system variables should be provided for normal operating conditions as well as for emergencies. There are six major components to a control system:

1. Sensing and measurement
2. Controller
3. Final control elements
4. Switches and alarms
5. Emergency shutdown and interlock systems
6. Computer control

Utilities

Failure of the utilities and ancillary systems occurs when one or more of the following is lost: electric power, cooling water or other heat removal systems, steam or other heat supply systems, fuel, air, inert gas, or effluent disposal facilities.

Generally, electricity is purchased from a local supplier. Power is required to generate motor drives for lighting and for general uses. Note that, in the typical distillation operation illustrated in Fig. 16.3.3, a power outage for the entire system would cause a less severe effect than failure of a reflux pump alone.

Cooling water required on site often is stored in towers; storage tank problems or piping and valve malfunctions could cause loss of this component. If seawater is used, materials of construction must be more resistant to salt. Loss of steam purchased or generated in water tube boilers could result from boiler tube failure, turbine failure, or piping or valve malfunction.

Fuel system outages may result for a variety of reasons, including running out of fuel or equipment failure.

Generally, rotary and reciprocating single-stage or two-stage compressors generate compressed air for the operation of the pneumatic controllers that may be used in chemical process plants. Compressor malfunctioning or air that is not clean (oil-free) and dry may cause loss of the system.

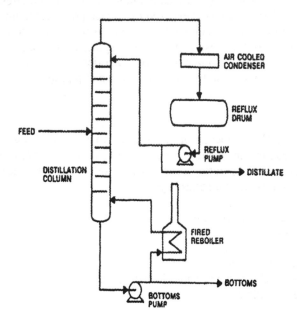

Figure 16.3.3. Utilities outage.

Inert gases such as nitrogen are used for purging and blanketing; hence, large quantities of these gases may be needed. Loss of the system could cause a major hazard if the blanketing effect were not available when an emission or leak occurred.

Ancillary Equipment

Most plant sites require waste disposal facilities. Some aqueous solutions may be disposed in public sewers or nearby bodies of waters, as permitted by the local, state, and federal laws. Flammable materials may be disposed of by burning in incinerators. Nonflammable liquid and solid waste can be dumped with approval. Toxic solid wastes must be chemically treated or drummed in tanks. Gaseous effluents containing toxic or noxious substances must be chemically treated before being discharged into the atmosphere. Often gases may be released in stacks tall enough to dilute and disperse the effluent. Scrubbers and adsorbers may be employed to remove pollutants before disposal. If one or more of these disposal systems should fail, the degree of the potential hazard will depend on the nature of the waste and the degree of dispersion into the environment.

Piping, Valves, and Pumps

Davenport has listed more than 60 major leaks of flammable materials, most, of which resulted in serious fires or *unconfined vapor cloud explosions* (UVCEs).[4] Table 16.3.1 classifies the leaks by point of origin and shows that if transport containers are excluded, pipe failures accounted for more than half the accidents. The biggest cause of these failures has been shown to be poor construction due to use of wrong specifications or failure to follow specifications established.

Piping is the most important transport facility in an industrial plant. In general, piping failures can result from

1. Insufficient or ineffective supports and neglected vibration effects
2. Poor weld quality
3. Temperature stresses
4. Overpressure
5. Inappropriate materials of construction

Typical pipe and valve failures and the reasons for their occurrence are discussed by Kletz.[5] The paragraphs that follow give some examples.

Many pipe failures are attributed to dead ends, where water, present in trace amounts in many oil streams, collects and freezes, resulting in pipe rupture. Corrosive materials may also dissolve in this stagnant water. For example, a dead-end branch 12 inches in diameter and 3 meters long was transporting natural gas at a gauge pressure of 550 psi. Water and impurities collected in the dead end, causing corrosion and ultimately failure. The escaping gas ignited at once, killing three men who were looking for the leak.

There are dead ends besides pipes that have been closed off. Valve branches, which are rarely used, are just as dangerous. As an example, the feed line to a furnace (Fig. 16.3.4) was provided with a permanent steam connection for use during decoking. The connection was on the bottom of the feed line and the steam valve was some distance away. Water collected above the steam valve, froze during cold weather, and ruptured the line, allowing oil at a gauge pressure of 450 psi to escape. If dead ends cannot be avoided, they should be connected to the top of the main pipeline. In another example, an unusual and unnecessary dead end was a length of 2-inch pipe welded onto a process line to provide a support for an instrument (Fig. 16.3.5). Water collected in the support, and 4 years after installation the process line corroded through, causing a liquefied gas leak.

TABLE 16.3.1   Origin of Leaks Causing Vapor Cloud Explosions[a]

Origin of Leak	Number of Incidents	Notes
Transport container	10	Includes 1 zeppelin
Pipeline (including valve, flange, sight-glass, etc.)	34	Includes 1 sight-glass and 2 hoses
Pump	2	
Vessel	5	Includes 1 internal explosion, 1 stopover, and 1 failure due to overheating
Relief valve or vent	8	
Drain valve	4	
Error during maintenance	2	
Unknown	2	
Total	67	

[a] From Davenport.[4]

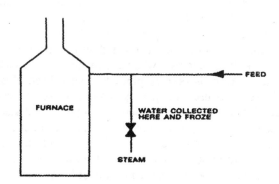

Figure 16.3.4.   Dead end formed at a steam connection.

Another serious failure occurred when water in a dead end suddenly vaporized. A heavy oil was dried by heating it to 120°C in a tank filled with steam coils. The oil was circulated while it was being dried. The suction line projected into the conical base of the tank, forming a dead end (see Fig. 16.3.6). As long as the circulation pump was kept running, water could not settle out in the dead end. The pump was turned off, however, and, as the water collected, it became warmed by the heating oil. The water soon reached boiling, vaporized, and exploded violently. The oil caught fire, the tank burst, and five lives were lost.

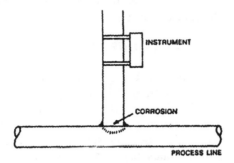

Figure 16.3.5. Corrosion resulting from collected water.

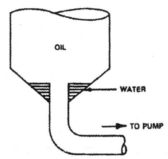

Figure 16.3.6. Water collected in a dead end.

Pipes, particularly small-bore pipes, often have failed by fatigue because their support was insufficient and the pipes were free to vibrate. On other occasions, pipes failed because their support was too rigid and they were not free to expand. Supports for pipes usually are constructed on site, and often it is not apparent until start-up that the supports are inadequate. Even at this point, the necessary reinforcement may not be done because it is not considered a priority at start-up.

The biggest hazard associated with pumps is failure of the gland, which may lead to a massive leak of flammable, toxic, or corrosive substances. Another common cause of pump accidents is *deadheading*, a condition in which the pump is allowed to run against a closed delivery valve. As a result, the temperature rises, seals are damaged, and fluid leaks out. Deadheading has caused explosions when the materials being pumped decomposed at high temperatures. Pumps fitted with an AutoSort will deadhead if they start when they should not. Such pumps should be fitted with a relief valve or one of the other protective devices discussed in Section 5.8. Many bearing failures and leaks have occurred as the result of lack of lubrication, or because of water in the lubricating oil. Pumps require maintenance more than any other piece of equipment. Many accidents have occurred because pumps under repair were not properly isolated from the rest of the plant.

16.4 MECHANICAL OPERATIONS

The mixing and separation of highly reactive chemicals presents many hazardous situations. These operations include spraying (misting or fogging), dispersing (diffusing), filtering, centrifuging, mixing, agitating, separating (layering and flotation), classifying (screening), and precipitating. Mechanical operations involve combinations of gases, solids, and liquids. Spraying, misting, or fogging of combustible liquids may create serious fire or explosion hazards. Proper mixing and agitation are essential to chemical operations because they assure uniformity and control of reactions.

Size reduction and enlargement operations include grinding, crushing, milling, machining, compacting, extruding, screening, molding, and dust collecting. A solid material that is reduced to small particles often becomes combustible and under certain conditions explosive (see Chapter 7). If the dust is permitted to disperse into the atmosphere, an explosion may occur.

The proper handling of chemicals in the process is critical for accident prevention. The hazard potential of an operation depends on the nature and physical state of the material. For example, accidents may result from

1. Vaporization and diffusion of flammable or toxic liquids and gases
2. Spraying, misting, or fogging of flammable, combustible, or toxic liquids
3. Dusting and dispersion of combustible materials and strong oxidizing agents
4. Mixing combustible and strong oxidizing agents
5. Temperature or pressure increases affecting unstable chemicals
6. Exposing unstable chemicals to friction or shock
7. Separation of hazardous chemicals from protective inerts or diluents

16.5 OTHER CAUSES

Human Error

Among the phenomena that lead to accidents, human error is the most unpredictable. This section describes accidents due to errors of judgment. Even well trained people occasionally make such errors, and one must either accept an occasional mistake or change the work situation to minimize or remove the opportunities for error.

Ironically, many errors occur because a person is well trained. Routine operations are controlled by lower levels of the brain and are not continuously monitored by the conscious mind. When the normal pattern of action is interrupted, an error is more likely to occur.

Some accidents, however, have occurred because employees were not adequately trained or lacked basic knowledge of scientific principles. Errors also occasionally occur because people deliberately fail to follow instructions. For example, workers sometimes neglect to wear protective clothing or to take other obvious precautions. Typical operator errors include the inadvertent opening or closing of a valve, the badly timed stopping or starting of rotating equipment, and the improper adjustment of a control loop set point.

The examples of accidents caused by human error discussed below are described in detail by Kletz .[5]

1. *Wrong Valve Opened.* A pump feeding an oil stream to the tubes of a furnace failed. The operator closed the oil valve and intended to open a steam valve to purge the furnace tubes. The wrong valve was opened, however, prohibiting flow to the furnace. The tubes overheated and collapsed. An investigation later showed that access to the steam valve was poor; moreover, there was no indication in the control room that flow through the furnace coils was not occurring, and there was no low-flow alarm or trip system on the furnace.

2. *Abnormal Reading Unnoticed on Panel Instrument.* A reactor starting up was being filled with the reaction mixture from another reactor already in service. As the operator added fresh feed and gradually increased its flow, he failed to notice that there was no indication that the temperature level was rising. The temperature actually rose, but the recorder did not show this change. The operator failed to question why the normally expected temperature rise was not occurring or was occurring without being registered. Thus the unnoticed recorder failure resulted in a runaway reaction.

3. *Wrong Valve Closed.* Figure 16.5.1 schematically shows a portion of a plant containing five reactors in parallel, plus two gas feed lines with cross connections between them. Oxygen lines (not shown) also led to the reactor. At the time of the accident, reactors 1 and 4 were on-line. Wrongly believing that valve B was open, an operator shut valve A. This stopped the gas flow to reactor 1. Although a ratio controller controlled the oxygen flow, a small amount continued to flow. Upon realization of his mistake, the operator restored the gas flow to the reactor, now containing an excess of oxygen. As a result, an explosion occurred downstream in a waste heat boiler and four lives were lost. This example shows how a simple mistake can have severe consequences. However, poor design and lack of protective equipment were more at fault in this case; in particular, the reactor design should have included relief valves. As in examples 1 and 3, accidental operation of a valve should not have been allowed to result in overpressure (explosion) or a runaway reaction.

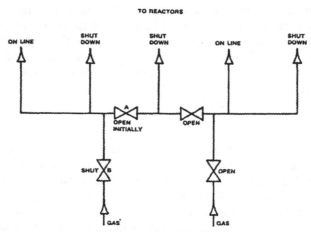

Figure 16.5.1. Schematic for accidental closing of a valve.

4. *Errors in Diagnosis.* The incident at Three Mile Island in 1979 (see Chapter 1) provides an example of error in diagnosis. Although two instrument readings indicated a high level in the primary water circuit, operators ignored other readings that were indicating a low level.

These examples illustrate incidents that led to accidents. Ignorance of readings and alarms, poor instructions, lack of proper training, poor design, and ignorance of basic scientific principles are often factors behind human error.

Human error analysis (HEA) is a systematic evaluation of the factors that influence the performance of human operators, maintenance staff, technicians, and other personnel in the plant. HEA involves the evaluation of one of several types of task analysis, which is a method for describing the physical and environmental characteristics of a task along with the skills, knowledge, and capabilities required of those who perform the task. This type of analysis can identify error-likely situations that can cause or lead to an accident.

Fire Exposure/Explosions

Many accidents result from fire exposure and explosions. This topic is covered in detail in Chapter 7.

Natural Causes

Accidents due to naturally occurring conditions resulting from the structure of the land or from the ravages of weather were reviewed briefly in Chapter 5. Outdoor processing, common in industries using hazardous chemicals, increases the hazards due to electrical storms (lightning), floods, hurricanes, or freezing temperatures. Earthquakes also should be considered. For example, if a piece of equipment is situated on a naturally occurring fault in the earth's crust, a tectonic shift could cause serious damage and hazards. Also, the topography must be studied; to be sure that adequate drainage is provided.

Plant Site Layout and Location

Most of the occurrences mentioned above maybe related to the location of the plant. Natural hazards, waste disposal hazards, and avoidance of congested areas, particularly those with inadequate emergency facilities, may eliminate or minimize the potential hazards of the plant itself. It is important to set an efficient and appropriate environment for the plant site, as was discussed in Chapter 5.

Inadequate plant layout and spacing may lead to congested process and storage areas, lack of isolation for extra hazardous operations and areas, sources of ignition that are too close to hazards, and lack of proper emergency facilities.

16.6 FAULT TREE ANALYSIS (FTA)

Generally a fault tree may be viewed as a diagram that shows the path that a specific accident takes. Fault tree analysis (FTA) begins with the ultimate consequence and works backward to the possible causes and failures. It is based on the most likely or most credible events that lead to the accident. FTA demonstrates the mitigating or reducing effects and can include causes stemming from human error as well as equipment failure. The task of constructing a fault tree is tedious and requires a probability background to handle common mode failures, dependent events, and time constraints. Chapters 20 and 21 discuss FTA in more detail.

Fault tree analysis seeks to relate the occurrence of an undesired event, the "top event," to one or more antecedent events, called "basic events." The top event may be, and usually is, related to the basic events via certain intermediate events. A fault tree diagram exhibits the casual chain linking the basic events to the intermediate events and the latter to the top event. In this chain, the logical connection between events is indicated by so-called "logic gates." The principal logic gates are the AND gate, symbolized on the fault tree by AND, and the OR gate symbolized by OR.

The fault tree symbols and their definitions are presented in Table 16.6.1. The construction of the fault tree for a tank overflow example is demonstrated in Figure 16.6.1.

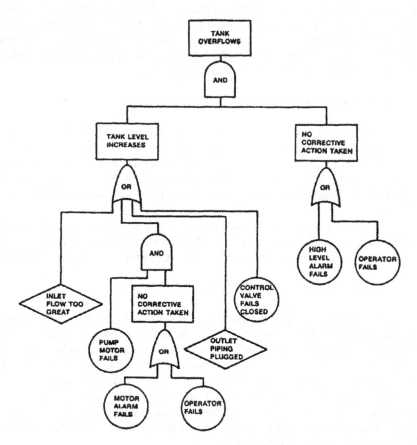

Figure 16.6.1. Example fault tree.

For another example of a fault tree, consider a water pumping system consisting of two pumps, A and B, where A is the pump ordinarily operating and B is a standby unit that automatically takes over if A fails. A control valve in both cases regulates flow of water through the pump. Suppose that the top event is no water flow, resulting from the following basic events: failure of a pump A *and* failure of pump B, *or* failure of the control valve. The fault tree diagram for this system is shown in Figure 16.6.2.

TABLE 16.6.1 Fault Tree Symbols[a]

Basic Event	Standard Usage: Basic initiating fault requiring no further development
	Modified ADL Usage: Represents initiating event and therefore has a yearly rate of occurrence
Undeveloped Event	Standard Usage: Event which is not developed any further as it is not required or data is unavailable
	Modified ADL Usage: Represents contributing events having taken place
External Event	Standard Usage: Event normally expected to occur
	Modified ADL Usage: Not used as even events normally expected to occur can lead to an undesired outcome and data may not be any more accurate than for any other type of event
Intermediate Event	Standard and ADL Usage: Intermediate level event caused by more primary events developed below
And Gate	Standard and ADL Usage: Logic gate where output fault occurs only if all input faults/events occur
Or Gate	Standard and ADL Usage: Logic gate where output fault occurs if at least one of the input faults/events occurs

[a]ADL=Alternate Digital Logic

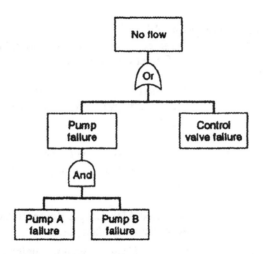

Figure 16.6.2. Fault tree diagram for a water pumping system consisting of two pumps (A and B).

16.7 ILLUSTRATIVE EXAMPLES

Example 16.1

List and briefly discuss human errors of a plant nature that lead to or cause accidents.

Solution 16.1

Some examples of human error of a plant nature are:

- Design error: Improper design for plant specific plant usage
- Construction mistakes: Problems caused by poor manufacturing
- Procedural mistakes: Not following proper steps
- Management error: Not paying attention to worker progress
- Maintenance: Leaving a tool in a reactor after maintenance

Example 16.2

Consider the release of a toxic gas from a storage tank. List and discuss possible causes for the release.

Solution 16.2

Some possible causes for a toxic gas release from a storage tank are:

- Rupture in storage tank
- Fire in tank farm; explosion of storage tank
- Collapse of tank due to earthquake
- Rupture in main line
- Leak in line or from tank

Example 16.3

Briefly discuss staffing and cost requirements for a FTA:

Solution 16.3

Normally, a single person can do the analysis. The analyst should create a fault tree based on consultations with operators and engineers along with other personnel who are familiar with the equipment/systems that are being analyzed. By having one person creating the fault tree, there is continuity in the analysis. The analyst must have full and complete access to all figures needed to define faults and failures, which contribute to the accident. If many fault trees are to be created, a team approach may be implemented. Here each member of the team should concentrate on a single fault tree. Members of the team must work together along with personnel to thoroughly complete the analysis.

The complexity of the system directly affects the time and cost requirements for the fault tree analysis. The larger the modeling processes the longer the time needed to determine a resolution of the analysis. Complex systems mean many potential accident events and larger problems.

Example 16.4

Briefly discuss the two primary sources for estimates of accident frequencies.

Solution 16.4

Sources include historical records and the application of fault tree analysis and related techniques, and they are not necessarily applied independently. For example specific historical data can sometimes be usefully applied as a check on frequency estimates of various sub events of a fault tree.

The use of historical data provides the most straightforward approach to the generation of incident frequency estimates but is subject to the applicability and the adequacy of the records. Care should be exercised in extracting data from long periods of the historical record over which design or operating standards or measurement criteria may have changed.

Example 16.5

Consider the following fault tree for a flammable storage tank fire. Identify whether the numbered boxes are (and) or (or) gates.

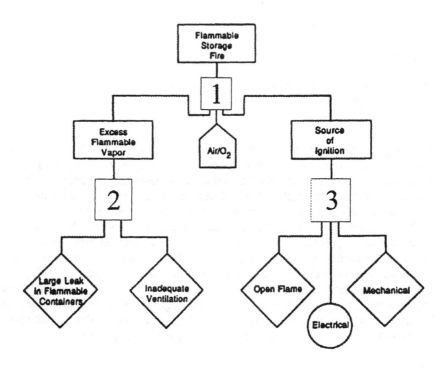

Solution 16.5

The numbered boxes above are as follows:

1. For a Flammable Storage Fire to occur, Excess Flammable Vapor and Air/O$_2$ and Source of Ignition must be present. Therefore, box 1 is an *and* gate since all situations must occur.
2. For Excess Flammable Vapor to occur, a Large Leak in Flammable Containers or Inadequate Ventilation must be present. Therefore, box 2 is an *or* gate since only one situation need occur.
3. For a Source of Ignition to occur, there must be an Open Flame or an Electrical Ignition or Mechanical Ignition must be present. Therefore, box 3 is an *or* gate since only one situation need occur.

Example 16.6

Discuss the three major factors that often influence equipment failure rates.

Solution 16.6

The variation of equipment failure rate with time-in-service is usually represented by three regions:

1. At initial start up, the rate of equipment failure is high due to factors such as improper installation or problems as a result of defects from the manufacturer.
2. The rate of failure declines when the equipment is under normal operation. At this point, failures are chance occurrences.
3. The rate of failure increases as the equipment ages. This can be termed as wear out failure.

This failure rate with time was discussed earlier and will be discussed again in Chapters 20 and 21. This graphical representation is known as the "bathtub curve" (or Weibull distribution to statisticians) because of its shape. See Figure 20.7.1 for this relationship.

16.8 SUMMARY

1. Accidents in industry occur for many reasons. A few of which can be attributed to mechanical failure, operational error (human error), and process upset, and design error. In order to understand the root cause of an accident, system safety approaches have been put to use.
2. The chemical industry is involved in a variety of manufacturing processes ranging from plastics and explosives to biological preparations. The hazards accompanied with these processes vary with the basic plans and designs for each individual plant. The process must be evaluated and safeguards must be put in place.
3. The equipment used in a processing system is designed under internal and external constraints. Most equipment failures occur when these constraints are exceeded. Each piece of equipment has its own set of constraints, which must be followed to avoid an accident.
4. The handling of chemicals can cause hazardous situations. If not properly mixed and agitated, uniformity and control of the reactions is very difficult.
5. Human error is a common and unpredictable cause of accidents. Inadequate training of employees results in error because of unfamiliarity of the process. Training alone does not eliminate error. Many times those with training make errors because their tasks are routine.

6.   The specific path that an accident takes can be represented graphically by a fault tree. The tree begins with the ultimate consequence and works backwards to the possible causes and failures.

PROBLEMS

1.   Discuss when ancillary equipment needs to be reviewed when attempting to reduce the causes of accidents.
2.   Discuss the advantages and disadvantages of sophisticated (automated) controls at a chemical plant.
3.   Assume that a leak has occurred from an H_2S storage tank. Discuss some potential causes.
4.   Discuss the various factors that can influence equipment failure rates.
5.   Describe the difference between catastrophic failure, degraded failures, incipient failure.
6.   List some of the operational failures that may occur at a chemical plant.
7.   Discuss human error that can arise due to an error caused solely by the worker.

REFERENCES

1.   J. Crowl and J. Louvar, *"Chemical Safety Fundamentals with Applications"*, Prentice-Hall. Saddle River, NJ, 1990.
2.   AIDUE, *"Guidelines of Hazard Evaluation Procedures"*, CCPS, AICHE, New York City, 1982.
3.   US Department of Labor, *"System Safety Engineering"*, Safety Manual No. 15, Mine Safety and Health Administration, Washington, D.C., 1982.
4.   *Hazard Survey of the Chemical and Allied Industries,* Engineering and Safety Service, Technical Survey No. 3, American Insurance Association, New York, 1979.
5.   T. A. Kletz, *What Went Wrong? Case Histories of Process Plant Disasters,* Gulf Publishing, Houston, TX, 1985.
6.   C. R. Burklin, *"Safety Standards, Codes and Practices for Plant Design,"* Chem. Eng., **79**, 56-63, October 2, 1972.
7.   J. A. Davenport, *"Survey of Vapor Cloud Incidents,"* Chem. Eng. Prog., **73** (9) 54 (1977).

17

Accident Consequences and Evaluation

17.1 INTRODUCTION

Consequences of accidents can be classified qualitatively by the degree of severity. A quantitative assessment is beyond the scope of this chapter. However, information is provided in Chapters 20 and 21. Factors that help to determine the degree of severity are the concentration in which the hazard is released, the relative toxicity of the hazard, and in the case of a chemical release, the length of time that a person is within the exposed environment. The worst-case consequence of a scenario is defined as a conservatively high estimate of the severe accident identified.[1] On this basis, one can qualitatively rank the consequences of accidents into low, medium, and high degrees of severity.[2] A low degree of severity means that the hazard consequence is nearly negligible, and the injury to person, property, or the environment may be observed only after an extended period of time. The degree of severity is considered to be medium when the accident is serious, but not catastrophic. An example of this could include a case where there is a release of low-level concentration of a chemical that is considered to be highly toxic. Another example of a medium degree of severity could be a highly concentrated release of a less toxic chemical, large enough to cause injury or death to persons and damage to the environment unless immediate action is taken. There is a high degree of severity when the accident is catastrophic or the concentrations and toxicity of a chemical hazard are large enough to cause injury or death to many individuals, and there is long-term damage to the surrounding environment.

The authors gratefully acknowledge the assistance of David James Braun in researching, reviewing, and editing this chapter.

Event trees are diagrams that evaluate the consequences of a specific hazard. The safety measures and the procedures designed to deal with the event are presented here in Chapter 17. The consequences of each specific event that led to the accident are also presented. An event tree is drawn (sequence of events that led up to the accident). The accident is described. This allows the path of the accident to be traced. It shows possible outcomes that could have arisen had a single event in the sequence been changed. Thus, an event tree provides a diagrammatic representation of event sequences that begin with the so-called initiating event and terminate in one or more undesirable consequences. In contrast to a fault tree, which works backward from an undesirable consequence to possible causes, an event tree works forward from the initiating event to possible undesirable consequences. As described earlier in Chapter 15, the initiating event may be equipment failure, human error, power failure, or some event that has the potential for adversely affecting an ongoing process.

Potential consequences of other hazardous conditions can include:

1. Flying shrapnel
2. Rocketing tank parts
3. Fireball created by mechanically atomized drops of burning liquid and vapor
4. Secondary fires and explosions caused by flaming tank contents
5. Release of toxic or corrosive substances to the surrounds

Perhaps the key to determining the consequences of an accident is the study of accident minimization/prevention. This topic receives extensive treatment in Section 17.2. The estimation (not calculation) of consequences is treated in Section 17.3, which is followed by evacuation procedures (Section 17.4). The next section examines failure modes, effects and critical analysis (FMECA). The chapter concludes with vulnerability analysis (Section 17.6) and event tree analysis (Section 17.7).

17.2 ACCIDENT MINIMIZATION/PREVENTION

The first step in minimizing accidents in a chemical plant is to evaluate the facility for potential fires, explosions, and vulnerability to other hazards, particularly those of a chemical nature. This calls for a detailed study of plant site and layout, materials, processes, operations, equipment, and training, plus an effective loss prevention program. The technical nature of industry requires detailed data and a broad range of experience. This complex task, today becoming the most important in plant design, is facilitated by the safety codes, standards, and practice information available. The technical approach to evaluating the consequences of hazards is discussed later in this chapter and in Part V (Chapters 20 and 21).

A good design:

1. Is intrinsically safe
2. Has adequate design margins
3. Ensures sufficient system reliability
4. Includes fail-safe controls
5. Provides for fault detection and alarms
6. Incorporates protection instrumentation

Safety Standards, Codes, and Practices

Coordinating safety, engineering, and industrial standards is principally the responsibility of the American National Standards Institute (ANSI), a federation of industrial companies, government agencies, trade organizations, technical societies, and consumer organizations. Burklin cites the following National Fire Protection Association (NFPA) definitions of the types of document that have been developed for safety design.[3]

1. *Code.* A code is a document containing only mandatory provisions using the word "shall" to indicate requirements. Explanatory material is only included in notes, footnotes, or an appendix.
2. *Standard.* This is a document containing both mandatory provisions ("shall" rules) and advisory provisions ("should" recommendations).
3. *Recommended Practice.* This term describes a document containing only advisory provisions using the word "should."
4. *Manual.* A manual is a document that is informative in nature and does not contain requirements or recommendations.

Table 17.2.1 lists the major organizations providing these documents. Table 17.2.2 is a cross-reference to the same information listing the design areas. An extremely useful and detailed description of these organizations is provided by Burklin.[3]

Plant Site

The most logical starting point in the safety design approach is to select a site where the number of undesirable weather and topographic conditions is limited. Adequate utilities and support systems plus fire protection service are also required for a safe environment. Chapter 5 presented a detailed account of plant site selection and layout. These features will now be considered from a safety point of view. The following guidelines should be observed in determining a site that is favorable for the efficient and economical operation of the process.

1. A fairly level site is needed to prevent spills (e.g., flammable liquids) from flowing. Slight inclines may be useful for drainage purposes. Firm soil well above the water level is also necessary.
2. Adequate water supplies for fire protection are vital. Other utility services, such as electricity, must be reliable and well maintained during emergencies. Plants depending on outside electricity should have two separate feeder circuits whenever possible, and possible failures of these utilities should be evaluated before construction. When one system fails, the other system must provide suitable switching or shutdown to prevent serious hazards.
3. Roadways should allow easy approach of emergency vehicles.
4. Sites near existing plants containing hazardous materials and near congested communities should be avoided.
5. Waste disposal systems containing flammable, corrosive, or toxic materials should be at least 250 feet from plant equipment.
6. Climate and natural hazards should be evaluated. Lightning arrestors should be installed to prevent lightning from directly striking flammable areas. For outdoor operations, temperatures must be considered.
7. Providers of emergency services, such as the fire department, should be well-trained, and well-equipped, and able to respond rapidly.

Plant Layout

Plant layout should be planned to provide efficient operation, proper working conditions, constant flow of materials, inventory control, and mechanisms for ensuring safety. Plant site and plans for future expansion are also considerations. Congestion and concentration of operations should be covered in the initial evaluation to determine that the control of fires, explosions, toxic chemical releases, and environmental hazards will be possible. Localizing such accidents may minimize the damage and permit control of the hazard. Guidelines for a general layout are as follows.

1. Process units, especially critical equipment and storage areas, should be adequately spaced. When this is impossible, barricades, fire protection units, and protective construction should be provided.
2. A rectangular block layout should be followed to provide adequate roadways between process equipment for emergency vehicle access and minimize the spread of fire.
3. The water system layout should provide an adequate supply to all parts of the plant. Piping, pumps, tanks, valves, hydrants, and auxiliary equipment should follow acceptable standards.
4. Utilities should be protected by dual supplies.

TABLE 17.2.1 Major Organizations Providing Codes and Standards, Recommended Practices, Design Criteria, or Guidelines for Equipment in Chemical and Allied Industry Process Plants

Technical and Trade Groups	Organization Abbreviation
Air Conditioning & Refrigeration Institute	ARI
Air Moving and Conditioning Association	AMCA
American Association of Railroads	AAR
American Gas Association	AGA
American Petroleum Institute	API
American Water Works Association	AWWA
Chemical Manufacturers Association (formerly Manufacturing Chemists Association)	CMA
Chlorine Institute	CI
Compressed Gas Association	CGA
Cooling Tower Institute	CTI
Manufacturers Standardization Society	MSS
National Electrical Manufacturers Association	NEMA
Pipe Fabrication Institute	PFI
Scientific Apparatus Makers Association	SAMA
Society of the Plastics Industry	SPI
Steel Structure Painting Council	SSPC
Tubular Exchanger Manufacturers Association	TEMA
U.S. Government Agencies	
Bureau of Mines	BM
Department of Transportation	DOT
U.S. Coast Guard	USCG
Hazardous Materials Regulation Board	HMRB
Federal Aviation Administration	FAA
Environmental Protection Agency	EPA
National Bureau of Standards	NBS
Occupational Safety and Health Administration	OSHA
Testing Standards and Safety Groups	
American National Standards Institute	ANSI
American Society for Testing and Materials	ASTM
National Fire Protection Association	NFPA
Underwriters Laboratories, Inc.	UL
National Safety Council	NSC

TABLE 17.2.1 (Continued)

Technical and Trade Groups	Organization Abbreviation
Insuring Associations	
American Insurance Association	AIA
Factory Insurance Association	FIA
Factory Mutual System	FM
Oil Insurance Association	OIA
Organization	Organization Abbreviation
Professional Societies	
American Conference of Governmental Industrial Hygienists	ACGIH
American Industrial Hygiene Association	AIUA
American Institute of Chemical Engineers	AIChE
American Society of Mechanical Engineers	ASME
American Society of Heating, Refrigeration, and Air-Conditioning Engineers	ASHRAE
Illuminating Engineering Society	IES
Institution of Chemical Engineers (Britain)	ICHE
Institute of Electrical and Electronics Engineers	IEEE
Instrument Society of America	ISA

5. Offices, the cafeteria, laboratories, and other populated places should be located on the periphery of the site (displaced from hazardous areas), and upwind from stacks and the center of the plant.
6. Storage, loading, and transportation facilities should also be located on the periphery to minimize traffic through operating areas.
7. Waste disposal should be located downwind to minimize exposure to plant workers and the community.
8. Sewers and drains should provide safe and efficient removal of liquids. The drainage system must provide sufficient capacity to prevent any likely flooding from severely damaging equipment.
9. Fire protection equipment and emergency facilities must be readily available and sufficient.
10. Plant security must provide efficient supervision of boarders and entrances.

TABLE 17.2.2 **Areas Covered by Codes, Standards, Recommended Practices, Design Criteria, or Guidelines of Designated Organizations** [a]

Design Area	Organization
Accident case history	ACA, AIA, AIChE, API, FIA, FM, NFPA, NSC, OIA, OSHA, USCG
Air compressors	AIA, ANSI, FM, USCG
Air-fin coolers	ARI, ASURAE, OIA, USCG
Boilers	ANSI, NFPA, NSC, UL
Combustion equipment and controls	ANSI, FIA, FM, NEPA, NSC, OIA, UL, USCG
Compressors	AIA, ARI, ASHRAE, ASME, FM, OIA, USCG
Cooling towers	M, FM, NFPA, OIA
Drain and waste systems	AIChE, AWWA, CMA, USCG
Dust collection equipment	FIA, FM, NFPA, USCG
Dust hazards	ACGIH, AIHA, ANSI, BM, FIA, FM, NFPA, NSC, UL, USCG
Electric motors	ANSI, CMA, IEEE, NFPA, UL, USCG
Electrical area classification	AIA, ANSI, API, CMA, FIA, FM, NFPA, NSC, OIA, OSHA, USCG
Electrical control and enclosures	AIA, ANSI, ARI, CMA, FIA, FM, IEEE, ISA, NEMA, NFPA, NSC, OIA, OSHA, UL, USCG
Emergency electrical systems	AGA, AIA, FM, IEEE, NEMA, NFPA, USCG
Fans and blowers	ACGIH, AIHA, AMCA, ARI, ASME, FM, USCG
Fire protection equipment	AIA, ANSI, API, AWWA, BM, CGA, CMA, FIA, NEMA, NFPA, NSC, OIA, OSMA, UL, USCG
Fire pumps	ANSI, FM, IEEE, NFPA, UL, USCG
Fired heaters	ANSI, ASME, FIA, FM, NFPA, OIA, UL, USCG
Gas engines	FM, NFPA, OIA, USCG
Gas turbines	AGA, FIA, FM, NFPA, OIA, USCG
Gear drives, power transmission	AIA, ANSI, NSC, USCG

TABLE 17.2.2 (Continued)

Grounding and static electrical	AIA, ANSI, API, FIA, FM, IEEE, NEMA, NFPA, NSC, OIA, OSHA, UL, USCG
Inspection and testing	AIChE, AMCA, API, ASHRAE, ASTM, AWWA, CGA, CM, DOT, IEEE, MSS, NFPA, NSC, PFI, USCG
Instrumentation	AIA, ANSI, API, ARI, ASTM, AWWA, CGA, FIA, FM, HMRB, IEEE, ISA, NBS, NFPA, OIA, SAMA, UL, USCG
Insulation and fireproofing	AIA, ANSI, ASHRAE, ASTM, FM, OIA, UL, USCG
Jets and ejectors	USCG
Lighting	ANSI, FM, IEEE, IES, NEMA, NFPA, NSC,UL,USCG
Lubrication	AMCA, ANSI, ASME, NFPA
Material handling	MCA, NFPA, NSC, OSHA
Materials of construction	AIA, ANSI, ASTM, AWWA, CGA, CMA, CI, CTI, FM, HMRB, ISA, NBS, NFPA, NSC, OIA, TEMA, UL, USCG
Noise and vibration	AGA, AIChE, AIUA, AMCA, ANSI, API, ARI, ASHRAE, ASTM, EPA, ISA, NFPA, NSC, OSHA, UL
Painting and coating	AIChE, ANSI, ASTM, AWWA, HMRB, OSHA, NBS, SSPC, UL
Piping materials and systems	AGA, AIA, ANSI, API, ARI, ASHRAE, ASTM, AWWA, CGA, CI, FIA, FM, HMRB, IES, MSS, NBS, NFPA, NSC, PFI, SPI, UL, USCG
Plant and equipment layout	AAR, AIA, API, AWWA, CGA, CMA, FIA, FM, HMRB, NFPA, NSC, OIA, USCG
Pneumatic conveying	ANSI, NFPA, USCG
Power wiring	ANSI, API, FIA, FM, IEEE, NEMA, NFPA, OIA, OSHA, UL, USCG
Pressure relief equipment systems	AIA, API, ASME, CGA, CI, FIA, FM

TABLE 17.2.2 (Continued)

Pressure vessels	AIA, API, ASME, CGA, DOT, NFPA, NSC, OSHA, HMRB, OIA, OSHA, USGC
Product storage and handling	AAR, AIA, AIChE, ANSI, API, CCA, Cl, CMA, FIA, FM, NFPA, OIA, OSHA, USCG
Pumps	AIChE, ANSI, AWWA, Cl, NFPA, OIA, UL, USCG
Refrigeration equipment	ANSI, API, ASHRAE, FM, NFPA, UL, USCG
Safety equipment	ACGIH, AIHA, ANSI, BM, CGA, Cl, CMA, FM, NSC, OSHA, UL, USCG
Shell and tube exchangers	AGA, AIChE, API, ASHRAE, ASME, CGA, PFI, USCG
Shutdown systems	ATA, API, FIA, NFPA, OIA, UL, USCG
Solids conveyors	CMA
Stacks and flares	FAA, OIA, USCG
Steam turbines	AIA, FM, IEEE, OIA, USCG
Storage tanks	AWWA, Cl, NBS, NFPA, OIA, OSHA, UL, USCG
Ventilation	ACGIH, AIHA, ANSI, BM, FIA, FM, NFPA, NSC, UL, USCG
Venting requirements	API, FIA, FM, HMRB, NFPA, USCG

[a] See Table 17.2.1 for Abbreviations

To contain the hazards in a critical area, process units, in general, should be located in the same vicinity instead of scattered throughout the plant. Spacing requirements of the units are determined by:

1. Fire, explosive, and health hazards of the process and the nature of the materials
2. The quantity of materials contained in any one unit or area
3. The exposure possibility of equipment to damage from potential nearby fires, explosives, and hazards
4. The value of the area
5. The importance of the unit to the continuity of production
6. The availability of adequate fire fighting and rescue operations
7. The climate and topographical conditions

Chemical plants also consist of process buildings, storage and warehouse buildings, control houses, laboratories, and general offices. Depending on the nature of activity and the quality of the contents, the structural requirements and protection features will vary. Building standards are defined by the National Code of the American Insurance Association.

Buildings erected must be adequate with respect to explosion venting and ventilation, firewalls, exits, drainage, and electrical wiring. The safety of the equipment and the structures is often a function of their age. The degree of "adequacy" must be evaluated based on this as well as the factors above.

Materials

All chemicals to be handled in a process plant must be evaluated during the design phase. Once an accident has occurred, knowledge of properties of any hazardous materials will allow appropriate handling.

An inventory list of the physical properties and hazardous nature (flammability, toxicity, reactivity, corrosiveness, etc.) of all chemicals should be composed. Also, to protect personnel and the community, the exposure effects of the chemical should be determined.

Unit Operations

Depending on the nature of the change, physical changes occurring during operations may increase the potential for fire, explosion, and health hazards of a product. Some general factors that increase the hazard potential are:

1. Vaporization and diffusion of flammable or toxic liquids or gases
2. Spraying, misting, or fogging of flammable, combustible, or toxic liquids
3. Dusting and dispersion of combustible or toxic solids
4. Mixing combustible materials and strong oxidizing agents
5. Increasing the temperature or pressure of unstable chemicals
6. Exposing unstable chemicals to friction or shock
7. Separation of hazardous chemicals from protective inert materials or diluents

The environmental conditions contemplated for the operating equipment also must be evaluated. Many of the unit operations are used in conjunction with or in close proximity to process equipment and, therefore, the safety of the overall operation is involved.

Vaporization and diffusion of flammable or toxic liquids or gases is a primary consideration with distillation, evaporation, extraction, and absorption operations. The basic principle of safety for these unit operations is containment of the materials within the system. These operations should be conducted outdoors whenever possible. In this way, any accidental release of flammable or toxic materials will have a lower chance of creating a hazardous situation. Suitable detection and warning devices should also be installed.

Size reduction and enlargement operations present dust problems; standards for control involve many safety factors, including control of dust dispersion, inerting, and elimination of sources of ignition. To minimize the possibility of an explosion, it is essential to prevent unstable materials or mixtures of combustible materials and strong oxidizing agents from entering size reduction equipment or from being subjected to pressure or friction. The toxic properties of some dusts also necessitate special ventilation, dust collection systems, and protective equipment for operators.[1]

Separatory operations may isolate the more hazardous components from the diluent or inert. Screening of solids, in effect, separates the more hazardous dusts (small particles and fines) from the less hazardous larger pieces of material. Centrifuges, like all high-speed equipment, must be securely positioned to provide protection to all personnel working in the vicinity.

Gases can be the cause of excessive foaming when blown through flammable liquids. The rapid expansion tendencies of foams can spread hazards throughout the entire area. To prevent this foaming, controlling agents and extraneous materials must be excluded from all mixing operations.

Heat transfer is a very widely used operation in chemical processing. Both heating and cooling affect the controllability of the hazards inherent in chemical processing. Heat transfer equipment can produce temperatures ranging from -455 to 4000°F.

Equipment

The equipment and systems of the processing plant are designed to contain the chemicals under processing conditions and to provide the controlled environment required for production. This equipment is designed to function under both specific process conditions and upset conditions. Upset conditions that are considered in design include fire, explosions, and accidental chemical releases.

A well-designed piece of equipment has safety and loss prevention features built into it. The following typical design considerations are important from the standpoint of loss prevention:

1. Reliability of the unit
2. Ease of operation
3. Flexibility of the unit
4. Amenability to future expansion
5. Provisions for inspection and maintenance
6. Adequate emergency shutdown facilities
7. Standardization of equipment for rapid replacement
8. Design to anticipate pressure range-with overpressure controls
9. Design to anticipate temperature range-with overtemperature controls

Equipment should be checked for:

1. Compliance with existing recognized good industry practices, safety codes, standards, and governmental regulations
2. Construction according to specifications
3. Operation without undue strain on the operator
4. Reliable operation under all situations, including a difficult start-up
5. Provision of adequate safeguards and protection equipment
6. An inspection and maintenance program sufficient to provide reliable operation

The factors that enter into the design of vessels include type of material, configuration, method of construction, design stresses, and thickness of the metal. As with any equipment, design pressures and temperatures should take into consideration the most severe combination of conditions anticipated. Vessels must be completely drainable. Liners and wear plates may be required to prevent corrosion.

Vessels should be provided with overpressure protection as required. Vents and relief valve vent piping should be so arranged that the vented vapors will not constitute a hazard. Relief valves must be kept free from corrosion or fouling and should be operable at all times.

All pressure vessels should be tested, inspected, and marked in accordance with code requirements. Further inspection of field fabrication vessels may include the radiography of all seams and other pertinent tests.

Standards and codes have very specific safety requirements for the design of tanks for flammable liquids or gases and pressurized chemicals such as ammonia. Overpressure devices and vents are included in the design. Tank and vessel supports should be built for the maximum intended load.

Mixers and agitators designed for flammable liquids or dusts should be constructed to minimize fire and explosive possibilities. Electrical equipment should follow the requirements of the National Electrical Code. Equipment should be bonded and grounded to prevent the accumulation of static electricity.

The most common heat transfer unit in chemical production is the tubular heat exchanger. Tube materials must be selected to resist corrosion and fouling. Whenever a heat exchanger services a normal or an emergency cooling facility, it is recommended that its reliability as well as the adequacy of the supply of the coolant be evaluated. Reactive heat exchanger fluids require carefully engineered units to ensure containment. Interlocking of agitation equipment and heat exchangers is also necessary. Adequacy of controls and instrumentation is vital in the area of temperature control. Heat exchangers should be equipped with relief valves, bypass piping, and adequate drains. To prevent tube leakage or rupture, only seamless tubes should be used. Heat exchanger tubes and piping should be designed to withstand thermal expansion and contraction to prevent excessive stress on connections.

Pumps, compressors, turbines, drivers, and auxiliary machinery should be designed to provide reliable, rugged performance. Pump selection and performance depend on the capacity required and the nature of the fluids involved. Remotely controlled power switches and shutoff valves are necessary to control fluid flow during an emergency. The inlets for air compressors should be strategically located to prevent the intake of hazardous materials.

All piping valves and fittings should be designed according to recognized standards for the working pressures and structural stresses anticipated. The piping should be well supported and protected against physical and mechanical damage. Piping systems should be located, in serviceable areas. Nondestructive testing such as ultrasonic, radiographic, or eddy current examination is preferred to the spot-checking of batches of pipes and tubes by destructive means at the point of manufacture.

Evaluation of piping networks should be oriented toward the elimination of unnecessary pipe runs transporting hazardous materials. Pipelines should contain a sufficient number of well-placed valves to control the flow of fluids during fire emergencies. Piping should be color coded for rapid identification; the diamond symbol used in the National Fire Protection Association's S stem 704 for fire hazard information of materials may also be appropriate for proper pipe labeling.[4]

Piping should also be arranged to allow the shortest, most direct route without congestion. They should be installed to prevent the trapping of liquids and should contain adequate valving and provisions for contraction and expansion. Joints must be properly welded and secure.

Good foundations assure stability of the equipment. The foundation should extend below the soil frost line to eliminate settling of equipment. Major loads and equipment producing vibrations (e.g., pumps) should not be placed on filled ground. High towers and major vessels should have properly installed anchor bolts.

17.3 CONSEQUENCE ESTIMATION

For any specific incident there will be an infinite number of incident outcome cases that can be considered. There is also a wide degree of consequence models which can be applied. It is important, therefore, to understand the objective of the study tin order to limit the number of incident outcome cases to those which satisfy that objective. An example of variables that can be considered is as follows: [5]

- Quality, magnitude, and duration of the release
- Dispersion parameters
  - Wind speed
  - Wind direction
  - Weather stability
- Ignition possibility and flammable releases; if applicable

      -Ignition sources/location
      -Ignition strength
- Energy levels contributing to explosive effects; if applicable
- Impact of release on people, property, or environment
      -Thermal radiation
      -Projectiles
      -Shock-wave overpressure
      -Toxic dosage
- Mitigating effects
      -Safe havens
      -Evacuation
      -Daytime/nighttime population

Following an accident, the effects on individuals able to escape or remain in a shelter (or equivalent) differ from those for people in the open. Factors to consider in relation to building types and human behavior include:

1. *The nature of the hazard considering both intensity and duration.* Shelters vary in the degree of protection provided. For thermal and toxic hazards, shelters can have a beneficial effect. However, for explosions, the hazard maybe greater because of the possibility of the building collapsing.
2. *The nature of the hazard considering its degree of toxicity and its warning properties.* A release of carbon monoxide provides no warning while a release of some amine normally provides a strong odor at concentrations well below harmful levels.
3. *The nature of the surrounding population.* The distribution of the population indoors varies depending on the time of day and the season, the overall health of the population (senior citizens, infirm, etc.), and the type of clothing being worn (cotton, wool, polyester, etc.) by the personnel exposed to a possible heat radiation.
4. *The type of buildings and their construction.* Factors can include, ventilation rates, resistance to blast effects, the ability of overhead fixtures to remain intact, etc.
5. *The effectiveness of training and the availability of equipment for emergency response and medical treatment.* This applies to both the plant and among emergency response services. Trained personnel obviously can improve the chance of survival for those exposed.
6. *The prevailing weather conditions, topography, and physical obstructions.*
7. *The intensity and duration to which toxic gas can incapacitate exposed personnel.*

17.4 EVACUATION PROCEDURE [6]

An accidental release of hazardous materials sometimes necessitates evacuation of people from certain areas to prevent injury or death. These areas can include those directly affected by toxic fumes and gases or fire, and those areas that may be potentially affected during the course of the incident (e.g., through wind shift, a change in the site conditions). Evacuation is a complex undertaking. Rather than attempting to provide specific step-by-step guidance for each possible scenario, this section will discuss general considerations that should be addressed by the local emergency planning committee (LEPC).

Making a Decision on Evacuation

The first evacuation consideration, determining whether an evacuation is necessary, involves a comprehensive effort to identify and consider both the nature of and circumstances surrounding the released hazardous material and its effect on people. No safe exposure limits have been established for extremely hazardous substances (EHSs).

The Department of Transportation's (DOT's) *Emergency Response Guidebook* provides initial isolation and evacuation distances for transportation incidents. The evacuation distances given in this guidebook are preceded by the following advice: "The [initial isolation/evacuation] table is useful only for the first twenty two minutes of an incident... There are several good reasons for suggesting that the use of the table be limited specifically to the initial phase of a no-fire spill incident during transport. The best calculations for these tables are not reliable for long vapor travel times or distances. At their best the estimates are for a cool, overcast night with gentle and shifting winds moving in a non-reactive, neutrally-buoyant vapor." Note that specific dispersion calculation details are provided in Chapter 12. The DOT Emergency Response Guidebook is intended to help first responders make informed judgements during the initial phases of a hazardous materials transportation incident. LEPCs are cautioned not to use it as a substitute for a specific plan for responses to hazardous materials incidents.

Hazardous Conditions Affecting Evacuation Decisions

Numerous factors affect the spread of hazardous substances into the area surrounding a leaking/burning container or containment vessel. Evacuation decision-makers must carefully consider each of these factors in order to determine the conditions created by the release, the areas that have been or will be affected, and the health affects on people. The factors that affect evacuation include amount of released material(s), physical and chemical properties of the released material(s), health hazards, dispersion pattern, rate of release, and potential duration of release. Each of these factors is explained below.

To begin with, it is necessary to know the material's physical and chemical properties, including:

1. Physical state – solid, liquid, or gas

2.   Flammability: flashpoint, ignition temperature, flammability limits
3.   Specific gravity: whether material sinks or floats in water
4.   Vapor density: whether vapors rise or remain near ground level
5.   Solubility: whether material readily mixes with water
6.   Reactivity: whether material reacts with air, water, or other materials
7.   Crucial temperature: boiling point, freezing point

It is also necessary to know the health effects resulting from short-term exposure. These can include:

1.   Acute or chronic hazards
2.   Respiratory hazards
3.   Skin and eye hazards
4.   Ingestion hazards

Another consideration is the dispersion pattern of the released hazardous material, for example:

1.   Does the release follow the contours of the ground?
2.   Is it a plume (vapor cloud from a point source)?
3.   Does the release have a circular dispersion pattern (dispersing in all directions)?

Atmospheric conditions must also be addressed when determining the appropriate evacuation response to a hazardous material release. Atmospheric conditions that may affect the movement of material and evacuation procedures include:

1.   Wind (speed and direction)
2.   Temperature
3.   Moisture (precipitation, humidity)
4.   Air dispersion conditions (inversion or normal)
5.   Time of day (daytime or darkness)

Other considerations important in making evacuation decisions include:

1.   Whether the hazardous material is being released into air, land, and/or water and its concentration in the air and/or water
2.   Size and potential duration of the release
3.   Rate and release of the material, as well as the projected rate (the rate of release may change during the incident)

Conducting an Evacuation

Should it become necessary that an area be evacuated, the evacuation must be conducted in a well-coordinated, thorough, and safe manner. Evacuation involves a number of steps, which include assigning tasks of evacuation assistance personnel, informing potential evacuees, providing emergency medical care as necessary, providing security for evacuated area, and sheltering evacuees as necessary.

17.5 FAILURE MODES, EFFECTS AND CRITICAL ANALYSIS [7]

FMECA, also known as failure modes and effects analysis (FMEA), is a systematic qualitative method by which equipment and system failures and the resulting effects of these failures are determined. FMECA is an inductive analysis; that is, possible events are studied, but not the reasons for their occurrences. FMECA has some disadvantages: human error is not considered and the study concentrates on system components, not the system linkages that often account for system failures. FMECA provides an easily updated systematic reference listing of failure modes and effects that can be used in generating recommendations for equipment design improvement. Generally, this analysis is first performed on a qualitative basis; quantitative data can later be applied to establish a criticality ranking that is often expressed as probabilities of system failures.

Five steps are required for a thorough analysis:

1. The level of resolution of the study must be determined
2. A format must be developed
3. Problem and boundary conditions are then defined
4. The FMECA table is completed
5. The study results are reported [8,9,10,11]

The first step in FMECA is to determine a level of resolution. If a system-level hazard is to be addressed, equipment in the system must be studied; for a plant-level hazard, individual systems within the plant must be examined. Once the level of resolution has been determined, a format must be developed-one to be used consistently throughout the study. A minimal format should include each item, its description, failure modes, effects, and criticality ranking.

Defining problem and boundary conditions includes identifying the plant or systems that are to be analyzed and establishing physical system boundaries. In addition, reference information on the equipment and its function within the system must be obtained. This can be found in piping and instrumentation design drawings as well as in literature on individual components or equipment. The final step in the problem definition step is to provide a consistent criticality ranking definition. In a quantitative study, probabilities are often the method used for ranking. If the study is being conducted on a qualitative basis, relative scales (see Table 17.5.1) are usually used as ranking methods. Table 17.5.2 summarizes the hazard classes used in the aerospace industry that may be used on a relative scale.[7] If this type of scale is used, however, "negligible, marginal, critical, and catastrophic" should be defined more clearly. Another more specific criticality ranking scale (summarized in Table 17.5.2) is suggested by the American Institute of Chemical Engineers.

The FMECA table should be concise, complete, and well organized. This table should identify equipment and relate it to a system drawing or location. This is to prevent confusion when similar equipment is used in different locations. One of the limitations of FMECA is that the table must include ALL failure modes for each piece of equipment and effects of each failure along with the associated criticality ranking. Table 17.5.3 shows a sample chart that can be completed for the FMECA table.

The final step in conducting a FMECA is to report the results. If the prepared table (see Table 17.5.3) is complete, that may be sufficient. Often, however, a report of suggested design changes or alterations should also be included.

It should be noted that FMECA identifies single failure modes that either directly result in or contribute significantly to important accidents. Human/operator errors are generally not examined in a FMECA; however, the effects of a misoperation are usually described by an equipment failure mode. It should also be noted that FMECA is not efficient for identifying combinations of equipment failures that lead to accidents.

17.6 VULNERABILITY ANALYSIS

Vulnerability analysis identifies areas in the community that may be affected or exposed, individuals in the community who may be subject to injury or death from certain specific hazardous materials, and what facilities, property, or environment may be susceptible to damage should a hazardous materials release occur. A comprehensive vulnerability analysis provides information on: [6]

1. The extent of the vulnerable zones (i.e., an estimation of the area that may be affected in a significant way as a result of a spill or release of a known quantity of a specific chemical under defined conditions)
2. The population, in terms of numbers, density, and types of individuals (e.g., facility employees; neighborhood residents; people in hospitals, schools, nursing homes, prisons, day care centers) that could be within a vulnerable zone; the private and public property (e.g., critical facilities, homes, schools, hospitals, businesses, offices) that may be damaged, including essential support

TABLE 17.5.1 Suggested Criticality Rankings Based on Aerospace Hazard Classification [12]

Criticality Ranking	Effects on System and Surroundings
I	Negligible effects
II	Marginal effects
III	Critical effects
IV	Catastrophic effects

TABLE 17.5.2 Suggested Scale for Criticality Ranking for a Qualitative Failure Modes Effects and Criticality Analysis (FMECA) [10]

Criticality Ranking	Effects on System and Surroundings
1	None
2	Minor system upset
	Minor hazard to facilities
	Minor hazard to personnel
	Orderly process shutdown necessary
3	Major system upset
	Major hazard to facilities
	Major hazard to personnel
	Orderly process shutdown necessary
4	Immediate hazard to facilities
	Immediate hazard to personnel
	Emergency shutdown necessary

TABLE 17.5.3 Sample Chart that Can Be Completed for a Failure Modes Effects and Criticality Analysis (FMECA)

Equipment	Failure Mode	Effects on			
		Other Systems	System	Relative Ranking	Remarks

3. Systems (e.g., water, food, power, communication, medical) and transportation facilities and corridors
4. The environment that may be affected, and the impact of a release on sensitive natural areas and endangered species

The vulnerability zone is generally that geographical area in which the concentration of a substance accidentally released may reach a critical concentration that may cause serious health effects. Details on these dispersion calculations are provided in Chapter 12. This critical concentration is often the *threshold limit value* (TLV). The area encompassed by this zone, which may require evaluation during an emergency, depends on the amount of chemical that may become airborne in a release and the level of concern for the chemical released. Figure 17.6.1 is a graphical representation of the vulnerable zones for a release of two different but extremely hazardous and/or toxic substances from a plant or facility. Figure 17.6.2 represents similar vulnerable zones for the same two pollutants involved in a transportation release. The vulnerable zone for a transportation route is represented as a corridor, since the exact location of an accident may be difficult to predict. Once an accident has occurred, however, the actual vulnerable zone is assumed to be circular and similar to that for a chemical plant. The effect of different assumptions or the calculation of the radius of estimated vulnerability zones is provided in Figure 17.6.3. After estimating the vulnerable zone for a particular hazard, one should look for information in the following areas.

1. The number of people potentially exposed to this discharge at home, at work, or in recreational areas.
2. Places where more sensitive populations might be exposed, such as nursing homes, hospitals, and schools.
3. Emergency service facilities within and near the zone, including communication facilities (e.g., local radio stations). Personnel in these facilities may not be available to respond if the wind is blowing toward them.
4. Access and egress roads used for response, rescue, or evacuation. (Depending on wind direction, these roads may be unusable.)
5. Other characteristics unique to the community that may cause significant problems, such as high traffic flow roads, methods for accident control and diversion of traffic; communication centers in the potential path of a cloud; and recreational facilities such as sports or community centers.

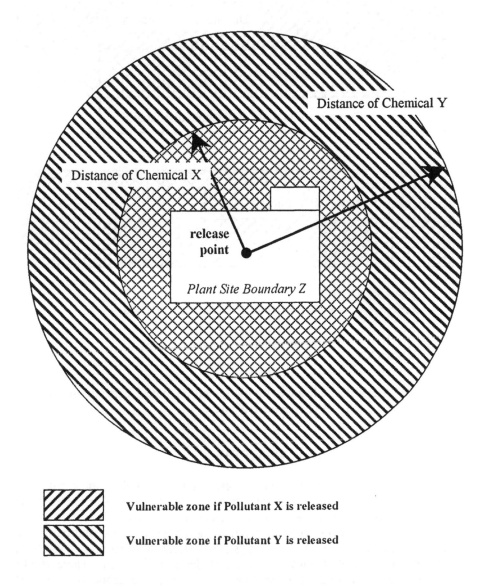

Figure 17.6.1. Vulnerable zones for atmospheric releases of chemicals X and Y from plant Z.

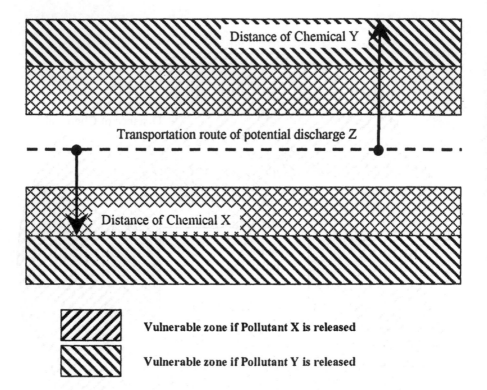

Figure 17.6.2. Vulnerable zones for transportation route Z with atmospheric releases of chemicals X and Y.

17.7 EVENT TREE ANALYSIS (ETA)

Event tree analysis is a technique for evaluating potential accident outcomes resulting from a specific equipment system failure or human error known as an initiating event. Event tree analysis considers operator response or safety system response to the initiating event in determining the potential accident outcomes. The results of the event tree analysis are accident sequences; that is, a chronological set of failures or errors that define an accident. These results describe the possible accident outcomes in terms of the sequence of events (successes or failures of safety functions) that follow an initiating event. Event tree analysis is well suited for systems that have safety systems or emergency procedures in place to respond to specific initiating events.[13]

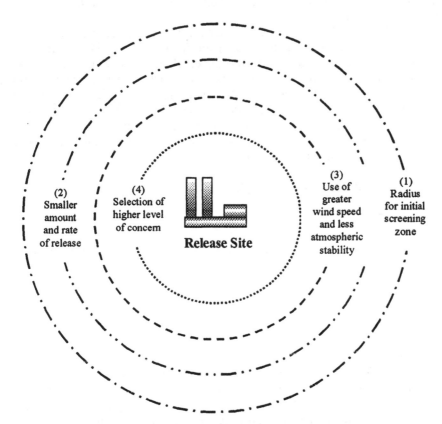

The effect of altering major assumptions on the downwind distance (radius) of the estimated vulnerable zone. Calculations made using (1) credible worst case assumptions for initial screening zone, (2) reevaluation and adjustment of quantity released and/or rate of release of chemical, (3) reevaluation and adjustment of wind speed (increase) and air stability (decrease), (4) selection of a higher level of concern. Note that adjustment of two or more variables can have an additive effect on reducing the size of the estimated vulnerable zone.

Note also that the relative sizes of the altered zones are not to scale (e.g., choosing a higher value for the level of concern does not always result in a smaller zone than the use of greater wind speed and less atmospheric stability).

Figure 17.6.3. The effect of different assumptions on the calculation of the radius of estimated vulnerable zones

The event tree model is started from the initial occurrence and built upon by sequencing the possible events and safety systems that come into play. The model displays at a glance, branches of events that relate the proper functioning or failure of a safety device or system and the ultimate consequence.

The model also allows quick identification of the various hazards that result from the single initial event.

The use of event trees is sometimes limiting for hazard analysis because it may lack the capability of quantifying the potential of the event occurring. The analysis may also be incomplete if all initial occurrences are not identified. Its use is beneficial in *examining*, rather than *evaluating*, the possibilities and consequences of a failure. For this reason, a fault tree analysis (FTA) should supplement this, to establish the probabilities of the event tree branches. This topic was introduced in a subsection of Chapter 16.

Figures 17.7.1 and 17.7.2 present a sample event tree analysis and an example of an event tree for a drum rupture.

17.8 ILLUSTRATIVE EXAMPLES

Example 17.1

Describe how the consequences of an accident at home could be minimized or eliminated.

Solution 17.1

Some possible ways to minimize the consequences of accidents in the home are listed below:

- All appropriate emergency services listed by the phone. (police, fire department, ambulance, poison control)
- Fire extinguishers should be available and easily accessible to areas where they are mostly likely to be needed. (kitchen, garage, workshop)
- Fire drills should be planned and practiced.
- Childproofing (cabinet hooks, plastic electrical outlet plugs, etc.)
- Carpeting should be secured in place to prevent tripping.

Example 17.2

Refer to Illustrative Example 2 in Section 16.8. Discuss some of the effects of the release.

Solution 17.2

Some possible effects of a toxic gas release are listed below:

- Warning system signals for system shutdown, plant evacuation, and emergency services response.
- Ignition of toxic gas kills all those unprotected within the blast radius.

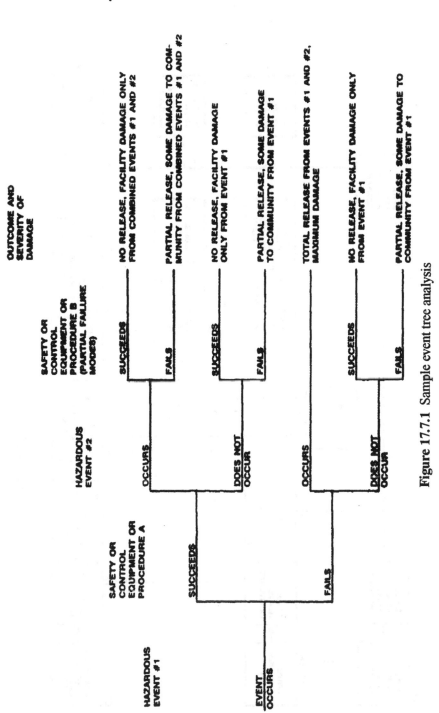

Figure 17.7.1 Sample event tree analysis

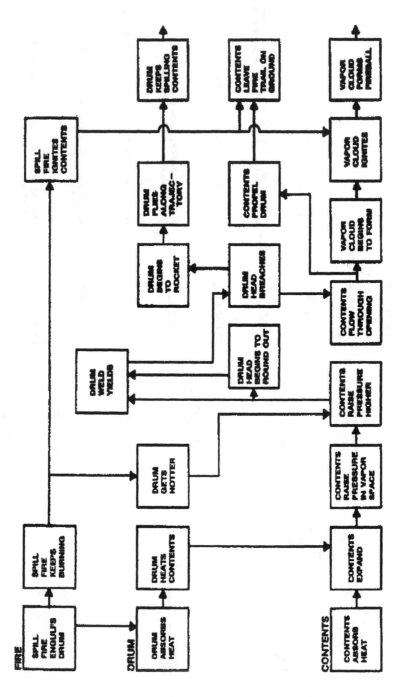

Figure 17.7.2 Event tree analysis

- Persons downwind of the release may be exposed. Effects will depend on the concentration and toxicity of the chemical. These can range from no adverse effect to death.
- Highly toxic release warrants the evacuation of persons in vulnerable zones downwind of the release. Releases of lesser toxicity may allow for other measures to be taken to minimize exposure, such as advising for persons to remain indoors.

Example 17.3

Describe the difference between an:

1. Event
2. External event
3. Initiating event
4. Intermediate event
5. Primary event

Solution 17.3

The above terms are described below.

Type of Event	Description
1. Event	An occurrence involving equipment performance or human action, or an occurrence external to the system that causes system upset. In this document an event is associated with an accident either as the cause or a contributing cause of the accident or as a response to the accident-initiating event.
2. External Event	An occurrence external to the system/plant, such as an earthquake or flood or an interruption of facilities such as electric power or process air.
3. Initiating Event	An event that will result in an accident unless systems or operations intervene to prevent or mitigate the accident.
4. Intermediate Event	An event of an accident event sequence that helps to propagate the accident or helps to prevent the accident or mitigate the consequences.
5. Primary Event	A basic independent event for which frequency can be obtained from experience or test.

Example 17.4

Describe staffing requirements for an *event tree* analysis.

Solution 17.4

A single analyst can perform an event tree analysis, but normally a team of 2 to 4 people is preferred. The team approach promotes "brainstorming" that results in a well defined event tree structure. The team should include at least one member with knowledge of event tree analysis, with the remaining members having experience in the operations of the systems and knowledge of the chemical processes that are to be of interest in the analysis.

Example 17.5

Outline the difference between an *event tree* and a *fault tree*.

Solution 17.5

The above terms are described below.

Event Tree: An *event tree* analysis begins with a specific initiating event and works forward to evaluate potential accident outcomes.

Fault Tree: A *fault tree* analysis begins with the ultimate consequence and works backward to the possible causes and failures.

17.9   SUMMARY

1. Consequences of accidents can be classified qualitatively by the degree of severity. Factors that help to determine the degree of severity are the concentration in which the hazard is released, the relative toxicity of the hazard, and in the case of a chemical release, the length of time that a person is within the exposed environment.
2. The minimizing/prevention of accidents in a chemical plant calls for a detailed study of plant site and layout, materials, processes, operations, equipment, and training, plus an effective loss prevention program.
3. It is important to understand that for any specific incident there will be an infinite number of incident outcome cases that can be considered. Therefore it is necessary to limit the number of possible incident outcome cases so that the proper consequence model can be applied.
4. An accidental release of hazardous materials sometimes necessitates evacuation of people from certain areas to prevent injury or death. Should it become necessary that an area be evacuated, the evacuation must be conducted in a well-coordinated, thorough, and safe manner.

5. Failure Modes, Effects and Criticality Evaluation (FMECA) is a systematic qualitative method by which equipment and system failures and the resulting effects of these failures are determined. FMECA studies possible events, but not the reasons for their occurrences.
6. A vulnerability analysis identifies those regions that may be affected or exposed, individuals who may be subject to injury or death, and what facilities, property, or environment may be susceptible to damage should a hazardous materials release occur.
7. Event tree analysis is a technique for evaluating potential accident outcomes resulting from a specific initiating event. The results of the event tree analysis are chronological sets of failures or errors that may define an accident.

PROBLEMS

1. List the types of process events that can result in a plant accident.
2. Refer to Problem 3 in Chapter 16. Discuss some of the effects of the leak.
3. List several guidelines that should be followed in selecting a site for a plant.
4. Outline the key ingredients that should be included in laying out a new plant.
5. List the key features (positive or negative) of Failure Modes, Effects, and Critical Analysis.
6. What equations are used for the estimation of vulnerable zones?
7. Describe the time and cost requirements for an event tree analysis.

REFERENCES

1. "Guidelines for Hazard Evaluation Procedures", CCPS, AIChE, New York City, 1992.
2. Author Unknown, "Technical Guidelines for Hazard Analysis", USEPA, Washington D.C., 1978.
3. C.R. Burklin, "Safety Standards, Codes and Practices for Plant Design," Chem. Eng., 79, 56-63, October 2, 1972.
4. J. Santoleri, J. Reynolds and L. Theodore, *Introduction to Hazardous Waste Incineration*, Wiley-Interscience, New York, 2000.
5. R. Perry and D. Green, *Perry's Chemical Engineers' Handbook*, 7th Edition, McGraw-Hill, New York City, 1997.
6. Adapted from: "Technical Guidelines for Hazards Analysis", USEPA, Washington, D.C., 1987.
7. Adapted from: H. Kavianian, J. Rao and G. Brown, "Application of Hazard Evaluation Techniques to the Dangers of Potentially Hazardous Industrial Chemical Processes", NIOSH, Cincinnati, OH, 1992.
8. U.S. Department of Labor, System Safety Engineering, Safety Manual No. 15, Mine Safety and Health Admistration, U.S. Department of Labor, Washington, D.C., 1986.

9.   Firenze, R.J., The Process of Hazard Control, 1st Edition, Kendall/Hurt Publishing Co., Dubuque, IA, 1978.

10. Battelle Columbus Division, Guidelines for Hazard Evaluation Procedures, The Center for Chemical Process Safety, AIChE, New York, NY, 1985.

11. Roland, H.E., and B. Moriarty, System Safety Engineering Management, 1st Edition, John Wiley & Sons, New York, NY, 1983.

12. Henley, E.J., and H. Kamamoto, Reliability and Risk Assessment, 1st Edition, Prentice Hall, Englewood Cliffs, NJ, 1981.

13. Adapted from: "Guidelines for Hazard Evaluation Procedures", CCPS, AIChE, New York City, 1992.

18

Hazard Risk Analysis

18.1 INTRODUCTION

This chapter serves to introduce the general subject of hazard risk assessment and analysis, including cause-consequence risk evaluation. The cause-consequence aspect of this topic is perhaps the key to understanding hazard risk. As such, it is treated in a separate section later in this chapter.

Risk analysis is an assessment of the likelihood (probability) of an accidental release of a hazardous material and the actual consequences that might occur, based on the estimated vulnerable zones. The risk analysis is a judgment of probability and severity of consequences based on the history of previous incidents, local experience, and the best available current technological information. It provides an estimation of:

1. The likelihood (probability) of an accidental release based on the history of current conditions and controls at the facility, consideration of any unusual environmental conditions (e.g., areas in flood plains), or the possibility of simultaneous emergency incidents (e.g., flooding or fire hazards resulting in the release of hazardous materials).
2. The severity of consequences of human injury that may occur (acute, delayed, and/or chronic health effects), the number of possible injuries and deaths, and the associated high-risk groups.
3. The severity of consequences on critical facilities (e.g., hospitals, fire stations, police departments, communication centers).
4. The severity of consequences of damage to property (temporary, repairable, permanent).

5. The severity of consequences of damage to the environment (recoverable, permanent).

Cause-consequence risk evaluation combines event tree and fault tree analysis to relate specific accident consequences to causes.[1] The process of this cause-consequence evaluation usually proceeds as follows:

1. Select an event to be evaluated.
2. Describe the safety system(s)/procedure(s)/factor(s) that interfere with the path of the accident.
3. Perform a fault tree analysis to determine the event or function that failed.
4. Perform an event tree analysis to find the path(s) an accident may follow.
5. Rank the results qualitatively on a basis of severity of consequences or perhaps quantitative calculations.

Following this introductory section, topics to be addressed include:

Section 18.2 Risk Characterization
Section 18.3 Cause-Consequence Analysis
Section 18.4 Qualitative Hazard Risk Analysis
Section 18.5 Quantitative Hazard Risk Analysis
Section 18.6 Uncertainties/Limitations
Section 18.7 Public Perception of Risk
Section 18.8 Risk Communication

18.2 RISK CHARACTERIZATION

Risk characterization estimates the health risk associated with the process under investigation. The result of this characterization is a number that represents the probability of adverse health effects from that process or from a substance released in that process. For instance, a risk characterization for all effects from an incineration process might be expressed as one additional cancer case per 1 million people.

Once a risk characterization is made, the meaning of that risk must be evaluated. Public health agencies generally only consider risk greater than 10 in 1 million (10^{-5} or 10×10^{-6}) to be significant risks warranting action.

Several major types of risk are detailed below.

1. *Individual Risk* – This provides a measure of the risk to a person in the vicinity of a hazard/accident, including the nature of the injury or other undesired outcomes, and the likelihood of occurrence. Individual risk is generally expressed in terms of a likelihood or probability of a specified undesired outcome per unit of time. For example (as indicated above), the

individual risk of a fatality at a particular location near a hazardous installation might be expressed as 1 in 100,000 per year, or 10^{-5} per year.

2. *Maximum Individual Risk (MIR)* – This is the maximum risk to an individual person. This individual is considered to have a 70-year lifetime of exposure to a process or a chemical. For discharge from a stack, for instance, the individual is considered to live downwind of the stack, never leaving this spot for every hour and every day of a 70-year life.

3. *Population Risk (PR)* – This is a risk to a population, expressed as a given number of deaths per thousand or per million people.

4. *Societal Risk* – This represents a measure of the risk to a group of people, including the risk of incidents potentially affecting more than one person. Individual risk (see above) is generally not significantly affected by the number of people involved in an incident. The risk to a person at a particular location depends on the probability of occurrence of the hazardous event, and on the probability of an adverse impact at that location should the event occur.

5. *Risk Indices* – A risk index is a single-number measure of the risk associated with a facility. Some risk indices are qualitative or semi-quantitative, ranking risks in various general categories. Risk indices may also be quantitative averages or benchmarks based on other risk measures.

Other general indices are provided in Table 18.2.1.

The reader should also note that the risk to people can be defined in terms of injury or fatality. The use of injuries as a basis of risk evaluation may be less disturbing than the use of fatalities. However, this introduces problems associated with degree of injury and comparability between different types of injuries. Further complications can arise in a risk assessment when dealing with multiple hazards. For example, how are second-degree burns, fragment injuries, and injuries due to toxic gas exposure combined? Even where only one type of effect (e.g., threshold toxic exposure) is being evaluated, different durations of exposure can markedly affect the severity of injury.

18.3 CAUSE-CONSEQUENCE ANALYSIS

Cause-consequence analysis serves to characterize the physical effects resulting from a specific incident and the impact of these physical effects on people, the environment, and property (causes are discussed throughout Chapter 16). Some consequence models or equations (see Chapter 17) used to estimate the potential for damage or injury fall into several categories.[2]

TABLE 18.2.1 Presentation of Measures of Risk

Risk Measure	Presentation Format
Equivalent Social Cost index:	A single-number index value representation
Fatal accident rate (discussed in Section 18.5):	A point estimate of fatalities/10^8 exposure hours
Individual hazard index:	An estimate of peak individual risk of FAR
Average rate of death:	A number representing the estimated average number of fatalities per unit time
Mortality index:	A single-value representation of consequence

1. *Source models* describe the release rate of material from the process equipment into the external environment, and the rate of release of spilled vapors and volatile liquids into the atmosphere.
2. *Dispersion models* describe the behavior of the released material in the atmosphere following its release.
3. *Fire and explosion models* describe the magnitude and physical effects (heat radiation, explosion overpressure) resulting from a fire or explosion.
4. *Effect models* describe the impact of the physical effects of a fire, explosion, or toxic gas release on exposed people, the environment or property, based on the results of the source, dispersion, and fire and explosion models.

Other models may be used to consider the effects of escape or evacuation, sheltering, protective equipment, or other factors (e.g., water contamination) that may be considered in a risk study.

Likelihood estimation, sometimes called *frequency estimation*, characterizes the probability of occurrence for each potential incident considered in the analysis. The major tools used for likelihood estimation are listed below.[3]

1. *Historical data* are used for facility types where there is extensive experience available from similar or identical installations.
2. *Failure sequence modeling techniques* such as fault tree analysis or event tree analysis are used to estimate the likelihood of incidents in facilities where historical data is unavailable, or is inadequate to accurately estimate the likelihood of the hazardous incidents of concern. Other modeling techniques may be required to consider the impact of external events (earthquakes, floods, etc.), common cause failures, and human factors and human reliability.
3. *Expert judgment* quantifies an expert's state of knowledge or perceptions of the likelihood of an incident. This knowledge may be based on historical data, insights gained from models, experience, or a combination of these factors.

In a more quantitative sense, cause-consequence analysis may be viewed as a blend of fault tree end event tree analysis (discussed in the two preceding chapters) for evaluating potential accidents. A major strength of cause-consequence analysis is its use as a communication tool. For example, a cause-consequence diagram displays the interrelationships between the accident outcomes (consequences) and their basic causes. The method can be used to quantify the expected frequency of occurrence of the consequences if the appropriate data are available.

Staffing and cost requirements are as follows: Cause-consequence analysis is best performed by a small team (2 to 4 people) with a variety of experience. One team member should be experienced in cause-consequence analysis (or fault tree and event tree analysis), with the remaining members having experience in the operations and interactions of the systems included in the analysis. Time and cost requirements for cause-consequence analysis are highly dependent on the number, complexity, and level of resolution of the events included in the analysis. Scoping-type analyses for several initiating events can usually be accomplished in a week or less. Detailed cause-consequence analyses may require two to six weeks, depending on the complexity of any supporting fault tree analyses.

18.4   QUALITATIVE HAZARD RISK ANALYSIS

Obviously, hazard risk information can be presented either qualitatively or quantitatively. This Section provides qualitative risk procedures and information. Several of these approaches are given below.

As described in the previous chapter, consequences of accidents can be classified qualitatively by the degree of severity. General factors that help to determine the degree of severity for chemical releases are the concentration in which the hazard is released, length of time that a person within the environment is exposed to the hazard, and the toxicity of the hazard. The worst-case consequence or scenario is defined as a conservatively high estimate of the most severe accident identified. On this basis, one can rank the consequences of accidents into low, medium, and high degrees of severity. A low degree of severity means that the hazard is nearly negligible, and the injury to person, property, or the environment is observed only after an extended period of time. The degree of severity or risk is considered to be medium when the accident is serious but not catastrophic, the toxicity of the chemical released is great, or the concentration of a less toxic chemical is large enough to cause injury or death to individuals and damage to the environment unless immediate action is taken. There is a high degree of risk when the accident is catastrophic or the concentration and toxicity of a hazard are large enough to cause injury or death to many individuals, and there is long-term damage to the surrounding environment. Figure 18.4.1 provides a graphical qualitative representation of the risk associated with both the probability of occurrence and consequences .[4] In line with the above, one might assign the following definitions to the terms in

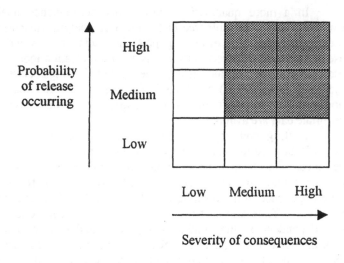

Major risk concerns

Figure 18.4.1. Qualitative probability consequence risk analysis

Figure 18.4.1 for a chemical release. Qualitative definitions regarding the probability of occurrence are as follows:

Low: The probability of occurrence is considered unlikely during the expected lifetime of the facility assuming normal operation and maintenance.

Medium: The probability of occurrence is considered possible during the expected lifetime of the facility.

High: The probability of occurrence is considered sufficiently high to assume the event will occur at least once during the expected lifetime of the facility.

Definitions of severity regarding the consequences are as follows:

Low: The chemical is expected to move into the surrounding environment in negligible concentrations. Injuries are expected only for exposure over extended periods or when individual personal health conditions create complications.

Medium: The chemical is expected to move into the surrounding environment in concentrations sufficient to cause serious injuries and/or deaths unless prompt and effective corrective action is taken. Death and/or injuries are expected only for exposure over extended periods or when individual personal health conditions create complications.

High: The chemical is expected to move into the surrounding environment concentrations sufficient to cause serious injuries and/or deaths upon exposure. Large numbers of people are expected to be affected.

Once the system components and their failure modes have been identified, the acceptability of risks taken as a result of such failures must be determined. The risk assessment process yields more comprehensive and better results when reliable statistical and probability data are available. In the absence of such data, the results are a strong function of the engineering judgment of the design team. The important issue is that both the severity and probability (frequency) of the accident must be taken into account.

Table 18.4.1 summarizes another method of risk assessment that can be applied to an accident system failure.[5] Both probability and consequence have been ranked on a scale of 0 to 1 with table entries being the sum of probability and consequence. The acceptability of risk is a major decision and can be described by dividing the situations presented in Table 18.4.1 into unacceptable, marginally acceptable, and acceptable regions. Figure 18.4.2 graphically represents this risk data.[6,7]

Each cell in the matrix (Table 18.4.2) is assigned a risk ranking as indicated by the letters. In this approach, an "A" level risk corresponds to a very severe consequence with a high likelihood of occurrence. Action must be taken, and it must be taken promptly. At the other end of the scale, a "E" level risk is of little or no consequence with a low likelihood of occurrence, and no action is needed or justified. For example, a level "C" risk might warrant mitigation with engineering and/or administrative controls or may represent risks that are acceptable with controls and procedures.

TABLE 18.4.1 Risk Data Summary

	Severity†									
Probability*	0.1	0.2	0.3	0.4	0.5	0.6	0.7	0.8	0.9	1.0
0	0.1	0.2	0.3	0.4	0.5	0.6	0.7	0.8	0.9	1.0
0.1	0.2	0.3	0.4	0.5	0.6	0.7	0.8	0.9	1.0	1.1
0.2	0.3	0.4	0.5	0.6	0.7	0.8	0.9	1.0	1.1	1.2
0.3	0.4	0.5	0.6	0.7	0.8	0.9	1.0	1.1	1.2	1.3
0.4	0.5	0.6	0.7	0.8	0.9	1.0	1.1	1.2	1.3	1.4
0.5	0.6	0.7	0.8	0.9	1.0	1.1	1.2	1.3	1.4	1.5
0.6	0.7	0.8	0.9	1.0	1.1	1.2	1.3	1.4	1.5	1.6
0.7	0.8	0.9	1.0	1.1	1.2	1.3	1.4	1.5	1.6	1.7
0.8	0.9	1.0	1.1	1.2	1.3	1.4	1.5	1.6	1.7	1.8
0.9	1.0	1.1	1.2	1.3	1.4	1.5	1.6	1.7	1.8	1.9
1.0	1.1	1.2	1.3	1.4	1.5	1.6	1.7	1.8	1.9	2.0

* Corresponds to ordinate in Figure 18.4.2
† Corresponds to abscissa in Figure 18.4.2

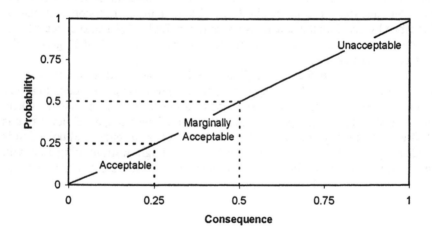

Figure 18.4.2. Graphic representation of risk data.

A qualitative approach is provided in Table 18.4.2

TABLE 18.4.2 Ranking of Severity and Consequence

Consequence	Frequency				
	1	2	3	4	5
1	A	AB	B	BC	C
2	AB	B	BC	C	CD
3	B	BC	C	CD	D
4	BC	C	CD	D	DE
5	C	CD	D	DE	E

Qualitative hazard risk policies for a company, e.g., Flynn and Theodore Enterprises (FATE), could take the form:

FATE will not knowingly pose a greater hazard risk to the public than it does to its own employees.
FATE will not expose its employees or neighbors to hazard risks that are considered unacceptable, based on industry practice and available technology.
FATE will comply with all applicable regulations and industry guidelines related to hazard risks, and will adopt its own standards where regulations do not exist or are inadequate.
FATE will neither undertake nor continue any operations whose associated hazard risks it does not understand or cannot control at a safe level.

Other possible hazard risk policies could include:

The average individual fatal risk level for the public should be less than _____.
The maximum individual fatal risk for employees should be less than _____.
The probability of one or more public fatalities should be less than _____.
The probability of 100 or more public fatalities should be less than _____.
The probability of one or more serious public injuries should be less than _____.
The probability of 100 or more serious public injuries should be less than _____.

Once again, the above can apply to either an annual or lifetime basis.

18.5   QUANTITATIVE HAZARD RISK ANALYSIS

Quantitative risk analyses usually produces single-number estimates. Although there are sufficient uncertainties associated with these quantitative numerical values (see next Section) they serve a valuable function. These may be used to compare one risk with another in a quantitative sense or occasionally employed in an absolute sense.

A simple procedure that can be employed in some estimates is to imagine all possible ways in which an accident can occur. This procedure is used with fault and event trees (see Chapters 20 and 21). For certain situations, the probability, of an accident can be calculated as the product of individual probabilities:

$$P = P_1 P_2 P_3 P_4 \cdots \qquad (18.5.1)$$

Even limited historical data and tests on the numbers P_1, P_2, P_3, and P_4 usually lead to results that give an extremely small overall probability.

In many applications the risk may be obtained by simply examining the frequency and consequence(s) associated with a hazard. The risk (R), consequence (C) and frequency (F) can be related through the equation:

$$R = (C)(F) \qquad (18.5.2)$$

One of the most popular risk policies employed by industry is the FAR Concept (Fatal Accident Rate). FAR represents the number of fatal accidents per 1,000 workers in a working lifetime (10^8 hr), where a working lifetime is assumed to be approximately 10^5 hrs. An acceptable FAR (by industries standards) is 4.0. This is made up of:

1. Ordinary industrial risks;
   FAR = 2
2. Chemical risks;
   FAR = 2

In addition, each individual chemical risk should not exceed 0.4.

Based on the above definition, if the hazard rate is the rate at which "dangerous" incidents occur, then it should not occur more often than:

0.4 occasion/10^8 working hrs

or

1.0 (once)/2.5×10^8 working hrs

or

once/30,000 yrs

This is approximately equivalent to 3×10^{-5} occasions/yr, i.e., the probability of occurrence should not exceed 3×10^{-5} $(yr)^{-1}$. If the worker is killed every 10th time the incident occurs, then the target hazard rate is:

once/3,000 yrs

or

$$3 \times 10^{-4} \text{ occasions/yr}$$

For workers in a chemical plant, the FAR can be calculated as follows:[8]

$$FAR = \frac{10^8}{8760} \times F \times \frac{D}{N} \qquad (18.5.3)$$

where: FAR = fatal accident rate
   F = frequency of the events in years
   D = expected number of fatalities, given the event
   N = average number of exposed individuals on each shift

The total individual risk at each location is equal to the sum of the individual risks at that location from all incident outcome cases.

$$IR_{x,y} = \sum_{i=1}^{n} IR_{x,y,i} \qquad (18.5.4)$$

where: $IR_{x,y}$ = total individual risk of fatality at geographical location x,y
   $IR_{x,y,i}$ = individual risk of fatality at geographical location x,y from incident outcome case i
   n = total number of incident outcome cases

The FAR concept application is expanded upon in Illustrative Example 6.

18.6   UNCERTAINTIES/LIMITATIONS

The reader should also note that this topic received treatment in Chapter 13, Part III. Obviously, knowledge about the present and the future is never completely accurate. Inadequate knowledge is usually the largest cause of uncertainty. The inadequacy of the knowledge means that the full extent of the uncertainty is also unknown. Uncertainty due to variability occurs when a single number (as often employed in risk analysis) is used to describe something that truly has multiple or variable values. Variability is often ignored by using values based on the mean of all the values occurring within a group.

Information or sources of uncertainties and limitations are available in the literature.[9] Some of this material is provided below.

System description:
- Process description of drawings are incorrect or out of date
- Procedures do not represent actual operation
- Site area maps and population data may be incorrect or out of date

- Weather data from the nearest available site may be inappropriate

Hazard identification:
- Recognition of major hazards may be incomplete
- Screening techniques employed for selection of hazards for further evaluation may omit important cases

Consequence techniques:
- Calculational burden (even with computers) due to the number of dispersion modeling variables
    Uncertainties in physical modeling -
    - Inappropriate model selection
    - Incorrect or inadequate physical basis for model
    - Inadequate validation
    - Inaccurate model parameters
    Uncertainties in physical model data -
    - Input data (composition, temperature, pressure)
    - Source terms for dispersion and other models
    Uncertainties in effects modeling -
    - Animal data inappropriate for humans (especially for toxicity)
    - Mitigating effects may be omitted
    - Lack of epidemiological data on humans of the same sex, age, education, etc.

Frequency techniques:
    Uncertainties in modeling -
    - Extrapolation of historical data to larger scale operations may overlook hazards introduced by scale up to larger equipment
    - Limitation of fault tree theory requires system simplification
    - Incompleteness in fault and event tree analysis
    Uncertainties in data -
    - Data may be inaccurate, incomplete, or inappropriate
    - Data from related activities might not be directly applicable
    - Data generated by expert judgment may be inaccurate
    - Improper or incomplete characterization of the general population

Risk estimation:
    Assumption of symmetry -
    - Uniform wind rose rarely occurs
    - Uniform ignition sources may be incorrect
    - Single point source for all incidents may be inaccurate
    Assumptions to reduce the depth of treatment -
    - A single condition of wind speed and stability may be too restrictive
    - A limited number of ignition cases can reduce accuracy

-   General problem with quality of data
-   Some risks cannot be quantified

The reader should note that since many risk assessments have been conducted on the basis of fatal effects, there are also uncertainties on precisely what constitutes a fatal dose of thermal radiation, blast effect, or a toxic chemical.   Where it is desired to estimate injuries as well as fatalities, the consequence calculation can be repeated using lower intensities of exposure leading to injury rather than death. In addition, if the adverse health effect (e.g. associated with a chemical release) is delayed, the cause may not be obvious. This applies to both chronic and acute emissions and exposures.

Another problem with risk estimates is that they are usually based on very conservative assumptions. Thus, the analyses may result in a calculation that presents too high a risk. Unnecessary equipment or procedures may have to be installed/instituted at a facility to reduce the calculated risk. In an effort to better understand the significance of risk analyses, it is often helpful to place the estimated risks in perspective with other risks.

In attempt to handle uncertainties and unknowns, Theodore has proposed a modified version of the standard Delphi panel approach that the authors have modestly defined as the FLTA (an acronym for the Flynn-Theodore Approach).[10]   In order to generate "better" estimates, several knowledgeable individuals within and perhaps outside the organization are asked to independently provide estimates, with explanatory details on these estimates. Each individual in the panel is then allowed to independently review all responses. The cycle is then repeated until the group's responses approach convergence.

18.7 PUBLIC PERCEPTION OF RISK

In making an effort to understand the significance of risk analyses, it is helpful to place the estimated risks in the same perspective as other everyday risks that have been determined by a similar methodology. Table 18.7.1 lists a number of risks for comparison.   These have been derived from actual statistics and reasonable estimates.[11,12] People often overestimate the frequency and seriousness of dramatic, sensational, dreaded, well-publicized causes of death and underestimate the risks from more familiar, accepted causes that claim lives one by one. Indeed, risk estimates by "experts" and lay people (or "the public") on many key environmental problems differ significantly. This problem and the reasons for it are extremely important because in our society the public generally does not trust experts to make important risk decisions alone.

To make wise judgments requires that individuals know what experts' estimates of the risks are, what it would cost (in terms of their other values) to reduce them, and how certain and free of bias the estimates are.   Scientific precision is not needed, but a sense of whether a risk is "big, " "medium," "small," or "infinitesimal" is. The challenge of risk communication is to provide

TABLE 18.7.1   Lifetime Risks to Life Community Faced by Individuals

Cause of Risk	Lifetime (70 yr) Risk, per Million Individuals
Cigarette smoking	252,000
All cancers	196,000
Construction	42,700
Agriculture	42,000
Police killed in line of duty	15,400
Air pollution (Eastern United States)	14,000
Motor vehicle accidents (traveling)	14,000
Home accidents	7,700
Frequent airplane traveler	3,500
Pedestrian hit by motor vehicle	2,900
Alcohol, light drinker	1,400
Background radiation at sea level	1,400
Peanut butter, 4 tablespoons per day	560
Electrocution	370
Tornado	42
Drinking water containing chloroform at maximum allowable EPA limit	42
Lightning	35
Living 70 years in zone of maximum impact from modern municipal airports	1
Smoking 1.4 cigarettes	1
Drinking 0.5 L of wine	1
Traveling 10 mi by bicycle	1
Traveling 30 mi by car	1
Traveling 1000 mi by jet plane (air crash)	1
Traveling 6000 mi by jet plane (cosmic rays)	1
Drinking water containing trichloroethylene at maximum allowable EPA Limit	0.1

this information in ways so that it can be properly incorporated in the view of people who have little time or patience for arcane scientific discourse. Success in risk communication is not to be measured by whether the public chooses the set of outcomes that minimizes risk as estimated by the experts; it is achieved instead when those outcomes are knowingly chosen by a well-informed public. This topic is treated in the next section.

18.8   RISK COMMUNICATION

Environmental risk communication is one of the more important problems that industry faces. Since the mid 1980s, public concerns about the environment have grown faster than concern about virtually any other national problem.[13] There are two major categories or risk: nonfixable and fixable. Nonfixable risks can never substantially be reduced, such as cancer-causing sunlight or cosmic radiation. Fixable risks can be reduced, and include those risks that are both large and small. There are so many of these fixable risks that all of them can never be successfully attacked, so choices must be made. When it comes to risk reduction, the outcome or end result should be to obtain the most reduction possible, taking into account that people fear some risks more than others. This essentially and often means that the technical community should concentrate on the big fixable targets, and leave the smaller ones to later.

Risk communication comes into play because citizens ultimately determine which risks government agencies attack. On the surface, it appears practical to remedy the most severe risks first, leaving the others until later or perhaps, if the risks are small enough, never remedying the others at all. However, the behavior of individuals in everyday life often does not conform with this view.

Two environmental issues that dramatized the need for intelligent and proper risk communication were:

1. Gasoline that contains lead
2. Ocean incineration

Although specific details are beyond the scope of this text, information on both types is provided in the literature.[14]

Seven Cardinal Rules of Risk Communication

There are no easy prescriptions for successful risk communication. However, those who have studied and participated in recent debates about risk generally agree on seven cardinal rules. These rules apply equally well to the public and private sectors. Although many of these rules may seem obvious, they are continually and consistently violated in practice. Thus, a useful way to read these rules is to focus on why they are frequently not followed.[15]

1. **Accept and involve the public as a legitimate partner.** A basic tenet of risk communication in democracy is that people and communities have a right to participate in decisions that affect their lives, their property, and the things they value.

   Guidelines: Demonstrate respect for the public and underscore the sincerity of effort involving the community early, before important decisions are made. Involve all parties that have an interest or stake in the issue under consideration. If you are a government employee, remember that you work for the public. If you do not work for the government, the public still holds you accountable.

Point to Consider: The goal in risk communication in a democracy should be to produce an informed public that is involved, interested, reasonable, thoughtful, solution-oriented, and collaborative. It should not be to diffuse public concerns or replace action.

2. **Plan carefully and evaluate your efforts.** Risk communication will be successful only if carefully planned.

Guidelines: Begin with clear, explicit risk communication objectives, such as providing information to the public, motivating individuals to act, stimulating response to emergencies, and contributing to the resolution of conflict. Evaluate the information you have about the risks and know its strengths and weaknesses. Classify and segment the various groups in your audience. Aim your communications at specific subgroups in your audience. Recruit spokespeople who are good at presentation and interaction. Train your staff, including technical staff, in communication skills; reward outstanding performance. Whenever possible, pretest your messages. Carefully evaluate your efforts and learn from your mistakes.

Points to Consider: There is no such entity as "the public"; instead, there are many publics, each with its own interests, needs, concerns, priorities, preferences, and organizations. Different risk communication goals, audiences, and media require different risk communication strategies.

3. **Listen to the public's specific concerns.** If you do not listen to the people, you cannot expect them to listen to you. Communication is a two-way activity.

Guidelines: Do not make assumptions about what people know, think or want done about risks. Take the time to find out what people are thinking. Use techniques such as interviews, focus groups, and surveys. Let all parties that have an interest or stake in the issue be heard. Identify with your audience and try to put yourself in their place. Recognize people's emotions. Let people know that you understand what they said, addressing their concerns as well as yours. Recognize the "hidden agendas," symbolic meanings, and broader economic or political considerations that often underlie and complicate the task of risk communication.

Point to Consider: People in the community are often more concerned about such issues as trust, credibility, competence, control, voluntariness, fairness, caring, and compassion than about mortality statistics and the details of quantitative risk assessment.

4. **Be honest, frank, and open.** In communicating risk information, trust and credibility are your most precious assets.

Guidelines: State your credentials; but do not ask or expect to be trusted by the public. If you do not know an answer or are uncertain, say so. Get back to people with answers. Admit mistakes. Disclose risk information as soon as possible (emphasizing any reservations about reliability). Do not minimize or exaggerate the level of risk. Speculate only with great caution. If in doubt, lean toward sharing more information, not less, or people may think you are hiding something. Discuss data uncertainties, strengths and weaknesses, including the ones identified by other credible sources. Identify worst-case estimates as such, and cite ranges of a risk estimate when appropriate.

Point to Consider: Trust and credibility are difficult to obtain. Once lost they are almost impossible to regain completely.

5. **Coordinate and collaborate with other credible sources.** Allies can be effective in helping you communicate risk information.

Guidelines: Take time to coordinate all inter-organizational and intra-organizational communications. Devote effort and resources to the slow, hard work of building bridges with other organizations. Use credible and authoritative intermediates. Consult with others to determine who is best able to answer questions about risk. Try to issue communications jointly with other trustworthy sources (for example, credible university scientists and/or professors, physicians, or trusted local officials).

Point to Consider: Few things make risk communication more difficult than conflicts or public disagreements with other credible sources.

6. **Meet the needs of the media.** The media are a prime transmitter of information on risks; they play a critical role in setting agendas and in determining outcomes.

   Guidelines: Be open and accessible to reporters. Respect their deadlines. Provide risk information tailored to the needs of each type of media (e.g., graphics and other visual aids for television). Prepare in advance and provide background material on complex risk issues. Do not hesitate to follow up on stories with praise or criticisms, as warranted. Try to establish long-term relationships of trust with specific editors and reporters.

   Point to Consider: The media are frequently more interested in politics than in risk; more interested in simplicity than in complexity; more interested in danger than in safety.

7. **Speak clearly and with compassion.** Technical language and jargon are useful as professional shorthand, but they are barriers to successful communication with the public.

   Guidelines: Use simple, non-technical language. Be sensitive to local norms, such as speech and dress. Use vivid, concrete images that communicate on a personal level. Use examples and anecdotes that make technical risk data come alive. Avoid distant, abstract, unfeeling language about deaths, injuries, and illnesses. Acknowledge and respond (both in words and with action) to emotions that people express that can include anxiety, fear, anger outrage, and helplessness. Acknowledge and respond to the distinctions that the public views as important in evaluating risks, e.g., voluntariness, controllability, familiarity, dread, origin (natural or man-made), benefits, fairness, and catastrophic potential. Use risk comparisons to help put risks in perspective, but avoid comparisons that ignore distinctions which people consider important. Always try to include a discussion of actions that are under way or can be taken. Tell people what you cannot do. Promise only what you can do, and be sure to do what you promise.

Points to Consider: Regardless of how well you communicate risk information; some people will not be satisfied. Never let your efforts to inform people about risks prevent you from acknowledging and saying that any illness, injury, or death is a tragedy. And finally, if people are sufficiently motivated, they are quite capable of understanding complex risk information, even if they may not agree with you.

18.9 ILLUSTRATIVE EXAMPLES

Example 18.1

Briefly describe the differences between individual and societal risks.

Solution 18.1

Individual risk is a measure of the risk to one person regardless of the amount of people present. For example, the likelihood of an incident causing a fatality at an office building is subject to a certain individual risk. If this individual risk is independent of the number if people present, it is then the same for each of the 400 people in the building during office hours and for the single guard working in the building when no one else is present.

Societal risk is a measure of the risk to a group of people. It is most often expressed in terms of a frequency distribution of multiple casualty events and it can also be expressed in terms similar to individual risk. The calculation of societal risk requires the same frequency and consequence information as individual risk. For example, the likelihood of 10 fatalities at a specific location is a type of societal risk measure. Additionally, societal risk estimation requires a definition of the population at risk around the facility. This definition can include the population type (e.g., residential, industrial, school), the likelihood of people being present, or mitigation factors.

Example 18.2

Refer to Example 2 in Section 16.8. Discuss some of the major effects associated with the leak.

Solution 18.2

Major effects:
1. Fatalities; injuries

2. High release of gas into the community; fatality
3. High release of gas into the community; injury

Example 18.3

Discuss the problems in valuing life.

Solution 18.3

There are various estimates of the value of life, ranging at the high end from about $2 million per life to about $200,000 per life. The choice of value in this range depends a great deal on how one subscribes to the various ethical bases upon which the estimates are based. Nonetheless, the whole range spans approximately an order of magnitude.

Example 18.4

List the four principal agencies that regulate risk issues.

Solution 18.4

1. The EPA, an independent agency, administers air, water, and toxic substance legislation.
2. The OSHA, a part of the Department of Labor, sets exposure standards and safety rules for work places.
3. The Food and Drug Administration (FDA) regulates foods, drugs, and cosmetics; it is housed in the Department of Health and Human Services, and reports to the Assistant Secretary for Health.
4. The Consumer Products Safety Commission (CPSC), an independent agency, controls the packaging, labeling, and distribution of a broad range of toys, clothes, electronics, and other products.

Example 18.5

Discuss the significance of the following figure.

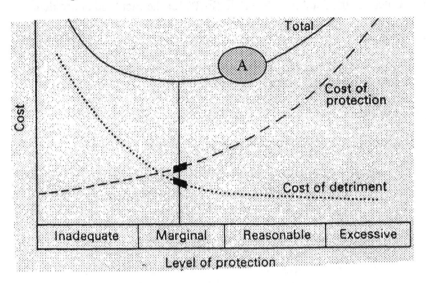

Solution 18.5

The figure above allows one to compare the cost of the detriment (associated with an accident) with the cost of improved protection. As seen on the graph, for low levels of cost protection, the risk of detriment costs are unreasonably high. However, for high levels of cost protection, the cost of detriment are significantly low. Therefore, a cost-benefit analysis should be performed in order to determine a reasonable cost for an acceptable level of protection while keeping the detriment costs to a minimum. From a plant's perspective, the level of protection should be set at point A. From a purely economic point of view, this point represents the minimum cost. Other factors such as regulation requirements, good will, etc., can change this.

Example 18.6

You have been hired as a consultant to a company administrator who has a limited budget for the mitigation of hazards in a certain chemical plant. The plant employs two kinds of workers: day employees that work one 8 hour shift daily, and shift employees that rotate through three 8 hour shifts each day. A HAZOP-HAZAN report reveals two kinds of accidents are possible during plant operation.

Accidents of the first kind result in the death of one-day employee per incident and occur with a frequency of 2.92×10^{-5} accidents per year. Accidents of the second kind result in the death of 100 shift workers per incident and occur with a frequency of 8.76×10^{-7} accidents per year.

1. Calculate the Fatal Accident Rate (FAR) for the first kind of accident.

2. Calculate the Fatal Accident Rate (FAR) for the second kind of accident.

3. What considerations would influence your recommendation on the allocation of funds to reduce these hazards?

Solution 18.6

1. Calculate the FAR for the first type of accident.

    $$FAR = (1 \text{ fatality/accident}) (2.92 \times 10^{-5} \text{ accidents/year})$$
    $$(1 \text{ yr/365 days}) (1 \text{ day/8 hr}) (10^8 \text{ hours/1000 lifetimes})$$
    $$= 1 \text{ fatality/1000 worker lifetimes}$$

2. Calculate the FAR for the second type of accident.

    $$FAR = (100 \text{ fatalities/accident}) (8.76 \times 10^{-7} \text{ accidents/year})$$
    $$(1 \text{ yr/365 days}) (1 \text{ day/24 hr}) (10^8 \text{ hours/1000 lifetimes})$$
    $$= 1 \text{ fatality/1000 worker lifetimes}$$

3. Provide advice on the allocation of funds.

    (While recognizing that any answer to this question is likely to be incomplete, the following is suggested as thought provoking.)
    FARs for the two types of accidents are equal and it is likely that over long periods of time an equal number of deaths may be expected from both types of accidents. This does not mean that equal consequences will result to the company from both types of accidents. Accidents of the first type involve a low (but, perhaps, steady) loss of life. Accidents of the second type, however, are sure to attract a great deal of attention in the media. Adverse public relations is nearly certain, as well as unfavorable attention from legislators and other public officials.
    Secondly, accidents of the second type have a catastrophic effect on production. Not only will the entire facility be demolished (in all likelihood), but a large fraction of the pool of trained personnel will be lost all at once. Who will train replacement personnel if everyone is lost in the disaster? Similar concerns make disruption in the community much greater for accidents of the second type. Some of these considerations can be factored into decision making as direct economic losses that increase the burden to both the company and the community for accidents of the second type.

One view (the only acceptable view to many people) is to give priority to the prevention of both types of accidents.

Considerations for the allocation of funds should include economic and public relations factors. Identical FARs do not necessarily result in equal consequences; effects on plant production, public relations, and other factors must also be considered. Finally, FARs in excess of 4.0 are not acceptable.

18.10   SUMMARY

1. Risk analysis is an assessment of the likelihood (probability) of an accidental release of a hazardous material and the actual consequences that might occur, based on the estimated vulnerable zones. It provides an estimation of the likelihood (probability) of an accidental release, the severity of consequences of human injury that may occur, the severity of consequences on critical facilities, the severity of consequences of damage to property, and the severity of consequences of damage to the environment.

2. Risk characterization estimates the health risk associated with the process under investigation. The result of this characterization is a number that represents the probability of adverse health effects from that process or from a substance released in that process. The major types of risk include: Individual Risk, Maximum Individual Risk (MIR), Population Risk (PR), Societal Risk, and Risk Indices.

3. Cause-consequence analysis serves to characterize the physical effects resulting from a specific incident and the impact of these physical effects on people, the environment, and property. Some consequence models or equations used to estimate the potential for damage or injury are as follows: Source Models, Dispersion Models, Fire & Explosion Models, and Effect Models. Likelihood estimation (frequency estimation), characterizes the probability of occurrence for each potential incident considered in the analysis. The major tools used for likelihood estimation are as follows: Historical Data, Failure sequence modeling techniques, and Expert Judgment.

4. Consequences of accidents can be classified qualitatively by the degree of severity. General factors that help to determine the degree of severity for chemical releases are the concentration in which the hazard is released, length of time that a person within the environment is exposed to the hazard, and the toxicity of the hazard.

5. Quantitative risk analyses usually produces single-number estimates. These may be used to compare one risk with another in a quantitative sense or occasionally employed in an absolute sense. One of the most popular risk policies employed by industry is the FAR Concept (Fatal Accident Rate). FAR represents the number of fatal accidents per 1,000 workers in a working lifetime (10^8 hr), where a working lifetime is assumed to be approximately 10^5 hrs. An acceptable FAR (by industries standards) is 4.0.

6. Knowledge about the present and the future is never completely accurate. Inadequate knowledge is usually the largest cause of uncertainty. Uncertainty due to variability occurs when a single number is used to describe something that truly has multiple or variable values. Variability is often ignored by using values based on the mean of all the values occurring within a group.

7. In making an effort to understand the significance of risk analyses, it is helpful to place the estimated risks in the same perspective as other everyday risks that have been determined by a similar methodology.

8. There are two major categories or risk: nonfixable and fixable. Nonfixable risks can never substantially be reduced, such as cancer-causing sunlight. Fixable risks can be reduced, and include those risks that are both large and small. The outcome or end result of risk reduction should be to obtain the most reduction possible. This essentially and often means that the technical community should concentrate on the big fixable targets, and leave the smaller ones to later.

PROBLEMS

1. Why should risk procedures be employed?
2. List some considerations that should be taken into account when attempting to reduce risks arising because of a chemical release.
3. Refer to Problem 3 in Chapter 16. Discuss some corrective/protective measures that can be taken to reduce the risk associated with the leak.
4. Is zero risk possible or can all risk be eliminated?
5. Consider the following two financial risk investment schemes.
   - There is a 50% probability that the investment will succeed. If successful, the profit is $100,000; if unacceptable, there is no gain.
   - There is a 2% probability that the investment will succeed. If successful, the profit is $2,000,000; if unacceptable, there is no gain.
6. The Fatal Accident Rate (FAR) is the number of fatal accidents per 1,000 workers in a working lifetime (10^8 hr). A responsible chemical company typically displays a FAR equal to 2 for chemical process risks such as fires, toxic releases or spillage of corrosive chemicals. Identify potential problem areas that may develop for a company if acceptable FAR numbers are exceeded.

REFERENCES

1. L. Theodore and K. Morris, *"Accident and Emergency Management"*, Theodore Tutorials, East Williston, NY, 1998.
2. C.A. Wentz, "Hazardous Waste Management", McGraw Hill, New York City, 1989.
3. "Guidelines for Hazard Evaluation Procedure", CCPS, AICHE, 1992.
4. Author Unknown, "Technical Guidance For Hazard Analysis", USEPA,

Washington DC, 1979.

5. H.R. Kavranian, J. Rao and G. Brown, "Application of Hazard Evaluation Techniques & the Design of Potentially Hazardous Industrial Chemical Processes", NIOSH, Cincinnati, OH, 1992.

6. L. Slote, "Handbook of Occupational Safety and Health", John Wiley & Sons, New York City, 1987.

7. E.J. Henley and H. Kumamoto, "Reliability Engineering and Risk Assessment", Prentice Hall, Saddle River, 1981.

8. R. Perry and D. Green, "Perry's 'Chemical Engineers' Handbook", McGraw Hill, New York City, 1997.

9. Adapted from "Guidelines for Chemical Process Quantitative Risk Analyses", 2nd ed., CCPS, AICHE, New York City, 2000.

10. L. Theodore: personal notes, 2001.

11. T. Main, Inc., "Health Risk Assessment for Air Emission of Metals and Organic Components for the Perc Municipal Waste & Energy Facilities", PERC, Boston, MA, 1985.

12. R. Wilson and E.A. Croush, "Risk Assessment and Comparisons: An Introduction", Science, April, 1987.

13. G. Burke, B. Singh and L. Theodore, "Handbook of Environmental Management and Technologies", 2nd ed. John Wiley & Sons, New York City, 2000.

14. M. Russel, "Communicating Risk & a Concerned Public, "USEPA Journal, Washington, D.C., 1989.

15. USEPA, "Seven Cardinal Rules of Risk Communication," EPA/OPA/8700, Washington, D.C., 1988.

Part V

Quantitative Hazard Risk Assessment

This part of the book reviews and develops quantitative methods for the analysis of hazard conditions in terms of the frequency of occurrence of unfavorable consequences. Uncertainty characterizes not only the transformation of a hazard into an accident, disaster, or catastrophe, but also the effects of such a transformation. Measurement of uncertainty falls within the purview of mathematical probability. Accordingly, Chapter 19 presents fundamental concepts and theorems of probability used in risk assessment. Chapter 20 discusses special probability distributions and techniques pertinent to risk assessment, and Chapter 21 presents actual case studies illustrating techniques in hazard risk assessment that use probability concepts, theorems, and special distributions.

19

Hazard Risk Assessment Fundamentals

19.1 INTRODUCTION

Hazard, risk, failure, and reliability are interrelated concepts concerned with uncertain events and therefore amenable to quantitative measurement via probability. "Hazard" is defined as a potentially dangerous event. For example, the release of toxic fumes, a power outage, or pump failure. Actualization of the potential danger represented by a hazard results in undesirable consequences associated with risk.

Risk is defined as the product of two factors: (1) the probability of an undesirable event and (2) the measured consequences of the undesirable event. Measured consequences may be stated in terms of financial loss, injuries, deaths, or other variables.[1] *Failure* represents an inability to perform some required function. *Reliability* is the probability that a system or one of its components will perform its intended function under certain conditions for a specified period. The reliability of a system and its probability of failure are complementary in the sense that the sum of these two probabilities is unity. This chapter considers basic concepts and theorems of probability that find application in the estimation of risk and reliability.

After defining fundamental terms used in probability and introducing set notation for events, we consider probability theorems facilitating the calculation of the probabilities of complex events. Conditional probability and the concept of independence lead to Bayes' theorem and the means it provides for revision of probabilities on the basis of additional evidence. Random variables, their probability distributions, and expected values provide the means for application of more sophisticated tools of mathematical analysis in probability. The chapter concludes with a consideration of the problem of estimating two important expected values, the *mean* and the *variance*, on the basis of a random sample of observations.

19.2 PROBABILITY DEFINITIONS AND INTERPRETATIONS

Probabilities are nonnegative numbers associated with the outcomes of so-called random experiments. A random experiment is an experiment whose outcome is uncertain. Examples include throwing a pair of dice, tossing a coin, counting the number of defectives in a sample from a lot of manufactured items, observing the time to failure of a tube in a heat exchanger, a seal in a pump, or a bus section in an electrostatic precipitator. The set of possible outcomes of a random experiment is called the *sample space* and is usually designated by S. Then P(A), the probability of event A, is the sum of the probabilities assigned to the outcomes constituting the subset A of the sample space S.

Consider, for example, tossing a coin twice. The sample space can be described as

$$S = \{HH, HT, TH, TT\}$$

If probability $\frac{1}{4}$ is assigned to each element of S and A is the event of at least one head, then

$$A = (HH, HT, TH)$$

The sum of the probabilities assigned to the elements of A is $\frac{3}{4}$. Therefore, $P(A) = \frac{3}{4}$.

The description of the sample space is not unique. The sample space S in the case of tossing a coin twice could be described in terms of the number of heads obtained. Then

$$S = (0, 1, 2\}$$

Suppose probabilities $\frac{1}{4}$, $\frac{1}{2}$, $\frac{1}{4}$ are assigned to the outcomes 0, 1, and 2, respectively. Then A, the event of at least one head, would have for its probability,

$$P(A) = P\{1, 2\} = \frac{3}{4}$$

How probabilities, are assigned to the elements of the sample space depends on the desired interpretation of the probability of an event. Thus P(A) can be interpreted as *theoretical relative frequency* – that is, a number about which the relative frequency of event A tends to cluster as n, the number of times the random experiment is performed, increases indefinitely. This is the objective interpretation of probability. Under this interpretation, saying P(A) is $\frac{3}{4}$ in the above example means that if a coin is tossed twice, n times, the proportion of times one or more heads occur clusters about $\frac{3}{4}$ as n increases indefinitely.

As another example, consider a single valve that can stick in an open (O) or closed (C) position. The sample space can be described as follows:

$$S = \{O, C\}$$

Suppose that the valve sticks twice as often in the open position as it does in the closed position. Under the theoretical relative frequency interpretation, the probability assigned to element O in S would be $\frac{2}{3}$, twice the probability assigned to the element C. If two such valves are observed, the sample space S can be described as

$$S = \{OO, OC, CO, CC\}$$

Assuming that the two valves operate independently, a reasonable assignment of probabilities to the elements of S as listed could be $\frac{4}{9}$, $\frac{2}{9}$, $\frac{2}{9}$ and $\frac{1}{9}$. The reason for this assignment will become clear after consideration of the concept of independence in Section 19.5. If A is the event of at least one valve sticking in the closed position, then

$$A = \{OC, CO, CC\}$$

The sum of the probabilities assigned to the elements of A is $\frac{5}{9}$. Therefore, $P(A) = \frac{5}{9}$.

Probability P/A can also be interpreted subjectively as a measure of degree of belief, on a scale from 0 to 1, that the event A occurs. This interpretation is frequently used in ordinary conversation. For example, if someone says, "The probability that I will go to the movies tonight is 90%", then 90% is a measure of the person's belief that he will go to the movies. This interpretation is also used when, in the absence of concrete data needed to estimate an unknown probability on the basis of observed relative frequency, the personal opinion of an expert is sought.[2] For example, an expert might be asked to estimate the probability that the seals in a newly designed pump will leak at high pressures. The estimate would be based on the expert's familiarity with the history of pumps of similar design.

19.3 SET NOTATION FOR EVENTS

Various combinations of the occurrence of any two events A and B can be indicated in set notation as follows:

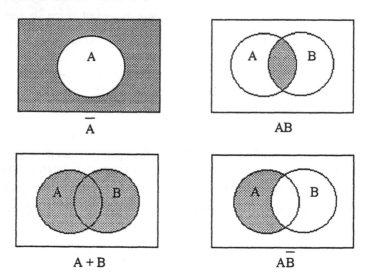

Figure 19.3.1. Venn diagrams.

$\overline{A}$	A does not occur.
$\overline{B}$	B does not occur.
A + B	A occurs or B occurs in the mutually inclusive sense to indicate the occurrence of A, B, or both A and B.
AB	A occurs and B occurs.
$A\overline{B}$	A occurs and B does not occur.

Venn diagrams (Fig. 19.3.1) provide a pictorial representation of these events.

In set terminology $\overline{A}$ is called the *complement* of A, that is, the set of elements in S that are not in A. Alternate notation for the *complement* of A is A^c or A'. AB is called the *intersection* of A and B, that is, the set of elements in both A and B. Alternate notation in the literature for the intersection of A and B is $A \cap B$. A + B is called the *union* of A and B, that is, the set of elements in A, B, or both A and B. Alternate notation for the union of A and B is $A \cup B$.

When events A and B have no elements in common they are said to be *mutually exclusive*. A set having no elements is called the null set and is designated by ϕ. Thus if events A and B are mutually exclusive then $AB = \phi$.

Note that the union of A and B consists of three mutually exclusive events: $A\overline{B}$, AB, $\overline{A}B$.

The algebra of sets-Boolean algebra-governs the way in which sets can be manipulated to form equivalent sets. The principal Boolean algebra laws used for this purpose are as follows.

Commutative Laws	$A + B = B + A$	(19.3.1)
	$AB = BA$	(19.3.2)
Associative Laws	$(A + B) + C = A + (B + C) = A + B + C$	(19.3.3)
	$(AB)C = A(BC) = ABC$	(19.3.4)
Distributive Laws	$A(B + C) = AB + AC$	(19.3.5)
	$A + BC = (A + B)(A + C)$	(19.3.6)
Absorption Laws	$A + A = A$	(19.3.7)
	$AA = A$	(19.3.8)
DeMorgan's	$\overline{(A + B)} = \overline{A}\,\overline{B}$	(19.3.9)
Laws	$\overline{AB} = \overline{A} + \overline{B}$	(19.3.10)

If the letter symbols for sets are replaced by numbers, the commutative and associative laws become familiar laws of arithmetic. In Boolean algebra the first of the two distributive laws, Eq. (19.3.5), has an analogous counterpart in arithmetic. The second, Eq. (19.3.6), does not. In risk analysis, Boolean algebra is used to simplify expressions for complicated events. For example, consider the event

$$T = (A + B + C)(\overline{C} + D)(E + F) \qquad (19.3.11)$$

where $\overline{C} = D + E$ and $\overline{D} = A + B$

Note that $C\overline{C} = \phi$, the null set.

After substituting

$$\overline{C} = D + E$$

and noting that

$$AD + AD = AD$$
$$BD + BD = BD$$
$$AD + BD = \phi$$

the expression for T in Eq. (19.3.11) reduces to:

Noting that:
$$T = (AE + BE + CD)(E + F) \qquad (19.3.12)$$

$$EE = E$$
$$CD = \phi$$

gives:

$$T = AE + AEF + BE + BEF \qquad (19.3.13)$$

The result indicates that the occurrence of any one of the events in the right member of Eq. (19.3.13) results in the occurrence of event T.

19.4 BASIC THEOREMS

The mathematical properties of $P(A)$, the probability of event A, are deduced from the following postulates governing the assignment of probabilities to the elements of a sample space, S.

1. $P(S) = 1$
2. $P(A) \geq 0$ for any event A
3. If $A_1, ..., A_n, ...$ are mutually exclusive, then
    $P(A_1 + A_2 + ... + A_n + ...) = P(A_1) + P(A_2) + ... + P(A_n) + ...$

In the case of a discrete sample space (i.e., a sample space consisting of a finite number or countable infinitude of elements), these postulates require that the numbers assigned as probabilities to the elements of S be nonnegative and have a sum equal to 1. These requirements do not result in complete specification of the numbers assigned as probabilities. The desired interpretation of probability must also be considered, as indicated in Section 19.2. The mathematical properties of the probability of any event are the same regardless of how this probability is interpreted. These properties are formulated in theorems logically deduced from the postulates above without the need for appeal to interpretation. Three basic theorems are:

Theorem 1. $P(\overline{A}) = 1 - P(A)$ $\qquad (19.4.1)$

Theorem 2. $0 \leq P(A) \leq 1$ $\qquad (19.4.2)$

Theorem 3. $P(A + B) = P(A) + P(B) - P(AB)$ $\qquad (19.4.3)$

Theorem 1 says that the probability that A does not occur is one minus the probability that A occurs. Theorem 2 says that the probability of any event lies between 0 and 1. Theorem 3, the addition theorem, provides an alternative way of calculating the probability of the union of two events as the sum of their probabilities minus the probability of their intersection. The addition theorem can be extended to three or more events. In the case of three events A, B, and C, the addition theorem becomes

$$P(A + B + C) = P(A) + P(B) + P(C) - P(AB) - P(AC) -$$
$$P(BC) + P(ABC) \qquad (19.4.4)$$

For four events A, B, C, and D, the addition theorem becomes

$$P(A + B + C + D) = \quad P(A) + P(B) + P(C) + P(D) - P(AB) - P(AC)$$
$$- P(AD) - P(BC) - P(BD) - P(CD) +$$
$$P(ABC) + P(ABD) + P(BCD) + P(ACD) -$$
$$P(ABCD) \qquad (19.4.5)$$

To illustrate the application of the three basic theorems (Eq. 19.4.1- 19.4.3), consider what happens when we draw a card at random from a deck of 52 cards. The sample space S may be described in terms of 52 elements, each corresponding to one of the cards in the deck. Assuming that each of the 52 possible outcomes would occur with equal relative frequency in the long run leads to the assignment of equal probability, $\frac{1}{52}$, to each of the elements of S. Let A be the event of drawing an ace and B the event of drawing a club. Thus A is a subset consisting of four elements, each of which has been assigned probability: $\frac{4}{52}$, and P(A) is the sum of these probabilities: $\frac{4}{52}$. Similarly the following probabilities are obtained:

$$P(B) = \frac{13}{52} ; \qquad\qquad P(AB) = \frac{1}{52}$$

Application of Theorem 1 gives

$$P(\overline{A}) = \frac{48}{52} ; \qquad\qquad P(\overline{B}) = \frac{39}{52}$$

Application of the addition theorem gives

$$P(A + B) = \frac{4}{52} + \frac{13}{52} - \frac{1}{52} = \frac{16}{52}$$

P(A + B), the probability of drawing an ace or a club, could have been calculated without using the addition theorem by calling A + B, the union of A and B, a set consisting of 16 elements. (We obtained the number 16 by adding the number of aces, 4, to the number of clubs, 13, and subtracting the card that is

---

counted twice – once as an ace and once as a club). Since each of the 16 elements in A + B has been assigned probability $\frac{1}{52}$, P(A + B) is the sum of the probabilities assigned, namely $\frac{16}{52}$.

19.5 CONDITIONAL PROBABILITY

The conditional probability of event B given A is denoted by P(B|A) and defined as

$$P(B|A) = P(AB)/P(A) \tag{19.5.1}$$

where P(B|A) can be interpreted as the proportion of A occurrences that also feature the occurrence of B.

For example, consider the random experiment of drawing two cards in succession from a deck of 52 cards. Suppose the cards are drawn without replacement (i.e., the first card drawn is not replaced before the second is drawn). Let A denote the event that the first card is an ace and B the event that the second card is an ace. The sample space S can be described as a set of 52 times 51 pairs of cards. Assuming that each of these (52)(51) pairs has the same theoretical relative frequency, assign probability 1/(52)(51) to each pair. The number of pairs featuring an ace as the first and second card is (4)(3). Therefore,

$$P(AB) = \frac{(4)(3)}{(52)(51)}$$

The number of pairs featuring an ace as the first card and one of the other 51 cards as the second is (4)(51). Therefore,

$$P(A) = \frac{(4)(51)}{(52)(51)}$$

Applying the definition of conditional probability Eq. (19.5.1) yields

$$P(B|A) = P(AB)/P(A)$$
$$P(B|A) = \frac{3}{51} = 0.0588$$

as the conditional probability that the second card is an ace, given that the first is an ace. The same result could have been obtained by computing P(B) on a new sample space consisting of 51 cards, three of which are aces. This illustrates the two methods for calculating a conditional probability. The first method calculates the conditional probability in terms of probabilities computed on the original sample space by means of the definition in Eq. (19.5.1). The second method uses the given event to construct a new sample space on which the conditional probability is computed.

  Conditional probability also can be used to formulate a definition for the independence of two events A and B. Event B is defined to be independent of event A if and only if

$$P(B|A) = P(B) \tag{19.5.2}$$

Similarly, event A is defined to be independent of event B if and only if

$$P(A|B) = P(A) \tag{19.5.3}$$

From the definition of conditional probability in Eq. (19.5.1), one can deduce the logically equivalent definition that event A and event B are independent if and only if

$$P(AB) = P(A)\,P(B) \tag{19.5.4}$$

  To illustrate the concept of independence, consider again the random experiment of drawing two cards in succession from a deck of 52 cards. This time suppose that the cards are drawn with replacement (i.e., the first card is replaced in the deck before the second card is drawn). As before, let A denote the event that the first card is an ace, and B the event that the second card is an ace. Then

$$P(B|A) = \frac{4}{52}$$

and since $P(B|A) = P(B)$, B and A are independent events.
From the definition of $P(B|A)$ and $P(A|B)$ one can deduce the multiplication theorem,

$$P(AB) = P(A)\,P(B|A) \tag{19.5.5}$$

$$P(AB) = P(B)\,P(A|B) \tag{19.5.6}$$

The multiplication theorem provides an alternate method for calculating the probability of the intersection of two events.

  The multiplication theorem can be extended to the case of three or more events. For three events A, B, C, the multiplication theorem states

$$P(ABC) = P(A)\,P(B|A)\,P(C|AB) \tag{19.5.7}$$

For four events A, B, C, and D, the multiplication theorem states:

$$P(ABCD) = P(A)\,P(B\,|\,A)\,P(C\,|\,AB)\,P(D\,|\,ABC) \qquad (19.5.8)$$

19.6 BAYES' THEOREM

Consider n mutually exclusive events A_1, A_2, ..., A_n whose union is the sample space S. Let B be any given event. Then Bayes' theorem states

$$P(A_i\,|\,B) = \frac{P(A_i)\,P(B|A_i)}{\sum\limits_{i=1}^{n} P(A_i)\,P(B|A_i)}; \quad i = 1,...n$$

$$(19.6.1)$$

where $P(A_1)$, $P(A_2)$, ..., $P(A_n)$ are called the "prior probabilities" of A_1, A_2, ..., A_n and $P(A_1\,|\,B)$, $P(A_2\,|\,B)$, ..., $P(A_n\,|\,B)$ are called the "posterior probabilities" of A_1, A_2, ..., A_n. Bayes' theorem provides the mechanism for revising prior probabilities, i.e., for converting them into posterior probabilities on the basis of the observed occurrence of some given event.[4]

As a example of the application of Bayes' theorem, suppose that 50% of a company's manufactured output comes from a New York plant, 30% from a Pennsylvania plant, and 20% from a Delaware plant. On the basis of plant records it is estimated that defective items constitute 1% of the output of the New York plant, 3% of the Pennsylvania plant, and 4% of the Delaware plant. If an item selected at random from the company's manufactured output is found to be defective, what are the revised probabilities that the item was produced, by each of the three plants?

Let A_1, A_2, A_3 denote, respectively, the events that the item was produced in the New York, Pennsylvania, and Delaware plants. Let B denote the event that the item was found to be defective. Then

$$P(A_1) = 0.50; \qquad P(B\,|\,A_1) = 0.01$$
$$P(A_2) = 0.30; \qquad P(B\,|\,A_2) = 0.03$$
$$P(A_3) = 0.20; \qquad P(B\,|\,A_3) = 0.04$$

Substituting in Eq.. (19.6.1), we obtain:

$$P(A_1|B) = \frac{P(A_1)P(B|A_1)}{P(A_1)P(B|A_1) + P(A_2)P(B|A_2) + P(A_3)P(B|A_3)}$$

$$P\left(A_1\middle|B\right)=\frac{\left(0.50\right)\left(0.01\right)}{\left(0.50\right)\left(0.01\right)+\left(0.30\right)\left(0.03\right)+\left(0.20\right)\left(0.04\right)}=0.23$$

Similarly,

$$P\left(A_2\middle|B\right)=0.41;\quad P\left(A_3\middle|B\right)=0.36$$

Therefore the information that the item selected at random was defective revises the probability that the item was produced in the New York plant downward from 0.50 to 0.23 and the probabilities for Pennsylvania and Delaware upward, respectively, from 0.30 to 0.41 and from 0.20 to 0.36.

For another example of the use of Bayes' theorem, suppose that the probability is 0.80 that an airplane crash due to structural failure is diagnosed correctly. Suppose, in addition, that the probability is 0.30 that an airplane crash not due to structural failure is incorrectly attributed to structural failure. If 35% of all airplane crashes are due to structural failure, what is the probability that an airplane crash was due to structural failure, given that it has been so diagnosed? Let A_1 be the event that structural failure is the cause of the airplane crash. Let A_2 be the event that the cause is other than structural failure. Let B be the event that the airplane crash is diagnosed as being due to structural failure. Then

$$P(A_1) = 0.35; \qquad P(B\,|\,A_1) = 0.80$$
$$P(A_2) = 0.65; \qquad P(B\,|\,A_2) = 0.30$$

Substituting in Eq. (19.6.1), we obtain

$$P\left(A_1\middle|B\right)=\frac{P\left(A_1\right)P\left(B\middle|A_1\right)}{P\left(A_1\right)P\left(B\middle|A_1\right)+P\left(A_2\right)P\left(B\middle|A_2\right)}$$

$$P\left(A_1\middle|B\right)=\frac{\left(0.35\right)\left(0.80\right)}{\left(0.35\right)\left(0.80\right)+\left(0.65\right)\left(0.30\right)}=0.59$$

Therefore the diagnosis of structural failure revises its probability as the cause of the airplane crash upward from 0.35 to 0.59.

19.7 RANDOM VARIABLES

A random variable is a real-valued function defined over the sample space S of a random experiment (Note that this application of probability theorem to plant and equipment failures, i.e., accidents, requires that the failure occurs randomly, i.e., by chance). The domain of the function is S, and the real numbers associated with the various possible outcomes of the random experiment constitute the range of the function. If the range of the random variable consists of a finite number or countable infinitude of values, the random variable is classified as *discrete*. If the range consists of a non-countable infinitude of values, the random variable is classified as *continuous*. A set has a countable infinitude of values if they can be put into one-to-one correspondence with the positive integers. The positive even integers, for example, consist of a countable infinitude of numbers. The even integer 2n corresponds to the positive integer n for n = 1, 2, 3, ... The real numbers in the interval (0, 1) constitute a non-countable infinitude of values.

Defining a random variable on a sample space S amounts to coding the outcomes in real numbers. Consider, for example, the random experiment involving the selection of an item at random from a manufactured lot. Associate X = 0 with the drawing of a non-defective item and X = 1 with the drawing of a defective item. Then X is a random variable with range (0, 1) and therefore discrete.

Let X denote the number of the throw on which the first failure of a switch occurs. Then X is a discrete random variable with range {1, 2, 3, ..., n, ...}. Note that the range of X consists of a countable infinitude of values and that X is therefore discrete.

Suppose that X denotes the time to failure of a bus section in an electrostatic precipitator. Then X is a continuous random variable whose range consists of the real numbers greater than zero.

19.8 PROBABILITY DISTRIBUTIONS

The probability distribution of a random variable concerns the distribution of probability over the range of the random variable. The distribution of probability is specified by the pdf (*probability distribution function*). This section is devoted to general properties of the pdf in the case of discrete and continuous random variables. Special pdf's finding extensive application in hazard and risk analysis are considered in Chapter 20.

The pdf of a discrete random variable X is specified by f(x) where f(x) has the following essential properties:

1. $f(x) = P(X = x)$
    $= $ probability assigned to the outcome corresponding to the number x in the range of X

2. $f(x) \geq 0$, and

3. $\sum_{x} f(x) = 1.$

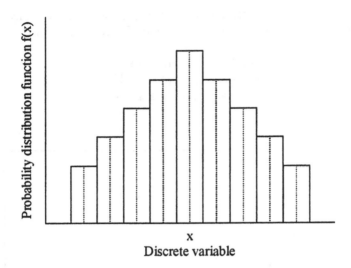

Figure 19.8.1. A discrete probability distribution function.

Property 1 indicates that the pdf of a discrete random variable generates probability by substitution. Properties 2 and 3 restrict the values of f(x) to nonnegative real numbers whose sum is 1. An example of a discrete probability distribution function (approaching a normal distribution – to be discussed in the next chapter) is provided in Figure 19.8.1.

Consider, for example, a box of 100 transistors containing five defectives. Suppose that a transistor selected at random is to be classified as defective or non-defective. Let X denote the outcome, with X = 0 associated with the drawing of a non-defective and X = 1 associated with the drawing of a defective. Then X is a discrete random variable with pdf specified by

$$f(x) = 0.5; \qquad x = 1$$
$$f(x) = 0.95; \qquad x = 0$$

The pdf of a continuous random variable X has the following properties:

1. $\int_a^b f(x)\,dx = P(a < X < b),$

2. $f(x) \geq 0,$ and

3. $\int_{-\infty}^{+\infty} f(x)\,dx = 1.$

Property 1 indicates that the pdf of a continuous random variable generates probability by integration of the pdf over the interval whose probability is required. When this interval contracts to a single value, the integral over the interval becomes zero. Therefore the probability associated with any particular value of a continuous random variable is zero. Consequently if X is continuous:

$$P(a \leq X \leq b) = P(a < X \leq b)$$
$$P(a \leq X \leq b) = P(a < X < b)$$
$$P(a \leq X \leq b) = P(a \leq X < b) \qquad (19.8.1)$$

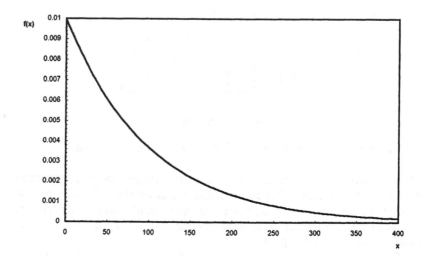

Figure 19.8.2. The pdf of time, in hours, between successive failures of an aircraft air conditioning system.

Property 2 restricts the values of f(x) to nonnegative numbers. Property 3 follows from the fact that:

$$P(-\infty < X < \infty) = 1$$

Inspection of the graph in Fig. 19.8.2 indicates that intervals in the lower part of the range of X are assigned greater probabilities than intervals of the same length in the upper part of the range of X because the areas over the former are greater than the areas over the latter. The expression P(a < X < b) can be interpreted geometrically as the area under the pdf curve over the interval (a, b). Integration of the pdf over the interval yields the probability assigned to the interval. For example, the probability that the time in hours between successive failures of the aircraft air conditioning system is greater than 6 but less than 10 is:

$$P(6 < X < 10) = \int_{6}^{10} 0.01e^{-0.01} dx = 0.04$$

Another function used to describe the probability distribution of a random variable X is the *cumulative distribution function* (cdf). If f(x) specifies the pdf of a random variable X, then F(x) is used to specify the cdf. For both discrete and continuous random variables, the cdf of X is defined by:

$$F(x) = P(X \le x); \quad -\infty < x < \infty \qquad (19.8.2)$$

Note that the cdf is defined for all real numbers, not just the values assumed by the random variable.

To illustrate the derivation of the cdf from the pdf, consider the case of a random variable X whose pdf is specified by:

$$f(x) = 0.2; \qquad x = 2$$
$$f(x) = 0.3; \qquad x = 5$$
$$f(x) = 0.5; \qquad x = 7$$

Applying the definition of cdf in Eq. (19.8.2), one obtains for the cdf of X:

$$f(x) = 0; \qquad x < 2$$
$$f(x) = 0.2; \qquad 2 \le x \le 5$$
$$f(x) = 0.5; \qquad 5 \le x < 7$$
$$f(x) = 1; \qquad x \ge 7$$

It is helpful to think of F(x) as an accumulator of probability as x increases through all real numbers.

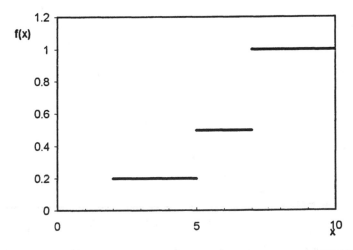

Figure 19.8.3. Graph of the cdf of a discrete random variable X.

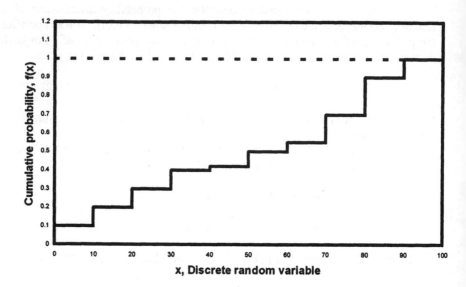

Figure 19.8.4. Discrete random variable

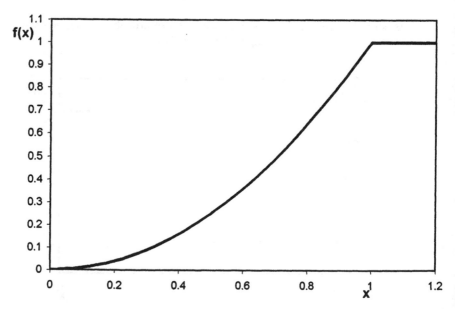

Figure 19.8.5. Graph of the cdf of a continuous random variable X.

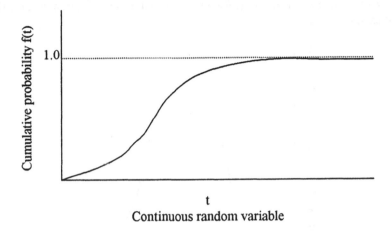

Figure 19.8.6. A continuous cumulative probability distribution function.

In the case of a discrete random variable, the cdf is a step function increasing by finite jumps at the values of x in the range of X. In the example above, these jumps occur at the values 2, 5, and 7. The magnitude of each jump is equal to the probability assigned to the value where the jump occurs. This is depicted in Fig. 19.8.3. Another form of representing the cdf of a discrete random variable is provided in Figure 19.8.4.

In the case of a continuous random variable, the cdf is a continuous function. Suppose, for example, that X is a continuous random variable with pdf specified by

$$f(x) = 2x; \quad 0 \le x < 1$$
$$f(x) = 0; \quad \text{elsewhere} \tag{19.8.3}$$

Applying the definition of the cdf in Eq. (19.8.2), leads to

$$F(x) = 0; \quad x < 0$$
$$F(x) = \int_0^x 2x\,dx = x^2; \; 0 \le x < 1$$
$$F(x) = 1; \quad x \ge 1 \tag{19.8.4}$$

Figure 19.8.5 displays the graph of this cdf. Another example of a cdf of a continuous random time variable is shown in Figure 19.8.6. A cdf of a continuous variable (a normal distribution – to be reviewed in the next chapter) is provided in Figure 19.8.7.

The pdf of a continuous random variable can be obtained by differentiating its cdf and setting the pdf equal to zero where the derivative of the cdf does not exist. For example, differentiating the cdf obtained in Eq.

(19.8.4) yields the pdf in Eq. (19.8.3). In this case the derivative of the cdf does not exist for x = 1.

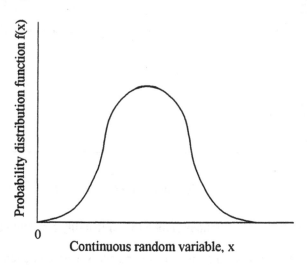

Figure 19.8.7.   A continuous probability distribution function.

The following properties of the cdf of a random variable X can be deduced directly from the definition of F(x) in Eq. (19.8.2).

1. $F(b) - F(a) = P(a < X \le b)$ (19.8.5)
2. $F(+\infty) = 1$ (19.8.6)
3. $F(-\infty) = 0$ (19.8.7)
4. F(x) is a non-decreasing function of x   (19.8.8)

These properties apply to the cases of both discrete and continuous random variables.

19.9 EXPECTED VALUES

The expected value of a random variable is the average value of the random variable. The expected value of a random variable X is denoted by E(X). The expected value of a random variable can be interpreted as the long-run average of observations on the random variable. The procedure for calculating the expected value of a random variable depends on whether the random variable is discrete or continuous.

If X is a discrete random variable with pdf specified by f(x), then:

$$E(X) = \sum_x xf(x) \qquad (19.9.1)$$

If X is a continuous random variable with pdf specified by f(x), then:

$$E(X) = \int_{-\infty}^{\infty} xf(x)\,dx \qquad (19.9.2)$$

Suppose, for example, that the pdf of a discrete random variable X is specified by:

$$f(x) = 0.3; \qquad x = 10$$
$$f(x) = 0.2; \qquad x = 20$$
$$f(x) = 0.5; \qquad x = 30$$

Then the expected value of X is given by:

$$E(X) = 10(0.3) + 20(0.2) + 30(0.5) = 22$$

Suppose, for example, that the pdf of a continuous random variable x is specified by:

$$f(x) = e^{-x}; \qquad x > 0$$
$$f(x) = 0; \qquad \text{elsewhere}$$

Application of Eq. (19.9.2) yields the following expected value for X:

$$E(X) = \int_0^{\infty} xe^{-x}\,dx = 1$$

The expected value of a random variable X is also called "the mean of X" and is often designated by μ. The expected value of $(X-\mu)^2$ is called the *variance* of X. The positive square root of the variance is called the *standard deviation*. The terms σ^2 and σ (sigma squared and sigma) represent variance and standard deviation, respectively. Variance is a measure of the spread or dispersion of the values of the random variable about its mean value. The standard deviation is also a measure of spread or dispersion. The standard deviation is expressed in the same units as X, while the variance is expressed in the square of these units.

The variance of X can be calculated directly from the definition

$$\sigma^2 = E(X - \mu)^2 \qquad (19.9.3)$$

However, it is usually calculated more easily by the equivalent formula

$$\sigma^2 = E(X^2) - \mu^2 = E(X^2) - [E(X)]^2 \qquad (19.9.4)$$

To illustrate the computation of variance and its interpretation in the case of discrete random variables, consider a random variable X having pdf specified by:

$$f(x) = 0.5; \qquad x = 1$$
$$f(x) = 0.5; \qquad x = -1$$

and a random variable Y having pdf specified by:

$$g(y) = 0.5; \qquad y = 10$$
$$g(y) = 0.5; \qquad y = -10$$

It is easily verified that both X and Y have the same expected value, zero. The expected value of X^2 is given by:

$$E(X^2) = \sum_x x^2 f(x)$$
$$E(X^2) = (1)(0.5) + (-1)^2(0.5) = 1$$

Therefore the variance of X is 1. The expected value of Y^2 is given by:

$$E(Y^2) = \sum_y y^2 g(y)$$
$$E(Y^2) = (10^2)(0.5) + (-10)^2(0.5) = 100$$

Therefore, the variance of Y is 100. The standard deviation of X is 1 and the standard deviation of Y is 10. The larger value for the standard deviation in the case of Y reflects the greater dispersion of Y values about their mean.

To illustrate the computation of variance and its interpretation in the case of continuous random variables, consider a random variable X having pdf specified by:

$$f(x) = \frac{1}{2}; \qquad 0 < x < 2$$
$$f(x) = 0; \qquad \text{elsewhere}$$

and a random variable Y having pdf specified by:

$$g(y) = \frac{1}{4}; \qquad -1 < y < 3$$
$$g(y) = 0; \qquad \text{Elsewhere}$$

The mean value of X is given by:

$$E(X) = \int_0^2 x\left(\frac{1}{2}\right)dx$$
$$E(X) = 1$$

The expected value of X^2 is given by:

$$E(X^2) = \int_0^2 x^2\left(\frac{1}{2}\right)dx$$

$$E(X^2) = \frac{4}{3}$$

Therefore the variance of X [see Eq. (19.9.4)] is:

$$\sigma^2 = E(X^2) - \mu^2$$

$$\sigma^2 = \frac{4}{3} - 1 = \frac{1}{3}$$

The mean value of Y is given by:

$$E(Y) = \int_{-1}^3 y\left(\frac{1}{4}\right)dy$$
$$E(y) = 1$$

The expected value of Y^2 is given by:

$$E(Y^2) = \int_{-1}^3 y^2\left(\frac{1}{4}\right)dy$$

$$E(y) = \frac{7}{3}$$

Therefore the variance of Y is:

$$\sigma^2 = E(Y^2) - \mu^2$$

$$\sigma^2 = \frac{7}{3} - 1 = \frac{4}{3}$$

Here X and Y have the same expected values but Y has the larger variance, reflecting the greater dispersion of Y values about their mean.

19.10 ESTIMATION OF MEAN AND VARIANCE

The mean μ and the variance σ^2 of a random variable are constants characterizing the random variable's average value and dispersion about its mean. The mean and variance can be derived from the pdf of the random variable. If the pdf is unknown, however, the mean and the variance can be estimated on the basis of a random sample of observations on the random variable. Let X_1, X_2, ..., X_n denote a random sample of n observations on X.

Then the sample mean $\overline{X}$ is defined by:

$$\overline{X} = \sum_{i=1}^{n} \frac{X_i}{n} \qquad (19.10.1)$$

and the sample variance S^2 is defined by:

$$S^2 = \sum_{i=1}^{n} \frac{(X_i - \overline{X})^2}{n-1} \qquad (19.10.2)$$

where $\overline{X}$ and S^2 are random variables in the sense that their values vary from sample to sample of observations on X. It can be shown that the expected value $\overline{X}$ is μ and that the expected value of S^2 is σ^2. Because of this, $\overline{X}$ and S^2 are called unbiased estimators of μ and σ^2, respectively.

The calculation of S^2 can be facilitated by use of the computation formula:

$$S^2 = \frac{n\sum_{i=1}^{n} X_i^2 - \left(\sum_{i=1}^{n} X_i\right)^2}{n(n-1)} \qquad (19.10.3)$$

For example, given the sample 5, 3, 6, 4, 7,

$$\sum_{i=1}^{5} X_i^2 = 135$$

$$\sum_{i=1}^{5} X_i = 25$$

$$n = 5$$

Substituting in Eq. (19.10.3) yields:

$$S^2 = \frac{(5)(135)-(25)^2}{(5)(4)} = 2.5$$

Chebyshev's theorem provides an interpretation of the sample standard deviation, the positive square root of the sample variance, as a measure of the spread (dispersion) of sample observations about their mean.[5] Chebyshev's theorem states that at least $(1 - 1/k^2)$, $k > 1$, of the sample observations lie in the interval $(\overline{X} - kS, \overline{X} + kS)$. For $k = 2$, for example, this means that at least 75% of the sample observations lie in the interval $(\overline{X} - 2S, \overline{X} + 2S)$. The smaller the value of S, the greater the concentration of observations in the vicinity of X.

In the case of a random sample of observations on a continuous random variable assumed to have a so-called normal pdf, the graph of which is a bell-shaped curve, the following statements give a more precise interpretation of the sample standard deviation S as a measure of spread or dispersion.

1. $\overline{X}$! S includes approximately 68% of the sample observations.

2. $\overline{X}$! 2S includes approximately 95% of the sample observations.

3. $\overline{X}$! 3S includes approximately 99.7% of the sample observations.

The source of these percentages is the normal probability distribution, which is studied in more detail in Chapter 20. Additional details are available in the literature.[6]

19.11 ILLUSTRATIVE EXAMPLES

Example 19.1

Suppose that the failure of either a generator or a switch will cause interruption of electrical power to a certain facility. If the probability of a generator failure is 0.02, the probability of switch failure is 0.01, and the probability of both failing is 0.0002, what is the probability of power interruption?

Solution 19.1

If A denotes the event of generator failure, and B the event of switch failure, then $A + B$ denotes the event of power interruption, the probability if which is given by:

$$P(A+B) = P(A) + P(B) - P(AB)$$
$$P(A+B) = 0.02 + 0.01 - 0.0002 = 0.0298$$

Example 19.2

Consider the case of a box of 100 transistors from which a sample of two items is to be drawn without replacement. If the box contains five defective transistors, what is the probability that the sample contains exactly two defectives?

Solution 19.2

Let A denote the event that the first transistor drawn is defective, and B, the event that the second is defective. Then the probability that the sample contains exactly two defectives is $P(AB)$. By application of the multiplication theorem, one obtains:

$$P(AB) = P(A)P(B|A)$$
$$P(AB) = \left(\frac{5}{100}\right)\left(\frac{4}{99}\right) = 0.002$$

Example 19.3

Suppose that an explosion at a chemical plant could have occurred as a result of one of three mutually exclusive causes: equipment malfunction, carelessness, or sabotage. It is estimated that such an explosion could occur with probability 0.20 as a result of equipment malfunction, 0.40 as a result of carelessness, and 0.75 as a result of sabotage. It is also estimated that the prior probabilities of the three possible causes of the explosion are, respectively, 0.50, 0.35, and 0.15. Using Bayes' theorem, determine the most likely cause of the explosion.

Solution 19.3

Let A_1, A_2, A_3 denote, respectively, the events that equipment malfunction, carelessness and sabotage occur. Let B denote the event of the explosion. Then

$$P(A_1) = 0.50; \quad P(B|A_1) = 0.20$$
$$P(A_2) = 0.35; \quad P(B|A_2) = 0.40$$
$$P(A_3) = 0.15; \quad P(B|A_3) = 0.75$$

$$P(A_1|B) = \frac{P(A_1)P(B|A_1)}{P(A_1)P(B|A_1) + P(A_2)P(B|A_2) + P(A_3)P(B|A_3)}$$

$$P(A_1|B) = \frac{(0.50)(0.20)}{(0.50)(0.20) + (0.35)(0.40) + (0.15)(0.75)} = 0.28$$

Similarly,

$$P(A_2|B) = 0.40; \quad P(A_3|B) = 0.32$$

Therefore, carelessness is the most likely cause of the explosion.

Example 19.4

Suppose that the probability that a switch fails on any throw is 0.001 and that successive throws are independent with respect to failure. If the switch fails for the first time on the throw x, it must have been successful on each of the preceding x-1 trials. What is the pdf of X given the above conditions?

Solution 19.4

The pdf of X is given by:

$$f(x) = (0.999)^{x-1}(0.001), \quad x = 1, 2, 3..., n,...$$

Example 19.5

Successive failures of a chemical reactor's cooling system in days is:

$$f(x) = 0.01e^{-0.01x}; \quad x > 0$$
$$f(x) = 0; \quad \text{elsewhere}$$

What is the probability that the time in days between successive failures is greater than 2 but less than 5?

Solution 19.5

The probability that the time in hours between successive failures is greater than 2 but less than 5 is:

$$P(2 < X < 5) = \int_{2}^{5} 0.005e^{-0.005x}dx = 0.015$$

19.12 SUMMARY

1.   Hazard, risk, failure, and reliability are interrelated concepts concerned with uncertain events and therefore amenable to quantitative measurement via probability.

2.   *Probabilities* are nonnegative numbers associated with the outcomes of so-called random experiments. A random experiment is an experiment whose outcome is uncertain.

3.   When events A and B have no elements in common they are said to be *mutually exclusive*. A set having no elements is called the null set and is designated by ϕ. Thus if events A and B are mutually exclusive then AB = ϕ.

4.   The mathematical properties of P(A), the probability of event A, are deduced from the following postulates governing the assignment of probabilities to the elements of a sample space, S. In the case of a discrete sample space (i.e., a sample space consisting of a finite number or countable infinitude of elements), these postulates require that the numbers assigned as probabilities to the elements of S be nonnegative and have a sum equal to 1.

5.   The conditional probability of event B given A is denoted by P(B|A) and defined as

$$P(B|A) = P(AB)/P(A)$$

(19.5.1)

where P(B|A) can be interpreted as the proportion of A occurrences that also feature the occurrence of B.

6.   Bayes' theorem provides the mechanism for revising prior probabilities, i.e., for converting them into posterior probabilities on the basis of the observed occurrence of some given event.[4]

7.   A random variable is a real-valued function defined over the sample space S of a random experiment (Note that this application of probability theorem to plant and equipment failures, i.e., accidents, requires that the failure occurs randomly, i.e., by chance). The domain of the function is S, and the real numbers associated with the various possible outcomes of the

random experiment constitute the range of the function. If the range of the random variable consists of a finite number or countable infinitude of values, the random variable is classified as *discrete*. If the range consists of a non-countable infinitude of values, the random variable is classified as *continuous*.

8.   The probability distribution of a random variable concerns the distribution of probability over the range of the random variable. The distribution of probability is specified by the pdf (*probability distribution function*).

9.   The expected value of a random variable is the average value of the random variable. The expected value of a random variable can be interpreted as the long-run average of observations on the random variable. The procedure for calculating the expected value of a random variable depends on whether the random variable is discrete or continuous.

10. In the case of a random sample of observations on a continuous random variable assumed to have a so-called normal pdf, the graph of which is a bell-shaped curve, the following statements give a more precise interpretation of the sample standard deviation S as a measure of spread or dispersion.

1. $\overline{X} \pm S$ includes approximately 68% of the sample observations.

2. $\overline{X} \pm 2S$ includes approximately 95% of the sample observations.

3. $\overline{X} \pm 3S$ includes approximately 99.7% of the sample observations.

The source of these percentages is the normal probability distribution.

PROBLEMS

1.   A coin is flipped. If heads occurs the coin is flipped again; otherwise, a die is tossed once.
     (a) List the elements in the sample space S.
     (b) Let A be the event of tails on the first flip of the coin.
     Let B be the event of a number less than 4 on the die. Find P(A), P(B), $P(\overline{B})$, $P(A\overline{B})$, $P(A + B)$.

2.   In a group of 100 engineering students 15 are studying to be chemical engineers, 70 are undergraduates, 10 of which are undergraduates studying to be chemical engineers. A student is selected at random from the group. Let C be the event that the student selected is studying to be a chemical engineer. Let U be the event that the student selected is an undergraduate.
     Find $P(\overline{C})$, $P(C + U)$, $P(\overline{C}U)$, $P(\overline{C}\ \overline{U})$, $P(\overline{C} + \overline{U})$, $P(C|U)$, $P(U|\overline{C})$.

3.   Three defective light bulbs are inadvertently mixed with seven good ones. If two bulbs are chosen at random for a lamp, what is the probability that

both are good?

4. From a list of 100 items, 10 of which are defective, 2 items are drawn in succession. Find the probability that both items are defective if
   (a) the first item is replaced before the second is drawn, and
   (b) the first item is not replaced before the second is drawn.

5. A plant received 75% of its voltage regulators from supplier A and 25% from supplier B; 90% of the regulators from A and 85% of the regulators from B perform according to specifications. What is the probability that a regulator came from supplier A, given that it performs according to specifications?

6. A casualty insurance company has high, medium, and low risk policy-holders who have probabilities 0.02, 0.01, and 0.0025, respectively, of filing a claim within any given year. The proportions of company policy - holders in the three risk groups are 0.10, 0.20 and 0.70. What proportion of claims filed each year come from the low risk group?

7. The probability that a light switch fails is 0.0001. Let X denote the number of the trial on which the first failure occurs. Find the probability that X exceeds 1000.

8. Let X denote the annual number of floods in a certain region. The pdf of X is specified as follows:

$$
\begin{aligned}
f(x) &= 0.25; &x &= 0 \\
f(x) &= 0.35; &x &= 1 \\
f(x) &= 0.24; &x &= 2 \\
f(x) &= 0.11; &x &= 3 \\
f(x) &= 0.04; &x &= 4 \\
f(x) &= 0.01; &x &= 5
\end{aligned}
$$

   (a) What is the probability of two or more floods in any year?
   (b) What is the probability of four or more floods in a year, given that two floods have already occurred?
   (c) What is the probability of at least one more flood in a year, given that one has already occurred?

9. A die is loaded so that the probability of any face turning up is directly proportional to the number of dots on the face. Let X denote the outcome of throwing the die once. Find the mean and variance of X.

10. The difference between the magnitude of a large earthquake, as measured on the Richter scale, and the threshold value of 3.25 is a random variable X having the pdf

$$f(x) = 1.7e^{-1.7x}; \quad x > 0$$
$$f(x) = 0; \quad \text{elsewhere}$$

   (a) Find $P(2 < X < 6)$.
   (b) Find the variance of X.

11. A random variable X denoting the years of college education of an employee selected at random from the personnel of an insurance company

dealing with accident claims has the pdf

$$f(x) = 0.6; \quad x = 2$$
$$f(x) = 0.4; \quad x = 4$$

12. A random variable X denoting the useful life of a battery in years has the pdf

$$f(x) = \left(\frac{3}{8}\right)x^2; \quad 0 < x < 2$$

$$f(x) = 0; \quad \text{elsewhere}$$

Find the cdf of X.

13. At a chemical plant, the probability that a new employee who has attended the safety-training program will have a serious accident in his or her first year is 0.001. The corresponding probability for a new employee who has not attended the safety-training program is 0.02. If 80% of all new employees attend the safety-training program, what is the probability that a new employee will have a serious accident during the first year?

REFERENCES

1. Ernst G. Frankel, *Systems Reliability and Risk Analysis,* Nijhoff, the Hague, 1984.
2. John E. Freund and Ronald E. Walpole, *Mathematical Statistics*, ed., Prentice-Hall, Englewood Cliffs, NJ, 1987.
3. Frank P. Lees, *Loss Prevention In the Process Industry*, Vol. 1, Butterworths, Boston, 1980.
4. Richard J. Larsen and Morris L. Marx, *An Introduction to Probability and Its Applications*, Prentice-Hall, Englewood Cliffs, NJ, 1985.
5. Cyrus Derman, Leon Glasser, and Ingram Olkin, *A Guide to Probability Theory and Its Applications*, Holt, New York, 1973.
6. L. Theodore, J. Reynolds, and K. Morris, "Health, Safety and Accident Prevention: Industrial Applications", Theodore Tutorials, East Willington, NY, 1997.

20

Hazard Risk Assessment Calculations

20.1 INTRODUCTION

This chapter is concerned with special probability distributions and techniques used in calculations of reliability and risk. Theorems and basic concepts of probability presented in Chapter 19 are applied to the determination of the reliability of complex systems in terms of the reliabilities of their components. The relationship between reliability and failure rate is explored in detail. Special probability distributions for failure time are discussed. The chapter concludes with a consideration of fault tree analysis and event tree analysis, two special techniques that figure prominently in hazard analysis and the evaluation of risk.

In Section 20.2, equations for the reliability of series and parallel systems are established. Various reliability relations are developed in Section 20.3. Sections 20.4 and 20.5 introduce several probability distribution models that are extensively used in reliability calculations in hazard and risk analysis. Section 20.6 deals with the Monte Carlo technique of mimicking observations on a random variable. Sections 20.7 and 20.8 are devoted to fault tree and event tree analyses, respectively.

20.2 SERIES AND PARALLEL SYSTEMS

Many systems consisting of several components can be classified as *series*, *parallel*, or a combination of both. However, the majority of industrial and process plants (units and systems) have series of parallel configurations.

A *series system* is one in which the entire system fails to operate if any one of its components fails to operate. If such a system consists of n components that function independently, then the reliability of the system is the product of the reliabilities of the individual components. If R_S denotes the reliability of a series system and R_i denotes the reliability of the ith component; i = 1, ..., n, then

$$R_S = R_1 R_2 ... R_n$$

$$R_S = \prod_{i=1}^{n} R_i \qquad (20.2.1)$$

A *parallel system* is one that fails to operate only if all its components fail to operate. If R_i is the reliability of the ith component, then $(1-R_i)$ is the probability that the ith component fails; i = 1, ..., n. Assuming that all n components function independently, the probability that all n components fail is $(1-R_1)(1-R_2)...(1-R_n)$. Subtracting this product from unity yields the following formula for R_P, the reliability of a parallel system.[1]

$$R_P = 1 - (1 - R_1)(1 - R_2)...(1 - R_n)$$

$$R_P = 1 - \prod_{i=1}^{n} (1 - R_i) \qquad (20.2.2)$$

The reliability formulas for series and parallel systems can be used to obtain the reliability of a system that combines features of a series and a parallel system. Consider, for example, the system diagrammed in Fig. 20.2.1. Components A, B, C, and D have for their respective reliabilities 0.90, 0.80, 0.80, and 0.90. The system fails to operate if A fails, if B and C both fail, or if D fails. Component B and C constitute a parallel subsystem connected in series to components A and D. The reliability of the parallel subsystem is obtained by applying Eq. (20.2.2), which yields

$$R_P = 1 - (1 - 0.80)(1 - 0.80) = 0.96$$

The reliability of the system is then obtained by applying Eq. (20.2.1), which yields

$$R_S = (0.90)(0.96)(0.90) = 0.78$$

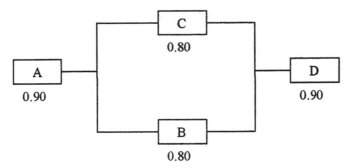

Figure 20.2.1. System with parallel and series components.

20.3 RELIABILITY RELATIONS

The reliability of a component will frequently depend on the length of time it has been in service. Let T, the time to failure, be a random variable having its pdf specified by $f(t)$. Then the probability that failure occurs in the time interval $(0, t)$ is given by

$$F(t) = \int_0^t f(t)\,dt \qquad (20.3.1)$$

Let the reliability of the component be denoted by $R(t)$, the probability that the component survives to time t. Therefore,

$$R(t) = 1 - F(t) \qquad (20.3.2)$$

Equation (20.3.2) establishes the relationship between the reliability of a component and the cdf of its time to failure.

     In accordance with the property expressed in Eq. (19.8.5), the probability that a component will fail in the time interval $(t, t + \Delta t)$ is given by

$$P(t < T < t + \Delta t) = F(t + \Delta t) - F(t) \qquad (20.3.3)$$

The conditional probability that a component will fail in the time interval $(t, t + \Delta t)$, given that it has survived to time t, is

$$P(t < T < t + \Delta t \mid T > t) = \frac{F(t + \Delta t) - F(t)}{P(T > t)} \qquad (20.3.4)$$

     Equation (20.3.4) is obtained by application of the definition of conditional probability in Eq. (19.5.1). Noting that

$$R(t) = P(T > t) \tag{20.3.5}$$

and substituting in Eq. (20.3.4) one obtains

$$P(t < T < t + \Delta t | T > t) = \frac{F(t + \Delta t) - F(t)}{R(t)} \tag{20.3.6}$$

Division of both sides of Eq. (20.3.6) by Δt yields

$$\frac{P(t < T < t + \Delta t | T > t)}{\Delta t} = \left[\frac{F(t + \Delta t) - F(t)}{\Delta t} \right] \left[\frac{1}{R(t)} \right] \tag{20.3.7}$$

Recall that $F'(t)$, the derivative of $F(t)$, is defined by

$$\lim_{\Delta t \to 0} \left[\frac{F(t + \Delta t)}{\Delta t} \right] = F(t) \tag{20.3.8}$$

By taking the limit of both sides of Eq. (20.3.7) as Δt approaches 0, we obtain

$$Z(t) = \frac{F'(t)}{R(t)} \tag{20.3.9}$$

where $Z(t)$ is defined by

$$Z(t) = \lim_{\Delta t \to 0} \frac{P(t < T < t + \Delta t | T > t)}{\Delta t} \tag{20.3.10}$$

Here $Z(t)$ is called the "failure rate" (also the "hazard rate") of the component. Equation (20.3.9) establishes the relationship among failure rate, reliability, and the cdf of time to failure.

Using Eqs. (20.3.2) and (20.3.9), one can obtain an expression for the reliability in terms of the failure rate. Differentiating both sides of Eq. (20.3.2) with respect to yields

$$R'(t) = -F'(t) \tag{20.3.11}$$

Recalling that the derivative of the cdf of a continuous random variable is equal to its pdf permits rewriting Eq. (20.3.11) as

$$R'(t) = -f(t) \tag{20.3.12}$$

Substitution in Eq. (20.3.9) yields

$$Z(t) = -\frac{R'(t)}{R(t)} \qquad (20.3.13)$$

Integrating both sides of Eq. (20.3.13) between 0 and t yields

$$\int_0^t Z(t)dt = -[\ln R(t) - \ln R(0)] \qquad (20.3.14)$$

Since $R(t) = P(T > t)$, $R(0) = 1$, and Eq. (20.3.14) becomes

$$\int_0^t Z(t)dt = -\ln R(t) \qquad (20.3.15)$$

Solving Eq. (20.3.15) for R(t) yields

$$R(t) = \exp\left[-\int_0^t Z(t)\, dt \right] \qquad (20.3.16)$$

the desired expression for reliability in terms of failure rate.
  The pdf of time to failure can also be expressed in terms of failure rate. Differentiating Eq. (20.3.116) with respect to t yields

$$R'(t) = -Z(t) \exp\left[-\int_0^t Z(t)\, dt \right] \qquad (20.3.17)$$

Substituting, once again,

$$R'(t) = -f(t)$$

from Eq. (20.3.12) produces

$$f(t) = Z(t) \exp\left[-\int_0^t Z(t)\, dt \right] \qquad (20.3.18)$$

the desired expression of the pdf in terms of failure rate.[2]
  Examples illustrating the application of these equations to real systems are provided in subsequent sections.

20.4   THE BATHTUB CURVE

Frequently, the failure rate of equipment exhibits three stages; a break-in stage with a declining failure rate, a useful life stage characterized by a fairly constant

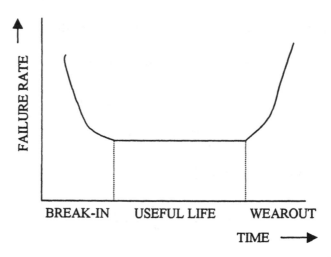

Figure 20.4.1. Bathtub curve.

failure rate, and a wearout period characterized by an increasing failure rate. Many industrial parts and components follow this path.   A failure rate curve exhibiting these three phases is called a "bathtub" curve (Figure 20.4.1).

In the cases of the bathtub curve, failure rate during useful life is constant. Letting this constant be α and substituting it for $Z(t)$ in Eq. (20.3.18) yields

$$f(t) = \alpha \exp\left[- \int_0^t \alpha \, dt \right]$$

$$f(t) = \alpha \exp(-\alpha t); \quad t > 0 \qquad (20.4.1)$$

as the pdf of time to failure during the useful life stage of the bathtub curve. Equation (20.4.1) defines an exponential pdf that is a special case of the pdf defining the Weibull distribution.

The Weibull distribution provides a mathematical model of all three stages of the bathtub curve. This is now discussed. An assumption about failure rate that reflects all three stages of the bathtub curve is

$$Z(t) = \alpha \beta t^{\beta - 1}; \quad t > 0 \qquad (20.4.2)$$

where α and β are constants. For $\beta < 1$ the failure rate $Z(t)$ decreases with time. For $\beta = 1$ the failure rate is constant and equal to α. For $\beta > 1$ the failure rate increases with time. Using Eq. (20.3.18) again to translate the assumption about failure rate into a corresponding assumption about the pdf of T, time to failure, one obtains

$$f(t) = \alpha\beta t^{\beta-1} \exp\left(-\int_0^t \alpha\beta t^{\beta-1}\, dt\right)$$

$$f(t) = \alpha\beta t^{\beta-1} \exp(-\alpha t^{\beta}); \quad t > 0; \quad \alpha > 0, \quad \beta > 0 \qquad (20.4.3)$$

Equation (20.4.3) defines the pdf of the Weibull distribution. The exponential distribution, whose pdf is given in Eq. (20.4.1), is a special case of the Weibull distribution with $\beta = 1$. The variety of assumptions about failure rate and the probability distribution of time to failure that can be accommodated by the Weibull distribution make it especially attractive in describing failure time distributions in industrial and process plant applications.

Estimates of the parameters α and β in Eq. (20.4.3) can be obtained by using a graphical procedure described by Bury.[1] This procedure is based on the fact that

$$\ln\left[\ln\frac{1}{1-F(t)}\right] = \ln\alpha + \beta\ln t \qquad (20.4.4)$$

where

$$F(t) = 1 - \exp(-\alpha t^{\beta}); \quad t > 0$$
$$F(t) = 0; \quad t < 0 \qquad (20.4.5)$$

defines the cdf of the Weibull distribution. In Eq. (20.4.4), the expression

$$\ln\left[\ln\frac{1}{1-F(t)}\right]$$

serves as a linear function of $\ln t$ with slope β and intercept $\ln \alpha$. The graphical procedure for estimating α and β on the basis of a sample of n observed values of t, time to failure, first involves the ordering of the observations from smallest $(i = 1)$ to largest $(i = n)$. The value of the ith observation varies from sample to sample. It can be shown that the average value of $F(t)$ for t equal to the value of the ith observation on T is $i/(n+1)$. One may then plot

$$\ln\left[\ln\frac{1}{1-\frac{i}{n+1}}\right]$$

against the natural logarithm of the ith observation $i = 1$ to $i = n$. Under the assumption that T has a Weibull distribution, the plotted points lie on a straight line whose slope is β and whose intercept is $\ln\alpha$. Special Weibull probability paper allows plotting the ith observation against $i/(n+1)$ to achieve the same result.

To illustrate this procedure, suppose that a sample of 10 observations on the time to failure of an electric component yields the observations in the first column of Table 20.4.1. Fitting a straight line to the derived values in columns 3 and 4 yields a line with slope 1.4 and intercept –6.5. Therefore, the estimated value of α is 0.0015 and the estimated value of β is 1.4.

To illustrate probability calculations involving the exponential and Weibull distributions introduced in conjunction with the bathtub curve of failure rate, consider first the case of a transistor having a constant rate of failure of 0.01 per thousand hours. To find the probability that the transistor will operate for at least 25,000 hours, substitute the failure rate

$$Z(t) = 0.01$$

into Eq. (20.3.18), which yields

$$f(t) = 0.01 \exp\left[-\int_0^t 0.01dt\right]$$
$$f(t) = 0.01 e^{-0.01t} dt; \quad t > 0$$

as the pdf of T, the time to failure of the transistor. Since t is measured in thousands of hours, the probability that the transistor will operate for at least 25,000 hours is given by

$$P(T > 25) = \int_{25}^{\infty} 0.01e^{-0.01t} dt = 0.78$$

As another example of probability calculations – this time involving the Weibull distribution – consider a component whose time to failure T in hours has a Weibull pdf with parameters $\alpha = 0.01$ and $\beta = 0.50$. To find the probability that the component will operate for at least 8100 hours, substitute $\alpha = 0.01$ and $\beta = 0.50$ in Eq. (20.4.3). This gives

$$f(t) = (0.01)(0.50)t^{0.5-1} e^{-(0.01)t^{0.5}}; \quad t > 0$$

TABLE 20.4.1 Data for Estimation of Weibull Parameters

(1) Time to Failure, t (days)	(2) Order of Failure, t	(3) ln t	(4) $\ln\left[\ln\dfrac{1}{1-\dfrac{i}{n+1}}\right]$
18	1	2.89	-2.36
36	2	3.58	-1.62
40	3	3.69	-1.16
53	4	3.97	-0.81
71	5	4.26	-0.51
90	6	4.50	-0.23
106	7	4.66	0.02
127	8	4.84	0.27
149	9	5.00	0.54
165	10	5.11	0.88

as the Weibull pdf of the failure time of the component under consideration. The probability that the component will operate at least 8100 hours is then given by

$$P(T > 8100) = \int_{8100}^{\infty} f(t)\, dt = \int_{8100}^{\infty} 0.005 t^{-0.5} e^{-(0.01)t^{0.5}}\, dt = 0 +$$
$$e^{-(0.01)(8100)^{0.5}} = 0.41$$

The Weibull distribution will again find application in the case study reviewed in Section 21.6.

20.5 OTHER SPECIAL DISTRIBUTIONS

In addition to the exponential and Weibull distributions, several other probability distributions figure prominently in reliability calculations and hazard risk analysis. Presented below are their pdfs, principal characteristics, and an indication of their application.

The Binomial Distribution

Consider n independent performances of a random experiment with mutually exclusive outcomes that can be classified "success" or "failure". The words "success" and "failure" are to be regarded as labels for two mutually exclusive categories of outcomes of the random experiment. They do not necessarily have the ordinary connotations of success or failure. Assume that p, the probability of success on any performances of the random experiment, is constant. Let q be the probability of failure, so that

$$q = 1 - p \tag{20.5.1}$$

The probability distribution of X, the number of successes in n performances of the random experiment, is the binomial distribution, with pdf specified by

$$f(x) = \frac{n!}{x!(n-x)!} p^x q^{n-x}; \quad x = 0, 1 \dots, n \tag{20.5.2}$$

Where f(x) is the probability of x successes in n performances. One can show that the expected value of the random variable X is np and its variance is npq.[2]

As a simple example of the binomial distribution, consider the probability distribution of the number of defectives in a sample of 5 items drawn with replacement from a lot of 1000 items, 50 of which are defective. Associate "success" with drawing a defective item from the lot. Then the result of each drawing can be classified success (defective item) or failure (non-defective item). The sample of items is drawn with replacement (i.e., each item in the sample is returned before the next is drawn from the lot; therefore the probability of success remains constant at 0.05. Substituting in Eq. (20.5.2) the values n = 5, p = 0.05, and q = 0.95 yields

$$f(x) = \frac{5!}{x!(5-x)!} (0.05)^x (0.95)^{5-x}; \quad x = 0, 1, 2, 3, 4, 5$$

as the pdf for X, the number of defectives in the sample. The probability that the sample contains exactly 3 defectives is given by

$$P(X = 3) = \frac{5!}{3!2!} (0.05)^3 (0.95)^2 = 0.0011$$

The binomial distribution can be used to calculate the reliability of a *redundant system*.[3] A redundant system consisting of n identical components is a system that fails only if more than r components fail. Familiar examples include single-usage equipment such as missile engines, short-life batteries, and flash bulbs, which are required to operate for one time period and are not reused. Associate "success" with the failure of a component. Assume that the n components are independent with respect to failure, and that the reliability of each is $1 - p$. Then X, the number of failures, has the binomial pdf of Eq. (20.5.2) and the reliability of the redundant system is

$$P(X \le r) = \sum_{x=0}^{r} \frac{n!}{x!(n-x)!} p^x q^{n-x} \qquad (20.5.3)$$

For example, consider the case of a standby redundancy system consisting of one operating pump and two on standby so that the system can survive two failures. Assume that the pumps are independent with respect to failure and that each has a probability of failure of 0.1. Thus, $n = 3$, $r = 2$, $1 - p = 0.9$, and the reliability of the system is found from Eq. (20.5.3)

$$P(X \le 6) = \sum_{x=0}^{2} \frac{3!}{x!(3-x)!} (0.10)^x (0.90)^{3-x} = 0.999$$

The Poisson Distribution

The pdf of the Poisson distribution can be derived by taking the limit of the binomial pdf as $n \to \infty$, $p \to 0$, and $np = \mu$ remains constant. The Poisson pdf is given by

$$f(x) = \frac{e^{-\mu} \mu^x}{x!}; \qquad x = 0, 1, 2, \dots \qquad (20.5.4)$$

Here, $f(x)$ is the probability of x occurrences of an event that occurs on the average μ times per unit of space or time. Both the mean and the variance of a random variable X having a Poisson distribution are μ.

The Poisson pdf can be used to approximate probabilities obtained from the binomial pdf given in Eq. (20.5.2) when n is large and p is small. In general, good approximations will result when n exceeds 10 and p is less than 0.05. When n exceeds 100 and np is less than 10, the approximation will generally be excellent. Table 20.5.1 compares the binomial and Poisson probabilities of the case of $n = 20$, $p = 0.05$ and $n = 100$, $p = 0.01$.

If λ is the failure rate of each component of a system, then λt is the average number of failures per unit of time. The probability of x failures in the specified unit of time is obtained by substituting $\mu = \lambda t$ in Eq. (20.5.4) to obtain

$$f(x) = \frac{e^{-\lambda t} (\lambda t)^x}{x!}; \qquad x = 0, 1, 2, \dots \qquad (20.5.5)$$

Suppose, for example that in a certain country the average number of airplane crashes per year is 2.5. What is the probability of 4 or more crashes during the next year? Substituting $\lambda = 2.5$ and $t = 1$ in Eq. (20.5.5) yields

TABLE 20.5.1 Binomial and Poisson Probabilities

x	n = 20, p = 0.05		n = 20, p = 0.05	
	Binomial	Poisson	Binomial	Poisson
0	0.3585	0.3679	0.3660	0.3679
1	0.3774	0.3679	0.3697	0.3679
2	0.1887	0.1839	0.1849	0.1839
3	0.0596	0.0613	0.0610	0.0613
4	0.0133	0.0153	0.0149	0.0153
5	0.0022	0.0031	0.0029	0.0031
6	0.0003	0.0005	0.0005	0.0005
≥7	0.0000	0.0001	0.0001	0.0001

$$f(x) = \frac{e^{-2.5}(2.5)^x}{x!}; \quad x = 0, 1, 2, \ldots$$

as the pdf of X, the number of airplane crashes in one year. The probability of 4 or more airplane crashes next year is then

$$P(X \geq 4) = 1 - P(X \leq 3) = 1 - \sum_{x=0}^{3} \frac{e^{-2.5}(2.5)^x}{x!}$$

$$P(X \geq 4) = 1 - 0.76 = 0.24$$

As another example suppose that the number of breakdowns of personal computers during 1000 hours of operation of a computer center is 3. What is the probability of no breakdowns during a 10-hour work period?

Substituting $\lambda = 11$ and $t = \frac{1}{100}$ in Eq. (20.5.5) yields

$$f(x) = \frac{e^{-0.03}(0.03)^x}{x!}; \quad x = 0, 1, 2, \ldots$$

as the pdf of X, the number of breakdowns in a 10-hour period. The probability of no breakdowns in a 10-hour work period is then

$$P(X = 0) = e^{-0.03} = 0.97$$

In addition to the applications cited above, the Poisson distribution can be used to obtain the reliability of a standby redundancy system,[3] in which one unit is in the operating mode and n identical units are in standby mode. Unlike a parallel system where all units in the system are active, in the standby redundancy system the standby units are inactive. If all units have the same failure rate in the operating mode, unit failures are independent, standby units have zero failure rate in the standby mode, and there is perfect switchover to a standby when the operating unit fails, the reliability R of the standby redundancy system is given by

$$R = \sum_{x=0}^{n} \frac{e^{-\lambda t}(\lambda t)^x}{x!} \qquad (20.5.6)$$

This is the probability if n or fewer failures in the time period specified by t. For example, consider the case of a standby redundancy system with one operating unit and one on standby (i.e., a system that can survive one failure). If the failure rate is 2 units per year, then the 6-month reliability of the system is obtained by substituting $n = 1$, $\lambda = 2$, and $t = \frac{1}{2}$ in Eq. (20.5.6), which gives

$$R = \sum_{x=0}^{1} \frac{e^{-1}(1)^x}{x!} = 0.74$$

as the 6-month reliability of the system.

The Normal Distribution

When T, time to failure, has a normal distribution, its pdf is given by

$$f(t) = \frac{1}{\sqrt{2\pi}\sigma} \exp\left[-\frac{1}{2}\left(\frac{t-\mu}{\sigma}\right)^2\right]; \quad -\infty < t < \infty \qquad (20.5.7)$$

where μ is the mean value of T and σ is its standard deviation. The graph of f(t) is the familiar bell-shaped curve shown in Fig. 20.5.1.

The reliability function corresponding to normally distributed failure time is given by

$$R(t) = P(T > t) = \frac{1}{\sqrt{2\pi}\sigma} \int_{t}^{\infty} \exp\left[-\frac{1}{2}\left(\frac{t-\mu}{\sigma}\right)^2\right] dt \qquad (20.5.8)$$

The corresponding failure rate (hazard rate) is obtained by substitution in Eq. (20.3.9) and Eq. (20.3.11), which state that the failure rate Z(t) is related to the reliability R(t) by

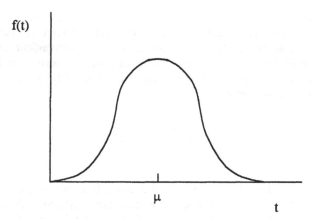

Figure 20.5.1 Normal pdf of time to failure.

$$Z(t) = \frac{R'(t)}{R(t)}$$

Recalling that Eq. (20.3.12) states

$$R'(t) = -f(t)$$

and substituting f(t) from Eq. (20.5.7) and R(t) from Eq. (20.5.8) yields

$$Z(t) = \frac{\exp\left[-\frac{1}{2}\left(\frac{t-\mu}{\sigma}\right)^2\right]}{\int_t^\infty \exp\left[-\frac{1}{2}\left(\frac{t-\mu}{\sigma}\right)^2\right]dt} \qquad (20.5.9)$$

as the failure rate corresponding to a normally distributed failure time.

    If T is normally distributed with mean μ and standard deviation σ, then the random variable $(T - \mu)/\sigma$ is normally distributed with mean 0 and standard deviation 1. The term $(T - \mu)/\sigma$ is called a "standard normal variable," and the graph of its pdf is called a "standard normal curve." Table 20.5.2 is a tabulation of areas under a standard normal curve to the right of z_0 of r nonnegative values of z_0.[4] Probabilities about a standard normal variable Z can be determined from the table. For example,

$$P(Z > 1.54) = 0.062$$

TABLE 20.5.2 Standard Normal, Cumulative Probability in Right-Hand Tail (for Negative Values of z, Areas Are Found by Symmetry)[a]

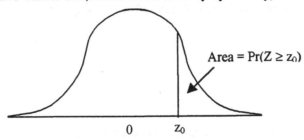

Area = $\Pr(Z \geq z_0)$

0   z_0

z_9	NEXT DECIMAL PLACE OF z_0									
	0	1	2	3	4	5	6	7	8	9
0.0	.500	.496	.492	.488	.484	.480	.476	.472	.468	.464
0.1	.460	.456	.452	.448	.444	.440	.436	.433	.429	.425
0.2	.421	.417	.413	.409	.405	.401	.397	.394	.390	.386
0.3	.382	.378	.374	.371	.367	.363	.359	.356	.352	.348
0.4	.345	.341	.337	.334	.330	.326	.323	.319	.316	.312
0.5	.309	.305	.302	.298	.295	.291	.288	.284	.281	.278
0.6	.274	.271	.268	.264	.261	.258	.255	.251	.248	.245
0.7	.242	.239	.236	.233	.230	.227	.224	.221	.218	.215
0.8	.212	.209	.206	.203	.200	.198	.195	.192	.189	.187
0.9	.184	.181	.179	.176	.174	.171	.189	.166	.164	.161
1.0	.159	.156	.154	.152	.149	.147	.145	.142	.140	.138
1.1	.136	.133	.131	.129	.127	.125	.123	.121	.119	.117
1.2	.115	.113	.111	.109	.107	.106	.104	.102	.100	.099
1.3	.097	.095	.093	.092	.090	.089	.087	.085	.084	.082
1.4	.081	.079	.078	.076	.075	.074	.072	.071	.069	.068
1.5	.067	.066	.064	.063	.062	.061	.059	.058	.057	.056
1.6	.055	.054	.053	.052	.051	.049	.048	.047	.046	.046
1.7	.045	.044	.043	.042	.041	.040	.039	.038	.038	.037
1.8	.036	.035	.034	.034	.033	.032	.031	.031	.030	.029
1.9	.029	.028	.027	.027	.026	.026	.025	.024	.024	.023
2.0	.023	.022	.022	.021	.021	.020	.020	.019	.019	.018
2.1	.018	.017	.017	.017	.016	.016	.015	.015	.015	.014
2.2	.014	.014	.013	.013	.013	.012	.012	.012	.011	.011
2.3	.011	.010	.010	.010	.010	.009	.009	.009	.009	.008
2.4	.008	.008	.008	.008	.007	.007	.007	.007	.007	.006
2.5	.006	.006	.006	.006	.006	.005	.005	.005	.005	.005
2.6	.005	.005	.004	.004	.004	.004	.004	.004	.004	.004
2.7	.003	.003	.003	.003	.003	.003	.003	.003	.003	.003
2.8	.003	.002	.002	.002	.002	.002	.002	.002	.002	.002
2.9	.002	.002	.002	.002	.002	.002	.002	.001	.001	.001

z_0	DETAIL OF TAIL ($._2135$, FOR EXAMPLE, MEANS 0.00135)									
2.	$._2228$	$._1179$	$._1139$	$._1107$	$._2820$	$._2621$	$._2466$	$._2347$	$._2256$	$._2187$
3.	$._2135$	$._3968$	$._3687$	$._3483$	$._3337$	$._3233$	$._3159$	$._3108$	$._4723$	$._4481$
4.	$._4317$	$._4207$	$._4133$	$._5854$	$._5541$	$._5340$	$._5211$	$._5130$	$._6793$	$._6479$
5.	$._6287$	$._6170$	$._7996$	$._7579$	$._7333$	$._7190$	$._7107$	$._8599$	$._8332$	$._8182$
	0	1	2	3	4	5	6	7	8	9

is obtained directly from the table as the area to the right of 1.54. The symmetry of the standard normal curve about zero implies that the area to the right of zero is 0.5 and the area to the left of zero is 0.5. Consequently;

$$P(0 < Z < 1.54) = 0.5 - 0.062 = 0.438$$

Also, because of symmetry

$$P(-1.54 < Z < 0) = 0.438 \quad \text{and} \quad P(Z < -1.54) = 0.062$$

Note that the area to the right of 1.54 is 0.062. The following probabilities can also be deduced from Fig. 20.5.2.

$$P(-1.54 < Z < 1.54) = 0.438$$
$$P(Z < 1.54) = 0.938; \quad P(Z > -1.54) = 0.938$$

Table 20.5.2 also can be used to determine probabilities concerning normal random variables that are not standard normal variables. The required probability is first converted to an equivalent probability about a standard normal variable. For example if T, the time to failure, is normally distributed with mean $\mu = 100$ and standard deviation $\sigma = 2$ then $(T - 100)/2$ is a standard normal variable and

$$P(98 < T < 104) = P\left[-1 < \frac{T-100}{2} < 2 \right] = P(-1 < Z < 2)$$
$$P(98 < T < 104) = 0.341 + 0.477 = 0.818$$

For any random variable X that is normally distributed with mean μ and standard deviation σ

$$P(\mu - \sigma < X < \mu + \sigma) = P(-1 < \frac{X-\mu}{\sigma} < 1)$$
$$P(\mu - \sigma < X < \mu + \sigma) = P(-1 < Z < 1) = 0.68 \qquad (20.5.10)$$

$$P(\mu - 2\sigma < X < \mu + 2\sigma) = P(-2 < \frac{X-\mu}{\sigma} < 2)$$
$$P(\mu - 2\sigma < X < \mu + 2\sigma) = P(-2 < Z < 2) = 0.95 \qquad (20.5.11)$$

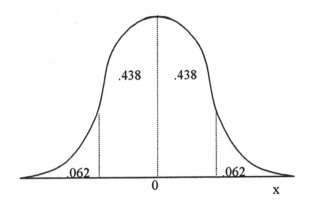

Figure 20.5.2 Areas under a standard normal curve.

$$P(\mu - 3\sigma < X < \mu + 3\sigma) = P(-3 < \frac{X - \mu}{\sigma} < 3)$$

$$P(\mu - 3\sigma < X < \mu + 3\sigma) = P(-3 < Z < 3) = 0.997 \qquad (20.5.12)$$

The probabilities given in Eqs. (20.5.10), (20.5.11), and (20.5.12) are the source of the percentages cited in statements 1, 2, and 3 at the end of Section 19.10. These can be used to interpret the standard deviation S of a sample of observations on a normal random variable, as a measure of dispersion about the sample mean $\overline{X}$.

The normal distribution is used to obtain probabilities concerning the mean $\overline{X}$ of a sample of n observations on a random variable X. If X is normally distributed with mean μ and standard deviation σ, then $\overline{X}$, the sample mean, is normally distributed with mean μ and standard deviation $\sigma/\sqrt{n}$. For example, suppose X is normally distributed with mean 100 and standard deviation 2. Then $\overline{X}$, the mean of a sample of 16 observations on X, is normally distributed with mean 100 and standard deviation 0.5. For example,

$$P(\overline{X} < 101) = P\left[\frac{\overline{X} - 100}{0.5} > \frac{101 - 100}{0.5}\right]$$

$$P(\overline{X} < 101) = P(Z > 2) = 0.023$$

If X is not normally distributed, then $\overline{X}$, the mean of a sample of n observations on X, is approximately normally distributed with mean μ and standard deviation $\sigma/\sqrt{n}$, provided the sample size n is large (>30). This result is based on an important theorem in probability called the *central limit theorem*. Suppose, for example, that the pdf of the random variable X is specified by

$$f(x) = \tfrac{1}{2}; \quad 0 < x < 2$$
$$f(x) = 0; \quad \text{elsewhere}$$

Application of Eq. (19.9.2) yields

$$\mu = E(X) = \int_0^2 x\left(\tfrac{1}{2}\right)dx = 1$$
$$E(X^2) = \int_0^2 x^2\left(\tfrac{1}{2}\right)dx = \tfrac{4}{3}$$

Therefore application of Eq. (19.9.4) yields

$$\sigma^2 = E(X^2) - \mu^2 = \tfrac{4}{3} - 1 = \tfrac{1}{3}$$

If $\overline{X}$ is the mean of a random sample of 48 observations on X, $\overline{X}$ is approximately normally distributed with mean 1 and standard deviation $\sqrt{1/(3)(48)}$ (i.e., 1/12). Therefore, for example,

$$P(\overline{X} > 9/8) = P\left[\frac{\overline{X}-1}{1/12} > \frac{9/8-1}{1/12}\right] = P(Z > 1.5) = 0.067$$

One of the principal applications of the normal distribution in reliability calculations and hazard and risk analysis is the distribution of time to failure due to wearout. Suppose, for example, that a production lot of a certain electronic device is especially designed to withstand high temperatures and intense vibrations has just come off the assembly line. A sample of 25 devices from the lot is tested under the specified heat and vibration conditions. Time to failure, in hours, is recorded for each of the 25 devices. Application of Eqs. (19.10.1) and (19.10.2) for the sample mean $\overline{X}$ and sample variance S^2 yields

$$\overline{X} = 125 \quad \text{and} \quad S^2 = 92$$

Past experience indicates that the wear-out time of this electronic device, like that of a large variety of products in many different industries, tends to be normally distributed. Using the calculated value of $\overline{X}$ and S as best estimates of μ and σ, we obtain for the 110-hour reliability of the electronic device

$$P(X > 110) = P\left[\frac{X - 125}{\sqrt{92}} > \frac{110 - 125}{\sqrt{92}}\right] = P(Z > -1.56) = 0.94$$

The Log-Normal Distribution

A nonnegative random variable X has a log-normal distribution whenever ln X, the natural logarithm of X, has a normal distribution. The pdf of a random variable x having a log-normal distribution is specified by

$$f(x) = \frac{1}{\sqrt{2\pi}\beta} x^{-1} \exp\left[-\frac{(\ln x - \alpha)^2}{2\beta^2}\right]$$

$$f(x) = 0; \quad \text{elsewhere} \tag{20.5.13}$$

The mean and variance of a random variable X having a log-normal distribution are given by

$$\mu = e^{\alpha + \beta^2 / 2} \tag{20.5.14}$$

$$\sigma^2 = e^{2\alpha + \beta^2}(e^{\beta^2} - 1) \tag{20.5.15}$$

Figure 20.5.3 plots the pdf of the log-normal distribution for $\alpha = 0$ and $\beta = 1$. Probabilities concerning random variables having a log-normal distribution can be calculated using tables of the normal distribution. If X has a log-normal distribution with parameters α and β, then ln X has a normal distribution with $\mu = \alpha$ and $\sigma = \beta$. Probabilities concerning X can therefore be converted into equivalent probabilities concerning ln X. Suppose, for example, that X has a log-normal distribution with $\alpha = 2$ and $\beta = 0.1$. Then

$$P(6 < X < 8) = P(\ln 6 < \ln X < \ln 8)$$

$$P(6 < X < 8) = P\left[\frac{\ln 6 - 2}{0.1} < \frac{\ln X - 2}{0.1} < \frac{\ln 8 - 2}{0.1}\right]$$

$$P(6 < X < 8) = P(-2.08 < Z < 0.79) = 0.78$$

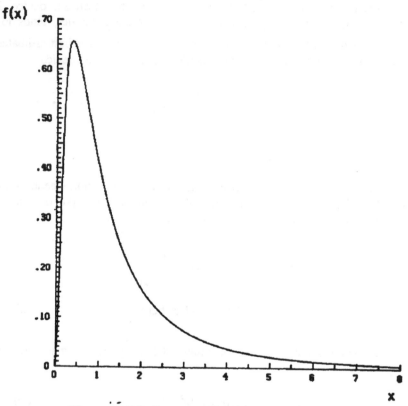

Figure 20.5.3 Log-normal pdf for $\alpha = 0$, $\beta = 1$

Estimates of the parameters α and β in the pdf of a random variable X having a log-normal distribution can be obtained from a sample of observations on X by making use of the fact that ln X is normally distributed with mean α and standard deviation β. Therefore, the mean and standard deviation of the natural logarithms of the sample observations on X furnish estimates of α and β. To illustrate the procedure, suppose the time to failure T, in thousands of hours, was observed for a sample of 5 electric motors. The observed values of T were 8, 11, 16, 22, and 34. The natural logs of these observations are 2.08, 2.40, 2.77, 3.09, and 3.53. Assuming that T has a log-normal distribution, the estimates of the parameters α and β in the pdf are obtained from the mean and standard deviation of the natural logs of the observations on T. Applying the Eqs. (19.10.1), and (19.10.2) yields 2.77 as the estimate of α and 0.57 as the estimate of β.

In addition to the probability distribution discussed above, there are several other well-known distributions that can be used in hazard risk analysis

TABLE 20.5.3 Probability Distributions

Statistical class	Distribution
Discrete	Binomial
	Geometric
	Multinomial
	Poisson
Continuous	Extreme Value
	Gamma
	Log-normal
	Normal
	Pareto
	Rayleigh
	Rectangular
	Weibull

TABLE 20.5.4 Statistical Distribution Details

Probability Distribution	Description	Application
Multinomial	Probability distribution of the number of failures in n independent demands in which at each trial there are more than two possible outcomes	Appropriate for situations similar to those for binomial distribution, except more than two outcomes can be found
Geometric	Probability distribution of the number of failures before the first success occurs. It is the discrete analog of the exponential distribution, where parameter p is analogous to λ_c. Distribution assumes "memoryless" property of independent trials	Can be applied to discrete failure on demand data in absence of other information
Exponential	Sometimes referred to as the "negative exponential" distribution. The distribution is characterized by a single parameter, λ_c, the failure rate assumed constant over time	Usually applied to data in the absence of other information; thus the most widely used in reliability work. Not appropriate for modeling burn-in or wearout failure rates

calculations. These distributions may be either discrete and continuous functions. A brief summary of them is provided in Table 20.5.3. A short description of each of these distributions not covered above is provided in Table 20.5.4.

20.6 MONTE CARLO SIMULATION OF FAILURE DISTRIBUTIONS

Monte Carlo simulation, a procedure for mimicking observations on a random variable, permits verification of results that ordinarily would require difficult mathematical calculations or extensive experimentation.

Monte Carlo simulation uses computer programs called *random number generators*. A random number may be defined as a number selected from the interval (0, 1) in such a way that the probabilities that the number comes from any two subintervals of equal length are equal. For example, the probability that the number is in the subinterval (0.1, 0.3) is the same as the probability that the number is in the subinterval (0.5, 0.7). Random numbers thus defined are observations on a random variable X having a uniform distribution on the interval (0, 1). This means that the pdf of X is specified by

$$f(x) = 1; \quad 0 < x < 1$$
$$f(x) = 0; \quad \text{elsewhere}$$

This pdf assigns equal probability to subintervals of equal length in the interval (0, 1).

Using random number generators, Monte Carlo simulation can generate observed values of a random variable having any specified pdf. To generate observed values of T, time to failure, when T is assumed to have a pdf specified by f(t), first use the random number generator to generate a value of X between 0 and 1. Substitute this value for the cdf F(t) and then solve for t. The solution is an observed value of the random variable T having pdf specified by f(t).

As a sample example of Monte Carlo simulation, consider a pump whose time to failure T, measured in years, has an exponential distribution with pdf specified by

$$f(t) = e^{-t}; \quad t > 0$$
$$f(t) = 0; \quad t \le 0$$

It is desired to estimate the average life of the pump on the basis of 15 simulated values of T.

The cdf of T is obtained from Eq. (19.8.2) as follows.

$$F(t) = P(T \le t) = \int_0^t f(t)\, dt$$

Therefore,

$$F(t) = 1 - e^{-t}; \quad t > 0$$
$$F(t) = 0; \quad t \le 0$$

If the first number generated by the random number generator is 0.93, the corresponding simulated value of t is obtained by solving the equation

$$0.93 = 1 - e^{-t}$$

to obtain 2.66.

The process is repeated with another random number until the desired 15 simulated values of T have been obtained. The results are shown in Table 20.6.1. The average value of the 15 simulated values of T is 1.02 years, a Monte Carlo estimate of the average life of the pump. The true average life of the pump is E(t), the expected value of T, obtained by application of Eq. (19.9.2):

$$E(t) = \int_{-\infty}^{\infty} tf(t)\,dt = \int_{0}^{\infty} te^{-t}\,dt = 1$$

A more accurate estimate of the true value of the average life of the pump can be obtained by increasing the number of simulated values in which the estimate is based. In this example the expected value of T, obtained by integration, provided a comparison with the estimated average value of T, obtained by Monte Carlo simulation. Generally Monte Carlo simulation is used to estimate some function of one or more random variables when direct calculation of this function would be difficult if not impossible.

In the case of random variables assumed to be normally distributed, Monte Carlo simulation is facilitated by use of a table of the normal distribution (Table 20.5.2). Consider, for example, a series system consisting of two electrical components, A and B. Component A has a time to failure T_A, assumed to be normally distributed with mean 100 hours and standard deviation 20 hours. Component B has a time to failure T_B, assumed to be normally distributed with mean 90 hours and standard deviation 10 hours. The system fails whenever component A or component B fails. Therefore, the time to failure of the system T_S is the minimum of time to failure of components A and B.

Suppose that using Monte Carlo simulation with 10 simulated values of T_A and 10 simulated values of T_B, it is desired to estimate an average value of T_S. First, 20 random numbers are generated. These are shown in columns 1 and 4 of Table 20.6.2. Regard each of the random numbers generated as the value of the cdf of a standard normal variable Z. Let z_1 be the simulated value of Z corresponding to 0.10, the first random number in column 1. Then, since 0.10 is the value of the cdf for $Z = z_1$,

$$P(Z < z_1) = 0.10$$

**TABLE 20.6.1   Data for Monte Carlo Estimation of Average Time to
                 Failure of a Pump**

Random Number	Simulated Time to Failure (years)
0.93	2.66
0.06	0.06
0.53	0.76
0.56	0.82
0.41	0.53
0.38	0.48
0.78	1.52
0.54	0.78
0.49	0.67
0.89	2.21
0.77	1.47
0.85	1.90
0.17	0.19
0.34	0.42
0.56	0.82

**TABLE 20.6.2   Data for Monte Carlo Estimation of Average Time to
                 Failure of Series System with Two Components, A, B**

(1) Random Numbers	(2) Simulated Z	(3) Simulated T_A	(4) Random Numbers	(5) Simulated Z	(6) Simulated T_B	(7) Simulated T_S
0.10	-1.28	74	0.92	1.41	104	74
0.54	0.10	102	0.86	1.08	101	101
0.42	-0.20	96	0.45	-0.13	89	89
0.02	-2.05	59	0.38	-0.31	87	59
0.81	0.88	118	0.88	-0.36	86	86
0.07	-1.48	70	0.21	-0.81	82	70
0.06	-1.56	69	0.26	-0.64	84	69
0.27	-0.61	88	0.51	0.03	90	88
0.57	0.18	104	0.73	0.61	96	96
0.80	0.84	117	0.71	0.56	96	96

From Table 20.6.2, z_1 is found to be -1.28. The corresponding simulated value of T_A is obtained by noting that $(T_A - 100)/20$ is a standard normal variable. Therefore, the simulated value of T_A corresponding to $z_1 = -1.28$ is

$$T_A = 100 - 1.28(20) = 74$$

The other simulated values of T_A listed in column 3 are obtained in similar fashion. The same procedure is used to obtain the simulated values of T_B, listed in column 6, after noting that $(T_B - 90)/10$ is a standard normal variable. For example, the simulated value of T_B corresponding to the random number 0.92 is

$$T_B = 90 + 1.41(10) = 104$$

The simulated value of T_S corresponding to $T_A = 74$ and $T_B = 104$ is the smaller of these two values (i.e., 74). The other simulated values of T_S are obtained in similar fashion and listed in column 7. The average of the simulated values of T_S, 83, is the estimated time to failure of the system.

Note that as previously mentioned, increasing the number of simulated values increases the accuracy of the estimate. Also note that Monte Carlo simulation provides an attractive alternative to solving the somewhat complicated mathematical problem of finding the expected value of the minimum of two normal random variables.

20.7 FAULT TREE ANALYSIS

A *fault tree* is a graphic technique used to analyze complex systems. The objective is to spotlight conditions that cause a system to fail. Fault tree analysis attempts to describe how and why an accident or other undesirable event has occurred. It may also be used to describe how and why an accident or other undesirable event could take place. Thus fault tree analysis finds wide application in hazard analysis and risk assessment of process and plant systems.[5]

Fault tree analysis seeks to relate the occurrence of an undesired event, the "top event," to one or more antecedent events, called "basic events". The top event may be, and usually is, related to the basic events via certain intermediate events. A fault tree diagram exhibits the causal chain linking the basic events to the intermediate events and the latter to the top event. In this chain the logical connection between events is indicated by so-called "logic gates". The principal logic gates are the AND gate, symbolized on the fault tree by ⌂, and the OR gate symbolized by ⌂.

As a simple example of a fault tree, consider a water pumping system consisting of two pumps, A and B, where A is the pump ordinarily operating and B is a standby unit that automatically takes over if A fails. Flow of water through the pump is regulated by a control valve in both cases. Suppose that the top event is no water flow, resulting from the following basic events: failure to pump A *and* failure of pump B, or failure of the control valve. The fault tree diagram for this system is shown in Fig. 20.7.1.

Since one of the purposes of a fault tree analysis is the calculation of the probability of the top event, let A, B, and C represent the failure of pump A, the failure of pump B, and the failure of the control valve, respectively. Then if T represents the top event, no water flow, we can write

$$T = AB + C$$

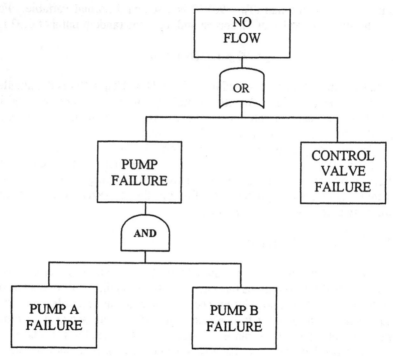

Figure 20.7.1.   Fault tree for a water pumping system.

This equation indicates that T occurs if both A and B occur or if C occurs. Assume that A, B, and C are independent and $P(A) = 0.01$, $P(B) = 0.005$, and $P(C) = 0.02$. Then application of the addition theorem in Eq. (19.4.3) yields

$$P(T) = P(AB) + P(C) - P(ABC) \qquad (20.7.1)$$

The independence of A, B, and C implies

$$P(AB) = P(A) P(B) \quad \text{and} \quad P(ABC) = P(A) P(B) P(C)$$

Therefore Eq. (20.7.1) can be written as

$$P(T) = P(A) P(B) + P(C) - P(A) P(B) P(C)$$
$$P(T) = (0.01)(0.005) + 0.02 - (0.01)(0.005)(0.02) = 0.020049$$

In connection with fault trees, *cut sets*, and *minimal cut sets* are defined as follows. A *cut set* is a basic event or intersection of basic events that will cause the top event to occur. A *minimal cut set* is a cut set that is not a subset of any other cut set; it may also be defined as the smallest combination of component and/or failures which, if they occur, will cause the top event to occur.[6] Minimal cut set analysis rearranges the fault tree so that any basic event that appears in different parts of the fault tree is not "double counted" in the quantitative evaluation. The result of minimal cut set analysis is a new fault tree, logically equivalent to the original, consisting of an OR gate beneath the top event, whose inputs are the minimal cut sets. Each minimal cut set is an AND gate containing a set of basic inputs necessary and sufficient to cause the top event. Additional details are available in the literaterature.[7]

In the example represented by Eq. (20.7.1), AB and C are cut sets because if AB occurs then T occurs, and if C occurs then T occurs. Also, AB and C are minimal cut sets, since neither is a subset of the other. The event T described in Eq. (19.3.13) can be regarded as a top event represented as a union of cut sets. In Eq. (19.3.13)

$$T = AE + AEF + BE + BEF$$

Here, AE, AEF, BE, and BEF are cut sets, since the occurrence of any one of the corresponding events results in the occurrence of T. AEF is not a minimal cut set because it is a subset of AE. Also BEF is not a minimal cut set because it is a subset of BE. The minimal cut sets in this case are, therefore, AE and BE.

Consider now the following hypothetical example of a more complicated fault tree involving eight basic events leading to the rupture of a pressure release disk, the top event T. The fault tree is shown in Fig. 20.7.2. Let the basic events be defined as follows:

A: Premature disk failure
B: Operator error
C: Pump motor failure
D: Reaction inhibitor system failure
E: Coolant system failure
F: Outlet piping obstruction
G: Motor alarm failure
H: Pressure sensor failure

The top event T can be represented in terms of the basic events as follows:

$$T = A + C(G + H) + B + DE + (F + G) \qquad (20.7.2)$$

Applying the Boolean algebra associative law and distributive law, given in Eqs. (19.3.3) and (19.3.5), to Eq. (20.7.2) yields

$$T = A + CG + CH + B + F + G \qquad (20.7.3)$$

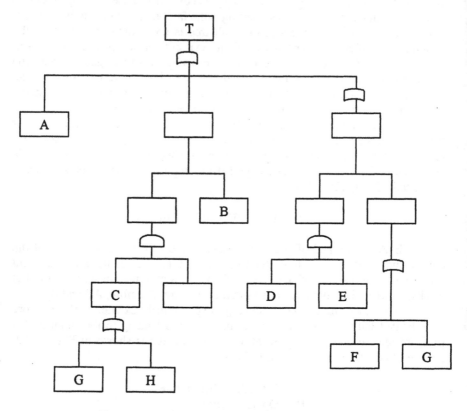

Figure 20.7.2. Fault tree with nine basic events.

The cut sets for this system are then:

  - A
  - CG
  - CH
  - B
  - DE
  - F
  - G

The only cut set that is a subset is CG, which is a subset of G. Eliminating CG yields the following minimal cut sets:

  - A
  - CH

- B
- DE
- F
- G

The probability of the top set T is the probability of the union of the events represented by the minimal cut sets. This probability usually can be approximated by the sum of the probabilities of these events. Suppose that all the basic events are independent and that their probabilities are

$P(A) = 0.03$
$P(B) = 0.01$
$P(C) = 0.005$
$P(D) = 0.04$
$P(E) = 0.009$
$P(F) = 0.07$
$P(G) = 0.02$
$P(H) = 0.05$

Then the probability of the disk rupturing can be approximated by

$P(T) = P(A) + P(CH) + P(B) + P(DE) + P(F) + P(G)$
$P(T) = 0.03 + (0.005)(0.05) + 0.01 + (0.04)(0.009) + 0.07 + 0.02$
$P(T) = 0.1306$

20.8 EVENT TREE ANALYSIS

An event tree provides a diagrammatic representation of event sequences that begin with a so-called initiating event and terminate in one or more undesirable consequences. In contrast to a fault tree, which works backward from an undesirable consequence to possible causes, an event tree works forward from the initiating event to possible undesirable consequences. The initiating event may be equipment failure, human error, power failure, or some other event that has the potential for adversely affecting an ongoing process.

The following illustration of an event tree analysis is based on one reported by Lees.[8] Consider a situation in which the probability of an external power outage in any given year is 0.1. A backup generator is available, and the probability that it will fail to start on any given occasion is 0.02. If the generator starts, the probability that it will not supply sufficient power for the duration of the external power outage is 0.001. An emergency battery power supply is available; the probability that it will be inadequate is 0.01.

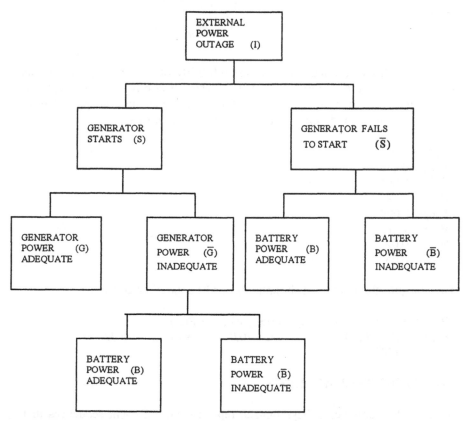

Figure 20.8.1. Event tree for a power outage.

Figure 20.8.1 shows the event tree with the initiating event and the external power outage, denoted by I. Labels for the other events on the event tree are also indicated. The event sequences I S $\overline{\text{G}}$ $\overline{\text{B}}$ and I $\overline{\text{S}}$ $\overline{\text{B}}$ terminate in the failure of emergency power supply. Applying the multiplication theorem in the form given in Eqs. (19.5.8) and (19.5.7) yields

$$P(I\ S\ \overline{G}\ \overline{B}) = (0.1)(0.98)(0.001)(0.01) = 0.998 \times 10^{-5}$$
$$P(I\ \overline{S}\ \overline{B}) = (0.1)(0.02)(0.01) = 2 \times 10^{-5}$$

Therefore the probability of emergency power supply failure in any given year is 2.098×10^{-5}, the sum of these two probabilities.

Additional details on these calculations are provided by Theodore et al.[9]

20.9 ILLUSTRATIVE EXAMPLES

Example 20.1

Refer to figure 20.2.1. If component A, B, C, and D have for their respective reliabilities 0.80, 0.80, 0.80, and 0.50, calculate the reliability of the system.

Solution 20.1

The reliability of the parallel subsystem is obtained by applying Eq. (20.2.2), which yields

$$R_P = 1 - (1 - 0.80)(1 - 0.80) = 0.96$$

The reliability of the system is then obtained by applying Eq. (20.2.1), which yields

$$R_S = (0.80)(0.96)(0.50) = 0.384$$

Example 20.2

Refer to the time and failure transistor example provided in Section 20.4. Determine the 10,000-hour reliability of a circuit of five such transistors connected in series.

Solution 20.2

The 10,000-hour reliability of one transistor is the probability that it will last at least 10,000 hours. This probability can be obtained by integrating the pdf of T, time to failure, which gives

$$P(T > 10) = \int_{10}^{\infty} 0.01e^{-0.01t} \, dt = 0.90$$

The same result can also be obtained directly from Eq. (20.3.16), which expresses reliability in terms of failure rate. Substituting the failure rate

$$Z(t) = 0.01$$

In Eq. (20.3.16) yields

$$R(t) = \exp-\left(\int_{0}^{t} 0.01 \, dt\right)$$

as the reliability function. The 10,000-hour reliability is therefore

$$R(10) = \exp-\left(\int_0^{10} 0.01\,dt\right) = e^{-0.01(10)} = 0.90$$

The 10,000-hour reliability of a circuit of five transistors connected in series is obtained by applying Eq. (20.2.1), the formula for the reliability of series system, to obtain

$$R_S = [R(10)^5] = 0.59$$

Example 20.3

Consider the case of a standby redundancy system consisting of one thermometer and 6 on standby. Assume that the thermometers are independent with respect to failure and that each has a probability of failure of 0.2. Determine the reliability of the above system.

Solution 20.3

For the above system, n = 7, r = 6, p = 0.2, and q = 1 − p = 0.8. Using Eq. (20.5.3), the reliability of the system may be determined.

$$P(X \le r) = \sum_{x=0}^{r} \frac{n!}{x!(n-x)!} p^x q^{n-x}$$

$$P(X \le 6) = \sum_{x=0}^{6} \frac{7!}{x!(7-x)!} (0.20)^x (0.80)^{7-x} = 0.9998$$

Example 20.4

Suppose that the number of breakdowns of personal computers during 2500 hours of operation of a computer center is 11. What is the probability of no breakdowns during a 50-hour work period?

Solution 20.4

Substituting $\lambda = 11$, $t = \frac{50}{2500} = \frac{1}{50}$, and $\lambda t = \frac{11}{50} = 0.22$, in Eq. (20.5.5) yields

$$f(x) = \frac{e^{-0.22}(0.22)^x}{x!}$$

The probability of no breakdowns in a 50-hour work period is then

$$P(X = 0) = e^{-0.22} = 0.80$$

Example 20.5

If T is the failure of a bag in a baghouse in 50 days with standard deviation 1.0, determine the probability of a bag failing between the 49th and 52nd day.

Solution 20.5

$$P(49 < T < 52) = P\left[-1 < \frac{T - 50}{1.0} < 2\right]$$

$$P(49 < T < 52) = 0.341 + 0.477 = 0.818$$

Example 20.6

Assume the time to failure in days, T, of a reactor cooling system has a log-normal distribution with $\alpha = 2.5$ and $\beta = 0.15$. Calculate $P(4 < X < 10)$.

Solution 20.6

$$P(4 < X < 10) = P(\ln 4 < \ln X < \ln 10)$$

$$P(4 < X < 10) = P\left[\frac{\ln 4 - 2.5}{0.15} < \frac{\ln X - 2.5}{0.15} < \frac{\ln 10 - 2.5}{0.15}\right]$$

$$P(4 < X < 10) = P(-7.42 < Z < -1.32)$$
$$= 0.000 + 0.093$$
$$= 0.093$$

20.10 SUMMARY

1. The basic concepts and theorems of probability presented in Chapter 19 may be applied to determine the reliability of complex systems in terms of the reliabilities of their components.
2. Many systems consisting of several components can be classified as *series*, *parallel*, or a combination of both. However, the majority of industrial and process plants (units and systems) have series of parallel configurations.
   A *series system* is one in which the entire system fails to operate if any one of its components fails to operate. A *parallel system* is one that fails to operate only if all its components fail to operate.
3. The reliability of a component will frequently depend on the length of time it has been in service.
4. Frequently the failure rate of equipment exhibits three stages; a break-in stage with a declining failure rate, a useful life stage characterized by a

fairly constant failure rate, and a wearout period characterized by an increasing failure rate. Many industrial parts and components follow this path. A failure rate curve exhibiting these three phases is called a "bathtub" curve.

5. Other important probability distributions include: the Binomial Distribution, the Polynomial Distribution, the Normal Distribution, and the Log-Normal Distribution.

6. Monte Carlo simulation, a procedure for mimicking observations on a random variable, permits verification of results that ordinarily would require difficult mathematical calculations or extensive experimentation.

7. A *fault tree* is a graphic technique used to analyze complex systems. Fault tree analysis attempts to describe how and why an accident or other undesirable event has occurred. It may also be used to describe how and why an accident or other undesirable event could take place.

8. An event tree provides a diagrammatic representation of event sequences that begin with a so-called initiating event and terminate in one or more undesirable consequences. In contrast to a fault tree, which works backward from an undesirable consequence to possible causes, an event tree works forward from the initiating event to possible undesirable consequences. The initiating event may be equipment failure, human error, power failure, or some other event that has the potential for adversely affecting an ongoing process.

PROBLEMS

1. Determine the reliability of the following electrical system, using the reliabilities indicated under the various components.

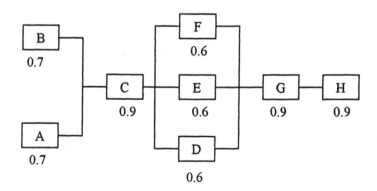

2. The life in hours of an electronic component is a random variable having a Weibull distribution with $\alpha = 0.025$ and $\beta = 0.50$.
    (a) What is the average life of the component?

(b) What is the probability that the component will last more than 4000 hours?

3. The life of an automobile seal has a Weibull distribution with failure rate $Z(t) = \dfrac{1}{\sqrt{t}}$, where t is measured in years. What is the probability that the seal will last at least 4 years?

4. A system has 20 independent components, each of which fails with probability 0.10. The system fails only if 4 or more of the components fail. What is the probability that the system will fail?

5. An engineer's ability to distinguish a natural from a synthetic diamond is tested independently on 10 different occasions. What is the probability of 7 correct identifications if the engineer is only guessing (i.e., has a probability of making a correct identification of 0.5)?

6. Microscopic slides of a certain culture of microorganisms contain on the average 20 microorganisms per square centimeter. After treatment by a chemical, one square centimeter is found to contain only 10 such microorganisms. If the treatment had had no effect, what would be the probability of finding 10 or fewer microorganisms in a given square centimeter?

7. Assume the number of particles emitted by a radioactive substance has a Poisson distribution with an average emission of one particle per second.
   (a) Find the probability that at most one particle will be emitted in 3 seconds.
   How low an emission rate would be required to make the probability of the emission of at most one particle in 3 seconds at least 0.80?

8. In a source of liquid, bacteria are known to be present with a mean number of 3 per cubic centimeter. Ten 1-cm^3 test tubes are filled with the liquid.
   (a) Calculate the probability that all 10 test tubes will show growth (i.e., contain at least one bacterium each).
   What is the probability that exactly 7 will show growth?

9. The measurement of the pitch diameter of the thread of a fitting is normally distributed with mean 0.4008 inch and standard deviation 0.0004 inch. The specifications are given as 0.4000 ± 0.0010 inch. What is the probability that a defective fitting will occur?

10. Let X denote the coded quality of bag fabric used in a particular utility baghouse. Assume that X is normally distributed with mean 10 and standard deviation 2.
    (a) Find c such that $P(|X - 10| < c) = 0.90$.
    (b) Find k such that $P(X > k) = 0.90$.

11. The failure rate per year, Y, of a coolant recycle pump has a log-normal distribution. If ln Y has mean, -2, and variance, 1.5, find $P(0.175 < Y < 1)$.

12. Three light bulbs (A, B, C) are connected in series. Assume that the bulb lifetimes are normally distributed, with the following means and standard deviations.

Mean (hours)	A	B	C
Standard deviation (hours)	100	90	80
	30	20	10

Using the following random numbers, simulate the lifetime of the system and estimate its mean and standard deviation.

0.52	0.01	0.77	0.67	0.14	0.90
0.80	0.50	0.54	0.31	0.39	0.28
0.45	0.29	0.96	0.34	0.06	0.51
0.68	0.34	0.02	0.00	0.86	0.56
0.59	0.46	0.73	0.48	0.87	0.82

13. If a building fire occurs, a smoke alarm sounds with probability 0.9. The sprinkler system functions with probability 0.7 regardless of whether the smoke alarm sounds. The range of consequences is minor fire damage, moderate fire damage with few injuries, moderate fire damage with many injuries, and major fire damage with many injuries. Construct an event tree and indicate the probabilities for each of the four consequences.

14. A runaway chemical reaction can occur if coolers fail (A) or if there is a bad chemical batch (B). Coolers fail only if both cooler 1 (C) and cooler 2 (D) fail. A bad chemical batch occurs if there is a wrong mix (E) or a process upset (F). A wrong mix occurs only if there is an operator error (G) and instrument failure (H). Construct a fault tree and obtain the minimum cut sets.

15. The top event in a fault tree occurs if A and B occur. A occurs if C and D occur. D occurs if E or F occur. B occurs if E or G occur. Assume that the probabilities of events A, B, C, D, E, F, and G are 0.003, 0.008, 0.0015, 0.07, 0.0025, 0.0065, and 0.0009, respectively.
    (a) Construct a fault tree and obtain the minimum cut sets.
    (b) Assuming that the basic events are independent, compute the probability of the top event.

REFERENCES

1. Karl V. Bury, *Statistical Models in Applied Science*, Wiley, New York, 1975.
2. Irwin Miller and John E. Freund, *Probability and Statistics for Engineers*, 3rd ed., Prentice-Hall, Englewood Cliffs, NJ, 1985.
3. Bertram L. Amstadter, *Reliability Mathematics*, McGraw-Hill, New York, 1971.
4. R. J. Woonacott, T. H. Woonacott, *Introductory Statistics*, 4th ed., Wiley, New York, 1985.
5. B. S. Dhillon, Chanan Singh, *Engineering Reliability*, Wiley, New York, 1981.

6. Chemical Manufacturers Association, *Risk Analysis in the Chemical Industry*, Government Institutes, Inc., Rockville, Maryland, 1985.
7. Guidelines for Chemical Process Quantitative Risk Analysis, CCPS, AIChE, New York City, 1989.
8. Frank P. Lees, *Loss Prevention in the Process Industries*, Vol. 1, Butterworths, Boston, 1980.
9. L. Theodore, J. Reynolds, and K. Harris, *"Health Safety and Accident Prevention: Industrial Applications"*, Theodore Tutorials, East Willington, NY, 1997.

21

Hazard Risk Analysis Applications

21.1 INTRODUCTION

This chapter presents case studies illustrating the application of basic principles and special techniques of hazard risk analysis introduced in Chapters 19 and 20. Section 21.2 begins by applying Bayes' theorem to revise the probability distribution of the failure rate of a coolant recycle pump on the basis of no observed failures over a 10-year period. The next case study, presented in Section 21.3, involves the risk of transporting hazardous chemicals by rail. The case study in Section 21.4 features an event tree analysis of a sequences of events leading to the release of toxic vapor and a fault tree analysis of a sequence producing moderate release of this vapor. Section 21.5 examines the catastrophic release of the contents of a holdup tank caused by a runaway chemical reaction induced by uncontrolled temperature increases. In Section 21.6, a case study on a bus section in an electrostatic precipitator illustrates the use of Monte Carlo methods in conjunction with the binomial and Weibull probability distributions for estimating out-of-compliance probabilities. The chapter concludes (Section 21.7) with preliminary hazard analysis and a fault tree presentation of an ethylene production plant.

Readers that are not completely at ease with the contents of Chapters 19 and 20 will have difficulty negotiating these case studies. They are real-world examples drawn from literature accounts of potential or actual events. The methods of quantitative analysis employed go beyond the level of elementary illustrative examples.

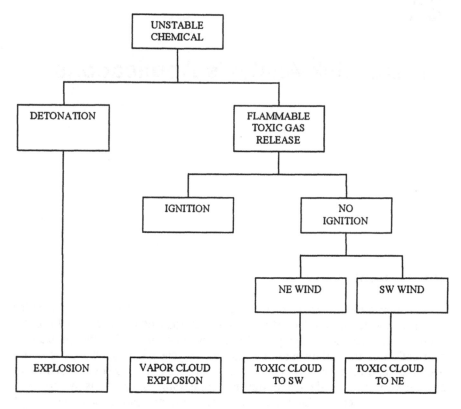

Figure 21.1.1 Event tree showing outcomes of two incidents involving an unstable chemical.

The following simplified example, constructed by Hendershot, will facilitate the transition to the case studies. [1] Suppose that a risk assessment is being conducted at a chemical plant to determine the consequences of two incidents (the initiating events of the event tree shown in Fig. 21.1.1) defined as

I.   an explosion resulting from detonation of an unstable chemical
II.   a release of a flammable toxic gas

Incident **I** has one possible outcome, an explosion, the consequences of which are assumed to be unaffected by weather conditions. Incident **II** has several possible outcomes which, for purposes of simplification, are reduced to

IIA.   vapor cloud explosion, caused by ignition of the gas released, centered at the release point, and unaffected by weather conditions
IIB.   toxic cloud extending downwind and affected by weather conditions

Again for purposes of simplification, only two weather conditions are envisioned, a northeast and a southwest wind. Associated with these two wind directions are events **IIB1** and **IIB2**.

IIB1. toxic cloud to the southwest
IIB2. toxic cloud to the northeast

The probabilities and conditional probabilities of the occurrence of the defined events in any given year are estimated as follows:

$$
\begin{aligned}
P(I) &= 10^{-6} \\
P(II) &= 1/33{,}333 \\
P(IIA \,|\, II) &= 0.33 \\
P(IIB \,|\, II) &= 0.67 \\
P(IIB1 \,|\, IIB) &= 0.5 \\
P(IIB2 \,|\, IIB1) &= 0.5
\end{aligned}
$$

Application of the multiplication theorem, Eq. (19.5.5), yields the probabilities

$$
\begin{aligned}
P(IIA) &= P(II)\,P(IIA \,|\, II) \\
&= (1/33{,}333)(0.33) = 10^{-5}
\end{aligned}
$$

$$
\begin{aligned}
P(IIB1) &= P(II)\,P(IIB \,|\, II)\,P(IIB1 \,|\, IIB) \\
&= (1/33{,}333)(0.67)(0.5) = 10^{-5}
\end{aligned}
$$

$$
\begin{aligned}
P(IIB2) &= P(II)\,P(IIB \,|\, II)\,P(IIB2 \,|\, IIB) \\
&= (1/33{,}333)(0.67)(0.5) = 10^{-5}
\end{aligned}
$$

The consequences of events I, IIA, IIB1, and IIB2 in terms of fatalities are estimated as follows.

I. All persons within 200 meters of the explosion center are killed; all persons beyond this distance are unaffected.
IIA. All persons within 100 meters of the explosion center are killed; all persons beyond this distance are unaffected.
IIB1. All persons downwind in a 22.5° segment of radius 500 meters are killed; all persons outside this area are unaffected.
IIB2. Same as IIB1.

Figure 21.1.2 shows the distribution of people in the impact zones of these events.

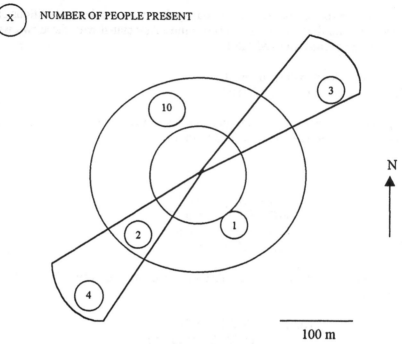

Figure 21.1.2. Population distribution in impact zones.

The total risk measured in terms of the average annual total number of people killed is obtained by multiplying the number of people in each impact zone by the sum of the probabilities of the events affecting the zone, and summing the results. Therefore,

$$
\begin{aligned}
\text{total risk} \quad &= 3[P(\text{IIB2})] + 4[P(\text{IIB1})] + 2[P(\text{IIB1}) + P(\text{I})] + 1[P(\text{I})] + \\
&\quad 10[P(\text{I})] \\
&= 3(10^{-5}) + 4(10^{-5}) + 2[10^{-5} + 10^{-6}] + 1(10^{-6}) + 10(10^{-6}) \\
&= 1.03 \times 10^{-4}
\end{aligned}
$$

The average annual individual risk for the 20 people in the impact zones shown in Fig. 21.1.2 is obtained by dividing the total risk by 20. Therefore,

$$
\text{average annual individual risk} = \frac{1.03 \times 10^{-4}}{20} = 5.2 \times 10^{-6}
$$

TABLE 21.1.1   Events and Probabilities for Incidents I and II

N	Events resulting in N or more deaths	Annual probability N or more deaths
0	I, IIA, IIB1, IIB2	3.1×10^{-5}
3	I, IIB1, IIB2	2.1×10^{-5}
6	I, IIB2	1.1×10^{-5}
13	I	10^{-6}

The result, 5.2×10^{-6}, may be interpreted as the average probability that a person in the impact zones will be killed in the course of a year, as a result of the occurrence of incident I or incident II.

Figure 21.1.3 depicts the risk to society in terms of the annual probability of N or more deaths as a result of the occurrence of incident I or incident II.   Note that the scales in Fig. 21.1.3 are logarithmic.   The plotted probabilities are obtained by summing the probabilities of the events resulting in N or more deaths for N = 0, 3, 6, 13.   Table 21.3.1 lists these events and probabilities.

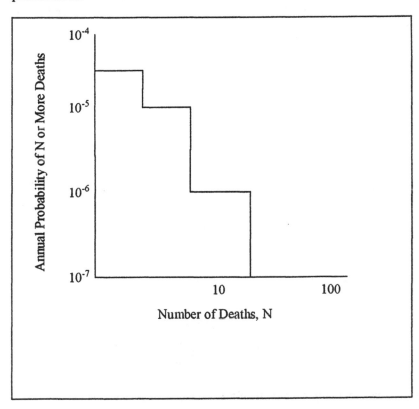

Figure 21.1.3.   Risk to Society

21.2 ESTIMATION OF FAILURE RATE OF COOLANT RECYCLE PUMP

Mosleh, Kazarians, and Gekler[2] obtained a Bayesian estimate of the failure rate, Z, of a coolant recycle pump in the hazard/risk study of a chemical plant. The estimate was based on evidence of no failures in 10 years of operation. Nuclear industry experience with pumps of similar types was used to establish the prior distribution of Z. This experience indicated that the 5th and 95th percentiles of the failure rate distribution developed for this category were 2.0 x 10^{-6} per hour (about one failure per 57 years of operation) and 98.3 x 10^{-6} per hour (about one failure per year). Extensive experience in other industries suggested the use of a log-normal distribution with the 5th and 95th percentile values as the prior distribution of Z, the failure rate of the coolant recycle pump.

Recall that if Z has a log-normal distribution, then ln Z has a normal distribution with mean μ and standard deviation σ equal to the parameters in the pdf of Z. This fact, in conjunction with the specified 5th and 95th percentiles of the probability distribution of Z, can be used to obtain the values of μ and σ, and thereby the parameters of the log-normal pdf of Z. If Z denotes the failure rate per year, the fact that the 5th percentile of the distribution of Z is 1/57 implies

$$P(Z < 1/57) = 0.05$$
$$P(\ln Z < -\ln 57) = 0.05$$
$$P\left(\frac{\ln Z - \mu}{\sigma} < \frac{-\ln Z - \mu}{\sigma}\right) = 0.05$$

Since $(\ln Z - \mu)/\sigma$ is a standard normal variable, we can refer to a table that gives areas under the standard normal curve (see Table 21.5.2) to find that

$$\frac{-\ln 57 - \mu}{\sigma} = -1.645 \tag{21.2.1}$$

The fact that the 95th percentile of the distribution of Z is 1 implies

$$P(Z < 1) = 0.95$$
$$P(\ln Z < 0) = 0.95$$
$$P\left(\frac{\ln Z - \mu}{\sigma} < \frac{-\mu}{\sigma}\right) = 0.95$$

Referring again to a published table of areas under the curve yields

$$\frac{-\mu}{\sigma} = 1.645 \qquad (21.2.2)$$

Solving Eqs. (21.2.1) and (21.2.2) for μ and σ yields

$$\mu = -2.02 \quad \text{and} \quad \sigma = 1.23$$

Substituting the values of μ and σ for the parameters in the log-normal pdf in Eq. (20.5.13) produces the following pdf of the prior distribution of Z

$$f(z) = \frac{1}{\sqrt{2\pi(1.23)^2}} z^{-1} \exp\left[-\frac{1}{2(1.23)^2}(\ln z + 2.02)^2\right]; \quad z > 0 \qquad (21.2.3)$$

The conditional probability of event B, no failures in 10 years, given that the failure rate is Z per year, is obtained by applying the Poisson distribution to give

$$P(B|z) = \frac{e^{-10z}(10z)^0}{0!} = e^{-10z} \qquad (21.2.4)$$

Bayes' theorem provides the mechanism for converting the prior pdf of Z to the posterior pdf of Z on the basis of the occurrence of event B (i.e., no failures in 10 years). Applying Bayes' theorem and denoting the posterior pdf of Z by $f(z|B)$ yields

$$P(z|B) = \frac{f(z)P(B|z)}{\int_0^\infty f(z)P(B|z)dz} ; \qquad (21.2.5)$$

where $f(z)$, the prior pdf, is given by Eq. (21.2.3).

The 5th percentile $P_{0.05}$ and the 95th percentile $P_{0.95}$ of the posterior distribution of z are defined by

$$\int_{-\infty}^{P_{0.05}} f(z|B)dz = 0.05 \qquad (21.2.6)$$

$$\int_{-\infty}^{P_{0.95}} f(z|B)dz = 0.95 \qquad (21.2.7)$$

Equations (21.2.6) and (21.2.7) yield

$$P_{0.05} = 1/134 \quad \text{and} \quad P_{0.95} = 1/7$$

Comparison of the 5th and 95th percentiles of the posterior distribution of Z with the 5th and 95th percentiles of the prior distribution of Z indicates that the posterior pdf lies to the left of the prior pdf. Therefore, the posterior pdf assigns higher probability to intervals in the lower part of the range of z than the prior pdf. This reflects the influence of the observed occurrence of no failures in 10 years.

The mean of the posterior distribution of Z is Bayesian estimate of the failure rate per year. If $E(Z|B)$ is the mean of the posterior distribution, then

$$E(Z|B) = \int_{-\infty}^{\infty} zf(z|B)dz = 1/15$$

provides the Bayesian estimate of the failure rate per year.

21.3 TRANSPORTATION OF HAZARDOUS CHEMICALS

Kazarians, Boykin, and Kaplan investigated the risk of transporting an acutely toxic and flammable chemical 400 miles in a specially equipped railroad tank car.[3] During its journey, the tank car changes trains in three major rail yards. Each change of trains is technically described as a "classification". The chemical is generally shipped one tank at a time. There are approximately 150 shipments a year.

Accident frequencies were evaluated separately for the two types of activity: mainline transit and rail yard classification. When an accident occurs and the tank car is damaged, the severity of public exposure depends on several factors, including the likelihood of a breach in the tank car, the severity level of the release (i.e., the rate or volume of spillage), the likelihood of an explosion, the magnitude of the explosion, and the dispersion pattern of the unignited vapors. Recall that Part II of this book deals with explosions and their effects; Part III treats this subject of dispersion. Table 21.3.1 summarizes the transportation risk data for the mainline and rail yard segments of the tank car journey.

Multiplying the average number of cars damaged per car mile, by the annual number of shipments, by the distance in miles traveled by each shipment gives:

$$(1.5)(10^{-7})(150)(400) = 9 \times 10^{-3}$$

as the average annual number of cars damaged. Multiplying this result by the likelihood of release following an accident, by the likelihood of no ignition, by the likelihood of a release of small severity gives

$$(9 \times 10^{-3})(0.20)(0.1)(0.4) = 7.2 \times 10^{-5}$$

TABLE 21.3.1 Transportation Risk Data

Factor	Mainline	Rail Yard
	Activity	
Average number of cars damaged	1.5×10^{-7}/ car mile	4.3×10^{-5}/ classification
Likelihood of release following accident	0.20	0.06
Severity of release likelihood		
Small	0.4	0.4
Medium	0.3	0.3
Large	0.2	0.2
Very large	0.1	0.1
Release ignition likelihood	0.9	0.9

TABLE 21.3.2 Average Annual Frequency of Release

Severity	Mainline	Rail Yard	Total
Small	7.2×10^{-5}	4.6×10^{-5}	11.8×10^{-5}
Medium	5.4×10^{-5}	3.5×10^{-5}	8.9×10^{-5}
Large	3.6×10^{-5}	2.3×10^{-5}	5.9×10^{-5}
Very Large	1.8×10^{-5}	1.2×10^{-5}	3.0×10^{-5}

TABLE 21.3.3 Health Impact Data for Various Release Severities

Release Severity	Number Affected	Likelihood of Exposure to Lethal Concentration	Average Annual Frequency of Exposure
Small	0	0.57	6.7×10^{-5}
Medium	200	0.34	3×10^{-5}
Large	500	0.07	4.1×10^{-6}
Very Large	800	0.02	6.0×10^{-7}

as the average annual frequency of a release of small severity due to mainline accidents. Similar computations give corresponding values for rail yard accidents and or releases of medium, large, and very large severity. Table 21.3.2 shows the results of these computations.

The health impact of a release of toxic vapor varies with the severity of the release, the population density along the route of the tank car, and weather conditions affecting dispersion. Table 21.3.3 shows, for each degree of severity, hypothetical estimates of the number of people affected, the likelihood of exposure to a potentially lethal concentration, and the product of this likelihood by the average annual frequency of each release severity. This latter product represents average annual frequency of exposure to a potentially lethal concentration. Average annual frequency of exposure is plotted against the number of people affected in the risk curve shown in Fig. 21.3.1.

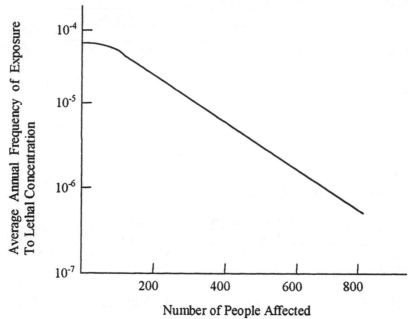

Figure 21.3.1   Risk curve for tank car transport of hazardous chemical.

21.4   RELEASE OF TOXIC VAPOR

Burns and Hazzan demonstrated the use of event tree and fault tree analysis in the study of a potential accident sequence leading to a toxic vapor release at an industrial chemical process plant.[4] The initiator of the accident sequence studied is event P, the failure of a plant programmable automatic controller. This event, in conjunction with the success or failure of a process water system (a glycol cooling system) and an operator-manual shutdown of the distillation system produced minor, moderate, or major release of toxic material as indicated in Fig. 21.4.1. The symbols W, G, O represent the events listed:

W: failure of process water system
G: failure of glycol cooling system
O: failure of operator-manual shutdown of the distillation system

The symbols $\overline{W}$, $\overline{G}$, and $\overline{O}$, the complements of W, G, and O, represent the corresponding successes.

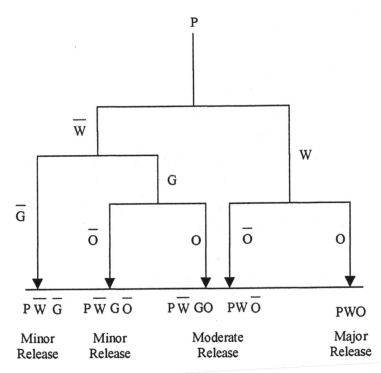

Figure 21.4.1. Event tree for toxic vapor release.

The three event sequences $P\overline{W}$ GO, $PW\overline{O}$, and PWO produced moderate or severe releases of toxic vapor.   One of these, $P\overline{W}$ GO, was subjected to a fault tree analysis. It was assumed that events P and O were independent events having no common link with $\overline{W}$ and G. Table 21.4.1 show the fault tree in Boolean equation format; that is, the top event $P\overline{W}$ GO is represented in terms of unions and intersections of intermediate and basic events. The event symbols are defined in Table 21.4.2. Computer analysis using the SETS program of the U.S. Nuclear Regulatory Commission identified 117 minimal cut sets: 7 single failures, 66 double failures, 28 triple failures, and 16 quadruple failures. The sum of the probabilities of the 117 minimal cut sets was 2.28×10^{-1}. The probabilities of events P and O were estimated to be 8.76×10^{-3} and 1×10^{-1} per plant-year, respectively. Multiplication of these three probabilities yields

$$(2.28 \times 10^{-1})(8.76 \times 10^{-3})(1 \times 10^{-1}) = 2.00 \times 10^{-4}$$

as the probability per plant-year of the accident sequence $P\overline{W}$ GO.

TABLE 21.4.1 Linked Fault Tree

$P\,\overline{W}\,\overline{GO}$ = P * $\overline{W}$ * G * O

= {P} * { $\overline{PWFS}$ * [$\overline{LS2912A}$ * $\overline{TLV2912A}$ * ($\overline{HP2D27}$ *

$\overline{LS2912A}$) * $\overline{PWSTIC}$ * $\overline{CNTL1}$

    * <($\overline{LONP}$ * $\overline{CNTL1}$ * $\overline{PWCTPAM}$) + ($\overline{LONP}$ * $\overline{CNTL1}$ *

$\overline{PWCTPAM}$ >)

   * [<($\overline{PWSAM}$ * $\overline{NPSHSAB}$ * $\overline{PWSAC2}$ * $\overline{PWSAO}$ * $\overline{LONP}$) +

   * [<($\overline{PWSBM}$ * $\overline{NPSHSAB}$ * $\overline{PWSBC2}$ * $\overline{PWSBO}$ * $\overline{LONP}$) >

   * <($\overline{PWCAM}$ * $\overline{NPSHCAB}$ * $\overline{PWCAC2}$ * $\overline{PWCAO}$ * $\overline{LONP}$) +

   * <($\overline{PWCBM}$ * $\overline{NPSHCAB}$ * $\overline{PWCBC2}$ * $\overline{PWCBO}$ * $\overline{LONP}$)}*

{[G5TV + G5TIC + G5MV + << G5CMB + LONP + G5CBO +
(CMMV + (GVSLLK + GPSLLK))
+ (GP1A * GHXTLK)) > * < G5CAM + LONP + GSCAO +
(GMMV + (GVSLLK + GPSLLK)
+ (GP1A * GHXTLK)) >> + < GVSLLK + GPSLLK + (GP1A *
GHXTLK) >
+ < GAGBYP + < IAS + G5TV + < (GCPFPN + GCPFPAM +
GCPFPAO + LONP)
* (GCPFPBM + GCPFPN + GCPFPBO + LONP) >
+ < (GCPAOIL + GRATIC + W + GPWMVA + GCPAM + LONP
+ GCPAC + GRLEAK + GRBVA + GCPPO
* (GCPBOIL + GCPBO + GCPBM + W + GPWMVB + GRBTIC +
LONP + GCPBC + GRLEAK + GRBVB >]
*[G15MV + < G15CAM + LONP + G15CAO + < TKV1413LK +
GL1438 + GL1440 + (GVSLLK + GPSLLK + (GP1A * GHXTLK))
>
+ < G15CBM + LONP + G15CBO + < TKV1413LK + GL1438 +
GL1440 + (GVSLLK + GPSLLK + (GP1A * GHXTLK)) >
+ < GVSLLK + GPSLLK + (GP1A * GHXTLK) >
+ < GAGBYP + < IAS + G5TV + < (GCPFPN + GCPFPAM +
GCPFPAO + LONP)
* (GCPFPBM + GCPFPN + GCPFPBO + LONP) >
+ < (GCPAOIL + GRATIC + W + GPWMVA + GCPAM + LONP
+ GCPAC + GRLEAK + GRBVA + GCPAO
* (GCPBOIL + GCPBO + GCPBM + W + GPWMVB + GRBTIC +
LONP + GCPBC + GRLEAK + GRBVB) >}
   * {0}

TABLE 21.4.2 Definition of P W̄ GO Variables, in Alphabetical Order

CNTL1	Controller 1 failure
G15CMB/G15CAM	- 15 C glycol cooling system (GC) circulation pumps B/A mechanical failure
G15CBO/G15CAO	- 15 C GC circulation pumps B/A stopped by operator
G15MV	-15 C GC 1 of 5 manual valves closed
G5CBM/G5CAM	+5 C GC circulation pumps B/A mechanical failure
G5CBO/G5CAO	+5 C GC circulation pumps B/A stopped by operator
G5MV	+5 C GC 1 of 5 manual valves closed
G5TIC	+5 C GC temperature instrument fails low
G5TV	+5 C GC temperature valve fails closed
GAGBYP	Glycol tanks bypass agitator failure
GCPBC/GCPAC	GC compressors B/A capacity controller failure
GCPBM/GCPAM	GC compressors B/A mechanical failure
GCPBO/GCPAO	GC compressors B/A stopped by operator
GCPBOIL/GCPAOIL	GC motors B/A oil supply failure
GCPFPBM/GCPFPAM	GC compressor feed pumps B/A stopped by operator
GCPFPN	GC compressor 10-inch feed line low NPSH
GHXTLK	Glycol heat exchanger tube leak
GL1438	Condenser fails
GL1440	Product cooler fails
GMMV	+5 C GC 1 of 2 manual valves closed
GPLA	Glycol system pressure indicator fails
GPSLLK	Glycol pump seal leaks undetected
GPWMVB/GPWMVA	GC compressors B/A process water system (PW) supply manual valve closed
GRBTIC/GRATIC	GC refrigerant systems B/A temperature instrument failure
GRBVB/GRBVA	GC refrigerant bypass control valves B/A fail open
GRLEAK	GC refrigerant medium leak
GVSLLK	Glycol header vessel ruptures
HP2D27	Drain open
IAS	Instrument air system failure
LONP	Loss of normal electrical power supply
LS2912A	Level switch fails high
NPSHCAB	PW circulation pumps low NPSH
NPSHSAB	PW supply pumps low NPSH
O	Operator fails to shut down distillation system
P	Automatic controller failure
PWCBC2/PWCAC2	PW circulation pumps B/A controller 2 failure

TABLE 21.4.2 (Continued)	
PWCBM/PWCAM	PW circulation pumps B/A mechanical failure
PWCBO/PWCAO	PW circulation pumps B/A stopped by operator
PWCTPBM/PWCTPAM	PW collecting tanks pumps B/A failure
PWFS	PW flow switch failure
PWSBC2/PWSAC2	PW supply pumps B/A controller 2 failure
PWSBM/PWSAM	PW supply pumps B/A mechanical failure
PWSBO/PWSAO	PW supply pumps B/A stopped by operator
PWSTIC	PW surge tank instrument failure
TKV1413LK	-15 C glycol storage tank leak
TLV2912A	PW level valve fails
W	Process water system failure

The seven single failures causing moderate release of toxic vapor resulting from the occurrence of the event $\overline{P W}$ GO identified the components targeted for most attention from the plant operators. The glycol cooling system circulation pumps contributed to 26% of the sequence failures appearing in the cut set list. It was concluded that, as the components occurring most frequently in the cut set list, they merited high priority in terms of inspection, maintenance, and/or possible replacement by more reliable components.

This particular case study does illustrate how event tree and fault tree analysis can be used in a hazard operability (HAZOP) study.

21.5 OPERATION OF A HOLDUP TANK CONTAINING A HAZARDOUS CHEMICAL

Kazarians, Bradford, and Abrams reported a risk analysis designed to evaluate the potential impact on public health from the operation of a holdup tank containing a product fluid that is unstable (i.e., decomposes and liberates heat) and toxic bother under normal conditions. [5] The fluid has to be kept cool, and an inhibitor must be added to it to slow the decomposition reaction. Hot product fluid from the process stream is cooled in a heat exchanger and after an inhibitor has been added to it from the inhibitor addition system (IAS), the hot product flows into the holdup tank. The contents of the tank are further cooled to remove the heat of decomposition through the recycle loop. If the temperature in the system increases above a threshold value, the emergency inhibitor system (EIS) is triggered. The EIS is a separate inhibitor addition system designed to dump its contents into the holdup tank by means of gravity plus a pressurized gas assist. This action results in the complete control and subsequent heat reduction of the decomposition reaction.

The hazards of this arrangement arise from the possibility of a catastrophic release of the contents of the holdup tank into the atmosphere. A runaway reaction in the holdup tank, which occurs only when the fluid temperature rises, can breach the tank and thereby produce the catastrophic release. The events identified as initiators of event sequences terminating in a runaway reaction are listed, along with an estimate of the mean annual frequency of each, in Table 21.5.1.

Seventeen event sequences resulted in a runaway reaction. These are listed in Table 21.5.2 (using Table 21.5.1 notation) along with the mean annual frequency of release produced by each sequence; G stands for failure of the EIS on demand. The average annual frequency of failure on demand is estimated as 0.592.

The method used for obtaining the mean annual frequencies in Table 21.5.2 is illustrated as follows for the first event sequence (ABCG$\overline{E}$F) only. The mean annual frequency of event A is 0.183. The conditional mean frequency of B given A is 1, the conditional mean frequency of C given AB is 1, the conditional mean frequency of G given ABC is approximately 1, the conditional mean frequency of $\overline{E}$ given ABCG is approximately 1, the conditional mean frequency of F given ABCG$\overline{E}$ is 0.598. The product

$$P = (0.183)(1)(1)(1)(1)(0.598) = 1.1 \times 10^{-1}$$

represents the mean annual frequency for the first event sequence. In the case of event sequences ABCG$\overline{E}$F, (B + D) G$\overline{E}$F, and F$\overline{A}$ B$\overline{G}$E D, an operator may forestall the release of toxic vapor by emptying the holdup tank, thus mitigating the runaway reaction. The mean conditional failure frequency of such an operator action is estimated as 1.6×10^{-3}. Multiplying the mean annual frequencies for these three sequences by 1.6×10^{-3} results in revision of the total mean annual frequency downward from 3.43×10^{-1} to 3.55×10^{-2}.

Combinations of weather conditions, wind speed and wind direction along with boiling point, vapor density, diffusivity, and heat of vaporization of the chemical released vary the health impact of the released chemical on the nearby population. To model a runaway reaction, the release of 10,000 gallons was assumed to occur over a 15-minute period. The concentration of the chemical released was estimated, using procedures described in Part III (Chapter 12) for each combination of weather condition, wind speed, and wind direction. The results, combined with population data for the area adjacent to the plant, led to probability estimates of the number of people affected. Table 21.5.3 summarizes the findings.

Frequency of exceedance per year is obtained by multiplying 0.036, the mean annual frequency of release, by the cumulative probabilities in Table 21.5.3. Frequency of exceedance plotted against the number of people affected produces a risk curve portraying health impact in terms of the frequency with which the number of people affected exceeds various amounts. The risk curve is

TABLE 21.5.1 Mean Annual Frequency of Initiator Events

Initiator Event	Symbol	Mean Annual Frequency
Coolant A unavailable	A	0.183
Heat exchanger 1 fails to remove heat	B	0.018
Heat exchanger 2 fails to remove heat	C	0.018
Pump 1 fails to operate	D	0.068
Electric power lost	E	0.0584
IAS fails to operate	F	0.241

TABLE 21.5.2 Mean Annual Frequency of Event Sequences Terminating in Runaway Reaction

Event Sequence	Mean Annual Frequency
$ABCG\overline{E}F$	1.1×10^{-1}
$ABCGEF$	6.06×10^{-6}
$ABCGE$	1.47×10^{-6}
$\overline{C}\,\overline{A}\,\overline{B}\,\overline{G}EF\overline{D}$	7.15×10^{-6}
$\overline{C}\,\overline{A}BGE\,\overline{F}\,\overline{D}$	5.34×10^{-7}
$\overline{C}\,\overline{A}BGEFD$	2.95×10^{-11}
$C\overline{G}\overline{E}\,\overline{F}D$	2.12×10^{-6}
$C\overline{G}EFD$	1.17×10^{-10}
$(B+D)\,\overline{G}\,\overline{E}\,\overline{F}$	5.36×10^{-2}
$(B+D)\,\overline{G}EF$	2.96×10^{-6}
$EABCGFD$	3.49×10^{-2}
$\overline{F}\,\overline{A}\,\overline{B}GED$	1.44×10^{-1}
$\overline{F}A\overline{B}C\overline{G}ED$	7.15×10^{-6}
$\overline{F}\,\overline{A}BCGE\,\overline{D}$	3.55×10^{-10}
$\overline{F}\,\overline{A}\,\overline{C}GED$	2.84×10^{-5}
$\overline{F}\,\overline{A}CGE\overline{D}$	1.41
$FABCG\overline{E}$	7.27×10^{-5}
Total	3.43×10^{-1}

shown in Fig. 21.5.1. (Note that this case study required the application of many of the principles presented in both Parts III, IV, and V).

TABLE 21.5.3 Distribution of Number of People Affected by Chemical Release

Number Affected	Probability	Cumulative Probability
825	0.02	0.02
673	0.02	0.04
653	0.02	0.06
529	0.03	0.09
303	0.08	0.17
297	0.08	0.25
281	0.07	0.32
264	0.07	0.39
234	0.70	0.46
229	0.14	0.60
195	0.06	0.66
191	0.14	0.80
0	0.20	1.00

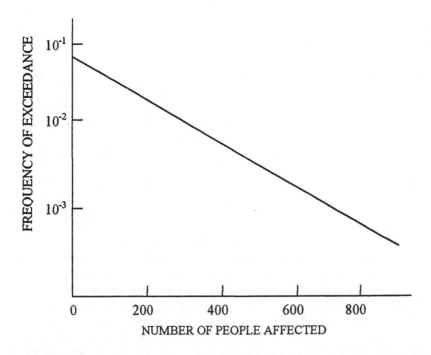

Figure 21.5.1. Risk curve for people affected by release of toxic chemical.

21.6   BUS SECTION FAILURES IN ELECTROSTATIC PRECIPITATORS

Theodore et al. employed Monte Carlo methods in conjunction with the binomial and Weibull distributions to estimate out-of-compliance probabilities for electrostatic precipitators on the basis of observed bus section failures. [6] The following definitions apply (see Fig. 21.6.1).

Chamber. One of many passages for gas flow
Field. One of several high voltage sections for the removal of particulates; these fields are arranged in series (i.e., the gas passes from the first field into the second, etc.)
Bus section. A region of the precipitator that is independently energized; a given bus section can be identified by a specific chamber and field

Thus an M x N electrostatic precipitator consists of M chambers and N fields. A precipitator is "out of compliance" when its overall collection efficiency falls below a designated minimum because of bus section failures. When several bus sections fail, the effect of the failures depends on where they are located. To determine directly whether a precipitator is out of compliance after a given number of bus sections have failed, it would be necessary to test all possible arrangements of the failure locations. The out of compliance probability is given by the percent of arrangements that result in overall collection efficiencies less than the prescribed minimum standard. The number of arrangements to be tested often makes the direct approach impractical. For example, Theodore et al. were requested (as part of a consulting assignment) to investigate a precipitator unit consisting of 64 bus sections; if 4 of these were to fail, there would be 15,049,024 possible failure arrangements.

Instead of the direct approach that would have required study of more than 15×10^6 potential failure arrangements, Theodore et al. used a Monte Carlo technique, testing only a random sample of the possible failure arrangements. The arrangements to be chosen are selected by use of random numbers. A set of random numbers is generated equal in quantity to the number of bus section failures assumed. Each of the random numbers is used to identify a bus section which, during the calculation of overall collection efficiency, is assumed to be out of commission. For all the out-of-compliance probabilities calculated, 5000 failure location arrangements were sampled. The random numbers used were generated by the power residue method.[7] These programs were written in FORTRAN and run on a VAX/ 780 (Digital Equipment Corporation) computer. The programs were later modified to run on a Hewlett-Packard microcomputer.

A leading utility in the northeast had requested out-of-compliance information to better schedule outages (plant shutdowns). The time to failure, T, of a bus section was assumed to have a Weibull distribution, the parameters of which were estimated by the method of maximum likelihood on the basis of observed bus section failures shown in Table 21.6.1 for the utility's 5 x 8

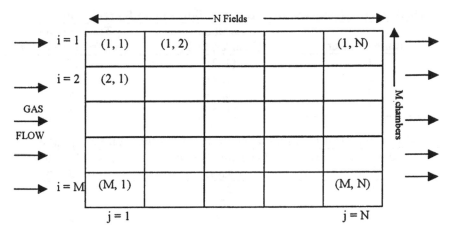

Figure 21.6.1. An M x N precipitator.

precipitator. [8] The probability that the precipitator would be out of compliance after 330 days, at which time a plant outage had been scheduled, was calculated (a) assuming that the last failure had just occurred and (b) assuming that 3 weeks had passed without the occurrence of another failure.

For each of the 36 bus sections that had not already failed, the Weibull distribution was used to determine the probability of failure before the next outage. Under assumption (a), this probability is $P(T < 330 | T > 209)$ i.e., the conditional probability of failure before 330 days, given that the bus section has survived 209 days. Under assumption (b), the corresponding probability is $P(T < 330 | T > 230)$. For part (b), the estimates of the Weibull distribution parameters used in part (a) were modified to take into consideration the absence of failures for 3 additional weeks.

Assuming a binomial distribution for the number of additional failures before the next outage, with probability of success equal to $P(T < 330 | T > 209)$ in part (a), probability of success equal to $P(T < 330 | T > 230)$ in part (b), and n = 36, the probabilities of 0, 1, 2, ..., 36 failures were calculated. On the basis of those probabilities, the Monte Carlo method was used to estimate out-of-compliance probability as follows. For each run, a random number was generated that selected a failure number between 0 and 36 with probability calculated as above. Then, a set of random numbers was generated equal in number to the failure number selected. As described previously, each of these random numbers was used to identify a bus section which, during the calculation of overall collection efficiency, was assumed to be out of commission. After 5000 runs, the percentage of runs yielding an overall collection efficiency less than the prescribed standard was the estimated out-of-compliance probability. Table 21.6.2 shows the results for assumptions (a) and (b) and also shows the effect of reducing the duration of the operating period (time between outages).

TABLE 21.6.1   Observed Bus Section Failures in 5 x 8 Precipitator

Failure Number	Location (chamber, Field)	Time to failure (days from last outage)
1	(4, 2)	62
2	(7, 4)	112
3	(8, 1)	153
4	(1, 3)	209

TABLE 21.6.2   Out-of-Compliance Probabilities for 5 x 8 Precipitator

Time (days) from Last Failure	Operating Period (days)	Out-of-Compliance Probability
0	330	0.504
	300	0.271
	270	0.095
21	330	0.181
	315	0.116
	300	0.066

21.7 ILLUSTRATIVE EXAMPLE

One illustrative example is presented in this final chapter. It has been adopted from the outstanding work of Kavianaian et al. and is concerned with an ethylene production plant. [9]   The solution involves a preliminary hazards analysis (PHA) and the development of a fault tree for the process.

Ethylene is the major feedstock used by the petrochemical industry for producing a variety of synthetic polymers. Ethylene is produced by steam cracking hydrocarbons such as ethane, propane, naptha, and gas oil. Steam pyrolysis of hydrocarbons produces ethylene along with a wide range of by-products such as hydrogen, methane, propylene, and butadiene. Since ethylene purity is a critical factor in polymerization units, the mixture of gases obtained as a result of steam cracking must be purified. Therefore, ethylene plants are composed of a pyrolysis section in which the feedstock is cracked in pyrolysis furnaces to produce ethylene and other gases, and in a purification section in which the pyrolysis are separated and recovered. [10, 11, 12]

Ethylene production involves high temperatures (1500°F) in the pyrolysis section and cryogenic temperatures in the purification section. [13] The feedstocks, products, and by-products of pyrolysis are flammable and pose severe fire hazards. Benzene, which is produced in small amounts as a by-product, is a known carcinogen.   Table 21.7.1 summarizes some of the properties of ethane (feedstock) and the product gases. Figure 21.7.1 shows a simplified schematic diagram of the pyrolysis and waste heat recovery section on an ethylene plant.

The feedstock is mixed with steam before entering the pyrolysis reactors. Steam reduces the hydrocarbon partial pressure, acts as a heat transfer media, and reduces coke laydown inside the reactor tubes. The tubular reactors are heated to reaction temperatures (1100°F to 1700°F) by means of direct fired heaters. The flow of steam and feedstock to the reactors can be adjusted to provide an optimum residence time, which is a function of the feedstock used.

After completing the cracking reactions in the tubular reactors, the gaseous mixture flows to a liquid quench tower where the gas temperature is lowered enough to stop the cracking reactions. Oil or water can be used as the cooling media. Transfer line heat exchangers can be used to recover the heat contained in the product gas, and this energy can be used to produce high pressure steam.

The cooled gaseous products are dried using an adsorbent such as molecular sieves and compressed to about 500 psig by a multistage compressor. The compressed gas is then sent to an acetylene converter where acetylene is selectively hydrogenated to ethane. The gaseous mixture then flows to the purification section of the plant where each component of the gas is recovered by means of cryogenic distillation.

Table 21.7.2 summarizes the results of application of a preliminary hazard analysis (PHA) to the pyrolysis section of an ethylene plant. The major hazard to the personnel and plant is the fire or explosion hazard of the gases used or produced in the process.

Figure 21.7.2 demonstrates the preliminary steps for a fault tree analysis (FTA); in addition, the TOP event, bounds, configurations, and unallowed events are specified, and the level of resolution is shown. Once all the limits have been determined, the fault tree is constructed (Figure 21.7.3). Note that every branch of the fault tree ends in a basic fault or cause leading to the TOP event.

As can be noted in Figure 21.7.2, steam and ethane are mixed before entering the reactor tubes where pyrolysis reactions take place. All feed and product lines must be equipped with appropriate control devices to ensure safe operation.[16] The FTA flow chart breaks down a TOP event (see description of fault tree in Unit II) into all possible basic causes. Although, this method is more structured than a PHA, it addresses only one individual event at a time. To use an FTA for a complete hazard analysis, all possible TOP events must be identified and investigated; this would be extremely time consuming and perhaps unnecessary in a preliminary design.

As can be noted from the analysis of the ethylene plant, one of the major disadvantages of the FTA is lack of recommendations for preventative and corrective measures. FTA, however, has the advantage of pinpointing the sequence of events that could lead to an undesired TOP event. Once these causes have been identified, an experienced design team can recommend solutions in the form of design alternatives and/or instrumentation. In recommending solutions, the probability, severity, and economics of each case must be taken into account. For example, the problem of temperature control

TABLE 21.7.1   Hazard and Toxic Properties of Materials in Ethylene
                Production[14, 15]

Feedstock	Toxicity	Fire Hazard	Explosion Hazard
Ethane	Low	Very dangerous	Moderate
Hydrogen	None	Dangerous	Dangerous
Acetylene	Moderate	Very dangerous	Moderate
Methane	Low	Very dangerous	Dangerous
Ethylene	Low	Very dangerous	Moderate
Carbon dioxide	Low	None	None

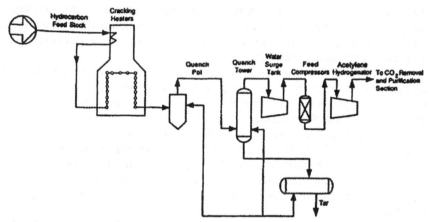

Figure 21.7.1 Pyrolysis and waste recovery section of an ethylene production
                 plant.

failure in the reactor tubes as a result of disruption in ethane flow can also be
solved by installing flow controls on the lines. Although flow controls on feed
and steam lines are installed for the purpose of controlling the residence time in
the reactor and product distribution, the flow controllers also contribute to
temperature control in the reactor. These interactions are important and their
effects must be taken into account as a conceptual design develops into a flow
diagram and finally into a piping and instrumentation diagram. The earlier
analysis of PHA indicates that PHA is not only capable of identifying major
hazards in the process, but it also recommends corrective measures at the very
early stages of design.   This is extremely important in developing new
technologies and in feasibility studies. The overall economic picture of a
process can change drastically as a result of instrumentation and/or procedures
to minimize risk or to bring the plant into compliance with regulations.

TABLE 21.7.2   Applying a Preliminary Hazard Analysis (PHA) to an
                          Ethylene Plant

Hazard	Cause	Major Effects	Corrective/ Preventative Measures
Damage to feed reactor tubes	Feed compressor failure (no endothermic reactions in reactor)	Capital loss, downtime Damage to the furnace coils due to high temperature	• Provide spare compressor with automatic switch-off control • Develop emergency response system
Explosion, fire	Pressure buildup in the reactor due to plug in transfer lines	Fatalities, injuries	• Provide pressure relief valve on the reactor tubes • Provide warning system for pressure fluctuations (high-pressure alarm) • Provide auxillary lines with automatic switch off
Explosion, fire	Violent reaction of H_2 to acetylene converter with air in presence of ignition source	Potential for injuries and fatalities due to fire or explosion	• Provide warning system (hydrogen analyzer) • Eliminate all sources of ignition near hydrogen storage area • Develop emergency fire response • Automatically shut off the H_2 feed • Provide fire fighting equipment

TABLE 21.7.2 (cont'd)

Hazard	Cause	Major Effects	Corrective/ Preventative Measures
Flammable gas release	Ethane storage tank ruptures	Potential for injuries and fatalities due to fire or explosion	• Provide warning control system (pressure control) • Minimize on-site storage • Develop procedure for tank inspection • Develop emergency response system • Provide gas monitoring system
Flammable gas release	CH_4 Storage tank (line) leak/rupture (fuel for the furnace)	Potential for injuries and fatalities due to fire or explosion	• Provide warning system • Minimize on-site storage • Develop procedure for tank inspection • Develop emergency response system • Provide gas monitoring system
Flammable gas release	Radiant tube rupture in the furnace	Potential for injuries and fatalities due to fire	• Improve reactor materials of construction • Monitor design vs. operating reactor temperature • Provide temperature control instrument

TABLE 21.7.2 (cont'd)

Hazard	Cause	Major Effects	Corrective/ Preventative Measures
Employee exposure to benzene (carcinogen)	Leak in knock-out pots or during handling benzene	Chronic health hazard	• Install warning signs in the area • Provide appropriate PPE • Develop safety procedures for handling and cleanup • Monitor concentration of benzene in area to meet TLV requirements
Fire/explosion in acetylene converter	Runaway reaction (exothermic)	Fatality, injury, or loss of capital	• Install temperature control on converter • Install pressure relief on reactor responding to temperature control
Flammable atmosphere	Leak in transfer lines	Fire/explosion	• Install combustible gas meter in sensitive areas • Provide adequate fire fighting equipment • Provide for emergency shutdown • Educate and train personnel on emergency procedures

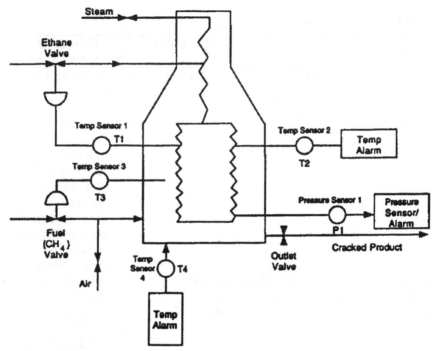

Figure 21.7.2 Fault tree analysis (FTA) preliminary steps, ethylene plant.

21.8 SUMMARY

1. This chapter has provided real-world hazard and risk analysis applications of the fundamental principles of probability and the various special techniques presented in Chapters 19 and 20.
2. Bayes' theorem and the log-normal distribution are used in the first case study (Section 21.2) to obtain an estimate of the failure rate of a coolant recycle pump.
3. The second case study (Section 21.3), focusing on the health impact of the release of toxic vapors during transportation of a hazardous chemical, illustrated the derivation of a risk curve showing the relation between number of people affected and average annual frequency of exposure to a potential lethal concentration of toxic vapors.
4. In Section 21.4 the effects of the release of toxic vapors were considered in connection with an accident sequence initiated by the failure of a plant programmable automatic controller. In this study, event tree analysis and fault tree analysis led to identification of the glycol cooling system circulation pumps as components meriting high priority for inspection,

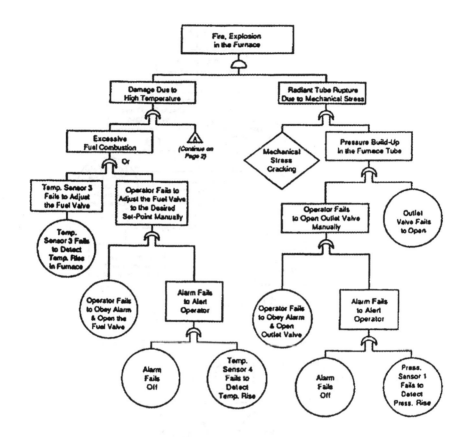

Figure 21.7.3 Fault tree analysis (FTA) for an ethylene plant design.

maintenance, and replacement to avoid even moderate release of toxic vapors.

5. The fourth case study (Section 21.5) was a hazard and risk analysis of the potential impact of the catastrophic release of the chemical contents of a holdup tank because of a runaway reaction. The study traced calculations leading to a risk curve portraying the health impact in terms of the frequency with which the number of people affected exceeded various amounts.

6. The final case study (Section 21.6) illustrated the use of Monte Carlo techniques and the binomial and Weibull distributions in estimating out-of-compliance probabilities for electrostatic precipitators.

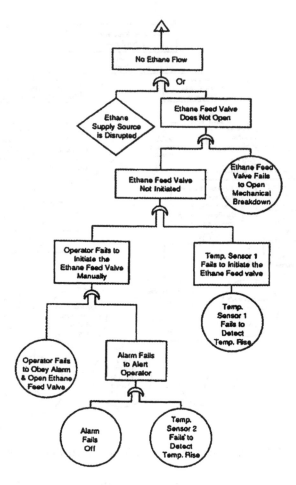

Figure 21.7.3 (cont'd)

PROBLEMS

1.   Using the procedure described in Section 20.4, obtain graphical estimates of the parameters α and β for the Weibull distribution of time to failure on the basis of the data in Table 21.6.1.

2.   The number of defective items in a sample of size n produced by a certain machine has a binomial distribution with parameters n and p, where p is the probability that an item produced is defective. For the case of 2 observed defectives in a sample of size 20, obtain the Bayesian estimate of p if the prior distribution of p is specified by the pdf

$$g(p) = 2(1 - p); \quad 0 < p < 1$$

3. The time to failure T of an electronic component has an exponential distribution with pdf specified by
$$f(t) = \theta e^{-\theta t}; \quad t > 0$$

Suppose that the prior distribution of θ has the pdf

$$g(\theta) = 2e^{-2\theta}; \quad \theta > 0$$

Find the Bayesian estimate of θ on the basis of the following observations on T: 2, 3, 5, 10, 17, 30.

4. Verify the total average annual frequency of release of medium severity given in Table 21.3.2 and the average annual frequency of exposure given in Table 21.3.3 in the case of a release of large severity.

5. Using the probabilities provided in Section 21.4, compute the probability of the accident sequence $P \overline{W} G \overline{O}$ leading to the minor release of toxic vapor.

6. Assume the time to failure T of a bus section has a Weibull distribution with $\alpha = 1.3 \times 10^{-3}$ and $\beta = 0.77$.
(a) Find P(T < 1000).
(b) Find the probability of x failures in 1000 hours in the case of 100 bus sections.
(c) Find the probability that a bus section fails between 70 and 1000 hours, given that it has not failed up to 70 hours.

REFERENCES

1. D. Hendershot, "A Simple Example Problem Illustrating the Methodology of Chemical Process Quantitative Risk Assessment," paper presented at AIChE Mid-Atlantic Region "Day in Industry" for Chemical Engineering Faculty, Apr. 15, 1988.

2. A. Mosleh, M. Kazarians, and W. C. Gekler, "Development of Risk and Reliability Data Base for Chemical Facilities," paper prepared for AIChE 1986 Summer Meeting, Boston, Aug. 25-27, 1986.

3. Mardyros Kazarians, Raymond F. Boykin, and Stan Kaplan, "Transportation Risk Management-A Case Study," paper presented at AIChE Loss Prevention Symposium, New Orleans, Apr. 6-10, 1986.

4. C. C. Bums, M. J. Hazzan, "Risk Analysis in the Chemical and Process Industries," paper presented at AIChE Summer Meeting, Seattle, Aug. 25-28, 1985.

5.   M. Kazarians, W. J. Bradford, and M. J. Abrams, "Risk Assessment of a Process Facility," paper presented at AIChE Orange County Section Annual Technical Meeting, Oct. 15, 1985.

6.   L. Theodore, J. Reynolds, F. Taylor, and S. Errico, "Electrostatic Precipitator Bus Section Failure: Operation and Maintenance," paper No. 84-96.10, presented at the 77th Annual Meeting of Air Pollution Control Association, San Francisco, 1984.

7.   B. J. Ley, *Computer Aided Analysis and Design for Electrical Engineers*, Holt, New York, 1970.

8.   A. Mann, R. Schafer, and N. Singpurwalla, *Methods for Statistical Analysis of Reliability and Life Data*, Wiley, New York, 1974, p. 76.

9.   H. R. Kavianian, J. Rao, and G. V. Brown, "Application of Hazard Evaluation Techniques to the Design of Potentially Hazardous Industrial Chemical Processes", NIOSH, Cincinnati, OH, 1992.

10.   D. J. Paustenbach, Should Engineering Schools Address Environment and Occupational Health Issues?, J. Professional Issues in Engineering, 113(2):93-111, ASCE, April 1987.

11.   R. A. Meyers, Handbook of Petroleum Refining Processes, McGraw-Hill, New York, NY, 1986.

12.   L. Kniel, O. Winter, and K. Stork, Ethylene, Keystone to the Petrochemical Industry, Marcel Dekker Inc., New York, NY, 1980.

13.   R. Perry and D. Grier, "Chemical Engineers' Handbook", 7th ed., McGraw-Hill, New York City, 1997.

14.   E. R. Plunket, Handbook of Industrial Toxicology, 3rd ed., Chemical Publishing Co., Inc., New York, NY, 1987.

15.   J. Spero, B. Devito, and L. Theodore, "Regulatory Chemicals Handbook", Marcel Dekker, New York City, 2000.

16.   M. Abdulmalik, E. Firoszabadi, H. Kavianian, and S. Panahshashi, "Safety and Hazard Assessment for Preliminary Design of Ethylene Plant", School of Engineering, California State University, Long Beach, CA, 1988.

Index

Milton Keynes UK
Ingram Content Group UK Ltd.
UKHW020003071024
449327UK00031B/2635

9 780367 396893